control of partial differential equations and applications

LECTURE NOTES IN PURE AND APPLIED MATHEMATICS

1. *N. Jacobson,* Exceptional Lie Algebras
2. *L.-Å. Lindahl and F. Poulsen,* Thin Sets in Harmonic Analysis
3. *I. Satake,* Classification Theory of Semi-Simple Algebraic Groups
4. *F. Hirzebruch, W. D. Newmann, and S. S. Koh, Differentiable Manifolds and Quadratic Forms*
5. *I. Chavel,* Riemannian Symmetric Spaces of Rank One
6. *R. B. Burckel,* Characterization of C(X) Among Its Subalgebras
7. *B. R. McDonald, A. R. Magid, and K. C. Smith,* Ring Theory: Proceedings of the Oklahoma Conference
8. *Y.-T. Siu,* Techniques of Extension on Analytic Objects
9. *S. R. Caradus, W. E. Pfaffenberger, and B. Yood,* Calkin Algebras and Algebras of Operators on Banach Spaces
10. *E. O. Roxin, P.-T. Liu, and R. L. Sternberg,* Differential Games and Control Theory
11. *M. Orzech and C. Small,* The Brauer Group of Commutative Rings
12. *S. Thomier,* Topology and Its Applications
13. *J. M. Lopez and K. A. Ross,* Sidon Sets
14. *W. W. Comfort and S. Negrepontis,* Continuous Pseudometrics
15. *K. McKennon and J. M. Robertson,* Locally Convex Spaces
16. *M. Carmeli and S. Malin,* Representations of the Rotation and Lorentz Groups: An Introduction
17. *G. B. Seligman,* Rational Methods in Lie Algebras
18. *D. G. de Figueiredo,* Functional Analysis: Proceedings of the Brazilian Mathematical Society Symposium
19. *L. Cesari, R. Kannan, and J. D. Schuur,* Nonlinear Functional Analysis and Differential Equations: Proceedings of the Michigan State University Conference
20. *J. J. Schäffer,* Geometry of Spheres in Normed Spaces
21. *K. Yano and M. Kon,* Anti-Invariant Submanifolds
22. *W. V. Vasconcelos,* The Rings of Dimension Two
23. *R. E. Chandler,* Hausdorff Compactifications
24. *S. P. Franklin and B. V. S. Thomas,* Topology: Proceedings of the Memphis State University Conference
25. *S. K. Jain,* Ring Theory: Proceedings of the Ohio University Conference
26. *B. R. McDonald and R. A. Morris,* Ring Theory II: Proceedings of the Second Oklahoma Conference
27. *R. B. Mura and A. Rhemtulla,* Orderable Groups
28. *J. R. Graef,* Stability of Dynamical Systems: Theory and Applications
29. *H.-C. Wang,* Homogeneous Branch Algebras
30. *E. O. Roxin, P.-T. Liu, and R. L. Sternberg,* Differential Games and Control Theory II
31. *R. D. Porter,* Introduction to Fibre Bundles
32. *M. Altman,* Contractors and Contractor Directions Theory and Applications
33. *J. S. Golan,* Decomposition and Dimension in Module Categories
34. *G. Fairweather,* Finite Element Galerkin Methods for Differential Equations
35. *J. D. Sally,* Numbers of Generators of Ideals in Local Rings
36. *S. S. Miller,* Complex Analysis: Proceedings of the S.U.N.Y. Brockport Conference
37. *R. Gordon,* Representation Theory of Algebras: Proceedings of the Philadelphia Conference
38. *M. Goto and F. D. Grosshans,* Semisimple Lie Algebras
39. *A. I. Arruda, N. C. A. da Costa, and R. Chuaqui,* Mathematical Logic: Proceedings of the First Brazilian Conference
40. *F. Van Oystaeyen,* Ring Theory: Proceedings of the 1977 Antwerp Conference
41. *F. Van Oystaeyen and A. Verschoren,* Reflectors and Localization: Application to Sheaf Theory
42. *M. Satyanarayana,* Positively Ordered Semigroups
43. *D. L Russell,* Mathematics of Finite-Dimensional Control Systems
44. *P. T. Liu and E. Roxin,* Differential Games and Control Theory III: Proceedings of the Third Kingston Conference, Part A
45. *A. Geramita and J. Seberry,* Orthogonal Designs: Quadratic Forms and Hadamard Matrices
46. *J. Cigler, V. Losert, and P. Michor,* Banach Modules and Functors on Categories of Banach Spaces

47. *P.-T. Liu and J. G. Sutinen,* Control Theory in Mathematical Economics: Proceedings of the Third Kingston Conference, Part B
48. *C. Byrnes,* Partial Differential Equations and Geometry
49. *G. Klambauer,* Problems and Propositions in Analysis
50. *J. Knopfmacher,* Analytic Arithmetic of Algebraic Function Fields
51. *F. Van Oystaeyen,* Ring Theory: Proceedings of the 1978 Antwerp Conference
52. *B. Kadem,* Binary Time Series
53. *J. Barros-Neto and R. A. Artino,* Hypoelliptic Boundary-Value Problems
54. *R. L. Sternberg, A. J. Kalinowski, and J. S. Papadakis,* Nonlinear Partial Differential Equations in Engineering and Applied Science
55. *B. R. McDonald,* Ring Theory and Algebra III: Proceedngs of the Third Oklahoma Conference
56. *J. S. Golan,* Structure Sheaves Over a Noncommutative Ring
57. *T. V. Narayana, J. G. Williams, and R. M. Mathsen,* Combinatorics, Representation Theory and Statistical Methods in Groups: YOUNG DAY Proceedings
58. *T. A. Burton,* Modeling and Differential Equations in Biology
59. *K. H. Kim and F. W. Roush,* Introduction to Mathematical Consensus Theory
60. *J. Banas and K. Goebel,* Measures of Noncompactness in Banach Spaces
61. *O. A. Nielson,* Direct Integral Theory
62. *J. E. Smith, G. O. Kenny, and R. N. Ball,* Ordered Groups: Proceedings of the Boise State Conference
63. *J. Cronin,* Mathematics of Cell Electrophysiology
64. *J. W. Brewer,* Power Series Over Commutative Rings
65. *P. K. Kamthan and M. Gupta,* Sequence Spaces and Series
66. *T. G. McLaughlin,* Regressive Sets and the Theory of Isols
67. *T. L. Herdman, S. M. Rankin III, and H. W. Stech,* Integral and Functional Differential Equations
68. *R. Draper,* Commutative Algebra: Analytic Methods
69. *W. G. McKay and J. Patera,* Tables of Dimensions, Indices, and Branching Rules for Representations of Simple Lie Algebras
70. *R. L. Devaney and Z. H. Nitecki,* Classical Mechanics and Dynamical Systems
71. *J. Van Geel,* Places and Valuations in Noncommutative Ring Theory
72. *C. Faith,* Injective Modules and Injective Quotient Rings
73. *A. Fiacco,* Mathematical Programming with Data Perturbations I
74. *P. Schultz, C. Praeger, and R. Sullivan,* Algebraic Structures and Applications: Proceedings of the First Western Australian Conference on Algebra
75. *L Bican, T. Kepka, and P. Nemec,* Rings, Modules, and Preradicals
76. *D. C. Kay and M. Breen,* Convexity and Related Combinatorial Geometry: Proceedings of the Second University of Oklahoma Conference
77. *P. Fletcher and W. F. Lindgren,* Quasi-Uniform Spaces
78. *C.-C. Yang,* Factorization Theory of Meromorphic Functions
79. *O. Taussky,* Ternary Quadratic Forms and Norms
80. *S. P. Singh and J. H. Burry,* Nonlinear Analysis and Applications
81. *K. B. Hannsgen, T. L. Herdman, H. W. Stech, and R. L. Wheeler,* Volterra and Functional Differential Equations
82. *N. L. Johnson, M. J. Kallaher, and C. T. Long,* Finite Geometries: Proceedings of a Conference in Honor of T. G. Ostrom
83. *G. I. Zapata,* Functional Analysis, Holomorphy, and Approximation Theory
84. *S. Greco and G. Valla,* Commutative Algebra: Proceedings of the Trento Conference
85. *A. V. Fiacco,* Mathematical Programming with Data Perturbations II
86. *J.-B. Hiriart-Urruty, W. Oettli, and J. Stoer,* Optimization: Theory and Algorithms
87. *A. Figa Talamanca and M. A. Picardello,* Harmonic Analysis on Free Groups
88. *M. Harada,* Factor Categories with Applications to Direct Decomposition of Modules
89. *V. I. Istrăţescu,* Strict Convexity and Complex Strict Convexity
90. *V. Lakshmikantham,* Trends in Theory and Practice of Nonlinear Differential Equations
91. *H. L. Manocha and J. B. Srivastava,* Algebra and Its Applications
92. *D. V. Chudnovsky and G. V. Chudnovsky,* Classical and Quantum Models and Arithmetic Problems
93. *J. W. Longley,* Least Squares Computations Using Orthogonalization Methods
94. *L. P. de Alcantara,* Mathematical Logic and Formal Systems
95. *C. E. Aull,* Rings of Continuous Functions
96. *R. Chuaqui,* Analysis, Geometry, and Probability
97. *L. Fuchs and L. Salce,* Modules Over Valuation Domains

98. *P. Fischer and W. R. Smith,* Chaos, Fractals, and Dynamics
99. *W. B. Powell and C. Tsinakis,* Ordered Algebraic Structures
100. *G. M. Rassias and T. M. Rassias,* Differential Geometry, Calculus of Variations, and Their Applications
101. *R.-E. Hoffmann and K. H. Hofmann,* Continuous Lattices and Their Applications
102. *J. H. Lightbourne III and S. M. Rankin III,* Physical Mathematics and Nonlinear Partial Differential Equations
103. *C. A. Baker and L, M. Batten,* Finite Geometrics
104. *J. W. Brewer, J. W. Bunce, and F. S. Van Vleck,* Linear Systems Over Commutative Rings
105. *C. McCrory and T. Shifrin,* Geometry and Topology: Manifolds, Varieties, and Knots
106. *D. W. Kueker, E. G. K. Lopez-Escobar, and C. H. Smith,* Mathematical Logic and Theoretical Computer Science
107. *B.-L. Lin and S. Simons,* Nonlinear and Convex Analysis: Proceedings in Honor of Ky Fan
108. *S. J. Lee,* Operator Methods for Optimal Control Problems
109. *V. Lakshmikantham,* Nonlinear Analysis and Applications
110. *S. F. McCormick,* Multigrid Methods: Theory, Applications, and Supercomputing
111. *M. C. Tangora,* Computers in Algebra
112. *D. V. Chudnovsky and G. V. Chudnovsky,* Search Theory: Some Recent Developments
113. *D. V. Chudnovsky and R. D. Jenks,* Computer Algebra
114. *M. C. Tangora,* Computers in Geometry and Topology
115. *P. Nelson, V. Faber, T. A. Manteuffel, D. L. Seth, and A. B. White, Jr.,* Transport Theory, Invariant Imbedding, and Integral Equations: Proceedings in Honor of G. M. Wing's 65th Birthday
116. *P. Clément, S. Invernizzi, E. Mitidieri, and I. I. Vrabie,* Semigroup Theory and Applications
117. *J. Vinuesa,* Orthogonal Polynomials and Their Applications: Proceedings of the International Congress
118. *C. M. Dafermos, G. Ladas, and G. Papanicolaou,* Differential Equations: Proceedings of the EQUADIFF Conference
119. *E. O. Roxin,* Modern Optimal Control: A Conference in Honor of Solomon Lefschetz and Joseph P. Lasalle
120. *J. C. Díaz,* Mathematics for Large Scale Computing
121. *P. S. Milojević,* Nonlinear Functional Analysis
122. *C. Sadosky,* Analysis and Partial Differential Equations: A Collection of Papers Dedicated to Mischa Cotlar
123. *R. M. Shortt,* General Topology and Applications: Proceedings of the 1988 Northeast Conference
124. *R. Wong,* Asymptotic and Computational Analysis: Conference in Honor of Frank W. J. Olver's 65th Birthday
125. *D. V. Chudnovsky and R. D. Jenks,* Computers in Mathematics
126. *W. D. Wallis, H. Shen, W. Wei, and L. Zhu,* Combinatorial Designs and Applications
127. *S. Elaydi,* Differential Equations: Stability and Control
128. *G. Chen, E. B. Lee, W. Littman, and L. Markus,* Distributed Parameter Control Systems: New Trends and Applications
129. *W. N. Everitt,* Inequalities: Fifty Years On from Hardy, Littlewood and Pólya
130. *H. G. Kaper and M. Garbey,* Asymptotic Analysis and the Numerical Solution of Partial Differential Equations
131. *O. Arino, D. E. Axelrod, and M. Kimmel,* Mathematical Population Dynamics: Proceedings of the Second International Conference
132. *S. Coen,* Geometry and Complex Variables
133. *J. A. Goldstein, F. Kappel, and W. Schappacher,* Differential Equations with Applications in Biology, Physics, and Engineering
134. *S. J. Andima, R. Kopperman, P. R. Misra, J. Z. Reichman, and A. R. Todd,* General Topology and Applications
135. *P Clément, E. Mitidieri, B. de Pagter,* Semigroup Theory and Evolution Equations: The Second International Conference
136. *K. Jarosz*, Function Spaces
137. *J. M. Bayod, N. De Grande-De Kimpe, and J. Martínez-Maurica*, *p*-adic Functional Analysis
138. *G. A. Anastassiou*, Approximation Theory: Proceedings of the Sixth Southeastern Approximation Theorists Annual Conference
139. *R. S. Rees*, Graphs, Matrices, and Designs: Festschrift in Honor of Norman J. Pullman
140. *G. Abrams, J. Haefner, and K. M. Rangaswamy*, Methods in Module Theory

141. *G. L. Mullen and P. J.-S. Shiue*, Finite Fields, Coding Theory, and Advances in Communications and Computing
142. *M. C. Joshi and A. V. Balakrishnan*, Mathematical Theory of Control: Proceedings of the International Conference
143. *G. Komatsu and Y. Sakane*, Complex Geometry: Proceedings of the Osaka International Conference
144. *I. J. Bakelman*, Geometric Analysis and Nonlinear Partial Differential Equations
145. *T. Mabuchi and S. Mukai*, Einstein Metrics and Yang–Mills Connections: Proceedings of the 27th Taniguchi International Symposium
146. *L. Fuchs and R. Göbel*, Abelian Groups: Proceedings of the 1991 Curaçao Conference
147. *A. D. Pollington and W. Moran*, Number Theory with an Emphasis on the Markoff Spectrum
148. *G. Dore, A. Favini, E. Obrecht, and A. Venni*, Differential Equations in Banach Spaces
149. *T. West*, Continuum Theory and Dynamical Systems
150. *K. D. Bierstedt, A. Pietsch, W. Ruess, and D. Vogt*, Functional Analysis
151. *K. G. Fischer, P. Loustaunau, J. Shapiro, E. L. Green, and D. Farkas*, Computational Algebra
152. *K. D. Elworthy, W. N. Everitt, and E. B. Lee*, Differential Equations, Dynamical Systems, and Control Science
153. *P.-J. Cahen, D. L. Costa, M. Fontana, and S.-E. Kabbaj*, Commutative Ring Theory
154. *S. C. Cooper and W. J. Thron*, Continued Fractions and Orthogonal Functions: Theory and Applications
155. *P. Clément and G. Lumer*, Evolution Equations, Control Theory, and Biomathematics
156. *M. Gyllenberg and L. Persson*, Analysis, Algebra, and Computers in Mathematical Research: Proceedings of the Twenty-First Nordic Congress of Mathematicians
157. *W. O. Bray, P. S. Milojević, and Č. V. Stanojević*, Fourier Analysis: Analytic and Geometric Aspects
158. *J. Bergen and S. Montgomery*, Advances in Hopf Algebras
159. *A. R. Magid*, Rings, Extensions, and Cohomology
160. *N. H. Pavel*, Optimal Control of Differential Equations
161. *M. Ikawa*, Spectral and Scattering Theory: Proceedings of the Taniguchi International Workshop
162. *X. Liu and D. Siegel*, Comparison Methods and Stability Theory
163. *J.-P. Zolésio*, Boundary Control and Variation
164. *M. Křížek, P. Neittaanmäki, and R. Stenberg*, Finite Element Methods: Fifty Years of the Courant Element
165. *G. Da Prato and L. Tubaro*, Control of Partial Differential Equations
166. *E. Ballico*, Projective Geometry with Applications
167. *M. Costabel, M. Dauge, and S. Nicaise*, Boundary Value Problems and Integral Equations in Nonsmooth Domains
168. *G. Ferreyra, G. R. Goldstein, and F. Neubrander*, Evolution Equations
169. *S. Huggett*, Twistor Theory
170. *H. Cook, W. T. Ingram, K. T. Kuperberg, A. Lelek, and P. Minc*, Continua: With the Houston Problem Book
171. *D. F. Anderson and D. E. Dobbs*, Zero-Dimensional Commutative Rings
172. *K. Jarosz*, Function Spaces: The Second Conference
173. *V. Ancona, E. Ballico, and A. Silva*, Complex Analysis and Geometry
174. *E. Casas*, Control of Partial Differential Equations and Applications
175. *N. Kalton, E. Saab, and S. Montgomery-Smith*, Interaction Between Functional Analysis, Harmonic Analysis, and Probability

Additional Volumes in Preparation

control of partial differential equations and applications

proceedings of the IFIP TC7/WG-7.2 international conference, Laredo, Spain

edited by
Eduardo Casas
University of Cantabria
Santander, Spain

Marcel Dekker, Inc. **New York • Basel • Hong Kong**

Library of Congress Cataloging-in-Publication Data

Control of partial differential equations and applications : proceedings of the IFIP WG 7.2 international conference, Laredo / edited by Eduardo Casas.
p. cm. — (Lecture notes in pure and applied mathematics ; v. 174)
"Lectures delivered at the IFIP TC7/WG-7.2 Conference on Control of Partial Differential Equations and Applications, held in Laredo (Cantabria, Spain) [in 1994]"—Pref.
ISBN 0-8247-9607-1 (alk. paper)
1. Differential equations, Partial—Congresses. 2. Control theory—Congresses. I. Casas, Eduardo. II. IFIP TC7/WG-7.2 Conference on Control of Partial Differential Equations and Applications (1994 : Laredo, Spain) III. Series.
QA377.C7653 1996
515'.353—dc20 95-34363
CIP

The publisher offers discounts on this book when ordered in bulk quantities. For more information, write to Special Sales/Professional Marketing at the address below.

This book is printed on acid-free paper.

MARCEL DEKKER, INC.
270 Madison Avenue, New York, New York 10016

Current printing (last digit):
10 9 8 7 6 5 4 3 2 1

PRINTED IN THE UNITED STATES OF AMERICA

Preface

These proceedings include the lectures delivered at the IFIP TC7/WG-7.2 Conference on Control of Partial Differential Equations and Applications, held in Laredo (Cantabria, Spain) within the Summer Courses of Universidad de Cantabria. The conference comprised 27 lectures. The topics, as well as the invited speakers, were selected by the organizing committee to cover not only theoretical aspects, but also numerical analysis and applications to industrial problems. The main topics discussed were

- Controllability and stabilization
- Optimality conditions
- Numerical methods
- Applications

The organizers are grateful to all participants for their contribution to the success of this conference. We also wish to express our cordial thanks to the following organizations for their financial support:

- Ayuntamiento de Laredo
- Dirección General de Investigación Científica y Técnica
- Fundación Leonardo Torres Quevedo
- STRIDE Programme (EU)
- Universidad de Cantabria

Eduardo Casas

Contents

Preface	*iii*
Contributors	*ix*
Committees	*xiii*
Conference Participants	*xv*
An Algorithm of Interior Point for Multidisciplinary Optimal Design *Franciso Javier Barón*	1
On Drag Differentiability for Lipschitz Domains *Juan Antonio Bello, Enrique Fernández-Cara, Jerome Lemoine and Jacques Simon*	11
Some Examples of Optimality Conditions for Convex Control Problems with General Constraints *Maïtine Bergounioux and Dan Tiba*	23
Cauchy Problem for Hamilton-Jacobi Equations and Dirichlet Boundary Control Problems of Parabolic Type *Piermarco Cannarsa and Maria Elisabetta Tessitore*	31
Fictitious Domain Approach Used in Shape Optimization: Neumann Boundary Condition *Jana Daňková and Jarosiav Haslinger*	43
Modelling and Control of Thin/Shallow Shells *Michel C. Delfour and Jean-Paul Zolésio*	51
Obstruction and Some Approximate Controllability Results for the Burgers Equation and Related Problems *Jesús Ildefonso Díaz*	63
Some Controllability Problems for the Stokes and Navier-Stokes Systems *Enrique Fernández-Cara*	77

On Some Weighting Methods to Improve Output Least Squares Estimation 89
Jacques Henry and Jean-Pierre Yvon

Boundary Control of a Schrödinger Equation with Nonconstant Principal Part 101
Mary Ann Horn and Walter Littman

Augmented Lagrangian Formulation of Nonsmooth Convex Optimization in Hilbert Spaces 107
Kazufumi Ito and Karl Kunisch

Analysis of the Control of Wave Generators in a Canal 119
Ghislaine Joly, Stephan Mottelet and Jean-Pierre Yvon

Boundary Stabilization of Isotropic Elasticity Systems 135
Vilmos Komornik

Time-Minimal Control of Rotating Beams 147
Werner Krabs

A Singular Control Approach to Highly Damped Second-Order Abstract Equations and Applications 157
Irena Lasiecka, Luciano Pandolfi and Roberto Triggiani

Uniform Stabilization of a Shallow Spherical Shell with Boundary Dissipation 171
Irena Lasiecka, Roberto Triggiani and Vanda Valente

SQP Interior Point Methods for Parabolic Control Problems 181
Friedemann Leibfritz and Ekkehard Sachs

Reverberation Analysis and Control of Networks of Elastic Strings 193
Günter Leugering

Remarks on the Control of Sensitive Distributed Systems 207
Jacques-Louis Lions

Stabilization and Periodic Solutions of a Hybrid System Arising in the Control of Noise 219
Sorin Micu and Enrique Zuazua

Differentiability of Projection and Applications 231
Michel Pierre and Jan Sokołowski

Applications of Exact Controllability to Some Inverse Problems for the Wave Equations 241
Jean-Pierre Puel and Masahiro Yamamoto

Approximation of the Optimal Control of Vibrations 251
John Read

A Static Formation Theory for Active Elastic Materials 263
David Lewis Russell

Optimal Control of Free Boundaries 277
Srdjan Stojanovic

A Stability Theorem for Linear-Quadratic Parabolic Control Problems 287
Fredi Tröltzsch

Contributors

FRANCISCO JAVIER BARÓN, Grupo de Análisis Matemático Aplicado, Universidad de Málaga, Campus de Teatinos, s/n, 29071 Málaga, Spain

JUAN ANTONIO BELLO, Departamento de Ecuaciones Diferenciales y Análisis Numérico, Facultad de Matemáticas, Universidad de Sevilla, c/ Tarfia, s/n, 41012 Sevilla, Spain

MAÏTINE BERGOUNIOUX, URA CNRS 1803, Département de Mathématiques, Université d'Orléans, BP 6759 - 45067 Orléans Cedex 2, France

PIERMARCO CANNARSA, Dipartimento di Matematica, Universitá di Roma "Tor Vergata", Via O. Raimondo, 00173 Roma, Italy

JANA DAŇKOVÁ, Mathematical Institute, Academy of Sciences, Prague, Czech Republic

MICHEL C. DELFOUR, Centre de Recherches Mathématiques et Département de Mathématiques et de Statistique, Université de Montréal, C.P. 6128 Succ. Centre-Ville, Montréal, Québec, Canada H3C 3J7

JESÚS ILDEFONSO DÍAZ, Departamento de Matemática Aplicada, Facultad de Matemáticas, Universidad Complutense de Madrid, Avda. Complutense, s/n, 28040 Madrid, Spain

ENRIQUE FERNÁNDEZ-CARA, Departamento de Ecuaciones Diferenciales y Análisis Numérico, Facultad de Matemáticas, Universidad de Sevilla, c/ Tarfia, s/n, 41012 Sevilla, Spain

JAROSIAV HASLINGER, Faculty of Mathematics and Physics, Charles University, Prague, Czech Republic

JACQUES HENRY, INRIA Rocquencourt, BP 105 - 78153 Le Chesnay Cedex, France

MARY ANN HORN, School of Mathematics, University of Minnesota, 127 Vincent Hall, 206 Church Street, SE Minneapolis MN 55455–0100, USA

KAZUFUMI ITO, Department of Mathematics, North Carolina State University, Raleigh, North Carolina 27695-8205, USA

GHISLAINE JOLY, Université de Technologie de Compiègne BP 649 - 60206 Compiègne, Francc

VILMOS KOMORNIK, Département de Mathématique et CNRS, Université Louis Pasteur, 7 rue René Descartes, 67084 Strasbourg Cédex, France

WERNER KRABS, Fachbereich Mathematik, Technische Hochschule Darmstadt, Schlossgartenstrasse 7, D-64289 Darmstadt, Germany

KARL KUNISCH, Fachbereich Mathematik, Technische Universität Berlin, Strasse des 17 Juni 136, D-10623 Berlin, Germany

IRENA LASIECKA, Department of Applied Mathematics, University of Virginia, Thornton Hall, Charlottesville, VA 22903, USA

FRIEDEMANN LEIBFRITZ, Fachbereich IV - Mathematik, Universität Trier, D-54286 Trier Germany

JEROME LEMOINE, Laboratoire de Mathématiques Appliquées, Université Blaise Pascal (Clermont Ferrand 2), Aubière, France

GÜNTER LEUGERING, Institut für Mathematik, Universität Bayreuth, D-95440 Bayreuth Germany

JACQUES-LOUIS LIONS, Collège de France, 3 rue d'Ulm, 75231 Paris Cedex 05, France

WALTER LITTMAN, School of Mathematics, University of Minnesota, 127 Vincent Hall, 206 Church Street, SE, Minneapolis MN 55455–0100, USA

SORIN MICU, Facultatea de Matematica-Informatica, Universitatea din Craiova, 1100, Romania

STEPHAN MOTTELET, Université de Technologie de Compiègne BP 649 - 60206 Compiègne, France

LUCIANO PANDOLFI, Politecnico di Torino, Corso Duca degli Abruzzi 24, 10129 Torino, Italy

MICHEL PIERRE, Université de Nancy I, Département de Mathématiques, BP 239 URA - CNRS 750, Projet NUMATH INRIA Lorraine, 54506 Vandoeuvre-Les-Nancy Cedex, France

JEAN-PIERRE PUEL, Université de Versailles Saint-Quentin et CMAP, Ecole Polytechnique, 91128 Palaiseau Cedex, France

JOHN READ, Fachbereich Mathematik, Technische Hochschule Darmstadt, Schlossgartenstrasse 7, D-64289 Darmstadt, Germany

DAVID LEWIS RUSSELL, Department of Mathematics, Virginia Polytechnic Institute, 418 McBryde Hall, Blacksburg, VA 24061, USA

Ekkehard Sachs, Fachbereich IV - Mathematik, Universität Trier, D-54286 Trier, Germany

Jacques Simon, Laboratoire de Mathématiques Appliquées, Université Blaise Pascal (Clermont Ferrand 2), Aubière, France

Jan Sokołowski, Université de Nancy I, Département de Mathématiques, BP 239 URA CNRS 750, Projet NUMATH INRIA Lorraine, 54506 Vandoeuvre-Les-Nancy Cedex, France

Srdjan Stojanovic, Department of Mathematical Sciences, University of Cincinnati, Cincinnati, OH 45221-0025, USA

Maria Elisabetta Tessitore, Dipartimento di Matematica, Universitá di Roma "La Sapienza", Piazzale A. Moro 2, 00185 Roma, Italy

Dan Tiba, Institute for Mathematics, Romanian Academy of Sciences, P.O. Box 1-764, RO–70700 Bucuresti, Romania

Roberto Triggiani, Department of Applied Mathematics, University of Virginia, Thornton Hall, Charlottesville, VA 22903, USA

Fredi Tröltzsch, Technische Universität Chemnitz-Zwickau, Fakultät für Mathematik, Reichenhainer Strasse 41, D-09107 Chemnitz, Germany

Vanda Valente, Istituto per le Applicazioni del Calcolo "M. Picone" C.N.R. viale Policlinico 137 - 00161 Roma, Italy

Masahiro Yamamoto, Department of Mathematical Sciences, University of Tokyo, 3-8-1 Kobama Meguro Tokyo 153, Japan

Jean-Pierre Yvon, Université de Technologie de Compiègne, BP 649 - 60206 Compiègne, France

Jean-Paul Zolésio, Institut Non Linéaire de Nice, CNRS, 1361 Route des Lucioles, 06904 Sophia Antipolis Cédex, France

Enrique Zuazua, Departamento de Matemática Aplicada, Facultad de Matemáticas, Universidad Complutense de Madrid, 28040-Madrid, Spain

Committees

International Program Committee

A. Bermúdez, Universidad de Santiago de Compostela, Spain
E. Casas, Universidad de Cantabria, Spain
G. Da Prato, Scuola Normale, Pisa, Italy
M.C. Delfour, Centre de Recherches Mathématiques, Montréal, Canada
R. Glowinski, University of Houston, USA
K. Hoffman, Technische Universität, München, Germany
W. Krabs, Technische Hochschule, Darmstadt, Germany
I. Lasiecka (Chairperson), University of Virginia, USA
J.L. Lions, Collège de France, Paris, France
J.-P. Puel, Ecole Polytechnique, Paris, France
J. Simon, Université Blaise Pascal, Clermont–Ferrand, France
F. Tröltzsch, Technische Universität, Chemnitz, Germany
J.-P. Yvon, Université de Technologie, Compiègne, France
J.-P. Zolésio, Université de Nice, France

Local Organizing Committee

E. Casas, Universidad de Cantabria, Spain
L.A. Fernández, Universidad de Cantabria, Spain
C. Pola, Universidad de Cantabria, Spain

Conference Participants

F.J. Barón, Univ. de Málaga, Spain.
B. Barón, Univ. de Cantabria, Spain.
M. Bergounioux, Univ. d'Orléans, France.
O. Bodart, Univ. d'Orléans, France.
P. Cannarsa, Univ. di Roma "Tor Vergata", Italy.
E. Casas, Univ. de Cantabria, Spain.
J.P. Del Molino Cortés, Spain.
M.C. Delfour, Univ. de Montréal, Canada.
J.I. Díaz, Univ. Complutense, Spain.
A. El Badia, Univ. de Technologie de Compiègne, France.
R. Echevarría, Univ. de Sevilla, Spain.
R. Escobedo Martínez, Spain.
C. Fabre, Ecole Polytechnique, France.
L.A. Fernández, Univ. de Cantabria, Spain.
E. Fernández-Cara, Univ. de Sevilla, Spain.
D. Franco Leis, Spain.
J. Haslinger, Charles Univ., Czech Republic.
J. Henry, INRIA, France.
M.A. Horn, Univ. of Minnesota, USA.
V. Komornik, Univ. Louis Pasteur et CNRS, France.
W. Krabs, Technische Hochschule Darmstadt, Germany.
K. Kunisch, Technische Univ. Berlin, Germany.
I. Lasiecka, Univ. of Virginia, USA.
G. Lebeau, Univ. de Paris–Sud, France.
G. Leugering, Univ. Bayreuth, Germany.
J.L. Lions, Collège de France, France.
B. Llanas Juárez, Spain.
M. Luna Laynez, Spain.
C.O. Menéndez, Univ. de Oviedo, Spain.
S. Mottelet, Univ. de Technologie de Compiègne, France.
J.A. Murillo Hernández, Spain.
G. Neumann, Univ. de Cantabria, Spain.
J. Pérez López, Spain.
C. Pola, Univ. de Cantabria, Spain.
J.-P. Puel, Univ. de Versailles et Ecole Polytechnique, France.
A.M. Ramos, Univ. Complutense, Spain.

J. Read, Technische Hochschule Darmstadt, Germany.
R. Ríos, Univ. de Oviedo, Spain.
D.L. Russell, Virginia Polytechnic Institute, USA.
E.W. Sachs, Univ. Trier, Germany.
A. Santamaría, Univ. de Oviedo, Spain.
J. Simon, Univ. Blaise Pascal (Clermont–Ferrand 2), France.
J. Sokołowski, Univ. de Nancy I, France.
J.C. Soto Merino, Spain.
S. Stojanovic, Univ. of Cincinnati, USA.
M.E. Tessitore, Univ. di Roma "La Sapienza", Italy.
F. Tröltzsch, Technische Univ. Chemnitz-Zwickau, Germany.
A. Valle, Univ. de Málaga, Spain.
D. Wildermann, Technische Hochschule Darmstadt, Germany.
J.-P. Yvon, Univ. de Technologie de Compiègne, France.
E. Zuazua, Univ. Complutense, Spain.

control of partial differential equations and applications

An Algorithm of Interior Point for Multidisciplinary Optimal Design

Francisco Javier Barón López
Grupo de Análisis Matemático Aplicado
Universidad de Málaga
Campus de Teatinos, s/n
29071 Málaga, Spain
e–mail: fjbaron@ccuma.sci.uma.es

1 INTRODUCTION

We consider the problem of minimizing a function submitted to inequality constraints to be applied to *engineering design*, when smooth nonlinear functions are involved, using a general approach for nonlinear constrained optimization given by J. Herskovits. It is based on the iterative solution of Karush–Kuhn–Tucker (KKT) first order optimality conditions.

Let us consider the nonlinear constrained optimization problem

$$\left.\begin{array}{rl} \text{minimize} & f(x) \\ \text{submitted to} & g(x) \leq 0, \end{array}\right\} \qquad (1)$$

where $f \in \mathrm{R}$, $g \in \mathrm{R}^m$ are smooth functions in R^n.

By Herskovits technique, the nonlinear equations in the primal and dual variables, included in KKT optimality conditions, are solved by fixed point iterations. To ensure convergence to KKT points, the inequations are satisfied at each iteration. Based on the present approach, first order, quasi Newton or Newton algorithms can be obtained. Given an initial estimate of x, at the interior of the inequality constraints, a sequence of interior points is obtained. At each point a descent feasible direction is obtained. Then, an inaccurate line search is done to get a new interior point with a reasonable decrease of the objective.

The present method is simple to codify, strong and efficient. It does not involve penalty functions, active set strategies or quadratic programming subproblems. It merely requires to solve two linear systems with the same matrix at each iteration and to perform an inaccurate line search.

2 THE INEQUALITY CONSTRAINED OPTIMIZATION PROBLEM

We consider:

$$\left.\begin{array}{rl} \text{minimise} & f(x) \\ \text{submitted to} & g(x) \leq 0. \end{array}\right\} \qquad (1)$$

being $\Omega \equiv \{x \in R^n / g(x) \leq 0\}$ the feasible set

DEFINITION 2.1 *$d \in \mathrm{R}^n$ is a descent direction of a real function ϕ at $x \in \mathrm{R}^n$ if $d^t \nabla \phi < 0$.*

DEFINITION 2.2 *$d \in R^n$ is a feasible direction at $x \in \Omega$, if for some $\theta > 0$ we have $x + td \in \Omega$ for all $t \in [0, \theta]$.*

DEFINITION 2.3 *A vector field $d(x)$ defined on Ω is an uniformly feasible directions field if exists $\tau > 0$ such that $x + td(x) \in \Omega$ for all $t \in [0, \tau]$.*

2.1 COMPUTING A SEARCH DIRECTION

KKT optimality conditions are:

$$\nabla f(x) + \nabla g(x)\lambda = 0, \tag{2}$$

$$G(x)\lambda = 0, \tag{3}$$

$$g(x) \leq 0 \text{ and} \tag{4}$$

$$\lambda \geq 0. \tag{5}$$

where $\lambda \in R^m$ are the dual variables and $G(x)$ a diagonal matrix with $G_{ii}(x) = g_i(x)$ A Fixed Point Iteration for (2)–(3) is:

$$\begin{bmatrix} B & \nabla g(x) \\ \Lambda \nabla g^t(x) & G(x) \end{bmatrix} \begin{bmatrix} x_0 - x \\ \lambda_0 - \lambda \end{bmatrix} = - \begin{bmatrix} \nabla f(x) + \nabla g(x)\lambda \\ G(x)\lambda \end{bmatrix} \tag{6}$$

where (x, λ) is the present point, (x_0, λ_0) is the new estimate and Λ diagonal matrix; $\Lambda_{ii} = \lambda_i$.

We can take $B = \nabla^2 f(x) + \sum_{i=1}^m \lambda_i \nabla^2 g_i(x)$ to have a Newton's Method, a quasi–Newton approximation or $B = I$, for a first order method. Iteration (6) is modified in a way to obtain, for a given interior pair (x, λ), a new interior estimate with the objective improved.

We define a direction d_0 in the primal space, as $d = x_0 - x$. Then, it is

$$Bd_0 + \nabla g(x)\lambda_0 = -\nabla f(x) \text{ and} \tag{7}$$

$$\Lambda \nabla g^t(x) d_0 + G(x)\lambda_0 = 0. \tag{8}$$

It is proved in [**Her 90**] that, if B Positive Definite, $\lambda > 0$ and $g(x) \leq 0$, then (7) has an unique solution and d_0 is a descent direction of f.
However, d_0 is not always a feasible direction. In fact, (8) is equivalent to

$$\lambda_i \nabla g_i^t(x) d_0 + g_i(x)\lambda_{0i} = 0 \; ; \; 1, m. \tag{9}$$

Then, d_0 is tangent to the active constraints.

To obtain a feasible direction a negative number is added in the right side of (9):

$$\lambda_i \nabla g_i^t(x) d + g_i(x)\bar{\lambda}_i = -\rho \lambda_i \omega_i \quad i = 1, m,$$

and we have the iterates:

$$Bd + \nabla g(x)\bar{\lambda} = -\nabla f(x)$$
$$\Lambda \nabla g^t(x)d + G(x)\bar{\lambda} = -\rho\Lambda\omega, \tag{10}$$

where d is the search direction and $\bar{\lambda}$ the new dual variable, $\rho > 0$ and $\omega > 0$, $\omega \in R^m$.

We can consider that the inclusion of a negative number in the right hand of (8), produces the effect of deflecting d_0 in the sense of the interior of the feasible region, being the deflection of d_0 relative to the i-th constraint proportional to $\rho\omega_i$.

Finally, as the deflection of d_0 is proportional to ρ and d_0 is descent, by establishing upper bounds on ρ it is possible to ensure that d is a descent direction too. Since $d_0^t \nabla f(x) < 0$, we can obtain these bounds by imposing

$$d^t \nabla f(x) \leq \alpha d_0^t \nabla f(x), \tag{11}$$

which implies $d^t \nabla f(x) < 0$. Then d is in a circular cone whose axis is $\nabla f(x)$. In general, the derivative of f in the direction of d will be smaller than in the direction of d_0. This is the price that we pay to obtain a feasible descent direction.

Note that d can be obtained by solving

$$Bd_1 + \nabla g(x)\lambda_1 = O \tag{12}$$

$$\Lambda \nabla g^t(x)d_1 + G(x)\lambda_1 = -\Lambda\omega, \tag{13}$$

and computing $d = d_0 + \rho d_1$. Substitution in (11) gives a bound on ρ:

If $d_1^t \nabla f(x) > 0$, we can take ρ such that

$$\rho \leq (\alpha - 1)d_0^t \frac{\nabla f(x)}{d_1^t \nabla f(x)}. \tag{14}$$

Otherwise any $\rho > 0$ can be adopted. For $d(x)$ to be an uniformly feasible directions field, ρ must be also bounded below.

To determine a new primal point, an inaccurate line search is done in the direction of d, requiring feasibility and a satisfactory decrease of the objective. Different updating rules can be adopted to define a new positive λ.

2.2 LINE SEARCH

We use an extension of Armijo's Line Search for this kind of algorithms. Compute t, the first number of the sequence $\{1, \nu, \nu^2, \nu^3, ...\}$ such that:

$$f(x + dt) < f(x) + t\eta \nabla f^t(x) \tag{15}$$

and

$$g_i(x + td) < 0, \text{ if } \bar{\lambda}_i \geq 0, \tag{16}$$

or

$$g_i(x + td) \leq g_i(x) \text{ otherwise.} \tag{17}$$

Conditions (15) ensures a reasonable decrease of the function. Conditions (16) and (17) that the new point is interior and condition (17) avoids saturation of non active constraints. It can be proved that the step-length is bounded above from zero. Other procedures more efficient are used in practice.

3 A NUMERICAL EXAMPLE: A BREAKWATER OPTIMIZATION

Breakwaters can be optimized with respect to their shape. We shall investigate this problem when the waves are modeled by the wave equation, using the method previously described. The technique used is a simple adaptation of a study made a year ago for the optimization of stealth aircrafts (Barón et al [**BPP 93**], [**BDCT 94**]). However certain modifications are necessary concerning the constraint set that defines admissible breakwaters. The wave lengths are different.

3.1 MATHEMATICAL MODELLING

Consider a monochromatic incident wave from infinity:

$$z(\vec{x}) = e^{i\vec{\omega}\cdot\vec{x}}. \tag{18}$$

In slow varying depth, the perturbation (reflected wave) y_S due to the boundary of the harbor $\mathcal{O}$ and its breakwater $\mathcal{S}$ satisfies a wave equation. The computational domain is called Ω. It is limited in the far field by a circle Γ_∞.

$$\nabla \cdot (cc_g \nabla y_S) + cc_g k^2 y_S = 0 \qquad \text{in } \Omega, \tag{19}$$

where c is the phase speed, c_g is the group speed, related to the wave number k and the water depth h by:

$$c_g = \frac{c}{2}(1 + \frac{2kh}{sinh(2kh)}) \tag{20}$$

The total amplitude is $\eta_S = z + y_S$.

Boundary conditions are a radiation condition at infinity and a reflection condition on the island and its breakwater:

$$\frac{\partial \eta_S}{\partial n} + a\eta_S = 0 \qquad \text{on } \partial\mathcal{O} \cup \partial\mathcal{S}. \tag{21}$$

The absorption coefficient a is $a = ik \sin\alpha(1 - K_r)/(1 + K_r)$ where α is the incident angle and K_r a reflection coefficient.

The radiation condition is implemented by a domain decomposition method whereby the solution inside the circle is matched with the analytical solution outside the circle. This gives an integro-differential condition of the type:

$$\frac{\partial y_S}{\partial n} + \mathcal{M} y_S = 0 \qquad \text{on } \Gamma_\infty, \tag{22}$$

which is also

$$\frac{\partial \eta_S}{\partial n} = -\mathcal{M}\eta_S + \mathcal{M}z + \frac{\partial z}{\partial n} \text{on } \Gamma_\infty. \tag{23}$$

The method is exact even if Γ_∞ in near $\mathcal{S}$.

Our goal is to minimize the amplitude of the waves in a given region $\mathcal{C}$ of Ω:

$$j(\mathcal{S}) = \frac{1}{2}\int_{\mathcal{C}} \mid \eta_{\mathcal{S}} \mid^2 \, d\Omega. \tag{24}$$

The parameters are the shape of the breakwater and the coefficient K_r which is controlled, within bounds, by the shape and amount of rocks in front of the breakwater. To solve it by the Herskovits' algorithm we must compute

$$f(\mathcal{S}) \, , \, g(\mathcal{S}) \, , \, \frac{\partial f}{\partial \mathcal{S}} \, , \, \frac{\partial g}{\partial \mathcal{S}}$$

3.2 DISCRETIZATION

To put it simple, we consider the case of constant c_g and of perfect reflection ($K_r = 1$). The variational formulation is:

$$\begin{cases} \int_\Omega [\nabla \eta_{\mathcal{S}} \cdot \nabla \phi - k^2 \eta_{\mathcal{S}} \cdot \phi] \, d\Omega + \int_{\partial D} \mathcal{M} \eta_{\mathcal{S}} \cdot \phi = \int_{\partial D} (\mathcal{M} z + \frac{\partial z}{\partial n}) \cdot \phi \\ \text{for all } \phi \in H^1(\Omega). \end{cases} \tag{25}$$

Discretization is done by the finite element method of degree 1 on triangles. So Ω is replaced by Ω_h,and H^1_h is replaced by H^1_h. If q_j denotes the hat functions, we seek:

$$\eta_{u_h} = \sum_{j=1}^{n_h} W_j \cdot q_j^{u_h}, \tag{26}$$

where W_j, $j = 1, \cdots, n_h$ are the n_h unknowns of the linear system

$$\int_{\Omega_h} \left[\nabla \eta_h \cdot \nabla q_j - k^2 \eta_h \cdot q_j\right] d\Omega_h + \int_{\partial D_h} \mathcal{M}_h \eta_h \cdot q_j = \int_{\partial D_h} \left(\mathcal{M}_h z + \frac{\partial z}{\partial n}\right)_h \cdot q_j, \tag{27}$$

which is also:

$$A^h_{ij} W_j = B^h_i \quad , i = 1, \ldots, n_h, \tag{28}$$

where $A^h = \left[A^h_{ij}\right] \in M_{n_h}(\mathbf{C})$, $B^h = \left[B^h_i\right] \in M_{n_h \times 1}(\mathbf{C})$, $W = [W_i] \in M_{n_h \times 1}(\mathbf{C})$,

$$A^h_{ij} = \int_{\Omega_h} \left[\nabla q_i \cdot \nabla q_j - k^2 q_i \cdot q_j\right] d\Omega_h + \int_{\partial D_h} \mathcal{M}_h q^h_i \cdot q^h_j \tag{29}$$

$$B^h_i = \int_{\partial D_h} \left(\mathcal{M}_h z + \frac{\partial z}{\partial n}\right)_h \cdot q^h_i. \tag{30}$$

3.3 DISCRETIZATION OF THE COST FUNCTION

Let σ, be the K- set of coordinates of the mesh points with respect to which the optimization is performed . To compute the cost function one must do:

$$\sigma \longrightarrow \eta_h \longrightarrow f_h(\sigma) = F\left(\eta_h(\sigma)\right), \tag{31}$$

$$f_h(\sigma) \stackrel{\text{def}}{=} F(\eta_h) \stackrel{\text{def}}{=} \int_{C_h} | \eta_h(\sigma) |^2 \, d\Omega. \tag{32}$$

Let I_{C_h} be the set of vertices which belong to $\mathcal{C}_h$:

$$I_{C_h} = \{i = 1, \ldots, n_h \, : \, q_i \cap C_h \neq \emptyset\}. \tag{33}$$

Then

$$\begin{aligned} f_h(\sigma) &= \frac{1}{2} \int_{C_h} | \eta_h(\sigma) |^2 \, d\theta = \frac{1}{2} \int_{C_h} | \sum_{i \in I_{C_h}} W_i(\sigma) \, q_i |^2 \, d\theta \\ &= \frac{1}{2} \sum_{i,j \in I_{C_h}} W_i(\sigma) \overline{W_j(\sigma)} \int_{C_h} q_i q_j \, d\Omega = \frac{1}{2} \overline{W}^t(\sigma) \cdot C \cdot W(\sigma). \end{aligned} \tag{34}$$

where $C = [C_{ij}] \in M_{n_h}(\mathrm{R})$. The coefficients of the matrix C are independent of σ,

$$C_{ij} = \begin{cases} \int_{C_h} q_i^h q_j^h \, d\Omega & \text{si } i, j \in I_{C_h} \\ 0 & \text{otherwise.} \end{cases} \tag{35}$$

The discrete state $W(\sigma)$ depends on σ because it is the solution of :

$$A(\sigma) \cdot W(\sigma) = B(\sigma). \tag{36}$$

3.4 DISCRETE GRADIENTS

Consider the Lagrangian

$$\mathcal{L} \, : \, \mathrm{R}^K \times M_{n_h \times 1}(\mathbf{C}) \times M_{n_h \times 1}(\mathbf{C}) \longrightarrow \Re \tag{37}$$

$$\mathcal{L}(\sigma, W, P) = \frac{1}{2} W^t \cdot C \cdot W + Re\left\{P^t \cdot [A(\sigma) \cdot W - B(\sigma)]\right\}. \tag{38}$$

Let us compute its derivatives in three steps:

i) Discrete state equation. Derivates of $\mathcal{L}$ with respect to P gives the state equation:

$$\frac{D\,\mathcal{L}}{D\,P}(\sigma, W(\sigma), P) = 0 \Longleftrightarrow A(\sigma) \cdot W(\sigma) = B(\sigma). \tag{39}$$

ii) Discrete adjoint equation. Derivatives of $\mathcal{L}$ with respect to W gives the adjoint state equation:

$$\frac{D\,\mathcal{L}}{D\,W}(\sigma, W(\sigma), P(\sigma)) = 0 \Longleftrightarrow A(\sigma)^t \cdot P(\sigma) = C \cdot \overline{W(\sigma)}. \tag{40}$$

iii) Discrete gradient of the cost function. Total derivative of $\mathcal{L}$ with respect to σ gives

$$\begin{aligned} \frac{\partial f_h}{\partial \sigma_k} &= \frac{\partial \mathcal{L}}{\partial \sigma_k}(\sigma, W, P) + \frac{\partial \mathcal{L}}{\partial W}(\sigma, W, P) \cdot \frac{\partial W}{\partial \sigma_k} + \frac{\partial \mathcal{L}}{\partial P}(\sigma, W, P) \cdot \frac{\partial P}{\partial \sigma_k} \\ &= \frac{\partial \mathcal{L}}{\partial \sigma_k}(\sigma, W, P) \\ &= Re\left\{P^t \cdot \left[\frac{\partial A}{\partial \sigma_k}(\sigma) \cdot W - \frac{\partial B}{\partial \sigma_k}(\sigma)\right]\right\}. \end{aligned} \tag{41}$$

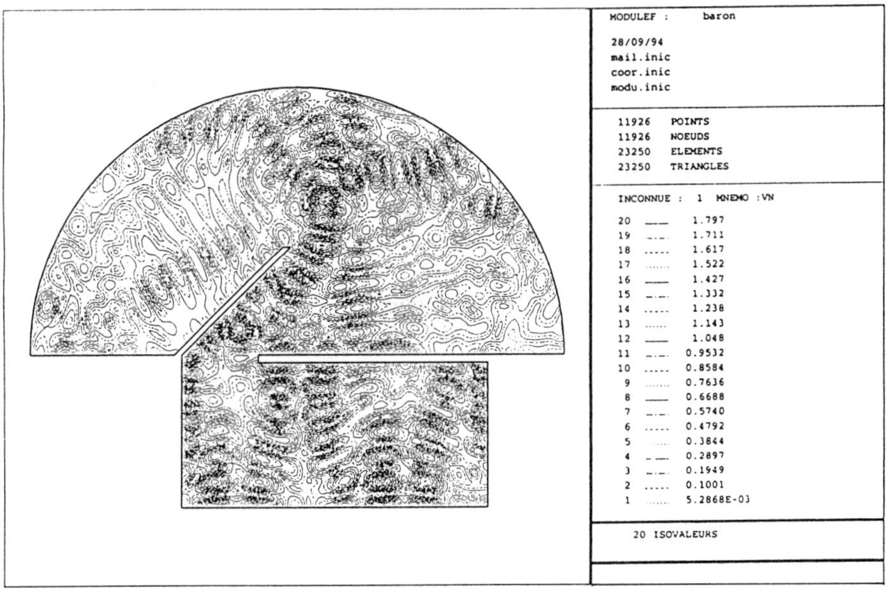

Figure 1: Simulation of the scattered waves for the initial breakwaters.

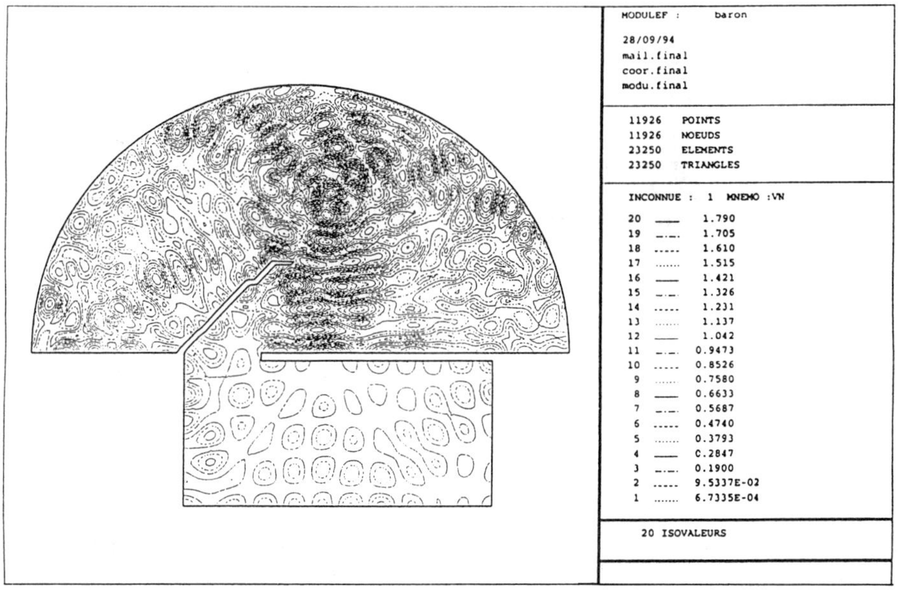

Figure 2: Simulation of the scattered waves for the optimized breakwaters.

3.5 NUMERICAL TESTS

To validate the method we take a harbor with a a breakwater whose length is prescribed but whose form can change provided that its curvature does not exceed a prescribed limit. The angle between the breakwater and the cost is bounded up and down.

One of the difficulties of the problem is that the water waves are small with respect to the size of the breakwater while the numerics require 12 points per wave length for a moderate precision. The triangulation of the domain 12,000 points. The linear system was solved by a Gauss factorization.

In this preliminary optimization the wave characteristics are fixed (wave speed, incident angle, wave length). The region of observation where waves should be small is the harbor.

The incident angle for z is $\alpha = 80°$. The constraints on the breakwater are box-type on mesh points that defines it. All mesh points of the breakwater can only move vertically. Optimization is performed with Herskovits [**Her 90**]. After 20 iterations the cost function goes from 1000 to 50. Observation domain is the bay limited by a horizontal line running from the lowest corner of the beginning of the breakwaters.

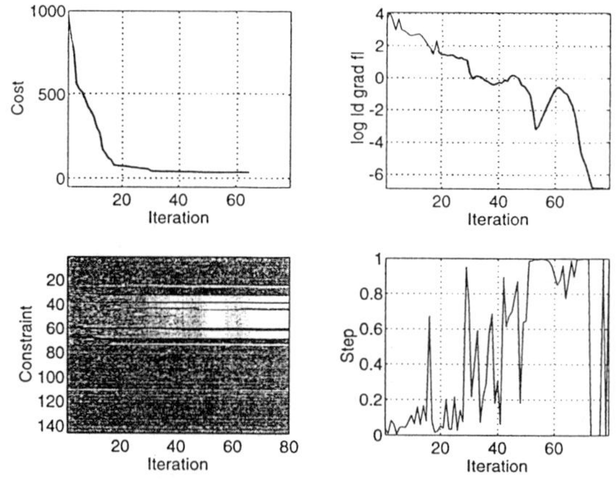

Figure 3: Historics of f, residual log $| d\nabla f |$, saturation $\log(g_i)$, steplength t.

References

[**BDCT 94**] F. Barón, G. Duffa, F. Carrere,P.Le Tallec, 'Optimization de forme en aerodinamique'. CHOCS n.12,Ed.Commisariat à l'Energie Atomique.,

[**BPP 93**] F. Barón, J. Periaux, O. Pironneau, 'Coating and shape optimization for Maxwell-Helmholtz equations'. *Proc 5th Conf on CFD*, Sendai, H. Daiguji ed., **I**, pp. 64,

[**Ber 74**] J.C. BERKHOFF, ' Computation of wave propagation by means of FEM'. *In Finite Element in Flow Problems*, J.T. Oden et al. UAH Press, Alabamma,

[**Her 90**] J. HERSKOVITS, J. ASQUIER, 'Quasi-Newton interior points algorithms for non-linear constrained optimization'. *DGOR-Proceedings of Operations Research 1990*, Vienna, (August 1990).

[**Mei 83**] C.C. MEI, *The applied dynamics of the ocean surface waves*. Wiley, New York, 1983.

[**SaK 93**] M. SAKURABA, K. KASHIYAMA, 'Stochastic finite element method analysis of wave diffraction-refraction considering the indeterminacy of input data'. *Finite Elements in Fluids*, CIMNE, Barcelona, Sept. 1993.

On Drag Differentiability for Lipschitz Domains

J. A. BELLO and E. FERNANDEZ-CARA Departamento de Ecuaciones Diferenciales y Análisis Numérico, Universidad de Sevilla, Sevilla, Spain

J. LEMOINE and J. SIMON Laboratoire de Mathématiques Appliquées, Université Blaise Pascal (Clermont-Ferrand 2), Aubière, France

INTRODUCTION

The behavior of the drag T associated to a body travelling at uniform velocity γ in a viscous incompressible fluid governed by the stationary Navier-Stokes equations is investigated. We are interested in viewing T as a function of the shape of the body. Let Ω be the fluid domain. The body variations are described by a field u and we search for a formula of the kind

$$T(\Omega + u) = T(\Omega) + T'(\Omega; u) + o(u),$$

where the modified fluid domain is:

$$\Omega + u = \{x \in \mathbb{R}^d;\ x = (I + u)(\xi),\ \xi \in \Omega\}.$$

We are thus led to an analysis of the differentiability of the function $u \mapsto T(\Omega + u)$.

We will mainly consider the case in which Ω is a Lipschitz domain and u is Lipschitz-continuous. This includes many interesting situations in which $\partial\Omega$ and/or $\partial(\Omega+u)$ have "corner" points.

The main result proved here is the differentiability of the function $u \mapsto T(\Omega + u)$, when the velocity γ is sufficiently small. More precisely, we show that it is a C^∞ mapping in a small ball $\mathcal{W}$. We also compute explicitly $T'(\Omega; u)$, i.e. the derivative at 0 in the direction u.

Results of this kind have been previously obtained for more regular domains by several authors. In particular, formal computations of the derivative were carried out by O. Pironneau [8] (see also [9]) using "normal" variations and also by F. Murat and J. Simon [6] using fields $u \in W^{2,\infty}(\mathbb{R}^d, \mathbb{R}^d)$.

Similar, but more detailed, results and proofs are given by the present authors in [1]. They include the linear case (Stokes equations) and the variations with respect to the vertices in a polygonal domain. In [1] the variation u is supposed to be 0 on the "outer" boundary of Ω.

Let us mention that in the more simple case of an elliptic equation, differentiability results have been established in [6] and in [7] without any regularity hypothesis on Ω. For the proofs, one introduces the change of variables $x = (I + u)(\xi)$, by means of which one is led to a fixed domain. This method has been used for many equations. It cannot be directly applied to the Stokes and Navier-Stokes cases, because of the incompressibility condition $\nabla \cdot y(u) = 0$. This difficulty is surmounted when Ω is a $W^{2,\infty}$ domain in [2] and [3], by introducing a family of isomorphisms which allow us to rewrite this equation appropriately. Here, the incompressibility equation is rewritten explicitly, see (22).

On the other hand, incompressibility forces Ω to be a Lipschitz domain, essentially because one has to be able to solve the equation $\nabla \cdot \chi = \phi$ in Ω, cf. (27). Otherwise, the same kind of results could be deduced without any regularity hypothesis on Ω.

Some other difficulties are related to weak regularity. The expression of the derivative given in [2] for a $W^{2,\infty}$ domain may be written, see (14), as the integral of a sum of terms. If Ω is only $W^{1,\infty}$, each of these terms does not necessarily belong to $L^1(\Omega)$. Nevertheless we prove that their sum is in $L^1(\Omega)$, by same method as in [7].

1 THE DEFINITION OF THE DRAG

Let D and B be bounded open connected sets in $\mathbb{R}^d$, $d = 2$ or 3, with $B \subset\subset D$. Let us set $\Omega = D \backslash \overline{B}$. In the sequel, it will be assumed that

$$\Omega \text{ is a Lipschitz domain,} \tag{1}$$

that is to say its boundary $\partial\Omega$ is locally the graph of a Lipschitz-continuous function and Ω is the corresponding epigraph.

Let $\gamma \in \mathbb{R}^d$ be a given vector. We consider the stationary Navier-Stokes problem

$$\begin{cases} y - g \in H_0^1(\Omega)^d, \ p \in L^2(\Omega), \ \int_\Omega p = 0, \\ -\nu \Delta y + (y \cdot \nabla) y + \nabla p = 0, \\ \nabla \cdot y = 0. \end{cases} \tag{2}$$

Here, $g \in H^1(\mathbb{R}^d)^d$ and satisfies

$$\nabla \cdot g = 0, \ g = \gamma \text{ in a neighborhood of } \partial D, \ g = 0 \text{ in a neighborhood of } B. \tag{3}$$

The requirement $\int_\Omega p = 0$ provides uniqueness for the pressure p that, otherwise, would be defined up to an additive constant.

If γ is sufficiently small, problem (2) possesses exactly one solution, which is "small" and does not depend on the choice of g. More precisely, theorem 2.1 in [6] gives:

LEMMA 1. *There exists a constant $\alpha > 0$ such that, if $|\gamma| < \alpha\nu$, then (2) possesses exactly one solution $(y,p) \in H^1(\Omega)^d \times L^2(\Omega)$. This solution does not depend on the choice of the function g satisfying (3). Furthermore, for each $\epsilon > 0$, the constant α can be chosen in such a way that*

$$\|y\|_{H^1(\Omega)^d} \leq \epsilon\nu.$$

Given $\mathcal{O} \subset\subset \mathcal{O}'$, one can also choose $\alpha = \alpha(\epsilon, \mathcal{O}, \mathcal{O}')$ not depending on B and D, provided $B \subset\subset \mathcal{O} \subset\subset \mathcal{O}' \subset\subset D$. Finally, if Ω is a $W^{2,\infty}$ domain, then $(y,p) \in H^2(\Omega)^d \times H^1(\Omega)$. □

Thus, at least when γ is small, one can associate to Ω a unique drag

$$T(\Omega) = \frac{\nu}{2} \int_\Omega \sigma(y)^2 \tag{4}$$

where $\sigma_{ij}(y) = \partial_i y_j + \partial_j y_i$ and $\sigma(y)^2 = \sum_{ij} \big(\sigma_{ij}(y)\big)^2$.

2 THE DOMAIN VARIATIONS

We introduce fixed open sets $\mathcal{O}$ and $\mathcal{O}'$ such that $B \subset\subset \mathcal{O} \subset\subset \mathcal{O}' \subset\subset D$, and

$$\mathcal{W} = \{u \in W^{1,\infty}(\mathbb{R}^d, \mathbb{R}^d);\ \|u\|_{W^{1,\infty}(\mathbb{R}^d,\mathbb{R}^d)} < c(\Omega)\}$$

where $c(\Omega)$ is such that the following properties (5) and (6) are satisfied.

LEMMA 2. *There exists $c(\Omega)$, $0 < c(\Omega) < 1$, such that : $\forall u \in \mathcal{W}$,*

$$\Omega + u \quad \textit{is a bounded Lipschitz domain in } \mathbb{R}^d \tag{5}$$

$$B + u \subset\subset \mathcal{O}, \quad \mathcal{O}' \subset\subset D + u\,. \ \square \tag{6}$$

The proof is similar to that of lemma 2 and 3 in [1] (in [1], $u|_{\partial D} = 0$).

It will also be assumed in the sequel that

$$|\gamma| < \alpha(\epsilon, \mathcal{O}, \mathcal{O}')\,\nu\,, \tag{7}$$

where α is furnished by lemma 1. The precise value of ϵ will be fixed below. Now, we choose g satisfying (3) and

$$g \equiv 0 \ \text{ in } \ \mathcal{O}\,, \quad g \equiv \gamma \ \text{ outside } \ \mathcal{O}'$$

(such a choice is always possible; for instance, one can take $g = a \wedge \nabla\psi$, where $a \in \mathbb{R}^3$, $a \cdot \gamma = 0$, $|a| = 1$ and $\psi \in C^\infty(\mathbb{R}^3)$, $\psi = 0$ in $\mathcal{O}$, $\psi(x) = (\gamma \wedge a) \cdot x$ outside $\mathcal{O}'$). If $u \in \mathcal{W}$, the Navier-Stokes problem in $\Omega + u$ can be written as follows

$$\begin{cases} y(u) - g \in H_0^1(\Omega + u)^d\,,\ p(u) \in L^2(\Omega + u),\ \int_\Omega p(u) \circ (I + u) = 0\,, \\ -\nu\,\Delta y(u) + (y(u) \cdot \nabla)\, y(u) + \nabla p(u) = 0\,, \\ \nabla \cdot y(u) = 0\,. \end{cases} \tag{8}$$

From lemma 1, we know that (8) possesses exactly one solution $(y(u), p(u))$. Accordingly, the drag associated to $B + u$ can be defined and is given by

$$T(\Omega + u) = \frac{\nu}{2} \int_{\Omega+u} \sigma\big(y(u)\big)^2 . \tag{9}$$

Remark. In principle, it seems more natural to normalize $p(u)$ by imposing $\int_{\Omega+u} p(u) = 0$. The choice we have made yields $\int_\Omega P(u) = 0$ for the transported pressure $P(u) = p(u) \circ (I + u)$; see (21). □

3 THE DIFFERENTIABILITY RESULTS

The most important result in this article is the following.

THEOREM 3. *There exists $\alpha > 0$ such that, if $|\gamma| < \alpha\nu$, then $u \mapsto T(\Omega + u)$ is a C^∞ mapping in the set $\mathcal{W}$.* □

In addition, the first derivative at 0, which is denoted by $T'(\Omega, \,.\,)$, can be obtained from any of the following expressions (10), (14) or (17):

THEOREM 4. *Assume $|\gamma| < \alpha\nu$.*

(i) *For all $u \in W^{1,\infty}(\mathbb{R}^d, \mathbb{R}^d)$, one has*

$$T'(\Omega; u) = \nu \int_\Omega \textstyle\sum_{ij} \sigma_{ij}(y) \Big(\sigma_{ij}(\dot{y}(u)) - \sum_k (\partial_i u_k \, \partial_k y_j + \partial_j u_k \, \partial_k y_i) + \frac{1}{2} \sigma_{ij}(y) \nabla \,.\, u \Big) \tag{10}$$

with $(\dot{y}(u), \dot{p}(u))$ being the unique solution to the linear problem

$$\begin{cases} \dot{y}(u) \in H_0^1(\Omega)^d \,, \ \dot{p}(u) \in L^2(\Omega), \ \int_\Omega \dot{p}(u) = 0 \,, \\ -\nu \, \Delta \dot{y}(u) + (\dot{y}(u) \,.\, \nabla)\, y + (y \,.\, \nabla)\, \dot{y}(u) + \nabla \dot{p}(u) = G(u, y, p) \,, \\ \nabla \,.\, \dot{y}(u) = \sum_{ij} \partial_i u_j \, \partial_j y_i \,. \end{cases} \tag{11}$$

Here, $y = y(0)$, $p = p(0)$ and $G_k(u, y, p) \in H^{-1}(\Omega)$ is given as follows for $1 \le k \le d$:

$$\begin{aligned} G_k(u, y, p) = & -\nu \textstyle\sum_{ij} \Big(\partial_j (\partial_i u_j \, \partial_i y_k) + \partial_j (\partial_j u_i \, \partial_i y_k) \Big) + \nu \sum_j \partial_j ((\nabla \,.\, u)\, \partial_j y_k) \\ & + \textstyle\sum_{ij} y_i \, \partial_i u_j \, \partial_j y_k - (y \,.\, \nabla)\, y_k \, \nabla \,.\, u + \sum_j \partial_j (\partial_k u_j \, p) - \partial_k ((\nabla \,.\, u)\, p) \,. \end{aligned} \tag{12}$$

Moreover, $y \in C^\infty(\Omega)^d$, $p \in C^\infty(\Omega)$ and, consequently,

$$G(u, y, p) = -\nu \, \Delta((u \,.\, \nabla)\, y) + \Big(((u \,.\, \nabla)\, y) \,.\, \nabla \Big)\, y + (y \,.\, \nabla)((u \,.\, \nabla)\, y) + \nabla (u \,.\, \nabla p) \,. \tag{13}$$

(ii) *One also has*

$$T'(\Omega; u) = \nu \int_\Omega \textstyle\sum_{ij} \Big(\sigma_{ij}(y)\, \sigma_{ij}(y'(u)) + \frac{1}{2} \nabla \,.\, (\sigma_{ij}(y)^2 u) \Big) \,, \tag{14}$$

with $(y'(u), p'(u))$ being the unique solution to

$$\begin{cases} y'(u) + (u \,.\, \nabla) y \in H_0^1(\Omega)^d\,, \\ (p'(u) + u \,.\, \nabla p) \in L^2(\Omega),\ \int_\Omega (p'(u) + u \,.\, \nabla p) = 0\,, \\ -\nu\, \Delta y'(u) + (y'(u) \,.\, \nabla)\, y + (y \,.\, \nabla)\, y'(u) + \nabla p'(u) = 0\,, \\ \nabla \,.\, y'(u) = 0\,. \end{cases} \tag{15}$$

Furthermore, $y'(u) \in H^1_{\text{loc}}(\Omega)^d$ and the sum in (14) satisfies

$$\textstyle\sum_{ij} \left(\sigma_{ij}(y)\, \sigma_{ij}(y'(u)) + \frac{1}{2} \nabla \,.\, (\sigma_{ij}(y)^2 u)\right) \in L^1(\Omega)\,. \tag{16}$$

(iii) *If B and D are $W^{2,\infty}$ domains and $u \in W^{2,\infty}(\mathbb{R}^d, \mathbb{R}^d)$, then $(y, p) \in H^2(\Omega)^d \times H^1(\Omega)$ and*

$$T'(\Omega; u) = \int_{\partial\Omega} u \,.\, n \left(\frac{\partial w}{\partial n} - \frac{\partial y}{\partial n}\right) \,.\, \frac{\partial y}{\partial n}\, ds\,, \tag{17}$$

with (w, q) being the unique solution to the "adjoint" problem

$$\begin{cases} w \in H_0^1(\Omega)^d \cap H^2(\Omega)^d\,, \quad q \in H^1(\Omega),\ \int_\Omega q = 0\,, \\ -\nu\, \Delta w_i + \sum_j \partial_i y_j\, w_j - \sum_j y_j\, \partial_j w_i + \partial_i q = -2\nu\, \Delta y_i\,, \quad 1 \le i \le d\,, \\ \nabla \,.\, w = 0\,.\ \square \end{cases} \tag{18}$$

In order to prove theorem 3, we will first show that $u \mapsto y(u)$ is, in a certain sense, a "differentiable" mapping. An important difficulty arises here, because $y(u)$ is a function only defined for $x \in \Omega + u$, a domain which depends on u. This is why we will introduce a suitable change of variables and we will rewrite the equations satisfied by $y(u)$ and $p(u)$ in the fixed domain Ω. Then, we will have to differentiate the transported variable $Y(u) = y(u) \circ (I + u)$, which is defined in Ω.

In the sequel, y and p stand for $y(0)$ and $p(0)$, respectively. We will check the following:

$$\dot{y}(u) = Y'(0) \,.\, u \equiv \lim_{t \to 0} \frac{y(tu) \circ (I + tu) - y}{t}\,.$$

This is the "total derivative" of $y(u)$ at 0, used in (10) to give an expression of $T'(\Omega; u)$. We will also have to use the "local derivative". In fact, we will check that

$$y'(u) = \frac{d}{dv} y(v)|_\omega(0) \,.\, u \equiv \lim_{t \to 0} \frac{y(tu)|_\omega - y|_\omega}{t} \quad \text{in } \omega\,.$$

This defines $y'(u)$ in each open set $\omega \subset\subset \Omega$ and, consequently, in the whole domain Ω. The previous local derivative was used in (14) to give an expression of $T'(\Omega; u)$. More precisely, the following result holds:

THEOREM 5. *There exists $\alpha > 0$ such that, if $|\gamma| < \alpha\nu$, then:*

(i) *The mapping $u \mapsto (y(u), p(u)) \circ (I + u)$ is C^∞ in $\mathcal{W}$, with values in the product space $H^1(\Omega)^d \times L^2(\Omega)$. Its derivative at 0 in the direction u is the unique solution $(\dot{y}(u), \dot{p}(u))$ to (11).*

(ii) *For all $\omega \subset\subset \Omega$, the mapping $u \mapsto y(u)|_\omega$ is differentiable in $\mathcal{W}$, with values in $L^2(\omega)^d$. Its derivative at 0 in the direction u is $y'(u)|_\omega$, where $y'(u)$ is uniquely defined by (15). One also has*

$$y'(u) = \dot{y}(u) - (u \cdot \nabla)\, y\,. \ \square \tag{19}$$

Theorems 3, 4 and 5 will be demonstrated in several steps, in the following sections.

4 PROOF OF THE DIFFERENTIABILITY AT 0 OF THE SOLUTION

The goal of this paragraph is to prove the following result:

LEMMA 6. *There exists $\alpha > 0$ such that, if $|\gamma| < \alpha\nu$, then the mapping $u \mapsto \big(y(u), p(u)\big) \circ (I + u)$, which is defined in $\mathcal{W}$ and takes values in $H^1(\Omega)^d \times L^2(\Omega)$, is differentiable at 0. Its derivative, denoted by $\big(\dot{y}(u), \dot{p}(u)\big)$, is uniquely determined by (11).*

The proof is based on the implicit function theorem. First, we will rewrite the equations (8) in the fixed domain Ω. For this, we have to "transport" all the terms, some of which belong to $H^{-1}(\Omega{+}u)$. But it is not clear for a distribution $f \in H^{-1}(\Omega{+}u)$ how $f \circ (I + u)$ can be defined. Contrarily, following [7] (definition 4.1), one can define $\big(f \circ (I + u)\big)\,\mathrm{Jac}(I + u) \in H^{-1}(\Omega)$, for $u \in \mathcal{W}$, such that: For any $\varphi \in H_0^1(\Omega)$, one has

$$\langle (f \circ (I + u))\,\mathrm{Jac}(I + u), \varphi\rangle_{H^{-1}(\Omega)\times H_0^1(\Omega)} = \langle f, \varphi \circ (I + u)^{-1}\rangle_{H^{-1}(\Omega+u)\times H_0^1(\Omega+u)}\,.$$

Therefore, we will "multiply" in some sense the transported equation by $\mathrm{Jac}(I + u)$. More precisely, we will use the following equalities, see [6] and [7].

We denote by $D(u)$ the operator whose components $D_i(u)$ are given as follows:

$$D_i(u) = \textstyle\sum_j M_{ij}(u)\, \partial_j\,, \quad M(u) = {}^t\left[\partial_j (I + u)_i\right]^{-1}. \tag{20}$$

Here, ${}^t\left[\partial_j(I + u)_i\right]^{-1}$ is the transposed of the inverse of the matrix of components $\partial_j(I + u)_i$. If $u \in \mathcal{W}$ and $f \in L^2(\Omega + u)$, then

$$((\partial_i f) \circ (I + u))\,\mathrm{Jac}(I + u) = \textstyle\sum_j \partial_j \Big(M_{ij}(u)\,(f \circ (I + u))\,\mathrm{Jac}(I + u)\Big)\,.$$

If $f \in H^1(\Omega + u)$. Then

$$(\partial_i f) \circ (I + u) = \textstyle\sum_j M_{ij}(u)\,\partial_j (f \circ (I + u)) = D_i(u)(f \circ (I + u))\,,$$
$$((\Delta f) \circ (I + u))\,\mathrm{Jac}(I + u) = \textstyle\sum_{ij} \partial_j \Big(M_{ij}(u)\,\mathrm{Jac}(I + u)\, D_i(u)(f \circ (I + u))\Big)\,.$$

Thus, the Navier-Stokes problem (8) can now be written as follows:

$$\begin{cases} Y(u) - g \in H_0^1(\Omega)^d\,, \quad P(u) \in L^2(\Omega),\ \int_\Omega P(u) = 0\,, \\ -\nu \sum_{ij} \partial_j \Big(M_{ij}(u)\,\mathrm{Jac}(I + u)\, D_i(u) Y_k(u)\Big) \\ \qquad + (Y(u) \cdot D(u)) Y_k(u)\,\mathrm{Jac}(I + u) \\ \qquad + \sum_j \partial_j \Big(M_{kj}(u)\, P(u)\,\mathrm{Jac}(I + u)\Big) = 0\,, \qquad 1 \le k \le d\,, \\ D(u) \cdot Y(u) = 0\,. \end{cases} \tag{21}$$

Here, we have set $Y(u) = y(u) \circ (I+u)$ and $P(u) = p(u) \circ (I+u)$. At present, we are going to replace the equation $D(u) \,.\, Y(u) = 0$ by

$$D(u) \,.\, Y(u) - \fint_\Omega D(u) \,.\, Y(u) = 0\,. \tag{22}$$

These two equations are equivalent for any $Y(u)$ such that $Y(u) - g \in H_0^1(\Omega)^d$. This is proved in [1], lemma 12.

We will also introduce the new variable $X(u) = Y(u) - g$. This leads to the following system, equivalent to (21) (which is, in turn, equivalent to (8)):

$$\begin{cases} X(u) \in H_0^1(\Omega)^d\,, \quad P(u) \in L^2(\Omega),\ \int_\Omega P(u) = 0\,, \\ -\nu \sum_{ij} \partial_j \Big(M_{ij}(u)\,\mathrm{Jac}(I+u)\, D_i(u) \big(X(u)+g\big)_k \Big) \\ \qquad + \big((X(u)+g) \,.\, D(u) \big) \big(X(u)+g\big)_k \,\mathrm{Jac}(I+u) \\ \qquad + \sum_j \partial_j \Big(M_{kj}(u)\, P(u)\, \mathrm{Jac}(I+u) \Big) = 0\,, \qquad 1 \le k \le d\,, \\ D(u) \,.\, (X(u)+g) - \fint_\Omega D(u) \,.\, (X(u)+g) = 0. \end{cases} \tag{23}$$

This equation can be written

$$H(u; X(u), P(u)) = 0 \tag{24}$$

where the function H is defined, from $\mathcal{W} \times H_0^1(\Omega)^d \times L_0^2(\Omega)$ into $H^{-1}(\Omega)^d \times L_0^2(\Omega)$, by

$$\begin{cases} H(u;\chi,\pi) = \big(F(u;\chi,\pi), R(u;\chi,\pi)\big), \quad F = (F_1, \dots, F_d)\,, \\ F_k(u;\chi,\pi) = -\nu \sum_{ij} \partial_j \Big(M_{ij}(u)\,\mathrm{Jac}(I+u)\, D_i(u)(\chi+g)_k \Big) \\ \qquad + \big((\chi+g) \,.\, D(u) \big)(\chi+g)_k \,\mathrm{Jac}(I+u) \\ \qquad + \sum_j \partial_j \Big(M_{kj}(u)\, \pi\, \mathrm{Jac}(I+u) \Big)\,, \qquad 1 \le k \le d\,, \\ R(u;\chi,\pi) = D(u) \,.\, (\chi+g) - \fint_\Omega D(u) \,.\, (\chi+g)\,. \end{cases} \tag{25}$$

Now, we check that the assumptions of the implicit function theorem are satisfied. First, H is C^1 in a neighborhood of $(0; X, P)$, where we have set $X = X(0) = y - g$, $P = P(0) = p$. Indeed, the coefficients in $D(u)$ and $M(u)$ are C^1 since, according to the results in [7], the mapping $u \mapsto M_{ij}(u)$ is C^1 in a neighborhood of 0 in $W^{1,\infty}(\mathbb{R}^d, \mathbb{R}^d)$, with values in $L^\infty(\mathbb{R}^d, \mathbb{R}^{d^2})$.

On the other hand, let us see that the differential operator $L = D_{(\chi,\pi)} H(0; X, P)$ is an isomorphism from $H_0^1(\Omega)^d \times L_0^2(\Omega)$ onto $H^{-1}(\Omega)^d \times L_0^2(\Omega)$. For each $(\chi, \pi) \in H_0^1(\Omega)^d \times L_0^2(\Omega)$, one has

$$\begin{aligned} L(\chi,\pi) &= \Big(-\nu\,\Delta\chi + (\chi \,.\, \nabla) y + (y \,.\, \nabla)\chi + \nabla\pi\,,\ \nabla \,.\, \chi - \fint_\Omega \nabla \,.\, \chi \Big) \\ &= \Big(-\nu\,\Delta\chi + (\chi \,.\, \nabla) y + (y \,.\, \nabla)\chi + \nabla\pi\,,\ \nabla \,.\, \chi \Big) \end{aligned} \tag{26}$$

because $\int_\Omega \nabla \cdot \chi = 0$. The operator L is linear and bounded from $H_0^1(\Omega)^d \times L_0^2(\Omega)$ into $H^{-1}(\Omega)^d \times L_0^2(\Omega)$. Hence, we have to check that, for each $f \in H^{-1}(\Omega)^d$ and $\phi \in L_0^2(\Omega)$, there exists a unique solution $(\chi, \pi) \in H_0^1(\Omega)^d \times L_0^2(\Omega)$ to the following system:

$$\begin{cases} -\nu\,\Delta\chi + (\chi \cdot \nabla)\,y + (y \cdot \nabla)\,\chi + \nabla\pi = f\,, \\ \nabla \cdot \chi = \phi, \end{cases} \tag{27}$$

and, also, that this solution depends continuously on the data. Since Ω is a Lipschitz domain, corollary 2.4 in [4] asserts :

$$\forall\phi \in L^2(\Omega) \text{ s. t. } \int_\Omega \phi = 0, \text{ there exists } \psi \in H_0^1(\Omega)^d \text{ s. t. } \nabla \cdot \psi = \phi\,. \tag{28}$$

Setting $\Phi = \chi - \psi$, the system (27) reduces to :

$$\begin{cases} \Phi \in V, \quad \pi \in L_0^2(\Omega), \\ -\nu\,\Delta\Phi + (\Phi \cdot \nabla)\,y + (y \cdot \nabla)\,\Phi + \nabla\pi = F \end{cases}$$

where $V = \{v \in H_0^1(\Omega)^d;\ \nabla \cdot v = 0\}$ and $F = f + \nu\Delta\psi - (\psi \cdot \nabla)y - (y \cdot \nabla)\psi$. This equation is elliptic with respect to Φ and possesses a unique solution depending continuously on the data if, for some appropriate $r = r(\mathcal{O}, \mathcal{O}') > 0$, one has:

$$\|y\|_{H_0^1(\Omega)^d} < r\,\nu\,. \tag{29}$$

Hence, if we choose $\epsilon < r$, $\alpha = \alpha(\epsilon, \mathcal{O}, D)$ as in lemma 1 and

$$|\gamma| < \alpha\nu,$$

this condition holds and L is an isomorphism.

This allows us to apply the implicit function theorem to the equation (24). We deduce that the mapping $u \mapsto \big(X(u), P(u)\big)$, which takes values in the space $H_0^1(\Omega)^d \times L_0^2(\Omega)$, is differentiable at 0. Since $y(u) \circ (I+u) = X(u) + g$ and $p(u) \circ (I+u) = P(u)$, the first part of lemma 6 is proven.

Finally, let us deduce the equations satisfied by $\big(\dot{y}(u), \dot{p}(u)\big)$. In accordance with the implicit function theorem,

$$L\big(\dot{y}(u), \dot{p}(u)\big) = -D_v H(0; X, P) \cdot u$$

for all admissible u. Taking into account (25) and also the identities

$$M'_{ik}(0) \cdot u = -\partial_i u_k \quad \text{and} \quad \frac{d}{dv}\mathrm{Jac}(I+v)(0) \cdot u = \nabla \cdot u \tag{30}$$

(see [7]), we find that $\big(\dot{y}(u), \dot{p}(u)\big)$ is a solution to (11). But this problem possesses exactly one solution, since L is an isomorphism. Consequently, lemma 6 is proven.

Remark. We can now explain why $D(u) \cdot Y(u)$ has been changed by $D(u) \cdot Y(u) - \fint_\Omega D(u) \cdot Y(u)$. Without this change, the last component of H is given by

$$R(u, \chi, \pi) = D(u) \cdot (\chi + g)\,.$$

In general, $D(u)\,.\,(\chi + g) \notin L_0^2(\Omega)$; therefore, $R(u;\chi,\pi)$ is not $L_0^2(\Omega)$-valued and, although $L = D_{(\chi,\pi)}H(0,X,P)$ remains the same, the implicit function theorem cannot be used. In short, this difficulty stems from the fact that the range of $H(u;\,.\,,\,.\,)$ is in this case a space $E(u)$ which is not included in the range of L. □

5 PROOF OF THE DIFFERENTIABILITY AT 0 OF THE DRAG

The goal of this paragraph is to prove theorem 4.

Proof of part (i): By definition, one has

$$T(\Omega+u) = \frac{\nu}{2}\int_{\Omega+u} \sum_{ij}\left(\partial_i y_j(u) + \partial_j y_i(u)\right)^2$$
$$= \frac{\nu}{2}\int_{\Omega} \sum_{ij}\Big(\sum_k \left(M_{ik}(u)\,\partial_k Y_j(u) + M_{jk}(u)\,\partial_k Y_i(u)\right)\Big)^2 \operatorname{Jac}(I+u)\,. \quad (31)$$

The map $u \mapsto z(u) = \sum_{ij}\Big(\sum_k \left(M_{ik}(u)\,\partial_k Y_j(u) + M_{jk}(u)\,\partial_k Y_i(u)\right)\Big)^2$ is differentiable at 0 with values in $L^1(\Omega)$. This is deduced from the differentiability at 0 of the $H_0^1(\Omega)$–valued mapping $u \mapsto Y(u)$. Then, by theorem 4.1 in [7], the map $u \mapsto T(\Omega+u)$ is differentiable at 0, and it's derivative at 0 in the direction u is given by

$$T'(\Omega;u) = \int_\Omega \left(\dot z(u) + z(0)\,\nabla\,.\,u\right)$$

where $\dot z(u)$ is the derivative of the map $u \mapsto z(u)$. From (30) and the fact that $M(0) = Id$, one has:

$$\dot z(u) = 2\sum_{ij}\left(\partial_i y_j + \partial_j y_i\right)\Big(\partial_i \dot y_j(u) + \partial_j \dot y_i(u) - \sum_k \partial_i u_k\,\partial_k y_j - \sum_k \partial_j u_k\,\partial_k y_i\Big)$$
$$= 2\sum_{ij}\sigma_{ij}(y)\Big(\sigma_{ij}(\dot y(u)) - \sum_k \partial_i u_k\,\partial_k y_j - \sum_k \partial_j u_k\,\partial_k y_i\Big)\,.$$

Since $z(0) = \sum_{ij}\sigma_{ij}(y)^2$, we have

$$T'(\Omega;u) = \nu\int_\Omega \sum_{ij}\sigma_{ij}(y)\Big(\sigma_{ij}(\dot y(u)) - \sum_k \left(\partial_i u_k\,\partial_k y_j + \partial_j u_k\,\partial_k y_i\right) + \frac{1}{2}\sigma_{ij}(y)\,\nabla\,.\,u\Big)\,.$$

This proves (10). The fact that $y \in C^\infty(\Omega)^d$ and $p \in C^\infty(\Omega)$ is well known (for instance, see [5]).

Proof of part (ii): Let us set

$$y'(u) = \dot y(u) - (u\,.\,\nabla)\,y, \quad p'(u) = \dot p(u) - u\,.\,\nabla p\,.$$

Using (13) we see that (11) and (15) are equivalent. On the other hand, these definitions provide the following identity:

$$\sum_{ij}\Big(\sigma_{ij}(y)\,\sigma_{ij}(y'(u)) + \frac{1}{2}\nabla\,.\,(\sigma_{ij}(y)^2 u)\Big)$$
$$= \sum_{ij}\sigma_{ij}(y)\Big(\sigma_{ij}(\dot y(u)) - \sum_k \left(\partial_i u_k\,\partial_k y_j + \partial_j u_k\,\partial_k y_i\right) + \frac{1}{2}\sigma_{ij}(y)\,\nabla\,.\,u\Big)\,.$$

Hence, (10) implies (16) and (14).

Proof of part (iii): Let us now suppose that Ω is a $W^{2,\infty}$ domain and $u \in W^{2,\infty}(\mathbb{R}^d, \mathbb{R}^d)$. According to lemma 1, one has $y \in H^2(\Omega)^d$ and $p \in H^1(\Omega)$. Consequently, one obtains from (14):

$$T'(\Omega; u) = \nu \int_\Omega \sum_{ij} \sigma_{ij}(y)\, \sigma_{ij}(y'(u)) + \frac{\nu}{2} \int_{\partial\Omega} \sum_{ij} \sigma_{ij}(y)^2\, u \,.\, n\, ds. \tag{32}$$

Since $\dot{y}(u) = 0$ and y is constant on $\partial\Omega$,

$$y'(u) = -(u \,.\, \nabla) y = -\, u \,.\, n \frac{\partial y}{\partial n} \qquad \text{on } \partial\Omega\,.$$

Therefore,

$$\begin{aligned}\nu \int_\Omega \sum_{ij} \sigma_{ij}(y)\, \sigma_{ij}(y'(u)) &= -2\nu \int_\Omega \Delta y \,.\, y'(u) - 2\nu \sum_{ij} \int_{\partial\Omega} u \,.\, n\, (\partial_i y_j + \partial_j y_i) \frac{\partial y_i}{\partial n}\, n_j\, ds \\ &= -2\nu \int_\Omega \Delta y \,.\, y'(u) - 2\nu \int_{\partial\Omega} \left|\frac{\partial y}{\partial n}\right|^2 u \,.\, n\, ds\end{aligned}$$

because

$$\begin{aligned}\sum_{ij} (\partial_i y_j + \partial_j y_i) \frac{\partial y_i}{\partial n} n_j &= \sum_{ij} \Big(\frac{\partial y_j}{\partial n} n_i + \frac{\partial y_i}{\partial n} n_j\Big) \frac{\partial y_i}{\partial n}\, n_j \\ &= \sum_j \partial_j y_j \sum_i \partial_i y_i + \sum_i \left|\frac{\partial y_i}{\partial n}\right|^2 \sum_j n_j^2 \\ &= \left|\frac{\partial y}{\partial n}\right|^2 .\end{aligned}$$

Finally, (32) gives

$$T'(\Omega; u) = -2\nu \int_\Omega \Delta y \,.\, y'(u) - \nu \int_{\partial\Omega} \left|\frac{\partial y}{\partial n}\right|^2 u \,.\, n\, ds\,.$$

If w and q are given by (18), after some manipulation, one obtains:

$$\begin{aligned}T'(\Omega; u) &= \int_\Omega \sum_i \Big(-\nu\, \Delta w_i\, y_i'(u) + \sum_j (\partial_i y_j\, w_j - y_j \partial_j w_i) y_i'(u) + \partial_i q\, y_i'(u)\Big) \\ &\qquad - \nu \int_{\partial\Omega} \left|\frac{\partial y}{\partial n}\right|^2 u \,.\, n\, ds \\ &= \langle -\nu\, \Delta y'(u) + (y'(u) \,.\, \nabla) y + (y \,.\, \nabla)\, y'(u) + \nabla p'(u), w \rangle_{H^{-1}(\Omega)^d \times H_0^1(\Omega)^d} \\ &\qquad + \nu \int_{\partial\Omega} u \,.\, n \left(\frac{\partial w}{\partial n} - \frac{\partial y}{\partial n}\right) . \frac{\partial y}{\partial n}\, ds\,.\end{aligned}$$

Using the equations (15) satisfied by $(y'(u), p'(u))$, the right hand side reduces to

$$\nu \int_{\partial\Omega} u \,.\, n \left(\frac{\partial w}{\partial n} - \frac{\partial y}{\partial n}\right) . \frac{\partial y}{\partial n}\, ds.$$

This proves (17). □

6 PROOF OF DIFFERENTIABILITY AT ANY POINT IN $\mathcal{W}$

In this paragraph, we prove the following result:

LEMMA 7. *The mapping* $u \mapsto \big(y(u), p(u)\big) \circ (I+u)$, *which takes values in* $H^1(\Omega)^d \times L^2(\Omega)$, *is differentiable at any point* $u_0 \in \mathcal{W}$. *The mapping* $u \mapsto T(\Omega + u)$ *is also differentiable at any* $u_0 \in \mathcal{W}$. □

Proof: Let $u_0 \in \mathcal{W}$ be given. We have

$$\Omega + (u_0 + v) = (\Omega + u_0) + v \circ (I + u_0)^{-1}. \tag{33}$$

for $v \in \mathcal{W}$ small enough in order to have $u_0 + v \in \mathcal{W}$. According to the results in § 6, the mapping $w \mapsto T\big((\Omega + u_0) + w\big)$ is differentiable at 0. The mapping $v \mapsto v \circ (I + u_0)^{-1}$ is linear and bounded (therefore differentiable) from $W^{1,\infty}(\mathbb{R}^d, \mathbb{R}^d)$ into itself. Consequently,

$$v \mapsto T\big((\Omega + u_0) + v \circ (I + u_0)^{-1}\big) \quad \text{is differentiable at } 0,$$

i.e. $u \mapsto T(\Omega + u)$ is differentiable at u_0.

Now we will apply the previous results to some new reference domains different from Ω. So we introduce the more explicit notation $(y(\Omega; v), p(\Omega; v))$ for the solution to the Navier-Stokes problem in $\Omega + v$. We see from (33) that, for small v,

$$y(\Omega; u_0 + v) \circ \big(I + (u_0 + v)\big) = y\big(\Omega + u_0; v \circ (I + u_0)^{-1}\big) \circ \big(I + u_0 + v\big). \tag{34}$$

On the other hand, from lemma 6, we know that the $H^1(\Omega + u_0)^d$-valued mapping $w \mapsto y(\Omega + u_0; w) \circ (I + w)$ is differentiable at 0. Thus, $v \mapsto y(\Omega; u_0 + v) \circ (I + u_0 + v)$ is differentiable at 0, i.e. $u \mapsto y(\Omega; u) \circ (I + u)$ is differentiable at u_0. A similar argument holds for the function $u \mapsto p(\Omega; u) \circ (I + u)$. □

Remark. Theorem 4.1 in [2] asserts that, when Ω is a $W^{2,\infty}$ domain, the mapping $u \mapsto \big(y(u), p(u)\big) \circ (I + u)$ is well defined for $u \in W^{2,\infty}(\mathbb{R}^d, \mathbb{R}^d) \cap \mathcal{W}$ and differentiable at 0, with values in $H^2(\Omega)^d \times H^1(\Omega)$. Adapting the previous argument, we can deduce differentiability at each point in a $W^{2,\infty}$-open ball centered at 0. □

7 HIGHER ORDER DIFFERENTIABILITY

In this paragraph, we will prove theorems 5 and 3.

Proof of part (i) of theorem 5: It remains to prove that $u \mapsto \big(Y(u), P(u)\big)$ is a C^∞ mapping (the remainder of part **(i)** has already been proved in section 4, lemma 6).

Observe that the mapping H, introduced in § 5 and defined from $\mathcal{W} \times H_0^1(\Omega)^d \times L_0^2(\Omega)$ into $H^{-1}(\Omega)^d \times L_0^2(\Omega)$, is C^∞. This is a consequence of the fact that $u \mapsto M_{ij}(u)$ and $u \mapsto \mathrm{Jac}(I + u)$ are C^∞ mappings. In turn, this stems from the following:

(a) The mapping $u \mapsto \mathrm{Jac}(I + u)$ is multilinear and, consequently, is of class C^∞.

(b) The mapping $u \mapsto M(u) - {}^t\left[\partial_i(I + u)_j\right]^{-1}$ is C^∞ on $\mathcal{W}$, because the inversion operator is indefinitely differentiable in the set of the nonsingular matrices.

From the implicit function theorem, we deduce that $u \mapsto (Y(u), P(u))$ possesses derivatives of all orders at 0. Using again (34), which can be written in the form

$$Y(\Omega; u_0 + u) = Y\big(\Omega + u_0; u \circ (I + u_0)^{-1}\big) \circ (I + u_0),$$

one also sees that $u \mapsto Y(\Omega; u)$ is C^∞ at each point $u_0 \in \mathcal{W}$. The same is true for $u \mapsto P(\Omega; u)$.

Proof of part (ii): The differentiability of the mapping $u \mapsto y(u)|_\omega$ at 0 in $L^2(\omega)^d$ and the identity (19) are consequences of the differentiability of $u \mapsto y(u)\circ(I+u)$ given in lemma 6. This is a consequence of general results on differentiation with respect to domains (see lemma 2.1 in [10]). On the other hand, (11) and (19) together imply (15). □

Proof of theorem 3: We have to check that $u \mapsto T(\Omega + u)$ is a C^∞ mapping. This is deduced from the above results and the equality (31). □

References

[1] J.A. Bello, E. Fernández-Cara, J. Lemoine, J. Simon, *The differentiability of the drag with respect to the variations of a Lipschitz domain in a Navier-Stokes flow,* to appear 1995.

[2] J.A. Bello, E. Fernández-Cara, J. Simon, *Optimal shape design for Navier-Stokes flow,* In "System modelling and optimization", ed. P. Kall, Lectures Notes in Control and Information Sciences, n° 180, p. 481–489, Springer-Verlag, Berlin 1992.

[3] J.A. Bello, E. Fernández-Cara, J. Simon, *Variation par rapport au domaine de l'énergie visqueuse dissipée dans un fluide de Navier-Stokes,* Note C. R. Acad. Sci. Paris, t. 313, Série I, p. 447–450, 1991.

[4] V. Girault, P.A. Raviart, *Finite element methods for Navier-Stokes equations,* Springer-Verlag, Berlin 1983.

[5] O.A. Ladyzhenskaya, *The mathematical theory of viscous incompressible flow,* Gordon and Breach, New York 1969.

[6] F. Murat, J. Simon, *Quelques résultats sur le contrôle par un domaine géométrique,* Rapport du L.A. 189 n° 74003, Université Paris VI, 1974.

[7] F. Murat, J. Simon, *Sur le contrôle par un domaine géométrique,* Rapport du L.A. 189 n° 76015, Université Paris VI, 1976.

[8] O. Pironneau, *On optimum design in fluid mechanics,* J. Fluid. Mech., Vol. 64, part I, p. 97–110, 1974.

[9] O. Pironneau, *Optimal shape design for elliptic systems,* Springer-Verlag, New-York 1984.

[10] J. Simon, *Differentiation with respect to the domain in boundary value problems,* Numer. Funct. Anal. and Optimiz., 2 (7 and 8), p. 649–687, 1980.

Some Examples of Optimality Conditions for Convex Control Problems with General Constraints

MAÏTINE BERGOUNIOUX URA 1803-Département de Mathématiques, U.F.R. Sciences, Université d'Orléans, Orléans, France

DAN TIBA Institute for Mathematics, Romanian Academy of Sciences , Bucuresti, Romania

1 SETTING THE PROBLEM

We are going to present some examples of optimal convex control problems, with mixed constraints on the state and the control. We give a condition which allows to set optimality conditions for such problems which are most of the time non qualified problems (in the Slater sense). These optimality conditions lead to generalized bang-bang results.

Let $V \subset H \subset V'$ compactly and densely be Hilbert spaces, $A(t) : V \to V'$, $B : U \to V'$ be linear bounded operators (U is another non trivial Hilbert space) and $L : L^2(0,T; H \times U) \to \mathbb{R}$, $l : H \to \mathbb{R}$ be convex, continuous and Gâteaux-differentiable mappings.
We consider the following optimal control problem :

$$(P) \quad \text{Min } \{ L(x,u) + l(x(T)) \}$$

subject to

$$x'(t) + A(t)x(t) = Bu(t) + f(t) \qquad \text{a.e. in }]0,T[\,, \tag{1}$$

$$[x,u] \in D \subset \mathcal{X} \times L^2(0,T;U)\ , \text{ closed convex subset}\,, \tag{2}$$

where

- $\mathcal{X} = L^2(0,T;V) \cap W^{1,2}(0,T;V')$,

- $f \in L^2(0, T; V')$,
- L is coercive in the sense

$$\exists c \text{ (a generic constant)} > 0 \text{ such that },\quad \forall [y, u] \in L^2(0, T; H \times U) \qquad L(y, u) > c\, \|u\|^2_{L^2(0,T;U)} - c, \tag{3}$$

- $\forall z \in V \quad t \mapsto A(t)z$ is V'-measurable on]0,T[, and

$$\forall z \in V \quad \|A(t)z\|_{V'} \leq c\|z\|_V \ , \tag{4}$$

$$\exists \alpha\ ,\ \exists \beta > 0\ ,\ \forall z \in V \quad \langle A(t)z, z\rangle_{V \times V'} + \alpha\|z\|^2_H \geq \beta\|z\|^2_V \ . \tag{5}$$

We assume the initial condition $x(0) = x^o \in H$ and the possible final restrictions are included in the definition of D. The evolution equation (1) has a unique solution $x \in \mathcal{X}$ for any $u \in L^2(0, T; U)$, by theorem 4.5, Barbu and Precupanu (1978), Ch. 1.

Moreover, by condition (3), it is a standard argument to show that (P) has at least one optimal pair (denoted $[x^*, u^*]$) in D if some admissibility assumption is fullfilled :

$$\exists\ [x, u] \in D \text{ such that } T(x, u) = 0\ , \tag{6}$$

with

$$\forall [x, u] \in \mathcal{X} \times L^2(0, T; U) \qquad T(x, u) = x' + A(t)x - Bu - f\ .$$

The problem (P) is a generalization of the Bolza optimal control problem studied by Barbu and Precupanu (1978), Ch.4, both with respect to the cost functional and with respect to the form of the mixed constraints. The continuity hypothesis on L and l is quite restrictive, but as we keep the constraints separately (i.e. we do not include them into the cost via the indicator function of D), then the class of examples is very large.

For state constrained control problems, one usually assume a Slater type interiority condition. In the general setting of (2), it takes the form :

$(\mathcal{S})$ $\quad \exists\ [\bar{x}, \bar{u}]$ feasible for (P) such that $\bar{x} \in \text{int}\ \{\ y \in \mathcal{X} \mid [y, \bar{u}] \in D\}$ in $\mathcal{C}(0, T; H)$,

and it has very severe implications on the set of possible applications.

It is our main concern to weaken this classical qualification constraint. Namely, instead of $(\mathcal{S})$, we shall suppose that

$(\mathcal{H})$ $\quad \exists \mathcal{M} \subset D$ bounded in $\mathcal{C}(0, T; H) \times L^2(0, T; U)$ such that $0 \in \text{int}\ T(\mathcal{M})$ in the W topology ,

where $W \subset L^2(0, T; V')$ continuously, densely is a Banach space. We also ask the following pairing compatibility condition, which is automatically fulfilled in many examples :

$$\langle v, w\rangle_{W \times W'} = \int_0^T \langle v, w\rangle_{V \times V'} \tag{7}$$

when both terms have sense.

Let us first notice that 0 appears naturally in $(\mathcal{H})$ since the problem constraint is expressed as $T(x, u) = 0$. Moreover, the elements (pairs) of the set $\mathcal{M}$ need not be feasible for (P).

REMARK 1 We have first to compare the two conditions $(\mathcal{S})$ and $(\mathcal{H})$ and it is easy to see that $(\mathcal{S})$ is always stronger than $(\mathcal{H})$. Moreover, assume that $U \subset V'$ continuously and $B : U \to V'$ is the canonical injection. Then one may ask an interiority assumption with respect to the control of the type :

$$(\mathcal{I}) \qquad \begin{array}{c} \exists\ [\tilde{x}, \tilde{u}] \text{ feasible for (P) such that} \\ \text{Int } \{\ u \in U \mid [\tilde{x}, u] \in D\ \} \ \text{ is non empty in } L^2(0,T;V'). \end{array}$$

This again implies $(\mathcal{H})$ by an argument as above. In this case the Slater condition need not be fulfilled, that is the condition $(\mathcal{H})$ is strictly weaker than $(\mathcal{S})$.

Condition $(\mathcal{H})$ may be mainly compared with the Zowe and Kurcyusz (1979) condition in the mathematical programming theory. This was previously used in abstract control problems by Tröltzsch (1983,1984), combined with interiority type assumptions at the level of applications. In the further examples, we show that the interior of the constraint set may be empty even in the uniform topology, but the argument still applies.

In the recent work of Barbu and Pavel (1993) another case of empty interior constraints is discussed for optimal control problems governed by periodic evolution equations and by a different method. Our approach is based on the penalization of the state system rather than of the constraints (as in Bonnans and Casas (1984)), which are kept explicitly throughout the proof. This is quite a classical philosophy in connection to Lagrange multipliers techniques (see for instance the monograph of Tikhomirov (1986)). In the setting of partial differential equations, it has been extensively exploited in the books of Lions (1983) or Tiba (1990), Ch.2, in connection to nonlinear singular control problems. Recently Bergounioux (1990,1992) has applied this method to control problems with state constraints governed by elliptic systems and Bergounioux, Männikkö and Tiba (1992) have studied some examples of parabolic control problems.
The main result is given by the following theorem:

THEOREM 1 The pair $[x^*, u^*]$ is an optimal pair for (P) if and only if there exists r^* in W' such that

$$\begin{aligned} \langle \nabla L_u(x^*, u^*), u^* - u\rangle_{L^2(0,T;U)} &+ \int_0^T \langle x^{*\prime} - x' + Ax^* - Ax, p^*\rangle_{V' \times V}\, dt \\ &- \langle x' + Ax - Bu - f, r^*\rangle_{W \times W'} \le 0 \,, \end{aligned} \tag{8}$$

for any $[x, u]$ such that $T(x, u) \in W$,
where p^* is the solution of the simplified adjoint system

$$-\frac{dp^*}{dt} + A^* p^* = \nabla L_x(x^*, u^*) \,, \tag{9}$$

$$p^*(T) = \nabla l(x^*(T)) \,. \tag{10}$$

2 A FIRST EXAMPLE : EMPTY INTERIOR CONSTRAINTS

We analyse in some detail the following example of optimal control problem governed by a parabolic partial differential equation :

$$(PP) \qquad \text{Min} \left\{ \frac{1}{2} \int_Q (y - z_d)^2 \, dx \, dt + \frac{N}{2} \int_Q u^2 \, dx \, dt \right\}$$

subject to :

$$\frac{\partial y}{\partial t} - \Delta y = f + u \quad \text{in } Q = \Omega \times]0, T[\,, \tag{11}$$

$$y(x,t) = 0 \quad \text{on } \Sigma = \partial\Omega \times]0, T[\,, \tag{12}$$

$$y(x,0) = y_o(x) \quad \text{in } \Omega\,, \tag{13}$$

and the constraints

$$e(x,t) \le y(x,t) \le g(x,t) \qquad \text{a.e. in } Q\,, \tag{14}$$

$$a(x,t) \le u(x,t) \le b(x,t) \qquad \text{a.e. in } Q\,. \tag{15}$$

Here Ω is a smooth, open and bounded domain of $\mathbb{R}^n$, $z_d \in L^2(Q)$, $N \ge 0$, $y_o \in L^2(\Omega)$, f, a, b are in $L^\infty(Q)$ and e, g in $\mathcal{C}(\bar{Q})$. We denote

$$K = \{ y \in L^2(0,T; H_o^1(\Omega)) \cap W^{1,2}(0,T; H^{-1}(\Omega)) \mid e \le y \le g,\ y(.,0) = y_o \}\,,$$

$$U_{ad} = \{ u \in L^2(Q) \mid a \le u \le b \text{ a.e. in } Q \}\,,$$

$$D = K \times U_{ad}\,,$$

which are closed convex sets. One has to assume the compatibility condition

$$e \le y^o \le g \quad \text{in } \Omega\,,$$

and some admissibility hypothesis. We ask

$$(\mathcal{E}) \qquad \exists \alpha > 0,\ \exists \tilde{u} \in U_{ad} \ \text{ such that } \ e \le Y(\tilde{u} - \alpha) \le Y(\tilde{u} + \alpha) \le g \quad \text{in } Q\,,$$

where Y is the solution operator $u \mapsto y$ defined by (11-13). By comparison, $(\mathcal{E})$ implies that the pair $[\tilde{u}, Y(\tilde{u})]$ is feasible for (PP). However, this is not an interiority assumption since e may be equal to g in some points. For instance, we may allow $e(x,t) = g(x,t) = 0$ on the border of the domain. Moreover, the mappings $\tilde{u} + \alpha, \tilde{u} - \alpha$ need not belong to U_{ad}, which may also have a void interior, i.e. $a = b$ on some subset.

In order to apply the abstract theory, we take the spaces $V = H_o^1(\Omega)$, $H = U = L^2(\Omega)$, the operators $A(t) : V \to V'$, $A(t)z = -\Delta z$, $B : H \to V'$, $B = i$, the canonical injection, and the mappings $l = 0$ and

$$L(y,u) = \frac{1}{2} \int_Q (y - z_d)^2 \, dx \, dt + \frac{N}{2} \int_Q u^2 \, dx \, dt\,.$$

The hypothesis $(\mathcal{H})$ is a clear consequence of $(\mathcal{E})$ with $W = L^\infty(Q)$, by fixing

$$y_\xi = Y(u_\xi) = Y(\tilde{u} + \alpha\xi),\ \ \xi \in W,\ \ \|\xi\|_W = 1\,.$$

A comparison argument shows that $[y_\xi, u_\xi] \in D$ for any ξ as above and we can choose in $(\mathcal{H})$ the bounded set

$$\mathcal{M} = \text{conv} \{ [y_\xi, u_\xi] \mid \xi \in W,\ \|\xi\|_W = 1 \}\,.$$

Then relation (8) may be rewritten as :

$$\int_Q Nu^*(u^* - u)\, dx\, dt + \int_Q (y^{*\prime} - y' - \Delta(y^* - y))p^*\, dx\, dt \\ - \langle y' - \Delta y - u - f, r^* \rangle_{W \times W'} \qquad \leq 0\,, \tag{16}$$

for any y in K, u in U_{ad} such that $T(y, u) \in W = L^\infty(Q)$.
Notice also that, since $f \in W$, $U_{ad} \subset W$, the last condition ($T(y, u) \in W = L^\infty(Q)$) is equivalent to a regularity condition on y : $y' - \Delta y \in L^\infty(Q)$, which is satisfied by y^*.
Here $[y^*, u^*]$ is the optimal pair of (PP) and p^* satisfies

$$\begin{aligned} -p^{*\prime} - \Delta p^* &= y^* - z_d && \text{in } Q \\ p^* &= 0 && \text{on } \Sigma \\ p^*(T) &= 0 && \text{in } \Omega\,. \end{aligned}$$

Choosing in turn $u = u^*$ and $y = y^*$ in (16), we get

$$\forall y \in K \text{ such that } y' - \Delta y \in L^\infty(Q) \quad \langle y^{*\prime} - y' - \Delta(y^* - y), p^* + r^* \rangle_{W \times W'} \leq 0\,, \tag{17}$$

$$\forall u \in U_{ad} \qquad \langle Nu^* - r^*, u^* - u \rangle_{W \times W'} \leq 0\,. \tag{18}$$

The relations (17) and (18) represent the decoupled form of the optimality condition (16).

Now, we are going to use these above relations to get some more precise information about the optimal pair. Let us define the sets

$$Q_e = \{\, (x,t) \in \bar{Q} \mid y^*(x,t) = e(x,t) \,\}\,,\ Q_g = \{\, (x,t) \in \bar{Q} \mid y^*(x,t) = g(x,t) \,\}\,,$$

$$Q^o = Q - (Q_e \cup Q_g)\,.$$

Thanks to (11) -(13) and the Sobolev embedding theorem, we infer that $y^* \in \mathcal{C}(\bar{Q})$ if y_o is regular. Then Q_e and Q_g are closed sets and Q^o is an open subset of Q. Let $d \in \mathcal{D}(Q)$ be a test function with compact support $supp\ d \subset Q^o$. By the continuity of y^*, e, g and the compactness of $supp\ d$, one can find $\rho > 0$ such that $y^* + \rho d$ and $y^* - \rho d$ remain in K . Obviously, they are also regular and we can use them in (17) as test functions to infer :

$$\left\langle p^* + r^*, \frac{\partial d}{\partial t} - \Delta d \right\rangle_{W \times W'} = 0\,,$$

for any $d \in \mathcal{D}(Q)$ with compact support in Q^o. Taking in account this relation and the equation satisfied by p^*, we see that there exists a distribution $j \in \mathcal{D}'(Q)$ with support included in $Q - Q^o$, such that

$$\frac{\partial r^*}{\partial t} - \Delta r^* + j = y^* - z_d \quad \text{in } \mathcal{D}'(Q)\,. \tag{19}$$

The previous equation is another familiar form of the adjoint system for state constrained control problems. In particular, it shows that $r^* \in W^{2,1,p}_{loc}(Q^o)$ for $p > 1$ if z_d belongs to $L^p(Q)$. Then $r^* \in \mathcal{C}(Q^o)$ by the Sobolev theorem if p is big enough.
We are now prepared to give a result on the structure of the optimal pair of (PP), which may be termed as a generalized bang-bang result, Tröltzsch (1983). We suppose that N=0.

THEOREM 2 We have :

$$Q^o \subseteq \{\, (x,t) \mid y^*(x,t) = z_d(x,t) \,\} \cup \{\, (x,t) \mid u^*(x,t) = a(x,t) \,\} \cup \{\, (x,t) \mid u^*(x,t) = b(x,t) \,\}.$$

Proof.- Choose $u = u^*$ in $Q - Q^o$ so that (18) yields

$$\forall u \in U_{ad} \qquad \int_{Q^o} r^*(u^* - u)\, dx\, dt \ \geq 0\,. \tag{20}$$

Since $r^* \in \mathcal{C}(Q^o)$, obviously

$$Q^o = \{\, (x,t) \in Q^o \mid r^*(x,t) > 0 \,\} \cup \{\, (x,t) \in Q^o \mid r^*(x,t) < 0 \,\} \cup \{\, (x,t) \in Q^o \mid r^*(x,t) = 0 \,\},$$

and (20) shows that $u^* = b$ on the first set and $u^* = a$ on the second set. If the last set has positive measure, then (19) and the maximal regularity of r^* on Q^o gives that $y^* = z_d$ on this subset.

REMARK 2 Taking into account the definition of Q^o, we see that at least one from y^* and u^* equals the extremal values $e,\ g,\ a,\ b$ or z_d in any point of Q. A similar analysis may be pursued when $N > 0$, but the structure of the optimal pair will be more complicated.

3 A SECOND EXAMPLE :"BOTTLENECK" PROBLEMS

We examine "bottleneck" type problems which were introduced by Bellman (1957) in connection with some models for industrial production processes. They were discussed by a different approach in the work of Miricà (1985).

We assume that the state equation and the cost functional are the same as in the previous example, with $f \in L^\infty(Q)$, $y_o \in W_o^{1,\infty}(\Omega) \cap W^{2,\infty}(\Omega)$, $y_o \geq 0$ a.e. in Ω, but the constraint has the form

$$|y| \leq u \quad \text{a.e. in } Q\,. \tag{21}$$

This is equivalent with : $-u \leq y \leq u$, so that the set D defined by (21) is convex.

REMARK 3 If $f \geq 0$, the maximum principle gives $y \geq 0$ (with (11)) and the constraint (21) is equivalent to

$$0 \leq y \leq u \quad \text{a.e. in } Q\,, \tag{21b}$$

which is similar to the original form considered by Bellman (1957). We also underline that the boundary condition (12) shows that the feasible pairs **are not** in the interior of D even in the $L^\infty(Q)$-topology, that is Slater-type conditions cannot be valid in (21b).

Take now $\tilde{u} = \alpha e^t$, $\alpha > 0$ and $\tilde{y}$ the solution of (3.4)-(3.6) associated to $\tilde{u}$. If α is great enough, then $f + \tilde{u} \geq 0$ a.e. in Q and $\tilde{y} \geq 0$ a.e in Q by the maximum principle. Let $w = \alpha e^t - \alpha \geq 0$ a.e in Q. We notice that w satisfies

$$\begin{aligned} \frac{\partial w}{\partial t} - \Delta w &= \alpha e^t && \text{in } Q \\ w &= (\alpha e^t - \alpha)_{|\Sigma} \ \geq 0 && \text{on } \Sigma \\ w(0) &= 0 && \text{in } \Omega\,. \end{aligned} \tag{22}$$

Let us denote y^v the solution of (3.4)-(3.6) associated to v (so that y^o is associated to $v = 0$). By comparison, it yields that $w \geq \tilde{y} - y^o$, that is $\tilde{u} - \alpha + y^o \geq \tilde{y} \geq 0$. There is some constant m such that $-m \leq y^o \leq m$ a.e in Q. Then, if $\alpha > 2m$ is large enough, we have

$$0 \leq \tilde{y} \leq \tilde{u} - m \qquad \text{a.e in } Q \,. \tag{23}$$

The pair $[\tilde{y}, \tilde{u}]$ is feasible and (23) shows that hypothesis $(\mathcal{H}')$ is satisfied. Indeed we take

$$\mathcal{M} = \{\ [y^v, \tilde{u}] \mid v \in B_{L^p(Q)}(\tilde{u}, \lambda)\ \},\ p > \frac{n+2}{2}\,,$$

where $B_{L^p(Q)}(\tilde{u}, \lambda)$ is the L^p-ball centered in $\tilde{u}$ with radius λ, and $\lambda > 0$ is small. By the continuity with respect of the right-hand side and the Sobolev embedding theorem, we can choose λ such that

$$\|\tilde{y} - y_v\|_{C(\bar{Q})} \leq m\,,$$

that is $\mathcal{M} \subset D$. Moreover $T(\mathcal{M}) = B_{L^p(Q)}(0, \lambda)$ and $(\mathcal{H})$ is fulfilled in $L^p(Q)$.

Finally, we also able to prove that a generalized bang-bang result remains valid in this case as well, under some regularity assumptions.

THEOREM 3 Let N be equal to 0 and assume that y^*, u^* exist and are continuous functions on $\bar{Q}$. Then

$$Q = \{\ (x,t) \in Q \mid |y^*(x,t)| = u^*(x,t)\ \} \cup \{\ (x,t) \in Q \mid y^*(x,t) = z_d(x,t)\ \}\,. \tag{24}$$

(The two sets above need not be disjoint. The first one corresponds to the case where the constraint is active.)

REFERENCES

V. Barbu - N. Pavel, Optimal control problems with two point boundary condition, *Journal of Optimization Theory and Applications, Vol.77*, pp. 51-78 (1993).

V. Barbu - Th. Precupanu, *Convexity and optimization in Banach spaces*, Sijthoff and Noordhoff, Leyden (1978).

R. Bellman, *Dynamic programming*, Princeton University Press (1957).

M. Bergounioux, Contrôle optimal de problèmes elliptiques avec contraintes sur l'état, *Comptes Rendus de l'Académie des Sciences, t 310, série I*, pp. 391-396 , (1990).

M. Bergounioux, A Penalization method for optimal control of elliptic problems with state constraints,*SIAM Journal on Control and Optimization - Vol 30 , n° 2* , pp. 305-323,(1992).

M. Bergounioux - T. Männikkö - D. Tiba, On nonqualified parabolic control problems, *Preprint n° 148*, University of Jyväskylä, Finland (July 1992).

J.F. Bonnans - E. Casas, On the choice of the function spaces for some state-constrained control problems, *Numerical Functional Analysis and Optimization 7 (4)*, pp. 333-348 (1984).

J.L. Lions , *Contrôle des systèmes distribués singuliers*, Gauthier-Villars, Paris (1983).

S. Mircà, Optimal feedback control for a class of bottleneck problems, *Journal Mathematical Analysis and Applications, 112, n° 1*, pp. 221-235 (1985).

D. Tiba, Optimal control of nonsmooth distributed parameter systems, *Lecture Notes n° 1459*, Springer-Verlag, Berlin (1990).
V. Tikhomirov, *Fundamental principles of the theory of extremal problems*, J.Wiley, Chichester, (1986).
F. Tröltzsch, A modification of the Zowe and Kurcyusz regularity condition with application to the optimal control of Noether operator equations with constraints on the control and the state, *Math. Operationforsch. Statist., Ser. Optimization, vol.14, n° 2*, pp.245-253 (1983).
F. Tröltzsch, *Optimality conditions for parabolic control problems and applications*, Teubner Texte, Leipzig, (1984).
J. Zowe - S. Kurcyusz, Regularity and stability for the mathematical programming problem in Banach spaces, *Applied mathematics and Optimization, 5*, pp.49-62 (1979).

Cauchy Problem for Hamilton–Jacobi Equations and Dirichlet Boundary Control Problems of Parabolic Type

PIERMARCO CANNARSA[1] Dipartimento di Matematica, Universitá di Roma "Tor Vergata", Via O. Raimondo, 00173 Roma, Italy

MARIA ELISABETTA TESSITORE Dipartimento di Matematica, Universitá di Roma "La Sapienza", Piazzale A. Moro 2, 00185 Roma, Italy

Abstract

The paper is concerned with the initial value problem for an infinite dimensional Hamilton–Jacobi equation related to boundary control problems of Dirichlet type for semilinear parabolic systems. The stationary version of this problem was studied by viscosity solution methods in [6]. Here we adapt the same techniques to the time dipendent case obtaining an existence and uniqueness result.

1 Introduction

In this paper we study the Cauchy problem related to the Hamilton–Jacobi equation

$$\begin{cases} -\dfrac{\partial v(t,x)}{\partial t} + H(t, A^{\beta}x, D_x v(t,x)) + < Ax + A^{-\beta}F(A^{\beta}x), D_x v(t,x) >= 0 \\ v(T,x) = g(A^{\beta}x) \end{cases} \tag{1.1}$$

where $t \in [0,T], x \in X$ and X is a real Hilbert space with scalar product $< \cdot, \cdot >$ and norm $|\cdot|$. In the above equation, $A : D(A) \subset X \to X$ is a self–adjoint strictly accretive

[1]The research of this author was supported in part by the Institute for Mathematics and its Application with funds provided by the National Science Foundation and in part by the Italian National Project MURST 40% "Problemi nonlineari...".

operator in X, A^β is the β– fractional power of A, $\frac{3}{4} < \beta < 1$, $F : X \to X$ is a bounded Lipschitz map and H is a real valued Lipschitz function.

To motivate this research let us consider the problem of minimizing the functional

$$J(t_0, x_0; \gamma) = \left\{ \int_{t_0}^{T} L(s, x(s; t_0, x_0, \gamma), \gamma(s))ds + g(x(T; t_0, x_0, \gamma)) \right\} \tag{1.2}$$

over all measurable functions $\gamma : [t_0, T] \to \tilde{U}$, where $\tilde{U}$ is closed and bounded subset of another real Hilbert space U. The functions $L : [t_0, T] \times X \times U \to \mathbb{R}$ and $g : X \to \mathbb{R}$ are assumed to be smooth and the state variable $x(\cdot; t_0, x_0, \gamma)$ is the mild solution of the infinite dimensional equation

$$\begin{cases} x'(t) + Ax(t) + F(x(t)) = A^\beta B\gamma(t) \\ x(t_0) = x_0. \end{cases} \tag{1.3}$$

The value function of problem (1.2)–(1.3) is defined as

$$u(t_0, x_0) = \inf_{\gamma(\cdot)\in\tilde{U}} J(t_0, x_0; \gamma) \tag{1.4}$$

Optimal control problems of type (1.4)–(1.3) are related to boundary control of parabolic equations. Indeed, if $B = A^{1-\beta}D$, and D is the Dirichlet map, see e. g. [2], then equation (1.3) is the abstract version of a semilinear parabolic equation, controlled at the boundary by a Dirichlet datum.

If we use in (1.3) the change of variable $y(t) = A^{-\beta}x(t)$, then problem (1.4)–(1.3) is transformed into another optimal control problem. The value function v of this new problem is given by

$$v(t_0, y_0) = \inf_{\gamma(\cdot)\in\tilde{U}} \left\{ \int_{t_0}^{T} L(s, A^\beta y(s; t_0, y_0, \gamma), \gamma(s))ds + g(A^\beta y(T; t_0, y_0, \gamma)) \right\} \tag{1.5}$$

where $y_0 = A^{-\beta}x_0$. Moreover the Dynamic Programming method suggests that v should be the unique viscosity solution of the Cauchy problem (1.1).

Infinite dimensional Hamilton–Jacobi equations have been studied by several authors, see [1], [7] Parts I to VI, [8] , [10], [11]. In all these papers, however, it is assumed that $\beta = 0$ in (1.1), while more general classes of operators A are considered. In [4] stationary Hamilton–Jacobi equations of the form

$$\lambda v(x) + H(x, A^\beta Dv(x)) + < Ax + F(x), Dv(x) >= 0, \tag{1.6}$$

have been studied using viscosity solutions for $\beta \in \left(\frac{1}{4}, \frac{1}{2}\right)$. In [5] the Cauchy problem related to equation (1.6) has been considered. In [6] we studied the stationary Hamilton–Jacobi equation

$$\lambda v(x) + H(A^\beta x, Dv(x)) + < Ax + A^{-\beta}F(A^\beta x), Dv(x) >= 0, \tag{1.7}$$

always using a viscosity approach. We also note that partial results about equation (1.7) have been proved in [7] Parts VII, using a different set–up then ours.

In Section 3 of the present paper, we adapt the definition of viscosity solution to problem (1.1) and we obtain a comparison result between sub and supersolutions of this problem.

Here we follow the approach of [6], but we only consider Lipschitz continuous viscosity solutions of (1.1) for semplicity. An additional difference with respect to the stationary case is the fact that one cannot expect solutions of (1.1) to be uniformly continuous on the whole strip $[0,T] \times X$. For this purpose, we require, as in [8], that the initial datum is assumed in the following sense

$$\lim_{R\to\infty} \lim_{t\downarrow 0} \sup_{|A^{\frac{1}{2}}x|\le R} [w(t,x) - g(A^{\beta}e^{-tA}x)] = 0.$$

Finally, in Section 4 we show that the value function v is a viscosity solution of (1.1), thus providing an existence result. In this way we also obtain a characterisation of u since $v(t, A^{-\beta}x) = u(t,x)$.

2 Preliminaries

Let X and U be two real Hilbert spaces and let $\tilde{U} \subset U$ be closed and bounded. Let $x_0 \in X$ and $t_0 \in [0,T]$. Now we consider the problem of minimizing (1.4) over all measurable functions $\gamma : [t_0, T] \to \tilde{U}$ (usually called controls). We recall that, in (1.4), $x(\cdot; t_0, x_0, \gamma)$ is the mild solution of (1.3), that is the solution of the integral equation

$$x(t) = e^{-(t-t_0)A}x_0 + \int_{t_0}^{t} e^{-(t-s)A}F(x(s))ds + A^{\beta}\int_{t_0}^{t} e^{-(t-s)A}B\gamma(s)ds \,. \qquad (2.1)$$

In (1.3), A^{β} denotes the fractional powers of the operator A, see [9]. Here we assume that L and g satisfy the following assumptions

$$\begin{array}{ll} (i) & L \in C([0,T]\times X \times \tilde{U}),\ |L(t,x,\gamma)| \le C_L\,,\ \forall (t,x,\gamma) \in [0,T]\times X\times\tilde{U}\,,\ C_L > 0; \\ (ii) & |L(t,x,\gamma) - L(t,y,\gamma)| \le K_L|x-y|,\ \forall \gamma \in \tilde{U},\ x,y \in X\,,\ \forall t \in [0,T]\,,\ K_L > 0; \\ (iii) & g \text{ is bounded from below and } \forall R > 0 \text{ and some constant } C_{g,R} > 0: \\ & |g(x) - g(y)| \le C_{g,R}|A^{-\beta}(x-y)|, \forall x,y \in X \text{ such that } |A^{-\beta}x|, |A^{-\beta}y| \le R. \end{array} \qquad (2.2)$$

Assumption (2.2) (iii) is equivalent to requiring that g possesses a Lipschitz continuous extension to the completion X_{β} of X with respect to the norm $|x|_{\beta} = |A^{-\beta}x|$. Such a condition is necessary in order to define the terminal cost in functional (1.2) when $x_0 \in X$.

Moreover we assume

$$\begin{array}{ll} (i) & A : D(A) \subset X \to X \text{ is a closed linear operator} \\ & \text{such that } A = A^* \text{ and } < Ax, x > \geq \omega |x|^2 \text{ for some } \omega > 0 \text{ and all } x \in D(A); \\ (ii) & \text{the inclusion } D(A) \subset X \text{ is dense and compact ;} \\ (iii) & F : X \to X \ , \ |F(x) - F(y)| \leq K_F |x - y| \ , \ |F(x)| \leq C_F, \ \ \forall x, y \in X; \\ (iv) & \beta \in \left(\frac{3}{4}, 1\right); \\ (v) & \text{there exists } \rho > 0, \ \text{ such that } B \in \mathcal{L}(U, D(A^\rho)). \end{array} \tag{2.3}$$

for some constants K_F, $C_F > 0$.

In assumption (v) above, we have denoted by $\mathcal{L}(U, D(A^\rho))$ the Banach space of all bounded linear operators $B : U \to D(A^\rho)$, where $D(A^\rho)$ is equipped with the graph norm.

We note that (i) and (ii) imply that $-A$ is the infinitesimal generator of an analytic semigroup satisfying $||e^{-tA}|| \leq e^{-\omega t}$ for some $\omega > 0$ and all $t \geq 0$. Moreover we recall that for every $\theta \in [0, 1]$ there exists a constant $M_\theta > 0$ such that

$$|A^\theta e^{-tA} x| \leq \frac{M_\theta}{t^\theta} |x|, \ \ \forall t > 0, \forall x \in X. \tag{2.4}$$

In addition, let $\theta \in (0, 1]$ and $\alpha \in (0, \theta)$. Then, a well known interpolation inequality, see e.g. [9], states that for every $\sigma > 0$ there exists $C_\sigma > 0$ such that

$$|A^\alpha x| \leq \sigma |A^\theta x| + C_\sigma |x|, \ \ \ \forall x \in D(A^\theta) \tag{2.5}$$

and there exists $C_{\alpha\theta} > 0$ such that

$$|A^\alpha x| \leq C_{\alpha\theta} |A^\theta x|^{\frac{\alpha}{\theta}} |x|^{1-\frac{\alpha}{\theta}}, \ \ \ \forall x \in D(A^\theta). \tag{2.6}$$

It is well known that, under the above assumptions, problem (2.1) has a unique solution in $L^2(t_0, T; X)$ for any $T > 0$.

Control processes as above are very important for applications. In fact, (1.3) describes the evolution of a system which is governed by a parabolic PDE and controlled by Dirichlet type boundary data.

We now transform (1.3) by the change of variable

$$y(t) = A^{-\beta} x(t). \tag{2.7}$$

More precisely, let $y_0 \in X$ and denote by $y(\cdot; t_0, y_0, \gamma)$ the mild solution of

$$\begin{cases} y'(t) + Ay(t) + A^{-\beta} F(A^\beta y(t)) = B\gamma(t) \\ y(t_0) = y_0 \in X \end{cases} \tag{2.8}$$

We recall a known regularity result, see e. g. [6] for a proof.

Proposition 2.1 *Assume that (2.3) holds and fix $T > 0$. Let $\gamma : [0, T] \to \tilde{U}$. Then for any $y_0 \in X$ there exists a unique solution*

$$y \in C([t_0, T]; X) \cap L^1(t_0, T; D(A^\beta)). \tag{2.9}$$

Moreover, if $y_0 \in D(A^{\frac{1}{2}})$, then

$$y \in C([t_0, T]; D(A^{\frac{1}{2}})) \cap L^2(t_0, T; D(A)) \cap W^{1,2}(t_0, T; X). \tag{2.10}$$

Finally, if $y_0 \in D(A)$ and $\gamma(\cdot)$ is bounded, then

$$y \in C([t_0, T]; D(A)). \tag{2.11}$$

By inserting the change of variable (2.7) in the cost functional J in (1.2), we obtain a new optimal control problem whose value function v is given by (1.5). It is easy to realize that value functions v and u are related by the formula $u(t, x) = v(t, A^{-\beta}x)$, $\forall (t, x) \in [t_0, T] \times X$. In particular, u is uniquely determined once v has been characterized. Therefore, we will study problem (2.8)–(1.5) instead of (1.3)–(1.4).

The Hamilton–Jacobi–Bellman equation associated with (1.5) is

$$\begin{cases} -\dfrac{\partial v(t,x)}{\partial t} + H(t, A^\beta x, D_x v(t,x)) + < x + A^{-\beta} F(A^\beta x), D_x v(t,x) >= 0 \\ v(T, x) = g(A^\beta x) \end{cases} \tag{2.12}$$

where

$$H(t, x, p) = \sup_{\gamma \in \tilde{U}} \left[- < B\gamma, p > - L(t, x, \gamma) \right].$$

3 Comparison of viscosity solutions

In this Section we study the Hamilton–Jacobi equation

$$\begin{cases} \dfrac{\partial v(t,x)}{\partial t} + H(t, A^\beta x, D_x v(t,x)) + < x + A^{-\beta} F(A^\beta x), D_x v(t,x) >= 0 \\ v(0, x) = g(A^\beta x) \end{cases} \tag{3.1}$$

We assume that (2.3) holds and that $H : [0, T] \times X \times X \to \mathbb{R}$ is a function satisfying

$$|H(t, x, p) - H(s, y, q)| \leq K_H (|t - s| + |x - y| + |p - q|) \text{ for some } K_H > 0. \tag{3.2}$$

Let $w, \varphi : (0, T) \times D(A^{\frac{1}{2}}) \to \mathbb{R}$ be given. For any $\delta > 0$ we define $M_\delta^+(w, \varphi)$ to be the set of all points $(\bar{t}, \bar{x}) \in (0, T) \times D(A^{\frac{1}{2}})$ at which the function

$$(t, x) \to w(t, x) - \varphi(t, x) - \frac{\delta}{2} |A^{\frac{1}{2}} x|^2,$$

defined on $(0,T) \times D(A^{\frac{1}{2}})$ attains a local maximum. Similarly, we denote by $M_\delta^-(w,\varphi)$ the set of all points $(\bar{t},\bar{x}) \in (0,T) \times D(A^{\frac{1}{2}})$ at which the function

$$(t,x) \to w(t,x) - \varphi(t,x) + \frac{\delta}{2}|A^{\frac{1}{2}}x|^2,$$

defined on $(0,T) \times D(A^{\frac{1}{2}})$ attains a local minimum. Moreover, we set $a^+ = \max\{a,0\}$ and $a^- = \min\{a,0\}$.

Definition 3.1 *We say that a bounded continuous function* $w : [0,T] \times X \to \mathbb{R}$ *is a viscosity subsolution of* (3.1) *if* w *is sequentially weakly upper semicontinuous on* $(0,T) \times X$, *and, for every* $\varphi \in C^1((0,T) \times D(A^{\frac{1}{2}}))$ *and* $\delta > 0$,

$$\begin{array}{ll} (i) & M_\delta^+(w,\varphi) \subset (0,T) \times D(A)\,; \\ (ii) & \dfrac{\partial \varphi(t,x)}{\partial t} + H(t, A^\beta x, D\varphi(t,x) + \delta A x) + \left\langle Ax + A^{-\beta}F(A^\beta x), D\varphi(t,x)\right\rangle \\ & +\delta|Ax|^2 + \delta\left\langle Ax, A^{-\beta}F(A^\beta x)\right\rangle \le 0\ , \forall (t,x) \in M_\delta^+(w,\varphi)\,; \\ (iii) & \lim\limits_{R\to\infty} \lim\limits_{t \downarrow 0} \sup\limits_{|A^{\frac{1}{2}}x| \le R} [w(t,x) - g(A^\beta e^{-tA}x)]^+ = 0. \end{array} \tag{3.3}$$

We say that w *is a viscosity supersolution of* (3.1) *if* w *is sequentially weakly lower semicontinuous on* $(0,T) \times X$, *and, for every* $\varphi \in C^1((0,T) \times D(A^{\frac{1}{2}}))$ *and* $\delta > 0$,

$$\begin{array}{ll} (i) & M_\delta^-(w,\varphi) \subset (0,T) \times D(A)\,; \\ (ii) & \dfrac{\partial \varphi(t,x)}{\partial t} + H(t, A^\beta x, D\varphi(t,x) - \delta A x) + \left\langle Ax + A^{-\beta}F(A^\beta x), D\varphi(t,x)\right\rangle \\ & -\delta|Ax|^2 - \delta\left\langle Ax, A^{-\beta}F(A^\beta x)\right\rangle \ge 0\ , \forall (t,x) \in M_\delta^-(w,\varphi)\,; \\ (iii) & \lim\limits_{R\to\infty} \lim\limits_{t \downarrow 0} \sup\limits_{|A^{\frac{1}{2}}x| \le R} [w(t,x) - g(A^\beta e^{-tA}x)]^- = 0. \end{array} \tag{3.4}$$

We say that w *is a viscosity solution of* (3.1) *if it is both a viscosity subsolution and a supersolution of such a problem.*

Remark 3.2 The definition of viscosity solution can be easily formulated for problems with a terminal condition like (2.12). More precisely, we say that a continuous function $v : (0,T) \times X \to \mathbb{R}$ is a viscosity subsolution of (2.12) if $w(t,x) = v(T-t,x)$ is a viscosity subsolution of (3.1). Supersolution and solution of (2.12) are similarly defined.

Now we give a comparison result between viscosity subsolutions and supersolutions of (3.1).

Theorem 3.3 *Assume that* (2.3) *and* (3.2) *hold true. Let* u *and* v *be a viscosity subsolution and supersolution of the Hamilton–Jacobi equation* (3.1) *respectively. If* u *and* v *are Lipschitz continuous, then*

$$u(t,x) \le v(t,x), \quad \forall (t,x) \in [0,T] \times X. \tag{3.5}$$

Proof–For the reader' s convenience we will give the proof in several steps.
Step I–It is enough to show that

$$u_\sigma(t,x) \le v_\sigma(t,x)\ \forall (t,x) \in (0,T) \times D(A^{\frac{1}{2}}),\ \forall \sigma > 0$$

where $u_\sigma(t,x) = u(t,x) - \dfrac{\sigma}{T-t}$ and $v_\sigma(t,x) = v(t,x) + \dfrac{\sigma}{T-t}$.
It is easy to see that $u_\sigma(t,x)$ is a viscosity subsolution of the problem

$$\begin{cases} \dfrac{\partial u_\sigma(t,x)}{\partial t} + H(t, A^\beta x, D_x u_\sigma(t,x)) + < Ax + A^{-\beta}F(A^\beta x), D_x u_\sigma(t,x) > \le -\dfrac{\sigma}{T^2} \\ u_\sigma(0,x) = g(A^\beta x) - \dfrac{\sigma}{T} \end{cases} \tag{3.6}$$

while $v_\sigma(t,x)$ is a viscosity supersolution of

$$\begin{cases} \dfrac{\partial v_\sigma(t,x)}{\partial t} + H(t, A^\beta x, (D_x v_\sigma(t,x)) + < Ax + A^{-\beta}F(A^\beta x), D_x v_\sigma(t,x) > \ge \dfrac{\sigma}{T^2} \\ v_\sigma(0,x) = g(A^\beta x) + \dfrac{\sigma}{T} \end{cases} \tag{3.7}$$

We will argue by contradiction.

If $u_\sigma \le v_\sigma$ does not hold, then there exists $(t_0, x_0) \in (0,T) \times D(A^{\frac{1}{2}})$ such that

$$u_\sigma(t_0,x_0) - v_\sigma(t_0,x_0) = \alpha > 0.$$

Step II–For $\varepsilon > 0$, let us define

$$\begin{aligned} &\phi_{\varepsilon,\delta}(t,s,x,y) \\ &= u_\sigma(t,x) - v_\sigma(s,y) - \frac{1}{2\varepsilon} < A^{\frac{1}{2}}(x-y), x-y > - \frac{\delta}{2}[< Ax, x > + < Ay, y >] \\ &- \frac{(t-s)^2}{2\varepsilon},\ \forall (t,x)\ ,\ (s,y) \in (0,T) \times D(A^{\frac{1}{2}}). \end{aligned} \tag{3.8}$$

Let $Q_\tau = [\tau, T-\tau] \times D(A^{\frac{1}{2}})$, $0 \le \tau \le \frac{T}{2}$; if τ is small enough then $(t_0, x_0) \in Q_\tau$.

Since A is a positive operator and since u and v are weakly semicontinuous, it follows that $\phi_{\varepsilon,\delta}$ is weakly upper semicontinuous on Q_τ. Moreover

$$\phi_{\varepsilon,\delta}(t,s,x,y) \le ||u||_0 + ||v||_0 - \delta[|A^{\frac{1}{2}}x|^2 + |A^{\frac{1}{2}}y|^2]$$

where $||u||_0 = \sup_{[0,T]\times X} u$ and $||v||_0 = \sup_{[0,T]\times X} v$. Therefore there exists

$$(t_{\varepsilon,\delta}, s_{\varepsilon,\delta}, x_{\varepsilon,\delta}, y_{\varepsilon,\delta}) \in Q_\tau \times Q_\tau$$

such that

$$\phi_{\varepsilon,\delta}(t_{\varepsilon,\delta}, s_{\varepsilon,\delta}, x_{\varepsilon,\delta}, y_{\varepsilon,\delta}) = \max_{Q_\tau \times Q_\tau} \phi_{\varepsilon,\delta}.$$

Since $\phi_{\varepsilon,\delta}(t_{\varepsilon,\delta}, s_{\varepsilon,\delta}, x_{\varepsilon,\delta}, y_{\varepsilon,\delta}) \ge \phi_{\varepsilon,\delta}(t_0, t_0, x_0, x_0)$, it is easy to show that if τ is small enough, say $0 < \tau < \tau_\sigma$, then $t_{\varepsilon,\delta}, s_{\varepsilon,\delta} < T - \tau$.

Step III–We will obtain some a priori estimates on $t_{\varepsilon,\delta}, s_{\varepsilon,\delta}, x_{\varepsilon,\delta}, y_{\varepsilon,\delta}$.

Since $\phi_{\varepsilon,\delta}(t_{\varepsilon,\delta}, t_{\varepsilon,\delta}, x_{\varepsilon,\delta}, x_{\varepsilon,\delta}) + \phi_{\varepsilon,\delta}(s_{\varepsilon,\delta}, s_{\varepsilon,\delta}, y_{\varepsilon,\delta}, y_{\varepsilon,\delta}) \leq 2\phi_{\varepsilon,\delta}(t_\varepsilon, s_{\varepsilon,\delta}, x_{\varepsilon,\delta}, y_{\varepsilon,\delta})$, from the boundedness of u_σ and v_σ we derive

$$|t_{\varepsilon,\delta} - s_{\varepsilon,\delta}| \leq \sqrt{M\varepsilon}, \quad \text{for some positive M.} \tag{3.9}$$

Since $\phi(t_{\varepsilon,\delta}, t_{\varepsilon,\delta}, x_{\varepsilon,\delta}, x_{\varepsilon,\delta}) + \phi(t_{\varepsilon,\delta}, t_{\varepsilon,\delta}, y_{\varepsilon,\delta}, y_{\varepsilon,\delta}) \leq 2\phi(t_{\varepsilon,\delta}, t_{\varepsilon,\delta}, x_{\varepsilon,\delta}, y_{\varepsilon,\delta})$, then the Lipschitz continuity of u and v yields

$$|x_{\varepsilon,\delta} - y_{\varepsilon,\delta}| \leq C_1\varepsilon\ , \quad \text{for some positive } C_1. \tag{3.10}$$

Recalling the interpolation inequality (2.6), we get

$$K_H|A^\beta(x_{\varepsilon,\delta} - y_{\varepsilon,\delta})| \leq C_2|A(x_{\varepsilon,\delta} - y_{\varepsilon,\delta})|^{4\beta-3}|A^{\frac{3}{4}}(x_{\varepsilon,\delta} - y_{\varepsilon,\delta})|^{4-4\beta}, \tag{3.11}$$

for some $C_2 > 0$. Moreover recall the following well known inequality

$$ab \leq \frac{\sigma^p}{p}a^p + \frac{1}{\sigma^q q}b^q \tag{3.12}$$

for every $a, b \in \mathbb{R}^+$, $p > 1$, $\frac{1}{p} + \frac{1}{q} = 1$ and $\sigma > 0$. Choosing $\sigma = \left(\frac{\delta}{8}\right)^{\frac{4\beta-3}{2}}$ and $p = \frac{2}{4\beta - 3}$ in (3.11) and applying it to (3.11) we derive

$$\begin{aligned} &C_2|A(x_{\varepsilon,\delta} - y_{\varepsilon,\delta})|^{4\beta-3}|A^{\frac{3}{4}}(x_{\varepsilon,\delta} - y_{\varepsilon,\delta})|^{4-4\beta} \\ &\leq \frac{\delta}{8}|A(x_{\varepsilon,\delta} - y_{\varepsilon,\delta})|^2 + \frac{C_3}{\delta^{\frac{4\beta-3}{5-4\beta}}}\left|A^{\frac{3}{4}}(x_{\varepsilon,\delta} - y_{\varepsilon,\delta})\right|^{\frac{8-8\beta}{5-4\beta}}, \end{aligned} \tag{3.13}$$

where C_3 is some positive constant. On the other hand, again applying (3.12) with $p = \frac{5-4\beta}{4-4\beta}$ and $\sigma = \left(\frac{1}{4\varepsilon}\right)^{\frac{4-4\beta}{5-4\beta}}$, we find

$$\frac{C_3}{\delta^{\frac{4\beta-3}{5-4\beta}}}\left|A^{\frac{3}{4}}(x_{\varepsilon,\delta} - y_{\varepsilon,\delta})\right|^{\frac{8-8\beta}{5-4\beta}} \leq \frac{1}{4\varepsilon}|A^{\frac{3}{4}}(x_{\varepsilon,\delta} - y_{\varepsilon,\delta})|^2 + \frac{C_4\varepsilon^{4-4\beta}}{\delta^{4\beta-3}}, \tag{3.14}$$

for $C_4 > 0$. From estimates (3.13) and (3.14), inequality (3.11) can be rewritten as

$$K_H|A^\beta(x_{\varepsilon,\delta} - y_{\varepsilon,\delta})| \leq \frac{\delta}{8}\left[|Ax_{\varepsilon,\delta}|^2 + |Ay_{\varepsilon,\delta}|^2\right] + \frac{1}{4\varepsilon}|A^{\frac{3}{4}}(x_{\varepsilon,\delta} - y_{\varepsilon,\delta})|^2 + \frac{C_4\varepsilon^{4-4\beta}}{\delta^{4\beta-3}}. \tag{3.15}$$

Moreover

$$K_H\delta\left[|Ax_{\varepsilon,\delta}| + |Ay_{\varepsilon,\delta}|\right] \leq \frac{\delta}{8}\left[|Ax_{\varepsilon,\delta}|^2 + |Ay_{\varepsilon,\delta}|^2\right] + C_5\delta, \tag{3.16}$$

where C_5 is a positive constant. On the other hand we get

$$\delta C_F||A^{-\beta}||[|Ax_{\varepsilon,\delta}| + |Ay_{\varepsilon,\delta}|] \leq \frac{\delta}{8}\left[|Ax_{\varepsilon,\delta}|^2 + |Ay_{\varepsilon,\delta}|^2\right] + C_6\delta, \tag{3.17}$$

for $C_6 > 0$. Finally from estimate (3.10), it follows

$$\begin{aligned} &K_F|A^\beta(x_{\varepsilon,\delta}-y_{\varepsilon,\delta})|\frac{|A^{\frac{1}{2}-\beta}(x_{\varepsilon,\delta}-y_{\varepsilon,\delta})|}{\varepsilon} \\ &\leq \frac{C_7|A^\beta(x_{\varepsilon,\delta}-y_{\varepsilon,\delta})||x_{\varepsilon,\delta}-y_{\varepsilon,\delta}|}{\varepsilon} \leq C_8|A^\beta(x_{\varepsilon,\delta}-y_{\varepsilon,\delta})|, \end{aligned} \tag{3.18}$$

where $C_7, C_8 > 0$. Applying estimate (3.15) to (3.18) we find

$$\begin{aligned} &K_F|A^\beta(x_{\varepsilon,\delta}-y_{\varepsilon,\delta})|\frac{|A^{\frac{1}{2}-\beta}(x_{\varepsilon,\delta}-y_{\varepsilon,\delta})|}{\varepsilon} \\ &\leq \frac{\delta}{8}\left[|Ax_{\varepsilon,\delta}|^2+|Ay_{\varepsilon,\delta}|^2\right]+\frac{1}{4\varepsilon}|A^{\frac{3}{4}}(x_{\varepsilon,\delta}-y_{\varepsilon,\delta})|^2+\frac{C_9\varepsilon^{4-4\beta}}{\delta^{4\beta-3}}. \end{aligned} \tag{3.19}$$

with C_9 positive constant.

Step IV–We want to show that $\tau = t_{\varepsilon,\delta} \wedge s_{\varepsilon,\delta}$ for ε small enuogh. Let us suppose that $t_{\varepsilon,\delta}, s_{\varepsilon,\delta} > \tau$ and define the following functions

$$\begin{aligned} \varphi(t,x) &= v_\sigma(s_{\varepsilon,\delta},y_{\varepsilon,\delta})+\frac{1}{2\varepsilon}< A^{\frac{1}{2}}(x-y_{\varepsilon,\delta}),x-y_{\varepsilon,\delta}> +\frac{\delta}{2}< Ay_{\varepsilon,\delta},y_{\varepsilon,\delta}> +\frac{(t-s_{\varepsilon,\delta})^2}{2\varepsilon} \\ \psi(s,y) &= u_\sigma(t_{\varepsilon,\delta},x_{\varepsilon,\delta})-\frac{1}{2\varepsilon}< A^{\frac{1}{2}}(x_{\varepsilon,\delta}-y),x_{\varepsilon,\delta}-y> -\frac{\delta}{2}< Ax_{\varepsilon,\delta},x_{\varepsilon,\delta}> -\frac{(t_{\varepsilon,\delta}-s)^2}{2\varepsilon} \end{aligned}$$

Notice that $\varphi, \psi \in C^1(0,T) \times D(A^{\frac{1}{2}}))$. Also, $(t_{\varepsilon,\delta}, x_{\varepsilon,\delta}) \in M_\delta^+(u,\varphi)$ and $(s_{\varepsilon,\delta}, y_{\varepsilon,\delta}) \in M_\delta^-(v,\psi)$ by construction. Writing the subsolution viscosity inequality for u_σ, using φ as a test function, and subtracting from it the supersolution inequality for v_σ, using ψ as a test function, we get

$$\begin{aligned} &\delta\left[|Ax_{\varepsilon,\delta}|^2+|Ay_{\varepsilon,\delta}|^2\right]+\frac{1}{\varepsilon}|A^{\frac{3}{4}}(x_{\varepsilon,\delta}-y_{\varepsilon,\delta})|^2 \\ &\leq H(s_{\varepsilon,\delta},A^\beta y_{\varepsilon,\delta},\frac{A^{\frac{1}{2}}(x_{\varepsilon,\delta}-y_{\varepsilon,\delta})}{\varepsilon}-\delta Ay_{\varepsilon,\delta})-H(t_{\varepsilon,\delta},A^\beta x_{\varepsilon,\delta},\frac{A^{\frac{1}{2}}(x_{\varepsilon,\delta}-y_{\varepsilon,\delta})}{\varepsilon}+\delta Ax_{\varepsilon,\delta}) \\ &-\delta\left[\left\langle Ax_{\varepsilon,\delta},A^{-\beta}F(A^\beta x_{\varepsilon,\delta})\right\rangle+\left\langle Ay_{\varepsilon,\delta},A^{-\beta}F(A^\beta y_{\varepsilon,\delta})\right\rangle\right] \\ &+\left\langle A^{-\beta}\left[F(A^\beta y_{\varepsilon,\delta})-F(A^\beta x_{\varepsilon,\delta})\right],\frac{A^{\frac{1}{2}}(x_{\varepsilon,\delta}-y_{\varepsilon,\delta})}{\varepsilon}\right\rangle-\frac{2\sigma}{T^2}. \end{aligned} \tag{3.20}$$

Recalling assumption (3.2) on H and assumption (2.3) on F, the above inequality yields

$$\begin{aligned} &\delta\left[|Ax_{\varepsilon,\delta}|^2+|Ay_{\varepsilon,\delta}|^2\right]+\frac{1}{\varepsilon}|A^{\frac{3}{4}}(x_{\varepsilon,\delta}-y_{\varepsilon,\delta})|^2 \\ &\leq -\frac{2\sigma}{T^2}+K_H\delta\left[|Ax_{\varepsilon,\delta}|+|Ay_{\varepsilon,\delta}|\right]+K_H|A^\beta(x_{\varepsilon,\delta}-y_{\varepsilon,\delta})|+K_H|t_{\varepsilon,\delta}-s_{\varepsilon,\delta}| \\ &+\delta C_F\|A^{-\beta}\|[|Ax_{\varepsilon,\delta}|+|Ay_{\varepsilon,\delta}|]+K_F|A^\beta(x_{\varepsilon,\delta}-y_{\varepsilon,\delta})|\frac{|A^{\frac{1}{2}-\beta}(x_{\varepsilon,\delta}-y_{\varepsilon,\delta})|}{\varepsilon}. \end{aligned} \tag{3.21}$$

Substituting estimates (3.9), (3.15), (3.16), (3.17) and (3.19) in inequality (3.21) and letting first $\varepsilon \to 0$ and then $\delta \to 0$ we derive a contradiction, therefore $\tau = t_{\varepsilon,\delta} \wedge s_{\varepsilon,\delta}$.

Step V–Conclusion. Adapting the reasoning of [5] we derive, if $x_0 \in D(A^{\frac{1}{2}})$, $|A^{\frac{1}{2}}x_0| \leq R$,

$$u_\sigma(t_0,x_0) - v_\sigma(t_0,x_0) - \delta|A^{\frac{1}{2}}x_0|^2 \leq \frac{\alpha}{2}$$

Letting $\delta \to 0$ we derive a contradiction which yields that

$$u_\sigma(t,x) \leq v_\sigma(t,x)\ ,\ \forall x \in D(A^{\frac{1}{2}})\ ,\ \forall t \in (0,T)\ .$$

Since $D(A^{\frac{1}{2}})$ is dense in X, we have $u_\sigma(t,x) \leq v_\sigma(t,x)$ for every $(t,x) \in (0,T)\times X$. ∎

As a straightforward consequence of the above comparison theorem we have the following uniqueness result.

Corollary 3.4 *Assume that (2.3) and (3.2) hold true. Then the Hamilton–Jacobi equation (3.1) has at most one viscosity solution in the sense of Definition (3.1).*

4 Properties of the value function and existence results

In this Section we prove that the value function v of problem (2.8)–(1.5) is the unique viscosity solution of the Cauchy problem (2.12). We first state a Lipschitz continuity result for v, it can be easily proved using the technique of [3], [4], [6].

Proposition 4.1 *Assume* (2.3), (2.2). *Then, the value function v defined in* (1.5) *is continuous on $[0,T]\times X$. Moreover, for any $R > \frac{1}{T}$ and $\theta \in [0, 1-\beta)$, there exists $C_{R,\theta} > 0$ such that*

$$|v(t_0,x_0) - v(t_0,y_0)| \leq C_{R,\theta}|A^{-\theta}(x_0-y_0)| \tag{4.1}$$

for all $t_0 \in \left[0, T-\frac{1}{R}\right]$ and all $x_0, y_0 \in X$ satisfying $|x_0|, |y_0| \leq R$.

We now give an existence result for (2.12).

Theorem 4.2 *Assume that* (2.3) *and* (2.2) *hold true. Then the value function v is a viscosity solution of* (2.12) *in the sense of Definition* 3.1.

Proof – Recalling the compactness assumption (2.3) (ii), from (4.1) we conclude that v is sequentially weakly continuous in $[0,T)\times X$. It remains to prove that v satisfies (3.3) and (3.4). To prove (i) and (ii) of (3.3) and (3.4), one can easily adapt the reasoning of [6]. Let us show that v satisfies the initial condition.

By the Dynamic Programming Principle we know that

$$v(T-t,x) = \inf_{\gamma(\cdot)\in\tilde{U}} \left\{ \int_{T-t}^{T} L(s, A^\beta x(s), \gamma(s))ds + g(A^\beta x(T)) \right\}$$

therefore, for any control $\gamma(\cdot)$

$$v(T-t,x)-g(e^{-tA}x)\leq\int_{T-t}^{T}L(s,A^{\beta}x(s),\gamma(s))ds+K_g|A^{\beta}(x(T)-e^{-tA}x)|. \tag{4.2}$$

Writing $x(T)$ in mild form in (4.2) we have

$$v(T-t,x)-g(A^{\beta}e^{-tA}x)\leq\int_{T-t}^{T}L(s,A^{\beta}x(s),\gamma(s))ds+$$
$$+K_g\left|\int_{T-t}^{T}e^{-(T-s)A}F(A^{\beta}x(s))ds+A^{\beta}\int_{T-t}^{T}e^{-(T-s)A}B\gamma(s)ds\right|.$$

Since L and F are bounded, choosing a constant control $\gamma(\cdot)=\gamma_0\in\tilde{U}$, as $t\to 0$ we conclude that

$$\lim_{R\to\infty}\lim_{t\downarrow 0}\sup_{|A^{\frac{1}{2}}x|\leq R}[w(t,x)-g(A^{\beta}e^{-tA}x)]\leq 0 \tag{4.3}$$

In order to prove the opposite inequality, let us choose a δ - optimal control $\gamma_\delta(\cdot)\in\tilde{U}$ such that

$$v(T-t,x)+\delta>\int_{T-t}^{T}L(s,A^{\beta}\tilde{x}(s),\gamma_\delta(s))ds+g(A^{\beta}\tilde{x}(T))$$

where $\tilde{x}(t)=x(t;x,\gamma_\delta)$. Since L is bounded for t small enough we have

$$\int_{T-t}^{T}L(s,A^{\beta}\tilde{x}(s),\gamma_\delta(s))ds\geq-\delta.$$

Proceeding as before we derive

$$v(T-t,x)-g(A^{\beta}e^{-tA}x)>-2\delta-K_g|A^{\beta}(\tilde{x}(T)-e^{-tA}x)|.$$

On the other hand we have

$$K_g|A^{\beta}(\tilde{x}(T)-e^{-tA}x)|\leq\left|\int_{T-t}^{T}F(A^{\beta}\tilde{x}(s))ds\right|+\left|\int_{T-t}^{T}\frac{M_\beta\|B\|\,\|\gamma_\delta\|}{(T-s)^{\beta}}ds\right|$$

Since the right–hand side of the above inequality tends to 0 as $t\downarrow 0$, we obtain that $\lim_{R\to\infty}\lim_{t\downarrow 0}\sup_{|A^{\frac{1}{2}}x|\leq R}[w(t,x)-g(A^{\beta}e^{-tA}x)]\geq-2\delta$. Now, δ is arbitrary, so the above inequality and (4.3) yield the conclusion. ∎

References

[1] V. Barbu and G. Da Prato, *Hamilton-Jacobi equations in Hilbert spaces*, Pitman, Boston, 1982.

[2] A. Bensoussan, G. Da Prato, M. C. Delfour and S. K. Mitter, *Representation and control of infinite dimensional systems*, Birkhäuser, Boston, 1992.

[3] P. Cannarsa and H. Frankowska, *Value function and optimality conditions for semilinear control problems*, Applied Math. Optim., 26 (1992), pp. 139–169.

[4] P. Cannarsa, F. Gozzi and H. M. Soner, *A dynamic programming approach to nonlinear boundary control problems of parabolic type*, J. Funct. Anal., 117 (1993), pp. 25–61.

[5] P. Cannarsa and M. E. Tessitore, *Cauchy problem for the dynamic programming equation of boundary control*, Proceedings IFIP Workshop on Boundary Control and Boundary Variation (to appear).

[6] P. Cannarsa and M. E. Tessitore, *Infinite dimensional Hamilton–Jacobi equations and Dirichlet boundary control problems of parabolic type*, (pre–print).

[7] M. G. Crandall and P. L. Lions, *Hamilton– Jacobi equations in infinite dimensions.*

Part I: *Uniqueness of viscosity solutions*, J. Funct. Anal. 62 (1985), pp. 379–396.

Part II: *Existence of viscosity solutions*, J. Funct. Anal. 65 (1986), pp. 368–405.

Part III, J. Funct. Anal. 68 (1986), pp. 214–247.

Part IV: *Hamiltonians with unbounded linear terms*, J. Funct. Anal. 90 (1990), pp. 237–283.

Part V: *Unbounded linear terms and B–continuous solutions*, J. Funct. Anal. 97/2 (1991), pp. 417–465.

Part VI: *Nonlinear A and Tataru's method refined*, pre–print.

Part VII: *The HJB equation is not always satisfied*, pre–print.

[8] H. Ishii, *Viscosity Solutions for a class of Hamilton– Jacobi Equations in Hilbert spaces*, J. Func. Anal., 105 (1992), pp. 301– 341.

[9] A. Pazy, *Semigroups of linear operators and applications to partial differential equations*, Springer–Verlag, New York–Heidelberg– Berlin, 1983.

[10] D. Tataru, *Viscosity Solutions of Hamilton–Jacobi Equations with unbounded linear terms*, J. Math. Anal. Appl., 163 (1992), pp. 345–392.

[11] D. Tataru, *Viscosity Solutions for the dynamic programming equation*, Applied Math. Optim., 25 (1992), pp. 109–126.

Fictitious Domain Approach Used in Shape Optimization: Neumann Boundary Condition

J. DAŇKOVÁ Mathematical Institute, Academy of Sciences, Prague, Czech Republic

J. HASLINGER Faculty of Mathematics and Physics, Charles University, Prague, Czech Republic

1 INTRODUCTION

A fictitious domain approach is now very popular when solving large systems arising from the discretization of partial differential equations. Recently an optimal control based approach was used by Atamian et al. (1991) and by Glowinski, Pan, Periaux (1994). Fictitious domain methods applied in shape optimization problems are also very attractive because of the specific character of the problem. The classical boundary variation technique does not seem to be very efficient: one has to create a new partition of the current domain, then to recompute the stiffness matrix, the load vector and to solve the corresponding system of algebraic equations. All this has to be done at each iteration. The use of a fictitious domain approach makes the numerical solution of the state problem more efficient: the problem is solved on a fixed domain $\widehat{\Omega}$ with a simple shape, which does not change during the computation. The mathematical analysis of the fictitious domain approach in shape optimization for the state problem given by the homogeneous Dirichlet boundary problem can be found in (Haslinger, Hoffmann, Kočvara, 1993), where *distributed* controls are used, and in (Haslinger, Klarbring, 1995), (Kunisch, Peichl, 1994), where the technique of *Lagrange multipliers* is presented. In this contribution we show how to apply the fictitious domain method when the Neumann boundary condition on a moving part of the boundary is considered.

2 SETTING OF THE PROBLEM

Let $\Omega(\alpha)$ be a domain defined by

$$\Omega(\alpha) = \{[x_1, x_2] \in R^2 |\ 0 < x_1 < \alpha(x_2),\ x_2 \in (0,1)\},$$

where α is a positive, Lipschitz-continuous function, which describes the variable part $\Gamma(\alpha)$ of $\partial\Omega(\alpha)$ with

$$\Gamma(\alpha) = \{[x_1, x_2] \in R^2 |\ x_1 = \alpha(x_2),\ x_2 \in (0,1)\}.$$

Next we shall suppose that the function α belongs to the set U_{ad}, where

$$\begin{aligned} U_{ad} = \{\alpha \in C^{0,1}([0,1]) |\ 0 < C_0 \leqq \alpha(x_2) \leqq C_1\ \forall x_2 \in (0,1), \\ |\alpha'(x_2)| \leqq C_2 \text{ a. e. in } (0,1),\ \text{meas } \Omega(\alpha) = C_3\}, \end{aligned}$$

i.e., U_{ad} contains all functions which are *uniformly bounded* and *uniformly Lipschitz-continuous* in $[0,1]$ and preserve the area of $\Omega(\alpha)$. The constants $C_0, ..., C_3$ are chosen in such a way that $U_{ad} \neq \emptyset$. Let $\widehat{\Omega}$ be a domain containing $\Omega(\alpha)$ for all $\alpha \in U_{ad}$, $\widehat{\Omega} = (0,d) \times (0,1)$, $d > C_1$, $f \in L^2(\widehat{\Omega})$, $g \in H^1(\widehat{\Omega})$.

On each $\Omega(\alpha)$, $\alpha \in U_{ad}$, we formulate the following state problem:

$$\begin{cases} -\triangle u(\alpha) = f \text{ in } \Omega(\alpha), \\ \dfrac{\partial u(\alpha)}{\partial n} = g \text{ on } \Gamma(\alpha), \\ u(\alpha) = 0 \text{ on } \partial\Omega(\alpha) \setminus \overline{\Gamma(\alpha)}, \end{cases}$$

or in the weak form:

$$\begin{cases} \text{find } u(\alpha) \in V(\alpha) \text{ such that} \\ (\nabla u(\alpha), \nabla\varphi)_{\Omega(\alpha)} = (f, \varphi)_{\Omega(\alpha)} + \displaystyle\int_{\Gamma(\alpha)} g\varphi ds\ \forall \varphi \in V(\alpha), \end{cases} \tag{2.1}$$

where

$$V(\alpha) = \{v \in H^1(\Omega(\alpha)) |\ v = 0 \text{ on } \partial\Omega(\alpha) \setminus \overline{\Gamma(\alpha)}\}$$

and the symbol $(.,.)_{\Omega(\alpha)}$ stands for the $L^2(\Omega(\alpha))$, $(L^2(\Omega(\alpha)))^2$ scalar product, respectively.

We shall use *the dual formulation* of the state problem (2.1) because it will enable us to treat the boundary condition on $\Gamma(\alpha)$ more easily. For any $\alpha \in U_{ad}$ we define the set

$$K_{f,g}(\alpha) = \{\mathbf{q} \in (L^2(\Omega(\alpha))^2,\ (\mathbf{q}, \nabla\varphi)_{\Omega(\alpha)} = (f,\varphi)_{\Omega(\alpha)} + \int_{\Gamma(\alpha)} g\varphi ds\ \forall\varphi \in V(\alpha)\}.$$

It is easy to see that $\mathbf{q} \in K_{f,g}(\alpha)$ if and only if

$$-\operatorname{div} \mathbf{q} = f \text{ in } \Omega(\alpha),$$
$$\mathbf{q}.\mathbf{n} = g \text{ on } \Gamma(\alpha)$$

in the sense of distributions. Here $\mathbf{n}$ denotes the outward unit normal vector to $\Gamma(\alpha)$.

The dual form of (2.1) now reads as follows:

$$\begin{cases} \text{find } \mathbf{p}(\alpha) \in K_{f,g}(\alpha) \text{ such that} \\ \widetilde{\mathcal{S}}(\mathbf{p}(\alpha)) \leqq \widetilde{\mathcal{S}}(\mathbf{q}) \ \forall \mathbf{q} \in K_{f,g}(\alpha), \end{cases} \tag{2.2}$$

where $\widetilde{\mathcal{S}}(\mathbf{q}) = \frac{1}{2}(\mathbf{q}, \mathbf{q})_{\Omega(\alpha)}$. It is known (see e.g. (Ekeland, Temam, 1976)) that the problem (2.2) has a unique solution $\mathbf{p}(\alpha)$ and $\mathbf{p}(\alpha) = \nabla u(\alpha)$, where $u(\alpha)$ is the solution of (2.1).

By $\mathbf{q}_f$ we denote a function from $(L^2(\widehat{\Omega}))^2$, such that $-\operatorname{div} \mathbf{q}_f = f$ in $\widehat{\Omega}$. Then $K_{f,g}(\alpha)$ can be splitted as follows:

$$K_{f,g}(\alpha) = \mathbf{q}_f + K_{0,\tilde{g}}(\alpha),$$

where

$$K_{0,\tilde{g}}(\alpha) = \{\mathbf{q} \in (L^2(\Omega(\alpha))^2 \,|\, \operatorname{div} \mathbf{q} = 0, \mathbf{q}.\mathbf{n} = \tilde{g} \text{ on } \Gamma(\alpha)\} \tag{2.3}$$

with $\tilde{g} = \mathbf{q}_f.\mathbf{n} + g$. For $\Omega \subseteq \widehat{\Omega}$ and $\mathbf{q} \in (L^2(\Omega))^2$ we define

$$\mathcal{S}_\Omega(\mathbf{q}) = \frac{1}{2}(\mathbf{q}, \mathbf{q})_\Omega + (\mathbf{q}_f|_\Omega, \mathbf{q})_\Omega. \tag{2.4}$$

The state problem now takes the form:

$$\begin{cases} \text{find } \mathbf{p}_0(\alpha) \in K_{0,\tilde{g}}(\alpha) \text{ such that} \\ \mathcal{S}_{\Omega(\alpha)}(\mathbf{p}_0(\alpha)) \leqq \mathcal{S}_{\Omega(\alpha)}(\mathbf{q}) \ \forall \mathbf{q} \in K_{0,\tilde{g}}(\alpha). \end{cases} \tag{$\mathcal{P}(\alpha)$}$$

Then $\mathbf{p}(\alpha) \equiv \mathbf{p}_0(\alpha) + \mathbf{q}_f|_{\Omega(\alpha)}$, where $\mathbf{p}_0(\alpha)$ solves $(\mathcal{P}(\alpha))$ and $\mathbf{p}(\alpha)$ is the solution of (2.2).

Finally, let $I : (\alpha, \mathbf{q}) \to R^1$, $\alpha \in U_{ad}$ a $\mathbf{q} \in (L^2(\Omega(\alpha))^2$ be an *objective functional* and define the following optimal shape design problem:

$$\begin{cases} \text{find } \alpha^* \in U_{ad} \text{ such that} \\ I(\alpha^*, \mathbf{p}(\alpha^*)) \leqq I(\alpha, \mathbf{p}(\alpha)) \ \forall \alpha \in U_{ad}, \end{cases} \tag{$\mathbf{P}$}$$

where $\mathbf{p}(\alpha) \equiv \mathbf{p}_0(\alpha) + \mathbf{q}_f|_{\Omega(\alpha)}$ with $\mathbf{p}_0(\alpha) \in K_{0,\tilde{g}}(\alpha)$ being the solution of $(\mathcal{P}(\alpha))$, $\alpha \in U_{ad}$.

The existence of at least one solution α^* of $(\mathbf{P})$ will be guaranteed provided the objective functional I is *lower semicontinuous* in the following sense (see Haslinger, Neitaanmäki, 1988):

$$\left\{ \begin{array}{l} \alpha_n \rightrightarrows \alpha \text{ (uniformly) in } [0,1], \ \alpha_n, \alpha \in U_{ad} \\ \hat{\mathbf{q}}_n \rightharpoonup \hat{\mathbf{q}} \text{ in } (L^2(\widehat{\Omega}))^2, \ \hat{\mathbf{q}}_n, \hat{\mathbf{q}} \in (L^2(\widehat{\Omega}))^2 \end{array} \right\} \Longrightarrow$$
$$\Longrightarrow \quad \liminf_{n \to \infty} I(\alpha_n, \hat{\mathbf{q}}_n|_{\Omega(\alpha_n)}) \geqq I(\alpha, \hat{\mathbf{q}}|_{\Omega(\alpha)}). \tag{$\mathcal{A}$}$$

3 FICTITIOUS DOMAIN APPROACH

Instead of the problem $(\mathcal{P}(\alpha))$, which is defined on $\Omega(\alpha)$, we consider a new problem which is formulated and solved on $\widehat{\Omega}$:

$$\begin{cases} \text{find } \hat{\mathbf{p}}_0(\alpha) \in \widehat{K}_{0,\tilde{g}}(\alpha) \text{ such that} \\ \mathcal{S}_{\widehat{\Omega}}(\hat{\mathbf{p}}_0(\alpha)) \leqq \mathcal{S}_{\widehat{\Omega}}(\hat{\mathbf{q}}) \ \forall \hat{\mathbf{q}} \in \widehat{K}_{0,\tilde{g}}(\alpha), \end{cases} \qquad (\widehat{\mathcal{P}}(\alpha))$$

where

$$\widehat{K}_{0,\tilde{g}}(\alpha) = \{\hat{\mathbf{q}} \in (L^2(\widehat{\Omega}))^2 |\ \operatorname{div} \hat{\mathbf{q}} = 0 \text{ in } \widehat{\Omega},\ \hat{\mathbf{q}}.\mathbf{n} = \tilde{g} \text{ on } \Gamma(\alpha)\},$$

where $\tilde{g}$ is the same as in (2.3).

We shall consider the following optimal shape design problem:

$$\begin{cases} \text{find } \alpha^* \in U_{ad} \text{ such that} \\ I(\alpha^*, \hat{\mathbf{p}}(\alpha^*)|_{\Omega(\alpha^*)})) \leqq I(\alpha, \hat{\mathbf{p}}(\alpha)|_{\Omega(\alpha)}) \ \forall \alpha \in U_{ad}, \end{cases} \qquad (\widehat{\mathbf{P}})$$

where $\hat{\mathbf{p}}(\alpha) \equiv \hat{\mathbf{p}}_0(\alpha) + \mathbf{q}_f$ with $\hat{\mathbf{p}}_0(\alpha)$ being the solution of $(\widehat{\mathcal{P}}(\alpha))$, $\alpha \in U_{ad}$.

We show that the solution of $(\widehat{\mathcal{P}}(\alpha))$ restricted to $\Omega(\alpha)$ solves $(\mathcal{P}(\alpha))$, i.e., that the problems $(\widehat{\mathbf{P}})$ and $(\mathbf{P})$ are equivalent. Indeed: $\hat{\mathbf{p}}_0(\alpha)|_{\Omega(\alpha)} \in K_{0,\tilde{g}}(\alpha)$. It remains to verify that:

$$\mathcal{S}_{\Omega(\alpha)}(\hat{\mathbf{p}}_0(\alpha)|_{\Omega(\alpha)}) \leqq \mathcal{S}_{\Omega(\alpha)}(\mathbf{q}) \ \forall \mathbf{q} \in K_{0,\tilde{g}}(\alpha),$$

where $\hat{\mathbf{p}}_0(\alpha)$ is the solution of $(\widehat{\mathcal{P}}(\alpha))$. Let $\mathbf{q} \in K_{0,\tilde{g}}(\alpha)$ be arbitrary and fixed. Setting

$$\begin{aligned} &\hat{\mathbf{q}} = \mathbf{q} \text{ in } \Omega(\alpha), \\ &\hat{\mathbf{q}} = \hat{\mathbf{p}}_0(\alpha) \text{ in } \Xi(\alpha) \equiv \widehat{\Omega} \setminus \overline{\Omega(\alpha)} \end{aligned}$$

we see that $\hat{\mathbf{q}}$ belongs to $\widehat{K}_{0,\tilde{g}}(\alpha)$. Since $\mathcal{S}_{\widehat{\Omega}}(\hat{\mathbf{q}}) = \mathcal{S}_{\Omega(\alpha)}(\mathbf{q}) + \mathcal{S}_{\Xi(\alpha)}(\hat{\mathbf{p}}_0(\alpha)|_{\Xi(\alpha)})$ and $\mathcal{S}_{\widehat{\Omega}}(\hat{\mathbf{p}}_0(\alpha)) \leqq \mathcal{S}_{\widehat{\Omega}}(\hat{\mathbf{q}})$, the desired inequality holds.

In order to release the constraint $\hat{\mathbf{q}}.\mathbf{n} = \tilde{g}$ on $\Gamma(\alpha)$, we introduce the lagrange multipliers. This will be useful when realizing $(\widehat{\mathbf{P}})$ numerically. We denote

$$H_0(\text{div}, \widehat{\Omega}) = \{\mathbf{q} \mid \mathbf{q} \in (L^2(\widehat{\Omega}))^2, \operatorname{div} \mathbf{q} = 0 \text{ in } \widehat{\Omega}\},$$
$$\Lambda(\alpha) = H^{1/2}(\Gamma(\alpha)) = \{\varphi : \Gamma(\alpha) \to R^1 \mid \exists w \in H_0^1(\widehat{\Omega}) \text{ such that } w = \varphi \text{ on } \Gamma(\alpha)\}.$$

Instead of $(\widehat{\mathcal{P}}(\alpha))$ we define the new problem

$$\begin{cases} \text{find } \hat{\mathbf{p}}_0(\alpha) \in H_0(\text{div }, \widehat{\Omega}),\ \hat{v}(\alpha) \in \Lambda(\alpha), \text{ such that} \\ (\hat{\mathbf{p}}_0(\alpha), \mathbf{q})_{\widehat{\Omega}} = -(\mathbf{q}_f, \mathbf{q})_{\widehat{\Omega}} + \langle \mathbf{q}.\mathbf{n}, \hat{v}(\alpha)\rangle_\alpha \ \forall \mathbf{q} \in H_0(\text{div}, \widehat{\Omega}), \\ \langle \hat{\mathbf{p}}_0(\alpha).\mathbf{n}, v\rangle_\alpha = \langle \tilde{g}, v\rangle_\alpha \ \forall v \in \Lambda(\alpha), \end{cases} \qquad (\widehat{\mathcal{P}}(\alpha))'$$

where $\langle .,. \rangle_\alpha$ denotes the duality between $H^{1/2}(\Gamma(\alpha))$ and $H^{-1/2}(\Gamma(\alpha)) = (H^{1/2}(\Gamma(\alpha)))'$. It is easy to show that the problem $(\widehat{\mathcal{P}}(\alpha))'$ has a unique solution $\hat{\mathbf{p}}_0(\alpha)$, $\hat{v}(\alpha)$ for all

$\alpha \in U_{ad}$. Moreover $\hat{\mathbf{p}}_0(\alpha)$ solves $(\widehat{\mathcal{P}}(\alpha))$, i.e., $(\hat{\mathbf{p}}_0(\alpha) + \mathbf{q}_f)|_{\Omega(\alpha)} = \nabla u(\alpha)$, where $u(\alpha)$ is the solution of (2.1), and $\hat{v}(\alpha) = u(\alpha)|_{\Gamma(\alpha)} - z(\alpha)|_{\Gamma(\alpha)}$ where $z(\alpha)$ solves the same Dirichlet-Neumann boundary value problem but on $\Xi(\alpha)$.

4 APPROXIMATION OF $(\widehat{\mathbf{P}})$

The aim of the present section is to describe briefly an approximation of the problem $(\widehat{\mathbf{P}})$. The approximation will be based on a suitable finite-element discretization of the state problem, solved on the fixed domain $\widehat{\Omega}$.

Let $\{\mathcal{T}_h\}$, $h \to 0+$, be a *regular family* of partitions of $\widehat{\Omega}$ in the standard sense. With any $\mathcal{T}_h$ we associate a finite-dimensional space $V_h \subset H_0^1(\widehat{\Omega})$, containing *piecewise polynomial* functions over $\mathcal{T}_h$. The space $H_0(\text{div}, \widehat{\Omega})$ will be approximated by means of the space

$$H_h = \{\mathbf{q}_h \mid \exists y_h \in V_h : \ \mathbf{q}_h = \operatorname{curl} y_h\},$$

where $\operatorname{curl} y_h = (-\partial y_h/\partial x_2, \partial y_h/\partial x_1)$. It is easy to see that $H_h \subset H_0(\text{div}, \widehat{\Omega})$.

Let $\{\mathcal{T}_H\}$, $H \to 0+$ be a family of partitions of $[0,1]$. By U_{ad}^H we denote the approximation of U_{ad} defined as follows:

$$U_{ad}^H = \{\alpha_H \in C([0,1]) |\ \alpha_H \text{ is piecewise linear on } \mathcal{T}_H\} \cap U_{ad}.$$

It remains to construct the approximation of the space $\Lambda(\alpha)$. Let $\alpha_H \in U_{ad}^H$ be given. The space $\Lambda_H(\alpha_H)$ will contain all continuous functions which are piecewise linear on $\Gamma(\alpha_H)$, $\alpha_H \in U_{ad}^H$.

Let $\alpha_H \in U_{ad}^H$ be given. The approximation of $(\widehat{\mathcal{P}}(\alpha))'$ reads as follows:

$$(\widehat{\mathcal{P}}(\alpha_H))_h \quad \begin{cases} \text{find } \hat{\mathbf{p}}_{0h}(\alpha_H) \in H_h, \ \hat{v}_H(\alpha_H) \in \Lambda_H(\alpha_H), \ \text{such that} \\ (\hat{\mathbf{p}}_{0h}(\alpha_H), \mathbf{q}_h)_{\widehat{\Omega}} = -(\mathbf{q}_f, \mathbf{q}_h)_{\widehat{\Omega}} + \displaystyle\int_{\Gamma(\alpha_H)} \hat{v}_H(\alpha_H)\mathbf{q}_h.\mathbf{n} ds \ \forall \mathbf{q}_h \in H_h, \\ \displaystyle\int_{\Gamma(\alpha_H)} v_H \hat{\mathbf{p}}_{0h}(\alpha_H).\mathbf{n} ds = \displaystyle\int_{\Gamma(\alpha_H)} \tilde{g} v_H ds \ \forall v_H \in \Lambda_H(\alpha_H). \end{cases}$$

In order to guarantee the existence and uniqueness of the solution of $(\widehat{\mathcal{P}}(\alpha_H))_h$ we assume that for any $\alpha_H \in U_{ad}^H$ the following condition is satisfied:

$$v_H \in \Lambda_H(\alpha_H) \ \& \int_{\Gamma(\alpha_H)} \mathbf{q}_h.\mathbf{n} v_H ds = 0 \ \forall \mathbf{q}_h \in H_h \Rightarrow v_H \equiv 0. \qquad (\mathcal{B})$$

The *approximation* of the whole problem $(\widehat{\mathbf{P}})$ now reads as follows:

$$\begin{cases} \text{find } \alpha_H^* \in U_{ad}^H \text{ such that} \\ I(\alpha_H^*, \hat{\mathbf{p}}_h(\alpha_H^*)|_{\Omega(\alpha_H^*)}) \leqq I(\alpha_H, \hat{\mathbf{p}}_h(\alpha_H)|_{\Omega(\alpha_H)}) \ \forall \alpha_H \in U_{ad}^H, \end{cases} \qquad (\widehat{\mathbf{P}})_{h,H}$$

where $\hat{\mathbf{p}}_h(\alpha_H) \equiv \hat{\mathbf{p}}_{0h}(\alpha_H) + \mathbf{q}_f|_{\Omega(\alpha_H)}$ with $\hat{\mathbf{p}}_{0h}(\alpha_H)$ being the solution of $(\widehat{\mathcal{P}}(\alpha_H))_h$. Using classical compactness arguments one can easily prove that the problem $(\widehat{\mathbf{P}})_{h,H}$ has at least one solution for all $h, H > 0$.

Next we shall formulate sufficient conditions, under which the problems $(\widehat{\mathbf{P}})$ and $(\widehat{\mathbf{P}})_{h,H}$ are closed on subsequences.

We assume that $h \to 0+$ iff $H \to 0+$ and that the following condition is satisfied:

$$\left.\begin{cases} \alpha_H \rightrightarrows \alpha \text{ in } [0,1], \ \alpha_H \in U_{ad}^H, \ \alpha \in U_{ad} \\ \mathbf{q}_h \rightharpoonup \mathbf{q} \text{ in } (L^2(\widehat{\Omega}))^2, \ \mathbf{q}_h \in H_h, \ \mathbf{q} \in (L^2(\widehat{\Omega}))^2 \end{cases}\right\} \Longrightarrow \tag{$\mathcal{C}$}$$

$$\Longrightarrow \lim_{h \to 0+} I(\alpha_H, \mathbf{q}_h|_{\Omega(\alpha_H)}) = I(\alpha, \mathbf{q}|_{\Omega(\alpha)}).$$

The mutual relation between $(\widehat{\mathbf{P}})_{h,H}$ and $(\widehat{\mathbf{P}})$ is given by the following theorem:

THEOREM 4.1 Let the conditions $\mathcal{A}$, $\mathcal{B}$ and $\mathcal{C}$ be satisfied. Let $\alpha_H^* \in U_{ad}^H$ be a solution of $(\widehat{\mathbf{P}})_{h,H}$ and $\hat{\mathbf{p}}_h^*(\alpha_H^*) \in H_h$ be the corresponding state. Then there exist subsequences $\{\alpha_{H'}^*\} \subset \{\alpha_H^*\}$, $\{\hat{\mathbf{p}}_{h'}^*(\alpha_{H'}^*)\} \subset \{\hat{\mathbf{p}}_h^*(\alpha_H^*)\}$ and elements $\alpha^* \in U_{ad}$, $\hat{\mathbf{p}}^* \in H_0(\text{div}, \widehat{\Omega})$ such that

$$\begin{cases} \alpha_{H'}^* \rightrightarrows \alpha^*, \ H' \to 0+ \text{ in } [0,1], \\ \hat{\mathbf{p}}_{h'}^*(\alpha_{H'}^*) \rightharpoonup \hat{\mathbf{p}}^*, \ h' \to 0+ \text{ in } (L^2(\widehat{\Omega}))^2 \end{cases}$$

Moreover α^* solves $(\widehat{\mathbf{P}})$ and $\hat{\mathbf{p}}^*$ is the solution of $(\widehat{\mathcal{P}}(\alpha^*))$.
Proof: The detailed proofs of these results as well as the particular choice of the space V_h are discussed in (Daňková, Haslinger, 1995).

REFERENCES

C. Atamian, G. V. Dinh, R. Glowinski, J. He and J. Periaux (1991). On some imbedding methods applied to fluid dynamics and electro-magnetics, *Comput. Methods Appl. Mech. and Engrg., 91:* pp. 1271-1299

Daňková J., Haslinger J. (1995). Fictitious domain approach used in shape optimization – Neumann boundary condition (in preparation)

Ekeland I., Temam R. (1976). Convex Analysis and Variational Problems, North Holland, Amsterdam

Glowinski R., Pan T. W. and Periaux J. (1994). A fictitious domain method for Dirichlet problem and applications,*Comput. Methods Appl. Mech. Engrg., 111:* pp. 283-303

Haslinger J., Hoffmann K. - H., Kočvara M. (1993). Control/fictitious domain method for solving optimal shape design problems, *RAIRO Modèl. Math. Anal. Numér., 27 (2):* pp. 157-182

Haslinger J., Klarbring A. (1995) Fictitious domain/mixed finite element approach for a class of optimal shape design problems, to appear in *RAIRO Modèl. Math. Anal. Numér.*

Haslinger J., Neittaanmäki P. (1988). *Finite Element Approximation of Optimal Shape Design: Theory and Applications,* J. Wiley & Sons, New-York

Kunisch, K., Peichl,G. (1994). Shape optimization for mixed boundary value problems based on an embedding domain method (preprint)

Modelling and Control of Thin/Shallow Shells

MICHEL C. DELFOUR
Centre de recherches mathématiques et
Département de mathématiques et de statistique,
Université de Montréal, C.P. 6128 Succ. Centre-ville,
Montréal, Québec, Canada, H3C 3J7

JEAN-PAUL ZOLÉSIO
Institut Non Linéaire de Nice
CNRS
1361 Route des Lucioles
06904 Sophia Antipolis Cédex, France

ABSTRACT. In this paper we provide a mathematical construction of a dynamical second order operational differential equation for thin/shallow shells from linear elasticity, the tangential differential calculus, and the oriented boundary distance function. This model is obtained by an intrinsic approach as in P. Germain. It extends to thin/shallow shells the "natural theory" and the theory of Love-Kirchhoff of plates. We specify the appropriate function spaces, and give existence and uniqueness theorems. The final models are in the standard V-H-V' setting and the corresponding control operator is constructed. This provides a mathematically more tractable model for the study of control and optimal design problems.

1. INTRODUCTION

Local bases and Christoffel symbols are naturaly associated with the study of partial differential equations over a submanifold of $\mathbb{R}^N$ as in the theory of thin/shallow shells (cf. for instance, M. BERNADOU [1]). The object of this paper is to present an intrinsic approach which makes use the oriented boundary (resp. algebraic or signed) distance function and the tangential differential calculus. From linear elasticity and a simple rheological law, the "natural theory" and the Love-Kirchhoff theory of static plates (cf. for instance P. GERMAIN [1]) are extended to thin/shallow shells. The corresponding dynamical models (in time) take the form of second order operational differential equations involving a positive V-H coercive elliptic operator for an appropriate choice of Hilbert spaces V and H as in J.L. LIONS [1, Chapter 4]. This intrinsic model seems to be mathematically more tractable than the ones using local coordinates systems. Even though we only consider linear models, the constructions are quite general and more complex rheologies can be considered.

This work was first presented for a static model in DELFOUR and ZOLÉSIO [2], [3] with two simplifying assumptions: the Lamé coefficient λ was zero in the rheological law (first paper), and the strain tensor was approximated by an affine expression in the thickness variable z, $|z| < h > 0$, of the shell (the thickness of the shell is $2h$) in both papers. In the present paper we give a more complete version of the static model. It includes the Lamé coefficient λ and uses an approximation of the strain tensor by a quadratic expression in the thickness variable z, $|z| < h$. This is motivated by the fact that in so doing we first obtain a *Korn inequality* for shells. Secondly the kernel of the "approximate" strain tensor coincides with the equivalent of the "rigid displacements"

The research of the first author has been supported in part by a Killam fellowship from Canada Council, National Sciences and Engineering Research Council of Canada operating grant A-8730 and infrastructure grant INF-7939 and by a FCAR grant from the Ministère de l'Education du Québec.

of 3-D elasticity. Our previous model only gave an approximation of this mechanical property. Finally the Love-Kirchhoff theory of shells comes out of the analysis as a special case of the natural model. We obtain the same variational equation but over some closed linear subspaces of the Hilbert spaces associated with the natural theory.

More results and details can be found in DELFOUR and ZOLÉSIO [4]: structure of the Neumann boundary conditions, shells with variable thickness, etc. This new approach seems particularily well suited for control and optimal design problems since the intrinsic geometric parameters of the shell are readily accessible.

2. NOTATION, DEFINITIONS AND MATHEMATICAL DESCRIPTION OF THE SHELL

2.1. Mathematical description of the shell without boundary.

Let $\mathbb{R}^N$ be the N-dimensional Euclidean space for some integer $N > 1$ (in practice $N = 3$). Let Ω be a subset of $\mathbb{R}^N$ with a boundary $\partial\Omega$ which is a C^2 $(N-1)$-dimensional submanifold of $\mathbb{R}^N$. First consider a *shell without boundary* by assuming that the *mean surface* Γ is equal to $\partial\Omega$ and that $\partial\Omega$ is compact. Associate with Ω the *oriented boundary distance function*

$$b_\Omega(x) = d_\Omega(x) - d_{\complement\Omega}(x), \tag{2.1}$$

where $\complement\Omega = \{x \in \mathbb{R}^N : x \notin \Omega\}$ and d_A is the usual distance function to the set A. The function b_Ω captures all the geometrical properties of the boundary $\partial\Omega$ of Ω. In particular its gradient ∇b_Ω coincides with the unitary exterior normal field n to the boundary $\partial\Omega$ of Ω and induces an orientation of $\partial\Omega$. The reader is referred to DELFOUR and ZOLÉSIO [1] for a detailed study of the properties of the function b_Ω and to GILBARG and TRUDINGER [1] for the study of curvature via distance functions. Since the domain Ω is fixed throughout this paper the function b_Ω will now be denoted by b. For $k \geq 2$ a domain Ω has a C^k boundary $\partial\Omega$ if and only if the oriented distance function b is C^k in a neighbourhood of $\partial\Omega$. Since we have assumed that $\partial\Omega$ is compact and of class C^2, there exists $h > 0$ sufficiently small such that $\Omega_h = \{x \in \mathbb{R}^N : |b(x)| < h\}$ is a neighbourhood of $\partial\Omega$ where b is C^2. Using the function b, the *shell* is the union of the level sets $\partial\Omega_z = \{x \in \mathbb{R}^N : b(x) = z\}, \quad -h < z < h$, "parallel" to $\partial\Omega$.

At each point X of $\partial\Omega$, the eigenvalues of the matrix of second order partial derivatives D^2b are 0 and the $(N-1)$ *principal curvatures* of the surface $\partial\Omega$. Its trace $H(X)$, up to a multiplying factor, is the *mean curvature* and the trace $K(X)$ of the matrix of cofactors $M(D^2b)$ is the *total* or *Gaussian curvature*

$$H(X) \stackrel{\text{def}}{=} \text{tr}\,(D^2b(X)) = \Delta b(X) \qquad K(X) \stackrel{\text{def}}{=} \text{tr}\, M(D^2b(X)). \tag{2.2}$$

2.2. Projection operators and flow of the gradient of b.

When $\partial\Omega$ is C^2, the *projection mapping* p onto $\partial\Omega$ is given by

$$p(x) = x - b(x)\,\nabla b(x). \tag{2.4}$$

The vector ∇b is C^1 in Ω_h and we can consider the *flow mapping* T_z, defined by

$$T_z(X) = x(z), \quad \frac{dx}{dz}(z) = \nabla b(x(z)), \quad |z| < h, \quad x(0) = X. \tag{2.5}$$

It is an homeomorphism from $\partial\Omega$ onto $\partial\Omega_z = \{x \in \mathbb{R}^N : b(x) = z\}$. In particular

(2.6) $$T_z(X) = X + z\,\nabla b(X)$$

for $|z| < h$. This induces a "curvilinear coordinate system" $(X, z) \in \partial\Omega \times]-h, h[$ in Ω_h. The points on the level set $\partial\Omega_z$ are given by $\{X + z\,\nabla b(X) : X \in \partial\Omega\}$ and for each $(X, z) \in \partial\Omega \times]-h, h[$, $\nabla b(T_z(X)) = \nabla b(X + z\,\nabla b(X)) = \nabla b(X)$. In particular $\det DT_z(X)$ is a polynomial of degree at most $N-1$ with coefficients $K_i(X)$, $X \in \Gamma$,

(2.7) $$\det DT_z(X) = \sum_{i=0}^{N-1} K_i\, z^i,$$

where $K_0 = 1$, $K_1 = H$ for $N \geq 2$, and $K_{N-1} = K$ for $N \geq 3$.

2.3. Mathematical description of the shell with boundary.

Assume the existence of a domain Ω in $\mathbb{R}^N$ with a boundary $\partial\Omega$ of class C^2 which is now not necessarily compact. The *mean surface* Γ of the shell is a bounded open domain in $\partial\Omega$ with a Lipschitzian boundary $\partial_{\partial\Omega}\Gamma$ in $\partial\Omega$. Since Γ is bounded and $\partial\Omega$ is C^2, there exist a (bounded) neighbourhood $V(\Gamma)$ of Γ and $h > 0$ such that

(2.8) $$b \in C^2(V(\Gamma)), \qquad S_h = \{T_z(X) : \forall X \in \Gamma,\ \forall z,\ |z| < h\} \subset V(\Gamma).$$

This means that for all $x \in S_h$, $p(x) \in \Gamma$ and that

(2.9) $$S_h = \{x \in \mathbb{R}^N : |b(x)| < h \text{ and } p(x) \in \Gamma\}.$$

In this construction the *lateral boundary*

(2.10) $$\Sigma_h = \{T_z(X) : X \in \partial_{\partial\Omega}\Gamma,\ |z| < h\}$$

of the shell is an $(N-1)$-surface normal to the mean surface Γ. When $\Gamma = \partial\Omega$, then $S_h = \Omega_h$, $\Sigma_h = \varnothing$, and we are back to the shell without boundary of §2.1.

2.4. Tangential differential operators.

For any scalar function $w : \Gamma \to \mathbb{R}$, denote by $\nabla_\Gamma w$ the *tangential gradient*

(2.11) $$\nabla_\Gamma w = \nabla W\big|_\Gamma - \frac{\partial W}{\partial n}\, n$$

defined in terms of an extension W of w to S_h. This definition is independent of the choice of the extension W. It is easy to check that

(2.12) $$\nabla(w \circ p) = Dp\,(\nabla_\Gamma w) \circ p = [I - b\,D^2 b]\,\nabla_\Gamma w \circ p$$

and that $\nabla(w \circ p) = \nabla_\Gamma w$ on Γ. In addition we define the *tangential Jacobian matrix* of a vector $v : \Gamma \to \mathbb{R}^N$ through its transposed

(2.13) $$^*D_\Gamma v = (\nabla_\Gamma v_1, \ldots, \nabla_\Gamma v_N)$$

in terms of the column tangential gradients. In particular

(2.14) $$D(v \circ p) = (D_\Gamma v) \circ p\, Dp = (D_\Gamma v) \circ p\,[I - b\,D^2 b], \qquad D(v \circ p)_{|\Gamma} = D_\Gamma v.$$

In the same way we define the *tangential divergence* and the *tangential strain tensor*

(2.15) $$\operatorname{div}_\Gamma v \stackrel{\text{def}}{=} \operatorname{tr} D_\Gamma(v), \qquad \operatorname{div}(v \circ p)_{|\Gamma} = \operatorname{div}_\Gamma v$$

(2.16) $$\varepsilon_\Gamma(v) \stackrel{\text{def}}{=} \frac{1}{2}(D_\Gamma v + {}^*D_\Gamma v), \qquad \varepsilon_\Gamma(v) = \varepsilon(v \circ p)_{|\Gamma}$$

Furthermore the composition of $\operatorname{div}_\Gamma$ and ∇_Γ yields the *Laplace-Beltrami* operator

(2.17) $$\Delta_\Gamma w \stackrel{\text{def}}{=} \operatorname{div}_\Gamma(\nabla_\Gamma w) \quad (= \Delta(w \circ p)_{|\Gamma}).$$

2.5. Decomposition of the integration over S_h along the level sets of b.
The next ingredient is Federer's decomposition of the measure on S_h: for any sufficiently smooth function f, say in $H^1(S_h)$,

$$\int_{S_h} f(x)\,dx = \int_{-\infty}^{\infty}\int_{\Gamma_z} f\,d\Gamma_z\,dz, \tag{2.18}$$

where $d\Gamma_z$ is the surface measure of $\Gamma_z = \{x \in \mathbb{R}^N : b(x) = z \text{ and } p(x) \in \Gamma\}$. This technique has been used by R. TEMAM [1] and J.-P. ZOLÉSIO [1] for Plasma Physics problems. After a change of variable using the transformation T_z we get

$$\int_{-\infty}^{\infty} dz \int_{\Gamma_z} d\Gamma_z\, f = \int_{-\infty}^{\infty} dz \int_{\Gamma} d\Gamma\, f \circ T_z\, |j(z)| = \int_{\Gamma} d\Gamma \int_{-h}^{h} dz\, f \circ T_z\, |j(z)|, \tag{2.19}$$

where using the notation (2.7) and $n = \nabla b$

$$j(z) = \det(DT_z)\, |{}^*DT_z^{-1}\, n| = \det(DT_z) \quad \text{on } \Gamma. \tag{2.20}$$

The area density on the boundary Γ_z of $\Omega_z = \{x \in \mathbb{R}^N : b_\Omega(x) \leq z\}$ is given by $d\Gamma_z = |j(z)|\,d\Gamma$. But $j(0) = 1$ and for $z\,D^2b(x)$ sufficiently small $|j(z)| = j(z)$. Since b is of class C^2 and Γ is bounded, this is always true for h sufficiently small.

3. DISPLACEMENT FIELD AND STRAIN TENSOR

First introduce the basic assumption on the displacement vector as in the "natural theory" of plates (cf. for instance P. GERMAIN [1]).

Assumption 1. At each point x of the shell the *displacement vector* $V(x)$ is of the form

$$V(x) = e \circ p(x) + b(x)\,\ell \circ p(x), \quad x \in S_h, \tag{3.1}$$

for vector-valued maps e and ℓ in $H^1(\Gamma)^N$ such that $\ell(X)$ is a tangential vector. □

3.1. The strain tensor.
From (3.1), a direct computation yields the following expressions in S_h

$$DV = [D_\Gamma e \circ p + b\,D_\Gamma \ell \circ p + \ell \circ p\,{}^*\nabla b]\,[I - b\,D^2 b], \qquad DV\,\nabla b = \ell \circ p \tag{3.2}$$

and the *strain tensor* $\varepsilon(V)$ over S_h associated with the displacement field $V(x)$

$$\begin{aligned} 2\,\varepsilon(V) = {} & [D_\Gamma(e) \circ p + b\,D_\Gamma(\ell) \circ p + \ell \circ p\,{}^*\nabla b]\,[I - b\,D^2 b] \\ & + [I - b\,D^2 b]\,{}^*[D_\Gamma(e) \circ p + b\,D_\Gamma(\ell) \circ p + \ell \circ p\,{}^*\nabla b] \end{aligned} \tag{3.3}$$

$$2\,\varepsilon(V)\,\nabla b = [I - b\,D^2 b]\,[2\,\varepsilon_\Gamma(e) \circ p\,\nabla b + \ell \circ p], \qquad \varepsilon(V)\,\nabla b \cdot \nabla b = 0. \tag{3.4}$$

The tensor $\varepsilon(V) \circ T_z$ is almost affine in z since

$$DV \circ T_z = [D_\Gamma e + z\,D_\Gamma \ell + \ell\,{}^*\nabla b]\,[I + z\,D^2 b]^{-1} \tag{3.5}$$

$$2\,\varepsilon(V)\circ T_z = [D_\Gamma(e) + z\,D_\Gamma(\ell) + \ell^*\nabla b]\,[I + z\,D^2b]^{-1} + [I + z\,D^2b]^{-1\,*}[D_\Gamma(e) + z\,D_\Gamma(\ell) + \ell^*\nabla b]. \tag{3.6}$$

$$2\,\varepsilon(V)\circ T_z\,\nabla b = [I + z\,D^2b]^{-1}\,[2\,\varepsilon_\Gamma(e)\,\nabla b + \ell]. \tag{3.7}$$

The nonlinear part is contained in the matrix $[I + z\,D^2b]^{-1}$. For $\|z\,D^2b\|$ small

$$[I + z\,D^2b]^{-1} = \sum_{i=0}^{\infty}(-D^2b)^i\,z^i, \tag{3.8}$$

and we get the strain tensor as the infinite sum

$$\varepsilon(V)\circ T_z = \sum_{i=0}^{\infty}\varepsilon^i\,z^i, \tag{3.9}$$

$$2\,\varepsilon^0 = D_\Gamma(e) + \ell^*\nabla b + {}^*D_\Gamma(e) + \nabla b^*\ell \tag{3.10}$$

$$2\,\varepsilon^1 = D_\Gamma(\ell) - D_\Gamma(e)\,D^2b + {}^*D_\Gamma(\ell) - D^2b\,{}^*D_\Gamma(e) \tag{3.11}$$

$$2\,\varepsilon^2 = [D_\Gamma(\ell) - D_\Gamma(e)\,D^2b]\,(-D^2b) + (-D^2b)\,[{}^*D_\Gamma(\ell) - D^2b\,{}^*D_\Gamma(e)] \tag{3.12}$$

$$2\,\varepsilon^n = [D_\Gamma(\ell) - D_\Gamma(e)\,D^2b]\,(-D^2b)^{n-1} + (-D^2b)^{n-1}\,[{}^*D_\Gamma(\ell) - D^2b\,{}^*D_\Gamma(e)] \tag{3.13}$$

for $n \geq 3$. The following result is central to the subsequent development of the model.

Theorem 3.1. *(i) We have the following equivalences:*

$$\varepsilon(V)\circ T_z = 0, \quad \forall |z| < h,\ \forall X \in \Gamma \tag{3.14}$$

if and only if

$$\varepsilon^0 = \varepsilon^1 = \varepsilon^2 = 0 \quad \text{on } \Gamma \tag{3.15}$$

if and only if there exists a vector $a \in \mathbb{R}^N$ and an $N \times N$ matrix B such that

$$\ell(X) = 0, \quad e(X) = a + BX, \quad \forall X \in \Gamma, \tag{3.16}$$

where B is such that

$$B + {}^*B = 0 \quad \text{and} \quad B\,\nabla b(X) = 0, \quad \forall X \in \Gamma. \tag{3.17}$$

(ii) For all z, $|z| < h$, $X \in \Gamma$,

$$\varepsilon(V)\circ T_z = \varepsilon^0 + [I + z\,D^2b]^{-1}\{z\,[\varepsilon^1 + \varepsilon^1\,z\,D^2b + z\,D^2b\,\varepsilon^1] + z^2\varepsilon^2\}[I + z\,D^2b]^{-1}.$$

Remark 3.1. Conditions (3.16) and (3.17) correspond to the rigid displacements as in three-dimensional elasticity. Our previous model only yielded an approximation of this mechanical property (cf. Delfour ad Zolésio [3, Remark 5.1]). □

3.2. Approximate strain tensor and Korn's inequality.

Assumption 2. The equivalence of the conditions $\varepsilon(V)\circ T_z = 0$ and $\varepsilon^0 = \varepsilon^1 = \varepsilon^2 = 0$ suggests to use an approximation of the strain tensor which is quadratic in z:

$$\tilde{\varepsilon}(V)\circ T_z = \varepsilon^0 + \varepsilon^1\, z + \varepsilon^2\, z^2. \tag{3.18}$$

That is to say that the dimensionless quantity $\|z\, D^2 b(X)\|$ is small or equivalently that the shell is either thin (h small) or shallow ($\|D^2 b(X)\|$ small) or both. □

It is natural to associate with the above approximation the Hilbert spaces

$$\mathcal{H} = \{(e,\ell) \in L^2(\Gamma)^N \times L^2(\Gamma)^N \,:\, \ell\cdot n = 0 \text{ on } \Gamma\} \tag{3.19}$$

$$\mathcal{V} = \left\{(e,\ell) \in \mathcal{H} \,:\, \varepsilon^i \in L^2(\Gamma)^{N\times N},\ 0 \le i \le 2\right\} \tag{3.20}$$

$$\mathcal{N} = \left\{(e,\ell) \in \mathcal{V} \,:\ \varepsilon^i = 0 \text{ on } \Gamma,\ 0 \le i \le 2\right\} \tag{3.21}$$

with norms

$$|(e,\ell)|^2_{\mathcal{H}} = |e|^2_{L^2(\Gamma)} + |\ell|^2_{L^2(\Gamma)} \tag{3.22}$$

$$\|(e,\ell)\|^2_{\mathcal{V}} = |(e,\ell)|^2_{\mathcal{H}} + \sum_{i=0}^{2} h^{2i}\, \|\varepsilon^i(e,\ell)\|^2_{L^2(\Gamma)}. \tag{3.23}$$

In view of Theorem 3.1 (i)

$$\begin{aligned}\mathcal{N} = \{(e,0) \in \mathcal{V} : e(X) = a + B\,X, \quad &\forall a \in \mathbb{R}^N,\ \forall B \text{ an } N\times N \text{ matrix}\\ &\text{such that } B + {}^*B = 0 \quad\text{and}\quad B\,\nabla b(X) = 0, \quad \forall X \in \Gamma\}.\end{aligned} \tag{3.24}$$

To complete the characterization of the space $\mathcal{V}$, we need the *Korn's inequality*.

Theorem 3.2. *Assume that Γ is a bounded open domain in the C^2 $(N-1)$-dimensional submanifold $\partial\Omega$ of $\mathbb{R}^N$ with a Lipschitzian boundary $\partial_{\partial\Omega}\Gamma$ in $\partial\Omega$. As h goes to zero, there exists a constant $c = c(S_h, D^2 b) > 0$ such that for all $(e,\ell) \in \mathcal{V}$*

$$\begin{aligned}&\int_\Gamma 2\,h\,[\,|\ell|^2 + \|D_\Gamma(e)\|^2\,] + 2\,\frac{h^3}{3}\|D_\Gamma(\ell)\|^2\, d\Gamma\\ &\le c^2 \int_\Gamma 2\,h\,|e|^2 + 2\,\frac{h^3}{3}\,|\ell|^2\\ &\qquad + 2\,h\,\|\varepsilon^0(e,\ell)\|^2 + 2\,\frac{h^3}{3}\,\|\varepsilon^1(e,\ell)\|^2 + 2\,\frac{h^5}{5}\,\|\varepsilon^2(e,\ell)\|^2\, d\Gamma,\end{aligned} \tag{3.25}$$

where for two $N\times N$ tensors A and B and an N-vector a

$$A\cdot\cdot\, B = \sum_{i,j=1}^{N} A_{ij}\, B_{ij}(V), \qquad \|A\|^2 = \sum_{i,j=1}^{N} A_{ij}\, A_{ji}, \quad |a|^2 = \sum_{i=1}^{N} a_i^2. \tag{3.26}$$

In particular

$$\mathcal{V} = \left\{(e,\ell) \in H^1(\Gamma)^N \times H^1(\Gamma)^N \,:\, \ell\cdot n = 0\right\} \tag{3.27}$$

$$H^1(\Gamma)^N = \left\{e \in L^2(\Gamma)^N \,:\, \begin{array}{l}\varepsilon_\Gamma(e) \in L^2(\Gamma)^{N\times N}\\ D_\Gamma(e)\, D^2 b + D^2 b^*\, D_\Gamma(e) \in L^2(\Gamma)^{N\times N}\end{array}\right\}. \tag{3.28}$$

3.3. Love-Kirchhoff theory.

From (3.18), (3.10) to (3.12) we have

$$2\,\tilde{\varepsilon}(V)\circ T_z\,\nabla b = [I - z\,D^2 b + z^2 (D^2 b)^2]\,[2\,\varepsilon_\Gamma(e)\,\nabla b + \ell]. \tag{3.29}$$

The generalization of the Love-Kirchhoff condition from plates to shell is given by

$$\ell = -2\,\varepsilon_\Gamma(e)\,\nabla b. \tag{3.30}$$

Identities (3.29) and (3.30) yield the conditions

$$\tilde{\varepsilon}(V)\circ T_z \nabla b = 0 \quad \text{in } S_h, \qquad \forall\,|z|<h, \quad \tilde{\varepsilon}(V)\circ T_z\, n_z = 0 \quad \text{on } \Gamma_z \tag{3.31}$$

where n_z denotes the unit exterior normal to the level curve Γ_z.

4. Computation of the energies and the work

The total energy is made up of the strain energy plus the work of the external force minus the kinetic energy. By external forces we mean forces and torques applied to the shell. For convenience we assume that they are *distributed* within S_h, but more general loadings can be envisioned. The solution of the dynamical shell equation corresponds to a stationary point of the total energy functional.

4.1. Strain energy and work of the external forces.

Define the *strain energy* $\mathcal{P}$ and the *work of the external forces* $\mathcal{W}$ as follows:

$$\mathcal{P} = \frac{1}{2}\int_{S_h} \sigma\cdot\cdot\,\tilde{\varepsilon}(V)\,dx \qquad \mathcal{W} = \int_{S_h} [F\cdot V + M\cdot(\ell\circ p)]\,dx, \tag{4.1}$$

where σ is the *stress tensor*, F the loading and M the moment applied to the shell. Assume that F and M belong to $H^1(S_h)^N$ so that the trace is well-defined on the surface Γ. Furthermore assume that S_h is made up of an homogeneous isotropic elastic material which obeys the rheological law

$$\sigma = \lambda\,\mathrm{tr}\,\tilde{\varepsilon}(V)\,I + 2\mu\,\tilde{\varepsilon}(V), \tag{4.2}$$

where $\lambda \geq 0$ and $\mu > 0$ are the Lamé coefficients. Hence

$$\sigma\circ T_z\cdot\cdot\,\tilde{\varepsilon}(V)\circ T_z = \lambda\,|\,\mathrm{tr}\,\tilde{\varepsilon}(V)\circ T_z|^2 + 2\,\mu\,\|\tilde{\varepsilon}(V)\circ T_z\|^2. \tag{4.3}$$

Recall expressions (3.1) and (3.18):

$$V\circ T_z = e + z\,\ell, \qquad \tilde{\varepsilon}(V)\circ T_z = \varepsilon^0 + z\,\varepsilon^1 + z^2\,\varepsilon^2 \tag{4.4}$$

Recall also the measure decomposition (2.18)-(2.19). From (4.1) and (4.2)

$$\mathcal{P} = \frac{1}{2}\int_{-h}^{h} dz \int_\Gamma d\Gamma\,\left[\lambda\,|\,\mathrm{tr}\,\tilde{\varepsilon}(V)\circ T_z|^2 + 2\,\mu\,\|\tilde{\varepsilon}(V)\circ T_z\|^2\right]\,j(z), \tag{4.5}$$

$$\mathcal{W} = \int_{-h}^{h} dz \int_\Gamma d\Gamma\,\left[F\circ T_z\cdot V\circ T_z + M\circ T_z\cdot(\ell\circ p\circ T_z)\right]\,j(z). \tag{4.6}$$

Using identities (4.3) (2.7) and (2.20) for the polynomial $j(z)$

$$\sigma(V) \circ T_z \cdot\cdot \tilde{\varepsilon}(V) \circ T_z \, j(z) = \sum_{n=0}^{4} \left(\sum_{i=0}^{N-1} K_i \, z^{i+n} \right) A_n, \tag{4.7}$$

where

$$\begin{aligned} A_0 &= 2\,\mu\, \|\varepsilon^0\|^2 + \lambda\, |\operatorname{tr} \varepsilon^0|^2 \quad A_1 = 2\,\mu\, 2\,\varepsilon^0 \cdot\cdot \varepsilon^1 + \lambda\, 2 \operatorname{tr} \varepsilon^0 \operatorname{tr} \varepsilon^1 \\ A_2 &= 2\,\mu\, [\|\varepsilon^1\|^2 + 2\,\varepsilon^0 \cdot\cdot \varepsilon^2] + \lambda\, [|\operatorname{tr} \varepsilon^1|^2 + 2 \operatorname{tr} \varepsilon^0 \operatorname{tr} \varepsilon^2] \\ A_3 &= 2\,\mu\, 2\,\varepsilon^1 \cdot\cdot \varepsilon^2 + \lambda\, 2 \operatorname{tr} \varepsilon^1 \operatorname{tr} \varepsilon^2 \quad A_4 = 2\,\mu\, \|\varepsilon^2\|^2 + \lambda\, |\operatorname{tr} \varepsilon^2|^2. \end{aligned} \tag{4.8}$$

Similarly if we assume that the force F and the moment M are of the form

$$F = f \circ p \quad M = m \circ p \text{ on } S_h, \quad f, m \in H^1(\Gamma)^N, \quad m \cdot \nabla b = 0 \tag{4.9}$$

then we get

$$\begin{aligned} &[(F \circ T_z) \cdot (V \circ T_z) + (M \circ T_z) \cdot ((\ell \circ p) \circ T_z)]\, j(z) \\ &= [f \cdot e + m \cdot \ell] \sum_{i=0}^{N-1} K_i \, z^i + f \cdot \ell \sum_{i=0}^{N-1} K_i \, z^{i+1} \end{aligned} \tag{4.10}$$

and we obtain expressions for $\mathcal{P}$ and $\mathcal{W}$

$$\mathcal{P} = \frac{1}{2} \int_\Gamma \int_{-h}^{h} \sum_{n=0}^{4} A_n \, z^n \sum_{i=0}^{N-1} K_i \, z^i \, dz \, d\Gamma = \frac{1}{2} \int_\Gamma \sum_{n=0}^{4} \alpha_n(h) \, A_n \, d\Gamma$$

$$\mathcal{W} = \int_\Gamma \alpha_0(h) \, (f \cdot e + m \cdot \ell) + \alpha_1(h) \, f \cdot \ell \, d\Gamma,$$

where, after integration with respect to z, the α_n's are polynomials in odd powers of h

$$\alpha_n(h) = \int_{-h}^{h} z^n \sum_{i=0}^{N-1} K_i \, z^i \, dz, \quad 0 \le n \le 4. \tag{4.11}$$

We now introduce the operator $A : \mathcal{V} \to \mathcal{V}'$ associated with $\mathcal{P}$ and the operator $\mathcal{B} : L^2(\Gamma)^N \times L^2(\Gamma)^N \to \mathcal{H}'$ associated with $\mathcal{W}$. For all (e, ℓ) and $(\bar{e}, \bar{\ell})$ in $\mathcal{V}$

$$< A(e, \ell), (\bar{e}, \bar{\ell}) >_{\mathcal{V}} = \sum_{n=0}^{4} \int_\Gamma \alpha_n a_n((e, \ell), (\bar{e}, \bar{\ell})) \, d\Gamma, \tag{4.12}$$

where

$$\begin{aligned} a_0((e, \ell), (\bar{e}, \bar{\ell})) &= 2\,\mu\, \varepsilon^0 \cdot\cdot \bar{\varepsilon}^0 + \lambda \operatorname{tr} \varepsilon^0 \operatorname{tr} \bar{\varepsilon}^0 \\ a_1((e, \ell), (\bar{e}, \bar{\ell})) &= 2\,\mu\, [\varepsilon^0 \cdot\cdot \bar{\varepsilon}^1 + \bar{\varepsilon}^0 \cdot\cdot \varepsilon^1] + \lambda\, [\operatorname{tr} \varepsilon^0 \operatorname{tr} \bar{\varepsilon}^1 + \operatorname{tr} \bar{\varepsilon}^0 \operatorname{tr} \varepsilon^1] \\ a_2((e, \ell), (\bar{e}, \bar{\ell})) &= 2\,\mu\, [\varepsilon^1 \cdot\cdot \bar{\varepsilon}^1 + \varepsilon^0 \cdot\cdot \bar{\varepsilon}^2 + \bar{\varepsilon}^0 \cdot\cdot \varepsilon^2] \\ &\quad + \lambda\, [\operatorname{tr} \varepsilon^1 \operatorname{tr} \bar{\varepsilon}^1 + \operatorname{tr} \varepsilon^0 \operatorname{tr} \bar{\varepsilon}^2 + \operatorname{tr} \bar{\varepsilon}^0 \operatorname{tr} \varepsilon^2] \\ a_3((e, \ell), (\bar{e}, \bar{\ell})) &= 2\,\mu\, [\varepsilon^1 \cdot\cdot \bar{\varepsilon}^2 + \bar{\varepsilon}^1 \cdot\cdot \varepsilon^2] + \lambda\, [\operatorname{tr} \varepsilon^1 \operatorname{tr} \bar{\varepsilon}^2 + \operatorname{tr} \bar{\varepsilon}^1 \operatorname{tr} \varepsilon^2] \\ a_4((e, \ell), (\bar{e}, \bar{\ell})) &= 2\,\mu\, \varepsilon^2 \cdot\cdot \bar{\varepsilon}^2 + \lambda \operatorname{tr} \varepsilon^2 \operatorname{tr} \bar{\varepsilon}^2, \end{aligned} \tag{4.13}$$

where $\varepsilon^i = \varepsilon^i(e,\ell)$, $\overline{\varepsilon}^i = \varepsilon^i(\overline{e},\overline{\ell})$. Note that $A_n = a_n((e,\ell),(e,\ell))$, $0 \leq n \leq 4$.
Similarily define the continuous linear operator

$$\text{(4.14)}\quad \begin{aligned} &\mathcal{B} : U \stackrel{\text{def}}{=} L^2(\Gamma)^N \times L^2(\Gamma)^N \to \mathcal{H}' \\ &< \mathcal{B}(f,m),(e,\ell) >_{\mathcal{H}} = \int_\Gamma \alpha_0\,[f\cdot e + m\cdot \ell] + \alpha_1\, f\cdot \ell\, d\Gamma. \end{aligned}$$

By construction A is symmetrical and positive

$$< A(e,\ell),(\overline{e},\overline{\ell}) >_{\mathcal{V}} = < A(\overline{e},\overline{\ell}),(e,\ell) >_{\mathcal{V}} \qquad < A(e,\ell),(e,\ell) >_{\mathcal{V}} \geq 0.$$

Lemma 4.1. *There exists $\overline{h} > 0$ and $\alpha > 0$ such that for all $0 < h < \overline{h}$ and all $(e,\ell) \in \mathcal{V}$*

$$\text{(4.15)}\quad < A(e,\ell),(e,\ell) >_{\mathcal{V}} \geq 2\mu h \alpha \sum_{n=0}^{2} h^{2n} \|\varepsilon^n(e,\ell)\|^2.$$

If we identify the elements of the dual $\mathcal{H}'$ of $\mathcal{H}$ with those of $\mathcal{H}$, then from the above lemma A is a $\mathcal{V}$-$\mathcal{H}$ coercive operator.

Theorem 4.1. *Given $\overline{h} > 0$ as specified in Lemma 4.1, $0 < h \leq \overline{h}$, and assuming that the following condition is verified for all $(e,\ell) \in \mathcal{N}$ (recall from Theorem 3.1 that $\ell = 0$)*

$$\text{(4.16)}\quad \int_\Gamma \alpha_0(h) f\cdot e\, d\Gamma = 0,$$

then for all h, $0 < h \leq \overline{h}$, there exists a unique solution $(\hat{e},\hat{\ell}) \in \mathcal{V}/\mathcal{N}$ to the variational equation: for all $(e,\ell) \in \mathcal{V}$

$$\text{(4.17)}\quad < A(\hat{e},\hat{\ell}),(e,\ell) >_{\mathcal{V}} + < \mathcal{B}(f,m),(e,\ell) >_{\mathcal{H}} = 0.$$

4.2. Kinetic energy and dynamical shell equation.

Assume that $V(x,t)$, $e(x,t)$, and $\ell(x,t)$ depend on the time t and denote by V', e', and ℓ' the derivatives with respect to the time t:

$$\text{(4.18)}\quad \begin{aligned} \mathcal{K} &= \int_{S_h} \frac{1}{2}\rho\,|V'|^2\, dx = \int_{S_h} \frac{1}{2}\rho\,|e'\circ p + b\,\ell'\circ p|^2\, dx \\ &= \frac{1}{2}\rho \int_\Gamma \alpha_0(h)\,|e'|^2 \;+ \alpha_1(h)\,2\,e'\cdot\ell' + \alpha_2(h)\,|\ell'|^2\, d\Gamma, \end{aligned}$$

where ρ is the volume density and e' and ℓ' are the time derivatives of e and ℓ. Define the operator $\mathcal{M} : \mathcal{H} \to \mathcal{H}'$ as follows:

$$\text{(4.19)}\quad < \mathcal{M}(e,\ell),(\overline{e},\overline{\ell}) >_{\mathcal{H}} = \rho \int_\Gamma \alpha_0(h)\, e\cdot\overline{e} + \alpha_1(h)\,[e\cdot\overline{\ell} + \overline{e}\cdot\ell] + \alpha_2(h)\,\ell\cdot\overline{\ell}\, d\Gamma.$$

For h sufficiently small $\mathcal{M}$ is symmetrical, positive and invertible.

We want to find a stationary point (e, ℓ) of the total energy $\mathcal{P} + \mathcal{W} - \mathcal{K}$. Identify the elements of the dual $\mathcal{H}'$ of $\mathcal{H}$ with those of $\mathcal{H}$ and denote by $\Lambda : \mathcal{H}' \to \mathcal{H}$ the corresponding canonical isomorphism. Then the linear operators

$$M = \Lambda\, \mathcal{M}, \quad B = \Lambda\, \mathcal{B} \tag{4.20}$$

are bounded from $\mathcal{H}$ to $\mathcal{H}$ and from U to $\mathcal{H}$. For h sufficiently small, M is a positive symmetrical continuous and invertible linear operator. The dynamical equation of the shell is of the form

$$\frac{d^2}{dt^2}(M \begin{bmatrix} e(t) \\ \ell(t) \end{bmatrix}, \begin{bmatrix} \varphi \\ \psi \end{bmatrix})_{\mathcal{H}} + < A \begin{bmatrix} e(t) \\ \ell(t) \end{bmatrix}, \begin{bmatrix} \varphi \\ \psi \end{bmatrix} >_{\mathcal{V}} + (B \begin{bmatrix} f(t) \\ m(t) \end{bmatrix}, \begin{bmatrix} \varphi \\ \psi \end{bmatrix})_{\mathcal{H}} = 0 \tag{4.21}$$

for all $(\varphi, \psi) \in \mathcal{V}$. Standard existence and uniqueness theorems (cf. for instance J.L. LIONS [1, Thm 1.1 and Rem. 1.3, p. 294]) can now be used.

Theorem 4.2. *Given $\overline{h} > 0$ as specified in Lemma 4.1, $0 < h \le \overline{h}$, there exists a unique solution*

$$(e, \ell) \in C^1([0, T]; \mathcal{H}) \cap C([0, T]; \mathcal{V}). \tag{4.22}$$

to equation (4.21) verifying the initial conditions

$$(e(0), \ell(0)) = (e_0, \ell_0) \quad \text{in } \mathcal{V}, \qquad (e'(0), \ell'(0)) = (e_1, \ell_1) \quad \text{in } \mathcal{H}. \tag{4.23}$$

4.3. Homogeneous Dirichlet boundary conditions.

We have the analogue of Theorems 4.1 and 4.2. In the static case

Theorem 4.3. *Given $\overline{h} > 0$ as specified in Lemma 4.1 and h, $0 < h \le \overline{h}$, there exists a unique solution $(e(h), \ell(h)) \in \mathcal{V}_0$ to the variational equation: for all $(e, \ell) \in \mathcal{V}_0$*

$$< A(e(h), \ell(h)), (e, \ell) >_{\mathcal{V}} + < \mathcal{B}(f, m), (e, \ell) >_{\mathcal{H}} = 0, \tag{4.24}$$

$$\mathcal{V}_0 = \left\{ (e, \ell) \in H_0^1(\Gamma)^N \times H_0^1(\Gamma)^N \, : \, \ell \cdot n = 0 \right\}. \tag{4.25}$$

It follows from the following Poincaré inequality for the shell.

Lemma 4.2. *Assume that Γ is a bounded open domain in the C^2 $(N-1)$-dimensional submanifold $\partial\Omega$ of $\mathbb{R}^N$ with a Lipschitzian boundary $\partial\Gamma = \partial_{\partial\Omega}\Gamma$ in $\partial\Omega$. For h sufficiently small, there exists a constant $c = c(S_h, D^2 b) > 0$ such that for all $(e, \ell) \in \mathcal{V}_0$*

$$\begin{aligned} &\int_\Gamma 2\,h\,|e|^2 + 2\,\frac{h^3}{3}\,|\ell|^2 + 2\,h\,[\,|\ell|^2 + \|D_\Gamma(e)\|^2\,] + 2\,\frac{h^3}{3}\|D_\Gamma(\ell)\|^2\, d\Gamma \\ &\le c^2 \int_\Gamma 2\,h\,\|\varepsilon^0(e, \ell)\|^2 + 2\,\frac{h^3}{3}\,\|\varepsilon^1(e, \ell)\|^2 + 2\,\frac{h^5}{5}\,\|\varepsilon^2(e, \ell)\|^2\, d\Gamma \end{aligned} \tag{4.26}$$

and $\|\varepsilon_\Gamma(e)\| + \|\varepsilon^1(e, 0)\|$ is an equivalent norm on $H_0^1(\Gamma)^N$.

5. The Love-Kirchhoff theory

We have seen in §3.3 that the Love-Kirchhoff theory is characterized by identity (3.30). We can repeat the constructions of §4 with the following subspaces of $\mathcal{H}$ and $\mathcal{V}$:

$$\mathcal{H}_1 = \{(e,\ell) \in L^2(\Gamma)^N \times L^2(\Gamma)^N \,:\, \ell + 2\,\varepsilon_\Gamma(e)\,\nabla b = 0 \text{ on } \Gamma\} \tag{5.1}$$

$$\mathcal{V}_1 = \{(e,\ell) \in \mathcal{H} \,:\, \varepsilon^i(e,\ell) \in L^2(\Gamma)^{N\times N},\ 0 \leq i \leq 2\}, \tag{5.2}$$

$$\mathcal{N}_1 = \{(e,\ell) \in \mathcal{V}_1 \,:\, \varepsilon^i(e,\ell) = 0 \text{ on } \Gamma,\ 0 \leq i \leq 2\}. \tag{5.3}$$

Endowed with the norms (3.22) and (3.23), they are also Hilbert spaces. The subspace $\mathcal{N}_1$ coincides with $\mathcal{N}$ as defined in (3.21) and corresponds to the rigid displacements. From the Korn's inequality (3.25) in Theorem 3.2,

$$\mathcal{V}_1 = \{(e,\ell) \in H^1(\Gamma)^N \times H^1(\Gamma)^N \,:\, \ell + 2\,\varepsilon_\Gamma(e)\,\nabla b = 0\}. \tag{5.4}$$

Lemma 4.1 remains true on $\mathcal{V}_1$ and A is $\mathcal{V}_1$-$\mathcal{H}_1$ coercive and we have the equivalent of Theorems 4.1 and 4.2.

Theorem 5.1. *Given $\overline{h} > 0$ as specified in Lemma 4.1, $0 < h \leq \overline{h}$, and assuming that the following condition is verified for all $(e,\ell) \in \mathcal{N}$ (recall from Theorem 3.1 that $\ell = 0$)*

$$\int_\Gamma \alpha_0(h)\, f \cdot e\, d\Gamma = 0, \tag{5.5}$$

then for all $0 < h \leq \overline{h}$ there exists a unique solution $(\hat{e},\hat{\ell}) \in \mathcal{V}_1/\mathcal{N}$ to the variational equation: for all $(e,\ell) \in \mathcal{V}_1$

$$< A(\hat{e},\hat{\ell}),(e,\ell) >_{\mathcal{V}} + < \mathcal{B}\,(f,m),(e,\ell) >_{\mathcal{H}} = 0. \tag{5.6}$$

As in §4, identify the elements of the topological dual $\mathcal{H}_1'$ of $\mathcal{H}_1$ with those of $\mathcal{H}_1$ and denote by $\Lambda_1 : \mathcal{H}_1' \to \mathcal{H}_1$ the corresponding canonical isomorphism. Let

$$M_1 = \Lambda_1\,\mathcal{M}_1, \quad B_1 = \Lambda_1\,\mathcal{B}. \tag{5.7}$$

Theorem 5.2. *Given $\overline{h} > 0$ as specified in Lemma 4.1, $0 < h \leq \overline{h}$, there exists a unique solution*

$$(e,\ell) \in C^1([0,T];\mathcal{H}_1) \cap C([0,T];\mathcal{V}_1). \tag{5.8}$$

to the equation

$$\frac{d^2}{dt^2}(M_1 \begin{bmatrix} e(t) \\ \ell(t) \end{bmatrix}, \begin{bmatrix} \varphi \\ \psi \end{bmatrix})_{\mathcal{H}_1} + < A \begin{bmatrix} e(t) \\ \ell(t) \end{bmatrix}, \begin{bmatrix} \varphi \\ \psi \end{bmatrix} >_{\mathcal{V}_1} + (B_1 \begin{bmatrix} f(t) \\ m(t) \end{bmatrix}, \begin{bmatrix} \varphi \\ \psi \end{bmatrix})_{\mathcal{H}_1} = 0 \tag{5.9}$$

for all $(\varphi,\psi) \in \mathcal{V}_1$, verifying the initial conditions

$$(e(0),\ell(0)) = (e_0,\ell_0) \quad \textit{in } \mathcal{V}_1, \qquad (\ell_0 = 2\,\varepsilon_\Gamma(e_0)\,n) \tag{5.10}$$

$$(e'(0),\ell'(0)) = (e_1,\ell_1) \quad \textit{in } \mathcal{H}_1, \qquad (\ell_1 = 2\,\varepsilon_\Gamma(e_1)\,n). \tag{5.11}$$

Remark 5.1. The Love-Kirchhoff identity (3.30) can also be written in the form

$$\ell = D^2 b\, e - \nabla_\Gamma(e \cdot n). \tag{5.12}$$

If we assume that $\partial\Omega$ is of class C^3, then

$$\mathcal{H}_1 = \{(e, D^2 b\, e - \nabla_\Gamma(e \cdot n)) \in L^2(\Gamma)^N \times L^2(\Gamma)^N \,:\, \nabla_\Gamma(e \cdot n) \in L^2(\Gamma)^N\} \tag{5.13}$$

$$\mathcal{V}_1 = \left\{(e, D^2 b\, e - \nabla_\Gamma(e \cdot n)) \in H^1(\Gamma)^N \times H^1(\Gamma)^N \,:\, D^2_\Gamma(e \cdot n) \in L^2(\Gamma)^{N\times N}\right\} \tag{5.14}$$

and $\mathcal{N}_1 = \mathcal{N}$ in (3.24), where $D^2_\Gamma(e \cdot n) = D_\Gamma(\nabla_\Gamma(e \cdot n))$. Spaces where the normal component of e is one degree smoother than e itself are typical in the theory of plates. The assumption that $\partial\Omega$ be of class C^3 is not a necessary mathematical assumption of the Love-Kirchhoff theory resulting from identity (3.30). It only arises from a rearrangement of the variables in the form (5.12). This suggests to define the spaces

$$H_1 = \left\{e \in L^2(\Gamma)^N \,:\, \nabla_\Gamma(e \cdot n) \in L^2(\Gamma)^N\right\} \tag{5.15}$$

$$V_1 = \left\{e \in H^1(\Gamma)^N \,:\, D^2_\Gamma(e \cdot n) \in L^2(\Gamma)^{N\times N}\right\} \tag{5.16}$$

$$\begin{aligned} N = \{e \,:\, e(X) = a + B\,X, \quad &\forall a \in \mathbb{R}^N,\ \forall B \text{ an } N \times N \text{ matrix} \\ &\text{such that } B + {}^*B = 0 \quad \text{and} \quad B\,\nabla b(X) = 0, \quad \forall X \in \Gamma\}. \end{aligned} \tag{5.17}$$

and use identity (5.12) to substitute for ℓ in the previous expressions. □

Theorems 5.1 and 5.2 remain true with obvious changes for homogeneous Dirichlet boundary conditions without condition (5.5).

Acknowledgement. The authors would like to thank M. Bernadou for bringing to our attention the work of R. Valid.

REFERENCES

M. BERNADOU [**1**], *Méthodes d'éléments finis pour les problèmes de coques minces*, Masson, Paris, Milan, Barcelone, 1994.

M.C. DELFOUR and J.P. ZOLÉSIO [**1**], *Shape analysis via oriented distance functions*, J. Functional Analysis **123** (1994), 129–201.

[**2**], *On a variational equation for thin shells*, Control and Optimal Design of Distributed Parameter Systems (J. Lagnese, D.L. Russell, L. White, eds.), Springer-Verlag, New York, 1994, pp. 25–37.

[**3**], *A boundary differential equation for thin shells*, J. Differential Equations (to appear).

[**4**], *Tangential differential equation for dynamical thin/shallow shells*, J. Differential Equations (submitted).

P. GERMAIN [**1**], *Mécanique, tomes I and II*, Ellipses éditeur, Paris, 1986.

D. GILBARG and N.S. TRUDINGER [**1**], *Elliptic Partial Differential Equations of Second Order*, Springer-Verlag, Berlin, Heidelberg, New York, Tokyo, 1983.

J.E. LAGNESE and J.-L. LIONS [**1**], *Modelling, analysis and control of thin plates*, RMA, Vol. 6, Masson, Paris, Milan, Barcelone, Mexico, 1988.

J.-L. LIONS [**1**], *Contrôle optimal de systèmes gouvernés par des équations aux dérivées partielles*, Dunod, Gauthier-Villars, Paris, 1968.

R. TEMAM [**1**], *Monotone rearangement of a function and the Grad-Mercier equation of plasmas physics*, Recent methods in non linear analysis and application (E. De Giorgi, E. Magenes and U. Mosco, eds.), Pitagora Editrice, Bologna, 1979.

R. VALID [**1**], *Mechanics of continuous media and analysis of structures*, North-Holland Series in Applied Mathematics and Mechanics, Amsterdam, New York, Oxford, 1981.

J.P. ZOLÉSIO [**1**], *Domain variational formulation for free boundary problems*, Optimization of Distributed Parameter Structures, vol II (E.J. Haug and J. Céa, eds.), Sijthoff and Noordhoff, Alphen aan den Rijn, The Netherlands, 1981, pp. 1152–1194.

Obstruction and Some Approximate Controllability Results for the Burgers Equation and Related Problems

J.I. DIAZ Departamento de Matemática Aplicada. Facultad de Matemáticas, Universidad Complutense de Madrid, Spain.

1 INTRODUCTION

It is well accepted that the Burgers equation provides a realistic simplification of the Navier-Stokes system in Fluid Mechanics. One of the most important common features among both equations is the presence of a quadratic nonlinear term and a linear diffusion operator. Some *optimal control* results associated to the Burgers equation have been obtained in the literature (see *e.g.*, Glowinski and Lions (1994) and its references). This work concerns with other type of control problem associated to the Burgers equation: the approximate controllability for a final observation. To fix ideas, given some positive numbers T, L and ν we consider the following boundary control problem

$$\left.\begin{array}{ll} y_t + yy_x - \nu y_{xx} = 0 & \text{in } (0,T)\times(0,L) \\ y(t,0) = 0, \quad y(t,L) = u(t), & t \in (0,T) \\ y(0,x) = y_0(x), & x \in (0,L) \end{array}\right\} \tag{1}$$

where the control $u(t)$ is assumed to be a function on $(0,T)$ and the initial datum is given. For the sake of simplicity in the exposition we shall assume that $y_0 \in L^\infty(0,L)$. Other controllability problems associated to the Burgers equation are also considered (see Remark 1.5). We recall that the approximate controllability property, with final observation in a Banach space of states, $\mathcal{X}$, of functions defined on $(0,L)$ (*e.g.* $\mathcal{X} = \mathcal{C}([0,L])$ or $\mathcal{X} = L^2(0,L)$) can be stated in the following terms: given a desired state $y_d \in \mathcal{X}$ and $\varepsilon > 0$ find a control $u_\varepsilon(t)$ such that the solution $y(t,\cdot;u_\varepsilon)$ of (1) corresponding to $u = u_\varepsilon$ satisfies that

$$\|y(T,\cdot : u_\varepsilon) - y_d\|_{\mathcal{X}} \leq \varepsilon. \tag{2}$$

In Section 2 we shall prove that this property can not holds under this general statement. To do that, we shall show that an *Obstruction Phenomenon* arises due to the presence of the superlinear term $(y^2)_x/2$ at the equation. This Obstruction Phenomenon was already exhibited for the case of semilinear parabolic problems in a series of works: Henry (1978), Díaz (1991a), (1991b), (1993a), (1994a) and (1994b), Díaz and Ramos (1993), (1994), Bernis, Díaz and Ramos (1995).The results here presented complete and generalize previous considerations on the Burgers equation made in Díaz (1991a) and (1991b). As an application we give a necessary condition for the approximate controllability of the Navier-Stokes system on a rectangle $\Omega = (0,L_1) \times (0,L_2)$. For $\Gamma = \{L_1\} \times [0,L_2]$, we consider the boundary control problem

$$\left.\begin{array}{ll} \mathbf{y}_t + (\mathbf{y}\cdot\nabla)\mathbf{y} - \nu\Delta\mathbf{y} = -\nabla p & \text{in } (0,T)\times\Omega \\ \operatorname{div}\mathbf{y} = 0 & \text{in } (0,T)\times\Omega \\ \mathbf{y} = \mathbf{0} & \text{on } (0,T)\times(\partial\Omega\setminus\Gamma) \\ \mathbf{y} = \mathbf{u} & \text{on } (0,T)\times\Gamma \\ \mathbf{y}(0,x) = \mathbf{y}_0(x) & \text{on } \Omega. \end{array}\right\} \tag{3}$$

We find a necessary condition on the pressure for the approximate controllability of the problem (see Theorem 2 and Remark 2). Finally, in Section 3, we study the approximate controllability of the Burgers problem under suitable constraints on the desired state $y_d(x)$. We recall some previous results by El Badia and Ain Seba (1992) and Fursikov and Imanuvilov (1993b) on the exact controllability for suitable desired states y_d. Using the last of those references we give a L^p-approximate controllability for a larger class of desired states. We conjecture that some sharper results can be found following the ideas of Díaz (1994b). We start here such a long programme by proving the L^p-approximate controllability for the general quasilinear problem

$$\left.\begin{array}{ll} y_t - \Delta y + \operatorname{div}\mathbf{B}(y) = u\chi_\omega & \text{in } (0,T)\times\Omega \\ y = 0 & \text{on } (0,T)\times\partial\Omega \\ y(0,x) = y_0(x) & \text{on } \Omega, \end{array}\right\} \tag{4}$$

where Ω denotes an open bounded regular set of $\mathbb{R}^N$ and ω is an open subset of Ω. Here $\mathbf{B} \in \mathcal{C}(\mathbb{R} : \mathbb{R}^N)$ is assumed to be differentiable at some $s_0 \in \mathbb{R}$ and sublinear at the infinity, *i.e.* there exists $M > 0$ such that

$$|\mathbf{B}(s)| \leq c_1 + c_2|s| \qquad \text{for any } s \in \mathbb{R},\ |s| > M. \tag{5}$$

The remaining parts of the programme introduced in Díaz (1994b) will be developped for the Burgers problem elsewhere.

2 OBSTRUCTION FOR THE BURGERS EQUATION

The main goal of this section is to show that the approximate controllability in $\mathcal{X} = L^p(0,L)$, $1 \le p \le \infty$, or $\mathcal{X} = \mathcal{C}([0,L])$ can not holds for the Burgers problem (1). To do that we shall prove the existence of an *universal obstruction function* which will provide an explicit bound for any element of the attainable set $R_T := \{y(T,\cdot;u) : u \in \mathcal{U},\ y \text{ solution of (1)}\}$, where $\mathcal{U}$ denotes the space of controls.

We start by recalling that the existence and uniqueness of solutions of problem (1) can be obtained in different ways. So, for instance, the results of Alt and Luckhaus (1983) can be applied showing the existence and uniqueness of a weak solution $y \in \mathcal{C}([0,T] : L^2(0,L)) \cap L^\infty((0,T)\times(0,L)) \cap L^p(0,T : W^{1,p}(0,L))$ for any $p \in [1,\infty)$ with $y_t \in L^{p'}(0,T : W^{-1,p'}(0,L))$ assumed $u \in \mathcal{U} := W^{1,1}(0,T)$. In fact, it is not difficult to show that this solution is much more regular.

The existence of the mentioned *universal obstruction function* is a consequence of the presence of the superlinear term $yy_x = (y^2)_x/2$ at the equation. Such a function will be built as solution of the following problem

$$\left.\begin{array}{ll} Y_t + YY_x - \nu Y_{xx} = 0 & \text{in } (0,T)\times(0,L) \\ Y(t,0) = 0, \quad Y(t,L) = +\infty, & t \in (0,T) \\ Y(0,x) = y_0(x), & x \in (0,L). \end{array}\right\} \tag{6}$$

THEOREM 1
(i) *Problem* (6) *has a minimal solution* $\underline{Y} \in \mathcal{C}([0,T] : L^2(0,L-\varepsilon)) \cap L^p(0,T : W^{1,p}(0,L-\varepsilon))$, *for any* $\varepsilon \in (0,L)$ *and for any* $1 \le p < \infty$. *Moreover, there exists some positive constants* C_1, C_2 *only dependent on* L, ν *and* $\|y_0\|_\infty$ *such that*

$$\underline{Y}(t,x) \le C_1\left(\frac{1}{L-x} - \frac{1}{L}\right) + C_2. \tag{7}$$

(ii) *Let* $y(t,\cdot;u)$ *be the solution of the associated problem* (1). *Then*

$$y(t,x;u) \le \underline{Y}(t,x) \quad \text{for any } t \in [0,T] \text{ and a.e. } x \in (0,L). \tag{8}$$

In particular the approximate controllability property can not be satisfied in the set of states $\mathcal{X} = L^q(0,L)$, $1 \le q \le \infty$.

PROOF. Given $n \in \mathbb{N}$ we consider the truncated problem

$$\left.\begin{array}{ll} Y_t + YY_x - \nu Y_{xx} = 0 & \text{in } (0,T)\times(0,L) \\ Y(t,0) = 0, \quad Y(t,L) = n, & t \in (0,T) \\ Y(0,x) = y_0^n(x) := \min\{y_0(x), n\}, & x \in (0,L). \end{array}\right\} \tag{9}$$

The existence and uniqueness of the solution Y_n of (9) holds as mentioned before. Furthermore, the comparison principle is satisfied (see *e.g.* Alt and Luckhaus (1983)) and so we have

$$Y_1 \le Y_2 \le \cdots \le Y_n \le Y_{n+1} \le \cdots \quad \text{on } (0,T)\times(0,L).$$

In order to construct a global barrier function we define

$$W(x) := K_1(L-x)^{-\alpha} + K_2, \quad x \in (0,L) \tag{10}$$

where $K_1, \alpha > 0$ and $K_2 \geq 0$ will be choosen later. Then

$$WW_x - \nu W_{xx} = \alpha K_1^2(L-x)^{-2\alpha-1} + K_1K_2\alpha(L-x)^{-\alpha-1} - \nu\alpha(\alpha+1)K_1(L-x)^{-\alpha-2}.$$

So, making $\alpha = 1$ we find that

$$WW_x - \nu W_{xx} = (-2\nu + K_1)K_1(L-x)^{-3} + K_1K_2(L-x)^{-2} := f(x). \tag{11}$$

Then, for any $n \in \mathbb{N}$, we have

$$\left.\begin{array}{ll} Y_{n,t} + Y_nY_{n,x} - \nu Y_{n,xx} = 0 \leq WW_x - \nu W_{xx} & \text{in } (0,T)\times(0,L) \\ Y_n(t,0) = 0 \leq W(0), \quad \limsup_{x\uparrow L}(Y_n(t,x) - W(x)) < 0, & t \in (0,T) \\ Y_n(0,x) \leq \|y_0\|_\infty \leq W(x), & x \in (0,L) \end{array}\right\}$$

assuming that

$$K_1 \geq 2\nu, \quad K_2 \geq 0 \quad \text{and} \quad \|y\|_\infty \leq K_1L^{-1} + K_2. \tag{12}$$

In particular, if

$$K_1 = 2\nu \quad \text{and} \quad K_2 \geq \left[\|y_0\|_\infty - 2\nu L^{-1}\right]_+ \tag{13}$$

then (12) holds, $f \in L^\infty(0, L-\varepsilon)$, for any $\varepsilon \in (0,L)$, and from the comparison principle we deduce that $Y_n(t,x) \leq W(x)$ for any $n \in \mathbb{N}$, $t \in [0,T]$ and *a.e.* $x \in (0,L)$. By the Beppo-Levi Theorem we have that $Y_n(t,\cdot) \to \underline{Y}(t,\cdot)$ in $L^p(0, L-\varepsilon)$, for any $\varepsilon \in (0,L)$ and any $1 \leq p \leq \infty$. Using straightforward arguments it is easy to see that $\underline{Y}$ is the minimal solution of problem (6) and that (7) holds. Part (ii) comes from the fact that if u is a bounded function then choosing $n_0 \in \mathbb{N}$ such that $n_0 \geq \max\{\|y_0\|_\infty, \|u\|_\infty\}$, we have $y(t,x;u) \leq Y_n(t,x)$ for any $n \geq n_0$, $t \in [0,T]$ and *a.e.* $x \in (0,L)$. Finally, if the desired state $y_d(x)$ is such that $y_d(x) > W(x)$ on an open subset of $(0,L)$ then condition (3) fails, for any $u \in \mathcal{U}$, for any $\varepsilon > 0$. ■

REMARK 1

1. The fact that (1) is not approximately controllable can also be proved by using an *universal integral estimate* as obtained in Fursikov and Imanuvilov (1993) by multiplying the equation by $(b-x)^n y_+^3(t,x)$ with $b \in (0,L)$ and $n > 5$, integrating by parts and applying Hölder and Young inequalities. Such a method was already introduced by A. Bamberger for the study of superlinear semilinear parabolic equations (see Henry (1978)). A more sophisticated energy method can also be applied to higher order superlinear parabolic equations: see Bernis, Díaz and Ramos (1995).
2. The obstruction phenomenon also holds for other nonlinear parabolic controls problems with superlinear terms in the equations such as $y_t - \Delta y + \lambda|y|^{p-1}y = 0$ ($p > 1$: superlinear semilinear equation), $y_t - \Delta|y|^{m-1}y = 0$ ($m > 1$: the porous media equation) and $y_t - \operatorname{div}(|\nabla y|^{p-2}\nabla y) = 0$ ($p > 2$: Non-Newtonian flows). See Díaz (1991a).
3. The study of boundary value problems blowing-up on a part of the boundary has been largely considered by many authors: Bieberbach, Rademacher, Keller, P.L.Lions and Lasry, *etc.* (see references in Bandle and Marcus (1990), G.Díaz and Letelier (1993) and Bandle, Díaz and Díaz (1994)).
4. We point out that the conclusion of Theorem 1 holds if u is not bounded (truncate u and pass to the limit). It also holds for solutions of the non-homogeneous equation

$$y_t + yy_x - \nu y_{xx} = f(t,x) \tag{14}$$

and the rest of conditions as indicated in (1), assuming that $f \in L^1((0,T) \times (0, L-\varepsilon))$ for any $\varepsilon \in (0, L)$ and

$$f(t,x) \le M_1(L-x)^{-3} + M_2(L-x)^{-2} \qquad \text{for any } (t,x) \in (0,T) \times (0,L) \tag{15}$$

for some nonnegative constants M_1 and M_2.
5. Similar results can be obtained for other control problems associated to the Burgers equation. So, for the case in which the control acts on the left boundary, *i.e.* $y(t,0) = u(t)$, $y(t,L) = 0$, $t \in (0,T)$ and the rest of conditions of (1) an inequality (analogous to (8)) holds

$$\frac{c_1}{x} - c_2 \le y(t,x;u) \quad \text{for any } t \in (0,T),\ a.e.\ x \in (0,L) \tag{16}$$

for any bounded function u and for some constants c_1 and c_2. Analogously, consider the controllability from the interior problem

$$\left.\begin{array}{ll} y_t + yy_x - \nu y_{xx} = u\chi_\omega & \text{in } (0,T) \times (0,L) \\ y(t,0) = y(t,L) = 0, & t \in (0,T) \\ y(0,x) = y_0(x), & x \in (0,L), \end{array}\right\} \tag{17}$$

where $\omega = (a,b)$ is an open subinterval of $(0,L)$ and $u \in L^2(\omega)$. An easy application of the arguments of Theorem 1 shows that

$$y(t,x;u) \le c_1 \left(\frac{1}{a-x} - \frac{1}{b} \right) + c_2 \qquad \text{for any } t \in [0,T] \text{ and } a.e.\ x \in (0,a) \tag{18}$$

for any $u \in L^2(0,T) \times \omega)$ and for some constants c_1 and c_2. □

We shall end this section by applying Theorem 1 to the study of the approximate controllability of the Navier-Stokes system. Such a question was already raised by Lions (1990) and still remains an open question. Our contribution will be limited to give a necessary condition on the pressure assuming that the approximate controllability holds. For the sake of simplicity in the exposition we shall merely deal with the case of a planar flow occupying a rectangle $\Omega = (0,L_1) \times (0,L_2)$ (see problem (3) in the Introduction). Many other domains and other control problems can be considered analogously.

THEOREM 2
Let $\mathbf{y_0} \in L^\infty(\Omega)^2$. *Consider* $\varepsilon > 0$ *and* $\mathbf{y}_d = (y_{d,1}, y_{d,2}) \in L^2(\Omega)^2$ *such that there exists a positively measured subset* P *of* Ω *and a constant* $C \ge [\|y_{0,1}\|_\infty - \nu L_1]_+$ *such that*

$$\|[y_{1,d}(x) - \nu(L-x)^{-1} - C]_+\|_{L^2(P)} > \varepsilon. \tag{19}$$

Let $\mathbf{u} = (u_1, u_2)$ *be any control such that* $\|\mathbf{y}(T,\cdot;\mathbf{u}) - \mathbf{y}_d\|_{L^2(\Omega)^2} \le \varepsilon$. *Then necesarily*

$$-\frac{\partial p}{\partial x_1}(t, x_1, x_2; \mathbf{u}) > 2\nu C(L_1 - x_1)^{-2} \tag{20}$$

on some positively measured set $M \subset (0,T) \times \Omega$.

The proof will use some comparison results between the solution of a Burgers problem (1) and the first component of the solution of the Navier-Stokes system (4). A first result in this direction is the following

LEMMA 1

Let $z \in \mathcal{C}^{1,2}_{t,x}((0,T)\times(0,L_1))\cap\mathcal{C}([0,T]\times[0,L_1])$ *and assume that* $\mathbf{y}=(y_1,y_2)$ *is a classical solution of* (3). *Let* $z\in\mathcal{C}([0,T]\times(0,L_1))$ *be such that*

$$\left.\begin{array}{ll} z_t + z z_{x_1} - \nu z_{x_1x_1} = f(t,x_1) & \text{in } (0,T)\times(0,L_1) \\ z(t,0)=0, \quad \liminf_{x\uparrow L_1} z(t,x) > \|u_1(t,\cdot)\|_{L^\infty(\Gamma)}, & t\in(0,T), \\ z(0,x_1)\geq \|y_{0,1}\|_{L^\infty(\Omega)}, & x_1\in(0,L_1) \end{array}\right\} \tag{21}$$

where $f\in\mathcal{C}((0,T)\times(0,L_1))$ *verifies*

$$f(t,x_1)\geq -\frac{\partial p}{\partial x_1}(t,x_1,x_2;\mathbf{u}) \qquad \text{for } (t,x_1,x_2)\in(0,T)\times\Omega \tag{22}$$

(*the control* $\mathbf{u}=(u_1,u_2)$ *in* (3) *is here assumed to be known*). *Finally, assume that*

$$z_{x_1}(t,x_1)>0 \qquad \text{for any } (t,x_1)\in(0,T)\times(0,L_1). \tag{23}$$

Then

$$z(t,x_1)\geq y_1(t,x_1,x_2;\mathbf{u}) \qquad \text{for any } t\in[0,T] \text{ and } (x_1,x_2)\in\Omega. \tag{24}$$

PROOF. From (21) and (23) we can assume, without loss of generality, that there exists $t_0\in(0,T]$ and $(x_{0,1},x_{0,2})\in\Omega$ such that

$$\sup_{[0,T]\times\overline{\Omega}}(y_1(t,x_1,x_2)-z(t,x_1)) = y_1(t_0,x_{1,0},x_{2,0})-z(t,x_{1,0})>0$$

(otherwise (24) holds). Define $w(t,x_1,x_2):=y_1(t,x_1,x_2)-z(t,x_1)$. Then

$$\begin{array}{ll} \dfrac{\partial y_1}{\partial x_1}(t_0,x_{0,1},x_{0,2}) = \dfrac{\partial z}{\partial x_1}(t_0,x_{0,1})>0 \quad \text{and} & \dfrac{\partial y_1}{\partial x_2}(t_0,x_{0,1},x_{0,2})=0, \\ \Delta w(t_0,x_{0,1},x_{0,2})\leq 0 & (\text{ i.e. } \Delta y_1(t_0,x_{0,1},x_{0,2})\leq z_{x_1x_1}(t_0,x_{0,1})) \end{array}$$

and

$$w_t(t_0,x_{1,0},x_{2,0})\geq 0.$$

On the other hand

$$w_t = -y_1y_{1,x_1} - y_2y_{1,x_2} + \nu\Delta y_1 - \nu z_{x_1x_1} - p_{x_1} - f + z z_{x_1}.$$

Thus

$$w_t(t_0,x_{0,1},x_{0,2})\leq -w(t_0,x_{0,1},x_{0,2})z_{x_1}(t_0,x_{0,1})<0$$

which is a contradiction. ■

A more general result, obtained by replacing (23) by a nonnegativeness condition, and valid for bounded weak solutions of the Navier-Stokes problem, can be proved by introducing an artificial constant in the Burgers equation.

LEMMA 2

Let $z\in\mathcal{C}([0,T]: L^2(0,L_1))\cap L^2(0,T:H^1(0,L_1))\cap W^{1,1}(0,T:L^1(0,L_1))$ *be a nonnegative function satisfying*

$$\left.\begin{array}{ll} z_t + 2z z_{x_1} - \nu z_{x_1x_1} = f(t,x_1) & \text{in } (0,T)\times(0,L_1) \\ z(t,0)=0, \quad z(t,L_1)\geq \|u_1(t,\cdot)\|_{L^\infty(\Gamma)}, & t\in(0,T) \\ z(0,x_1)\geq \|y_{0,1}\|_{L^\infty(\Omega)}, & x_1\in(0,L_1). \end{array}\right\} \tag{25}$$

Let $\mathbf{u} \in \mathcal{U} \subset \mathrm{L}^\infty((0,T)\times\Gamma)^2$ *be given such that* (3) *has a unique bounded weak solution* $\mathbf{y}(t,\cdot;\mathbf{u}) = (y_1(t,\cdot;\mathbf{u}), y_2(t,\cdot;\mathbf{u}))$ *of pressure* $p = p(t,\cdot;\mathbf{u})$. *Assume* $f \in \mathrm{L}^1((0,T)\times(0,L_1))$ *satisfying* (22). *Then*

$$z(t,x_1) \geq y_1(t,x_1,x_2 : \mathbf{u}) \quad \textit{for any } t \in [0,T] \textit{ and a.e. } (x_1,x_2) \in \Omega.$$

PROOF. We start by noting that from the incompressibility condition div $\mathbf{y} = 0$ we deduce that

$$(\mathbf{y}\cdot\nabla)y_1 = y_1 y_{1,x_1} + y_2 y_{1,x_2} = \left(\frac{1}{2}y_1^2\right)_{x_1} + (y_1y_2)_{x_2} - y_1 y_{2,x_2}$$

$$= \left(\frac{1}{2}y_1^2\right)_{x_1} + (y_1y_2)_{x_2} + y_1 y_{1,x_1} = \mathrm{div}(y_1^2, y_1y_2).$$

Then

$$\begin{aligned} &z_t + \mathrm{div}\ (z_1^2, y_2 z) - \nu\Delta z \geq y_{1,t} + \mathrm{div}\ (y_1^2, y_2 y_1) - \nu\Delta y_1 && \text{in } (0,T)\times\Omega, \\ &z \geq y_1 && \text{on } (0,T)\times\partial\Omega \\ &z(0,\cdot) \geq y_1(0,\cdot) && \text{on } \Omega. \end{aligned}$$

Finally the comparison theorem of Díaz and de Thelin (1994) (see Theorem 3) can be applied since the function $K : (0,T)\times\Omega\times\mathbb{R} \to \mathbb{R}^2$ defined by $K(t,x_1,x_2,r) = (r^2, y_2(t,x_1,x_2)r)$ generates a locally Lipschitz functional on $\mathrm{L}^1(\Omega)$ for any fixed $t \in [0,T]$.
■

PROOF of THEOREM 2. Assume, by contrary, that $-p_{x_1}(t,x_1,x_2 : \mathbf{u}) \leq 2\nu(L_1 - x)^{-2}$ *a.e.* on $(0,T)\times\Omega$. Taking $z(t,x_1) = \nu(L_1 - x)^{-1} + C$ and arguing as in Theorem 1 it is easy to see that Lemma 2 can be applied on $(0, L_1 - \delta)$ for any $\delta > 0$ small enough. In consequence we have $y_1(t,x_1,x_2;\mathbf{u}) \leq z(t,x_1)$ for any $t \in [0,T]$ and *a.e.* $(x_1,x_2) \in \Omega$. Therefore

$$\begin{aligned} \|\mathbf{y}(T,\cdot;\mathbf{u}) - \mathbf{y}_d\|_{\mathrm{L}^2(\Omega)^2} &\geq \|y_1(T,\cdot;\mathbf{u}) - y_{1,d}\|_{\mathrm{L}^2(\Omega)} \\ &\geq \|\,[y_1(T,\cdot;\mathbf{u}) - y_{1,d}]_+\,\|_{\mathrm{L}^2(P)} + \|\,[y_{1,d} - y_1(T,\cdot;\mathbf{u})]_+\,\|_{\mathrm{L}^2(P)} \\ &\geq \|\,\left[y_{1,d} - \nu(L-x)^{-1} - C\right]_+\,\|_{\mathrm{L}^2(P)} > \varepsilon \end{aligned}$$

which is a contradiction. ■

REMARK 2

Roughly speaking Theorem 2 says that if $y_{d,1}(x_1,x_2)$ is big enough near the boundary $x_1 = 0$ then necessarily $p_{x_1}(t,x_1,x_2)$ must be "very negative" on some part of $(0,T)\times\Omega$. Notice that although we do not know if this is our case, there are many explicit solutions of the Navier-Stokes system on special domains having $p_{x_i} = 0$ for some i and that this would contradict the necessary condition for the approximate controllability. □

3 SOME CONTROLLABILITY RESULTS: THE BURGERS EQUATION AND A RELATED QUASILINEAR EQUATION

Although the obstruction phenomenon shows that the approximate controllability can not hold, some partial results in this direction can be obtained. So, El Badia and Ain Seba (1992) use some variants of the Hopf-Cole transformation Φ to prove the exact controllability of problems (1) and (17) when the desired state y_d belongs to the transformed set $\Phi(E)$ where E is the set of attainable states associated to the linear heat equation.

A different approach was followed by Fursikov and Imanuvilov (1993b). They prove several results on the attainable set associated to the Burgers equation. Their main result, stated for the formulation given in (1) says the following: Let $y_0 \in \mathrm{H}^1(0,L)$ arbitrary and let $\hat{y}(t,x)$ be any solution of the Burgers equation satisfying $\hat{y}(t,0) = 0$ for $t \in (0,T)$ and $\hat{y}(0,x) = y_0(x)$ on $(0,L)$ but without any prescription at $x = L$. Define $y_d(x) := \hat{y}(T,x)$. Then there exists $\delta_0 > 0$ such that for any $\delta \in (0,\delta_0)$ there exists $y_0^* \in \mathrm{H}^1(0,L)$ with $\|y_0 - y_0^*\|_{\mathrm{H}^1} \leq \delta$ and a control $u \in \mathcal{C}(0,T)$ such that if $y^*(t,x)$ is the solution of (1) associated to the data y_0^* and u then $y^*(T,x) = \hat{y}(T,x)$ on $(0,L)$.

With the help of the above result we can prove a L^p-approximate controllability criterium for a larger class of desired states.

PROPOSITION 1

Let $y_0 \in \mathrm{L}^\infty(0,L)$ and let $y_d(\cdot) = \tilde{y}(T,\cdot)$ where $\tilde{y} \in \mathcal{C}([0,T] : L^\infty(0,L))$ is any solution of the Burgers equation satisfying $\tilde{y}(t,0) = 0$ for $t \in (0,T)$ and $\tilde{y}(0,x) = y_0(x)$ on $(0,L)$. Then, for any $\varepsilon > 0$ there exists a control $u \in \mathcal{C}(0,T)$ such that for any p, $1 \leq p \leq \infty$, we have

$$\|y(T,\cdot;u) - y_d\|_{\mathrm{L}^p(0,L)} \leq \varepsilon$$

PROOF. By regularizing y_0 we can find $\bar{y}_0 \in \mathrm{H}^1(0,L)$ such that $\|y_0 - \bar{y}_0\|_\infty \leq \varepsilon/2$. On the other hand, by applying the T-accretiveness of the operator $-y_{xx} + yy_x$ on $\mathrm{L}^\infty(0,L)$ (*i.e.* by a generalization of the classical maximum principle, see, *e.g.* Benilan (1981)) we know that if $y_i(t,x)$, $i = 1,2$, are solutions of problem (1) associated to the initial data $y_{0,i}(x)$ ($y_{0,i} \in \mathrm{L}^\infty(0,L)$) and to the same boundary datum $u(t)$ then we have

$$\|y_1(t,\cdot) - y_2(t,\cdot)\|_\infty \leq \|y_{0,1}(\cdot) - y_{0,2}(\cdot)\|_\infty \tag{26}$$

for any $t \in [0,T]$. Let now $\bar{y}(t,x)$ be the solution of the Burgers equation for initial datum $\bar{y}_0$ and boundary conditions $\bar{y}(t,0) = 0$, $\bar{y}(t,L) = \tilde{y}(t,L)$ for $t \in (0,T)$. Thus, by virtue of (26) we have $\|\bar{y}(t,\cdot) - \tilde{y}(t,\cdot)\|_\infty \leq \varepsilon/2$ for any $t \in [0,T]$. Now, by the mentioned result by Fursikov and Imanuvilov (1993b) we deduce the existence of a control function $u \in \mathcal{C}(0,T)$ and $y_0^* \in \mathrm{H}^1(0,L)$ with $\|\bar{y}_0 - y_0^*\|_{\mathrm{H}^1} \leq \min(\delta_0/2, \varepsilon/2)$ such that if $y^*(t,x)$ is the solution of (1) associated to u and y_0^* we have $y^*(T,x) = \bar{y}(T,x)$ on $(0,L)$. Finally, if $y(t,x;u)$ is the solution of (1) associated to $u(t)$ and $y_0(x)$, by virtue of (26) we have

$$\|y(T,\cdot) - y_d(\cdot)\|_\infty \leq \|y(T,\cdot) - \bar{y}(T,\cdot)\|_\infty + \|\bar{y}(T,\cdot) - y^*(T,\cdot)\|_\infty$$

$$+\|y^*(T,\cdot) - \tilde{y}(T,\cdot)\|_\infty \leq \varepsilon/2 + \min(\delta_0/2, \varepsilon/2) \leq \varepsilon. \quad \blacksquare$$

Notice that if we define the class of functions

$$\mathcal{Y} := \{w \in \mathcal{C}((0,L)) : \ w(x) \leq \underline{Y}(T,x) \quad \text{for any } x \in (0,L)\}$$

where $\underline{Y}$ is the universal obstruction function given in Theorem 1 then, arguing as in the proof of Theorem 1, it is clear that the desired state y_d considered in Proposition 1 satisfies that $y_d \in \mathcal{Y}$. Notice also that part (ii) of Theorem 1 proves that if, for instance, we know that $y_d \in \mathcal{C}(0,L)$ satisfies the $\mathcal{C}^0$-approximate controllability property then, necessarily, $y_d \in \mathcal{Y}$. In fact, we conjecture that this necessary condition is also sufficient. Such type of results was obtained in the case of superlinear semilinear problems in Díaz (1994b). The proof there was established in two different steps: a) truncation of the nonlinear term and application of approximate controllability results for *sublinear* semilinear equations, and b) obtention of a priori estimates on the control (associated to the truncated problem) independent of the truncation value $n \in \mathbb{N}$.

A first difficulty to apply such a programme for the case of the Burgers equation is that, as far as we know, there is not any controllability result for the associated sublinear case available in the literature. Due to that, we shall start here the mentioned programme by considering the question of the L^2-approximate controllability for a general class of quasilinear parabolic problems including the case mentioned above. For the sake of simplicity in the exposition we replace the boundary controllability considered in (1) for an internal controllability formulation. More precisely, let Ω be an open bounded regular set of $\mathbb{R}^N$ and ω be an open subset of Ω. Given $y_0 \in L^2(\Omega)$ we consider the control problem (4) mentioned in the Introduction.

The existence of a solution $y \in \mathcal{C}([0,T] : \ L^2(\Omega))$ (when the control $u \in L^2((0,T) \times \omega)$ is given) can be obtained by different methods (see *e.g.* Alt and Luckhaus (1983)). The uniqueness of solutions is a more delicate question. It was established under the additional assumption

$$\mathbf{B} \in \mathcal{C}^{0,\alpha}(\mathbb{R} : \mathbb{R}^N), \ \text{(the space of Hölder continuous functions), and } \alpha \geq \frac{1}{2}$$

by several authors: Alt and Luckhaus, Artola, Chipot and Rodrigues, Gagneux and Guerfi, Díaz and de Thelin (see references in Díaz and de Telin (1994)). More recently, the uniqueness of the solution has been established for merely continuous functions by Gagneux and Madaune-Tort (1994) using previous ideas introduced by Carrillo (1986) for the elliptic case. Concerning the approximate controllability we have

THEOREM 3

Under condition (5) *on* **B** *problem* (4) *is approximately controllable in* $L^2(\Omega)$.

As usual, we shall prove Theorem 3 through a fixed point argument applied to an operator associated to a linearised problem. So we consider, previously, the following problem

$$\left.\begin{array}{ll} y_t - \mathcal{L}(t)y = u\chi_\omega & \text{in } (0,T) \times \Omega \\ y = 0 & \text{on } (0,T) \times \partial\Omega \\ y(0,x) = y_0 & \text{on } \Omega \end{array}\right\} \tag{27}$$

where

$$\mathcal{L}(t)y = \Delta y - \text{div}\ (\mathbf{b}(t,\cdot)y) \tag{28}$$

assuming (for simplicity) $\mathbf{b} \in L^\infty((0,T) \times \Omega)$. The existence and uniqueness of the solutions $y \in \mathcal{C}([0,T] : \ L^2(\Omega)) \cap L^2(0,T : H_0^1(\Omega))$, for a given data $y_0 \in L^2(\Omega)$ and $u \in L^2((0,T) \times \omega)$

can be obtained, for instance, by applying the theory of abstract operators $\mathcal{A}(t)$ such that $\mathcal{A}(t)+\lambda\mathrm{I}$ are maximal monotone operators on $\mathrm{L}^2(\Omega)$ for some $\lambda>0$ and *a.e.* $t\in(0,T)$ (see *e.g.* Brezis (1973)). We point out that, in general, λ cannot be taken as $\lambda=0$ if $\mathbf{b}\neq\mathbf{0}$ (see Lemma 8.4. of Gilbarg and Trudinger (1977)).

PROPOSITION 2

When u spans $\mathrm{L}^2((0,T)\times\omega)$, $y(T,\cdot;u)$ spans an affine subspace which is dense in $\mathrm{L}^2(\Omega)$.

PROOF. We follows the arguments of Lions (1968). Without loss of generality we can assume that $y_0\equiv 0$. Let $\varphi_0\in\mathrm{L}^2(\Omega)$ such that

$$\int_\Omega y(T,x;u)\varphi_0(x)dx=0 \qquad \text{for any } u\in\mathrm{L}^2((0,T)\times\omega). \tag{29}$$

Let $\mathcal{L}^*$ be the *formal adjoint* operator of $\mathcal{L}$, *i.e.* $\mathcal{L}^*(t)\varphi=\Delta\varphi+\mathbf{b}(t,\cdot)\cdot\nabla\varphi$ (see Gilbarg and Trudinger (1977) p. 172). Define φ as the solution to the backwards problem

$$\left.\begin{array}{ll} -\varphi_t-\mathcal{L}^*(t)\varphi=0 & \text{in } (0,T)\times\Omega \\ \varphi=0 & \text{on } (0,T)\times\partial\Omega \\ \varphi(T,x)=\varphi_0(x) & \text{on } \ \Omega. \end{array}\right\} \tag{30}$$

Applying Green's formula we get

$$\int_\Omega y(T,x;u)\varphi_0(x)dx=\int_0^T\int_\omega\varphi(t,x)u(t,x)dxdt. \tag{31}$$

Therefore (29) implies that $\varphi=0$ in $(0,T)\times\omega$. Finally, applying the unique continuation property (see Corollary 1.2 of Saut and Scheurer (1987), valid for non necessarily selfadjoint operators) we conclude that $\varphi\equiv 0$ in $(0,T)\times\Omega$ and the conclusion holds by the Hahn-Banach theorem. ∎

In order to consider the nonlinear problem (4) we shall need more information on the application $y_d\mapsto u$ (for a fixed $\varepsilon>0$) found in Proposition 2. In fact this is a multivalued map since it is easy to see that there are infinitely many $u\in\mathrm{L}^2((0,T)\times\Omega)$ satisfying

$$\|y(T,\cdot;u)-y_d\|_{\mathrm{L}^2(\Omega)}\le\varepsilon, \tag{32}$$

where $y(t,\cdot;u)$ denotes the solution of (27). We shall follow now some direct methods introduced in Lions (1992a), (1992b) and later generalized and improved in Fabré, Puel and Zuazua (1992a), (1992b) leading to the existence of "quasi bang-bang controls". The results of this last reference can be modified easily to our context although in the mentioned work it is always assumed that $\mathcal{L}^*=\mathcal{L}=\Delta$. Given $\varphi_0\in\mathrm{L}^2(\Omega)$ we define the functional

$$\mathcal{J}(\varphi_0)=\frac{1}{2}\left(\int_0^T\int_\omega|\varphi(t,x)|dxdt\right)^2+\varepsilon\|\varphi_0\|_{\mathrm{L}^2(\Omega)}-\int_\Omega y_d(x)\varphi_0(x)dx.$$

Following Fabré, Puel and Zuazua (1992b) this functional is continuous from $\mathrm{L}^2(\Omega)$ into $\mathbb{R}$, strictly convex and satisfies the coerciveness condition

$$\liminf_{\|\varphi_0\|\to\infty}\frac{\mathcal{J}(\varphi_0)}{\|\varphi_0\|_2}\ge\varepsilon$$

(this last condition is proved from the unique continuation property obtained in Saut and Scheurer (1987)). Therefore there is a unique $\psi_0 \in L^2(\Omega)$ such that

$$\mathcal{J}(\psi_0) = \min_{\varphi_0 \in L^2(\Omega)} \mathcal{J}(\varphi^0). \tag{33}$$

It is easy to see that, in fact, $\psi_0 = 0$ *iff* $\|y_d\|_2 \leq \varepsilon$. By studying the associated Euler-Lagrange (multivalued) equation it is proved the following result.

PROPOSITION 3

Let ψ_0 be the solution of (33) and let $\psi(t,x)$ be the solution of (30) associated to $\varphi_0 = \psi_0$. Then there exists $v \in$ sign $(\psi)\chi_\omega$ such that the control $u := (\|\psi\|_{L^1((0,T)\times\omega)})v$ leads to a solution $y(t,\cdot;u)$ of (27) such that $\|y(T,\cdot;u) - y_d\| \leq \varepsilon$.

REMARK 3

The operator $\mathcal{L}$ defined in (28) can be taken, more generally, of the form

$$\mathcal{L}(t)y = \sum_{i,j=1}^{N} \frac{\partial}{\partial x_i}\left(a_{ij}(t,x)\frac{\partial y}{\partial x_j}\right) + \sum_{i=1}^{N} \frac{\partial}{\partial x_i}(b_i(t,x)y) \tag{34}$$

with $a_{ij} \in \mathcal{C}^1((0,T)\times\Omega)$ satisfying

$$\sum_{i,j=1}^{N} a_{ij}(t,x)\xi_i\xi_j \geq \alpha(x,t)|\xi|^2, \qquad \forall (t,x) \in (0,T)\times\Omega,\ \forall \xi \in \mathbb{R}^N$$

for some $a(x,t) > 0$ and $\mathbf{b} = (b_1,\ldots,b_N) \in L^\infty((0,T)\times\Omega)$. We point out that in Glowinski and Lions (1994) the nondivergential form operator

$$\mathcal{L}y = \sum_{i,j=1}^{N} \frac{\partial}{\partial x_i}\left(a_{ij}(x)\frac{\partial y}{\partial x_j}\right) + \mathbf{V}_0 \cdot \nabla y \tag{35}$$

was considered under the assumption $\mathbf{V}_0 \in L^\infty(\Omega)^N$ and div $\mathbf{V}_0 = \mathbf{0}$ on Ω. Notice that under this assumption the operator can be written as in (34) (use that div $(\mathbf{V}_0 y) =$ div $\mathbf{V}_0 y + \mathbf{V}_0\cdot\nabla y$). Finally, we point out that if $\mathbf{b}$ is a $W^{1,\infty}(\Omega)^N$ function, independent on t, then the result of Lin (1991) allows to know that the set $\{(t,x) : \psi(t,x) = 0,\ \psi$ solution of $(30)\}$ has zero Lebesgue measure and so the controls u, in Proposition 2, are of type bang-bang.

□

PROOF of THEOREM 3. Let us assume that $\mathbf{B}$ is differentiable at $s_0 = 0$. The case $s_0 \neq 0$ can be easily treated by an homogeneization argument. Define the function $\mathbf{g} : \mathbb{R} \to \mathbb{R}^N$ by

$$\mathbf{g}(s) = \frac{\mathbf{B}(s) - \mathbf{B}(0)}{s} \quad \text{if } s \neq 0 \text{ and } \mathbf{g}(0) = \mathbf{B}'(0).$$

From the regularity of $\mathbf{B}$ and (5) it is clear that $\mathbf{g} \in \mathcal{C}(\mathbb{R} : \mathbb{R}^N) \cap L^\infty(\mathbb{R} : \mathbb{R}^N)$. Given $z \in L^2((0,T)\times\Omega)$, we define $\mathbf{b} = -\mathbf{g}(z) \in L^\infty((0,T)\times\Omega)$ and so, by Proposition 2, the associated linear problem (27) is approximately controllable. More precisely, given $y_d \in L^2(\Omega)$ and $\varepsilon > 0$ let $u(z)$ be the control and $y^*(t,\cdot;z)$ the solution of (27) mentioned in Proposition 3. Define now the nonlinear mapping $\Lambda :\ L^2(\Omega) \to \mathcal{P}(L^2(\Omega))$ by $\Lambda(z) =$

$\{y^*(t,\cdot;u(z))$ for some $u(z) \in \|\psi\|_1 \text{sign}(\psi)\chi_\omega\}$. Notice that Λ can be multivalued since the uniqueness of $u(z)$ is not assured. It is easy to see that any fixed point of Λ allow us to obtain the conclusion of the theorem. In order to apply the Kakutani fixed point theorem (in the weak form given in Aubin (1984)) we need to check the following conditions: (i) $\forall z \in L^2((0,T)\times\Omega)$, the set $\Lambda(z)$ is non-empty, convex and compact in $L^2((0,T)\times\Omega)$, (ii) $\Lambda(z)$ is upper hemi-continuous. The proof of both properties can be obtained by adapting the arguments of Fabré, Puel and Zuazua (1992b). The convexity of $\Lambda(z)$ is a consequence of the linearity of (27) and the convexity of the set $\{v \in L^2((0,T)\times\Omega) : v \in \text{sign}(\psi)\chi_\omega\}$. As $\mathbf{g}(z)$ is uniformly bounded in $L^\infty((0,T)\times\Omega)$ we can prove the existence of a compact subset $\mathcal{K} \subset L^2((0,T)\times\Omega)$ such that $\Lambda(z) \subset \mathcal{K}$ for any $z \in L^2((0,T)\times\Omega)$. To prove this we first notice that $\{y^*(\cdot,\cdot;z)\}$ and $\{y_t^*(\cdot,\cdot;z)\}$ remain uniformly bounded in $L^2(0,T:H_0^1(\Omega))$ and $L^2(0,T:H^{-1}(\Omega))$ respectively when z runs $L^2((0,T)\times\Omega)$ (this can be obtained multiplying by $y^*(z)$, integrating by parts and applying the coerciveness of $\mathcal{L}(t)$). Then, by well-known results (see *e.g.* Corollary 4 of Simon (1987)), we conclude that $\{y^*(\cdot,\cdot;z)\}$ is relatively compact in $L^2((0,T)\times\Omega)$. Choosing as $\mathcal{K}$ the closure of this set in $L^2((0,T)\times\Omega)$ the proof of property (i) is reduced to show that $\Lambda(z)$ is a closed set. This is proved without any difficulty using that the multivalued (maximal monotone) graph sign($\cdot$) is strongly-weakly closed, the coerciveness of $\mathcal{L}(t)$ and the compactness of the Green operator associated to (27). The proof of (ii) follows as in Fabré, Puel and Zuazua (1992b) once that we already know the compactness of the set $\mathcal{K}$ and the continuity of function $\mathbf{g}$. ■

REMARK 4

Theorem 3 can be improved in several directions. First of all, following closely the arguments of Fabré, Puel and Zuazua (1992b), the approximate controllability property can be also obtained on the spaces $L^p(\Omega)$, $1 \le p < \infty$, and $\mathcal{C}^0(\overline{\Omega})$. On the other hand, the approximate controllability can be obtained for a more general class of nonlinear equations of the type

$$y_t - \Delta y + \text{div}\,(\mathbf{B}(y)) + f(y) = u\chi_\omega$$

assuming $\mathbf{B}$ as before and f be a continuous real function, differentiable at some $s_1 \in \mathbb{R}$ and sublinear at the infinity. □

References

[1] Alt, H.W. and Luckhaus, S. (1983). Quasilinear Elliptic-Parabolic Differential Equations, Math. Z. 183, pp. 311-341.

[2] Aubin, J.P. (1984). L'analyse non linéaire et ses motivations économiques, Masson, 1984.

[3] Bandle, C. and Marcus, M. (1992). "Large" solutions of semilinear elliptic equations: existence, uniqueness and assymptotic behaviour, Journal d'Analyse Math. 58, pp. 9-24.

[4] Bandle, C., Díaz, G. and Díaz, J.I. (1994). Solutions d'équations de réaction-diffusion nonlinéaires explosant au bord parabolique, C.R. Acad. Sciences. Paris, 318, Serie I, pp. 455-460.

[5] Benilan, Ph. (1981). Evolution equations and accretive operators, Lecture Notes, Univ. of Kentucky, Lexington.

[6] Bernis, F., Díaz, J.I. and Ramos, A. (1995). Obstruction and approximate controllability results for higher order semilinear parabolic equations. Manuscript.

[7] Brezis, H. (1973). Opérateurs maximaux monotones et semigroupes de contractions dans les spaces de Hilbert. North Holland, Amsterdam.

[8] Carrillo, J. (1986). Unicité des solutions du type Kruskov pour des problèmes elliptiques avec des termes de transport non lineaires. C.R. Acad. Sci, Paris, 303, serie I, pp. 189-192.

[9] Díaz, G. and Letelier, R. (1993). Explosive solutions of quasilinear elliptic equations: existence and uniqueness, Nonlinear Analysis, T.M.A. 20, pp. 97-125.

[10] Díaz, J.I. (1991a).Sur la contrôllabilité approchée des inéquations variationelles et d'autres problémes non-linéaires. C.R. Acad. Sci. de Paris. 312, serie I, pp.519-522.

[11] Díaz, J.I. (1991b). Sobre la controlabilidad aproximada de problemas no lineales disipativos. In the proceedings of Jornadas Hispano-Francesas sobre Control de Sistemas Distribuidos (A. Valle ed.). Univ. de Málaga, pp. 41-48.

[12] Díaz, J.I. (1993a).Mathematical analysis of some diffusive energy balance models in Climatology. In Mathematics, Climate and Environment, J.I. Díaz and J.I. Lions (eds.). Masson, pp. 28-56.

[13] Díaz, J.I. (1993b). Approximate controllability for some nonlinear parabolic problems. Proceedings of 16th IFIP-TC7 Conference on "System Modelling and Optimizacion", Compiegne (France), 5-9 July 1993. Lecture Notes in Control and Information Sciences. Springer-Verlag.

[14] Díaz, J.I. (1994a). On the controllability of some simple climate models. In Environment, Economics and their Mathematical Models, J.I. Díaz and J.L. Lions (eds.). Masson, pp. 29-44.

[15] Díaz, J.I. (1994b). Controllability and Obstruction for some nonlinear parabolic problems in Climatology. In Modelado de Sistemas en Oceanografía. (A. Valle and C. Pares eds.). Universidad de Málaga, pp. 43-58.

[16] Díaz, J.I. and de Thelin, F. (1994). On a nonlinear parabolic problem arising in some models related to turbulent flows. SIAM J. Math. Anal. 25, pp. 1085-1111.

[17] Díaz, J.I. and Ramos, A.M. (1993). Resultados positivos y negativos sobre la controlabilidad aproximada de problemas parabólicos semilineales. Actas de XIII CEDYA/III Congreso de Matemática Aplicada. Madrid.

[18] Díaz, J.I. and Ramos, A.M. (1994). Positive and Negative approximate controllability results for semilinear problems. Submitted.

[19] El Badia, A. and Ain Seba, B. (1992).Contrôlabilité exacte de l'équation de Burgers, C.R. Acad. Sci. Paris, 314, serie I, pp. 373-378.

[20] Fabré, C., Puel, J.P. and Zuazua, E. (1992a). Contrôlabilité approchée de l'équation de la chaleur semilinéaire, C.R. Acad. Sci. Paris, 315, serie I, pp. 807-812.

[21] Fabré, C., Puel, J.P. and Zuazua, E. (1992b). Approximate controlability of the semilinear heat equations. IMA Preprint Series, Minnesota.

[22] Fursikov, A.V. and Imanuvilov, O.Y. (1993a). On approximate controllability of the Stokes system, Annales de la Faculté des Sciences de Toulouse, Vol II n^{0}2, pp. 205-232.

[23] Fursikov, A.V. and Imanuvilov, O.Y. (1993b). On controllability of certain systems simulating a fluid flow. Manuscript.

[24] Gagneux, G. and Madaune-Tort, M. (1994). Unicité des solutions faibles d'équations de diffusion-convection. C.R. Acad. Sci. Paris, 318, Série I, pp. 919-924.

[25] Gilbarg, D. and Trudinger, N.S. (1977). Elliptic Partial Differential Equations of Second Order. Springer-Verlag, Berlin.

[26] Glowinski, R. and Lions, J.L. (1994). Exact and Approximate Controllability for Distributed Parameter System. Manuscript.

[27] Henry, J. (1978). Etude de la contrôlabilité de certains équations paraboliques. Thèse d'Etat. Université de Paris VI.

[28] Lin, F.H. (1991). Nodal sets of solutions of elliptic and parabolic equations. Comm. in Pure and Appl. Math. 44, pp. 287-308.

[29] Lions, J.L. (1968). Contrôle Optimal Systems Gouvernés par les Equations Derivées Partielles. Dunod, Paris.

[30] Lions, J.L. (1990). Are there connections between turbulence and controllability ?. Lect. Notes 144. Springer-Verlag (A. Bensoussan and J.L. Lions eds.).

[31] Lions, J.L. (1992). Remarks on approximate controllability, J. Analyse Math. 59, pp. 103-116.

[32] Saut, J.C. and Scheurer, B. (1987). Unique continuation for some evolution equations. J. Diff. Equations, 66, pp. 118-139.

[33] Simon, J. (1987). Compact sets in the space $L^p(0,T:B)$, Annali di Matematica Pura ed Applicata, CXLVI, pp. 65-96.

Some Controllability Problems for the Stokes and Navier–Stokes Systems

Enrique FERNÁNDEZ-CARA Dpto. Ecuaciones Diferenciales y Análisis Numérico, University of Sevilla, C/ Tarfia s/n, 41012 SEVILLA, SPAIN.

Abstract: Let us present the kind of questions considered in this paper. Roughly speaking, assuming that a final time $T > 0$, a velocity field $\mathbf{y}_d$ and a positive number α are given,

We ask ourselves whether a flow can be "driven" to a state $\mathbf{y}(T) \in B(\mathbf{y}_d; \alpha)$

We will refer to 2D and 3D Stokes and Navier-Stokes flows. To "drive" a flow will be to act on its particles by introducing (for instance) distributed control terms.

1 STATEMENT OF THE PROBLEM FOR STOKES FLOWS

Assume we are given a bounded and connected open set $\Omega \subset \mathbb{R}^N$ ($N = 2$ or 3) whose boundary is C^2, a "small" nonempty open subset $\omega \subset\subset \Omega$ and two positive numbers T and ν (Ω is the fluid domain, T is a final instant of time and ν is the kinematic viscosity of the fluid). Consider the Hilbert spaces H and V, which are given by the closures of

$$J_0(\Omega) = \{\, \mathbf{v};\ \mathbf{v} = (v_1, \ldots, v_N),\ v_i \in C_0^\infty(\Omega),\ \nabla \cdot \mathbf{v} = 0 \,\}$$

under the L^2 and H_0^1 norms respectively — recall that $V \hookrightarrow H \equiv H' \hookrightarrow V'$, with dense and compact embeddings. Let us also introduce the space $\mathcal{U} = L^2(\omega \times (0,T))^N$. Then for each $\mathbf{v} \in \mathcal{U}$ we consider the associated *Stokes* problem

$$\frac{\partial \mathbf{y}}{\partial t} - \nu\Delta\mathbf{y} + \nabla\Pi = \mathbf{v}\mathrm{X}_\omega\,, \qquad \nabla\cdot\mathbf{y} = 0 \qquad \text{in } \Omega\times(0,T)\,, \tag{1.1}$$

$$\mathbf{y} = 0 \quad \text{on } \partial\Omega\times(0,T)\,, \qquad \mathbf{y}(0) = 0\,, \tag{1.2}$$

where X_ω is the characteristic function of the set ω. The unique solution to (1.1)–(1.2) will be denoted by $(\mathbf{y}_\mathbf{v}, \Pi_\mathbf{v})$.

Let $\mathbf{y}_d \in H$ and $\alpha > 0$ be given. Our first problem concerns approximate controllability for (1.1)–(1.2). It is the following:

PROBLEM 1 Find $\mathbf{v} \in \mathcal{U}$ such that $|\mathbf{y}_\mathbf{v}(T) - \mathbf{y}_d| \leq \alpha$.

Here and in the sequel, $|\cdot|$ stands for the norm in H. Together with this problem, several interesting variants can be introduced. First, notice that the control space need not be the whole $L^2(\omega\times(0,T))^N$. For instance, $\mathcal{U}$ can be replaced by $\mathcal{U}^*$, where

$$\mathcal{U}^* = \{\,\mathbf{v}\,;\ \mathbf{v} = (v_1, v_2, 0),\ v_i \in L^2(\omega\times(0,T))\,\}\,.$$

Secondly, observe that similar boundary controllability questions can be thought. More precisely, if $\gamma \subset\subset \partial\Omega$ is a "small" part of the boundary, the state equation (1.1)–(1.2) can be replaced by

$$\frac{\partial \mathbf{y}}{\partial t} - \nu\Delta\mathbf{y} + \nabla\Pi = 0\,, \qquad \nabla\cdot\mathbf{y} = 0 \qquad \text{in } \Omega\times(0,T)\,,$$

$$\mathbf{y} = \mathbf{g}\mathrm{X}_\gamma \quad \text{on } \partial\Omega\times(0,T)\,, \qquad \mathbf{y}(0) = 0\,,$$

where $\mathbf{g}$ is chosen appropriately. Let us also mention that something similar to Problem 1 — or any of these variants — can be formulated in the more interesting framework of *Navier-Stokes* flows. This will be done in section 6.

2 THE PHYSICAL INTERPRETATION

In physical terms, one can easily interpret the previous situation: We fix the small set ω, the final time T and the velocity field $\mathbf{y}_d \in H$; then, we ask ourselves whether the mechanical system decribed by (1.1)–(1.2) can be conducted from an initial state $\mathbf{y} = 0$ to a state arbitrarily close to $\mathbf{y}_d$ acting from the outside and, exclusively, on those particles in ω. In the case of boundary controllability, we simply act on the velocity field on $\gamma\times(0,T)$.

Exact controllability for (1.1)–(1.2) is false, i.e. in general the system cannot be conducted to $\mathbf{y}_d$ (the final states $\mathbf{y}_\mathbf{v}(T)$ are regular functions). Consequently, to solve Problem 1 for arbitrary $\mathbf{y}_d$ and α is the best we can do.

In the formulation of Problem 1, the control function $\mathbf{v}$ is a field of exterior forces. Problem 1 possesses infinitely many solutions (this is a well known result; see [8]). Accordingly, it is natural to search for the "minimal" or "best" $\mathbf{v}$. One is thus led to

PROBLEM 2 Minimize $\iint_{\omega\times(0,T)} |\mathbf{v}|^2$ subject to $\mathbf{y}_\mathbf{v}(T) \in B(\mathbf{y}_d; \alpha)$.

In some situations, the choice we have made of the state equation is justified. In most cases, "real" incompressible fluids are viscous, can be described by the Navier-Stokes equations and satisfy no-slip boundary conditions on solid walls. When inertial forces are small with respect to viscous forces, i.e. $|(\mathbf{y}\cdot\nabla)\mathbf{y}| << |\nu\Delta\mathbf{y}|$, the former can be neglected and the Stokes system provides an acceptable model for the flow of the fluid.

3 THE MAIN KNOWN RESULTS

These are essentially due to J.-L. Lions (see [8] for a complete presentation). They provide existence for Problem 1 and existence and uniqueness for Problem 2.

THEOREM 1 Let us set $Z = \{\, \mathbf{y}_\mathbf{v}(T);\ \mathbf{v} \in \mathcal{U}\,\}$. Then the space Z is dense in H.

In other words, there is approximate controllability for flows governed by (1.1) – –(1.2). The proof of this result is nonconstructive; it relies on a known consequence of the Hahn-Banach theorem. More precisely, let $(\cdot,\cdot)$ denote the usual scalar product in H; then one proves the following:

If $\mathbf{w} \in H$ is such that $(\mathbf{y}_\mathbf{v}(T), \mathbf{w}) = 0$ for all $\mathbf{v} \in \mathcal{U}$, then $\mathbf{w} = 0$.

This is deduced by introducing an appropriate adjoint state and using unique continuation (see e.g. [12]). Indeed, if $\mathbf{w}$ is as before and $(\mathbf{q}, \Phi)$ is the unique solution to

$$-\frac{\partial \mathbf{q}}{\partial t} - \nu\Delta\mathbf{q} + \nabla\Phi = 0, \qquad \nabla\cdot\mathbf{q} = 0 \qquad \text{in } \Omega\times(0,T), \tag{3.1}$$

$$\mathbf{q} = 0 \quad \text{on } \partial\Omega\times(0,T), \qquad \mathbf{q}(T) = \mathbf{w} \tag{3.2}$$

(the adjoint system), then

$$0 = (\mathbf{y}_\mathbf{v}(T), \mathbf{w}) = \iint_{\omega\times(0,T)} \mathbf{v}\cdot\mathbf{q} \qquad \forall \mathbf{v} \in \mathcal{U}.$$

This implies $\mathbf{q} = 0$ in $\omega\times(0,T)$. The unique continuation property for (3.1)–(3.2) yields $\mathbf{q} \equiv 0$ and, finally, $\mathbf{w} = 0$.

The previous argument can be adapted to other *linear* similar situations (recall those in section 1). An interesting case corresponds to *quasi-Stokes* flows, i.e. those governed by

$$\frac{\partial \mathbf{y}}{\partial t} - \nu\Delta\mathbf{y} + (\mathbf{z}\cdot\nabla)\mathbf{y} + \mathbf{M}\mathbf{y} + \nabla\Pi = \mathbf{v}\mathrm{X}_\omega\,, \tag{3.3}$$

$$\nabla\cdot\mathbf{y} = 0\,, \tag{3.4}$$

together with appropriate initial and boundary conditions. If $\mathbf{z}$ and $\mathbf{M}$ are regular enough, the adjoint system corresponding to (3.3)–(3.4) satisfies the unique continuation property. As a consequence, the previous Hahn-Banach argument holds for this class of flows and, again, one obtains approximate controllability. In particular, this holds for *linearized Navier-Stokes* flows, whose velocity field is of the form $\mathbf{z}+\mathbf{y}$ where $\mathbf{z}$ is (together with some pressure) a regular solution to the Navier-Stokes system and $\mathbf{y}$ is small (for details, see [7]).

By the time I was preparing this paper, in july 1994, I was also interested in proving that unique continuation holds for (3.3)–(3.4) under hypotheses of the following kind:

$$\mathbf{z}\in L^\infty(\Omega\times(0,T))^N, \qquad \mathbf{M}\in L^\infty(0,T;L^N(\Omega)^{N\times N}) \tag{3.5}$$

(apparently, these are the "natural" hypotheses, since they suffice in the similar scalar case; see [13]). A result like this would be very interesting in order to demonstrate approximate controllability for some quasi-linear systems close to the Navier-Stokes problem. My colleagues and I found a "proof" based on an adequate Carleman's inequality: but unfortunately the proof was erroneous. I thank C. Fabre and J.-P. Puel for having clarified this point and also for giving some interesting ideas.

As mentioned above, it is quite easy to check that Problem 1 possesses infinitely many solutions. This is why we have introduced Problem 2, which can also be formulated as follows:

$$\inf_{\mathbf{v}\in\mathcal{U}}\{F_1(\mathbf{v}) + F_2(L\mathbf{v})\} \tag{3.6}$$

Here, we have used the following notation:

$$F_1:\mathcal{U}\to\mathbb{R}, \qquad F_1(\mathbf{v}) = \frac{1}{2}\iint_{\omega\times(0,T)}|\mathbf{v}|^2,$$

$$F_2: H\to(-\infty,+\infty], \qquad F_2(\mathbf{p}) = \begin{cases} 0 & \text{if } \mathbf{p}\in B(\mathbf{y}_d;\alpha)\,,\\ +\infty & \text{otherwise}\,,\end{cases}$$

$$L\in\mathcal{L}(\mathcal{U};H)\,, \qquad L\mathbf{v} = \mathbf{y}_{\mathbf{v}}(T)\,.$$

As in [8], we will also introduce the dual problem

$$\inf_{\mathbf{p}\in H}\{F_1^*(-L^*\mathbf{p}) + F_2^*(\mathbf{p})\}\,, \tag{3.7}$$

where F_i^* is the convex conjugate of F_i and L^* is the adjoint of L:

$$F_1^*:\mathcal{U}\to\mathbb{R}\,, \qquad F_1^*(\mathbf{v}) = F_1(\mathbf{v}) = \frac{1}{2}\iint_{\omega\times(0,T)}|\mathbf{v}|^2,$$

$$F_2^* : H \to \mathbb{R}, \qquad F_2^*(\mathbf{p}) = \alpha|\mathbf{p}| + (\mathbf{y}_d, \mathbf{p}),$$
$$L^* \in \mathcal{L}(H; \mathcal{U}), \qquad L^*\mathbf{p} = \mathbf{q}\mathrm{X}_\omega,$$

with $\mathbf{q}$ being, together with Φ, the unique solution to

$$-\frac{\partial \mathbf{q}}{\partial t} - \nu\Delta\mathbf{q} + \nabla\Phi = 0, \qquad \nabla\cdot\mathbf{q} = 0 \quad \text{in } \Omega\times(0,T), \tag{3.8}$$
$$\mathbf{q} = 0 \quad \text{on } \partial\Omega\times(0,T), \qquad \mathbf{q}(T) = \mathbf{p}. \tag{3.9}$$

With (3.7) in mind, we can apply usual convex duality techniques to the solution of Problem 2. This can be made as in [8] (see also [3] and [4] for other related questions) and leads to our second main result:

THEOREM 2 Assume $|\mathbf{y}_d| > \alpha$. Then:
(a) (3.6) possesses exactly one solution $\widehat{\mathbf{v}}$.
(b) (3.7) possesses exactly one solution $\widehat{\mathbf{p}}$.
(c) $\widehat{\mathbf{v}}$ and $\widehat{\mathbf{p}}$ are the unique functions satisfying

$$\begin{cases} \widehat{\mathbf{v}} \in \mathcal{U}, & \widehat{\mathbf{p}} \in H, \\ \widehat{\mathbf{v}} = -L^*\widehat{\mathbf{p}}, & \widehat{\mathbf{p}} \in \partial F_2(L\widehat{\mathbf{v}}). \end{cases} \tag{3.10}$$

The system (3.10) characterizes $(\widehat{\mathbf{v}}, \widehat{\mathbf{p}})$ and must be viewed as the optimality conditions for this couple. It can also be written as follows:

$$\frac{\partial \widehat{\mathbf{y}}}{\partial t} - \nu\Delta\widehat{\mathbf{y}} + \nabla\Pi = \widehat{\mathbf{v}}\mathrm{X}_\omega, \qquad \nabla\cdot\widehat{\mathbf{y}} = 0 \quad \text{in } \Omega\times(0,T), \tag{3.11}$$
$$-\frac{\partial \widehat{\mathbf{q}}}{\partial t} - \nu\Delta\widehat{\mathbf{q}} + \nabla\widehat{\Phi} = 0, \qquad \nabla\cdot\widehat{\mathbf{q}} = 0 \quad \text{in } \Omega\times(0,T), \tag{3.12}$$
$$\widehat{\mathbf{y}} = \widehat{\mathbf{q}} = 0 \quad \text{on } \partial\Omega\times(0,T), \qquad \widehat{\mathbf{y}}(0) = 0, \quad \widehat{\mathbf{q}}(T) = \widehat{\mathbf{p}}, \tag{3.13}$$
$$\widehat{\mathbf{v}} = -\widehat{\mathbf{q}}\mathrm{X}_\omega, \qquad \widehat{\mathbf{y}}(T) = \mathbf{y}_d + \frac{\alpha}{|\widehat{\mathbf{p}}|}\widehat{\mathbf{p}}. \tag{3.14}$$

In section 4, theorems 1 and 2 will be used to deduce convergent algorithms for the computation of $\widehat{\mathbf{v}}$ (and $\widehat{\mathbf{p}}$). Then, in section 5, in the framework of Navier-Stokes flows, we will present some tentative ideas suggested by these results.

4 SOME ITERATIVE ALGORITHMS

It is usual to deduce iterative algorithms for problems of the kind (3.6) by introducing appropriate regularized or penalized (or even regularized-penalized) approximations. Here, we will present several different possibilities.

Exact regularization. Our goal is to find a couple $(\widehat{\mathbf{v}}, \widehat{\mathbf{p}})$ such that

$$\begin{cases} \widehat{\mathbf{v}} \in \mathcal{U}, & \widehat{\mathbf{p}} \in H, \\ \widehat{\mathbf{v}} = -L^*\widehat{\mathbf{p}}, & \widehat{\mathbf{p}} \in \partial F_2(L\widehat{\mathbf{v}}). \end{cases} \tag{3.10}$$

For a problem like this, there is an argument due to A. Bermúdez and C. Moreno [2] which leads to exact regularization:

a) For any $\lambda > 0$, let us denote by $(\partial F_2)_\lambda$ the corresponding Yosida approximation to the maximal monotone operator ∂F_2, i.e.

$$(\partial F_2)_\lambda = \frac{1}{\lambda}\left(\mathbf{Id} - (\mathbf{Id} + \lambda \partial F_2)^{-1}\right).$$

It is well known that this is a single-valued Lipschitz-continuous mapping on H. Then, for each $\lambda > 0$, $(\widehat{\mathbf{v}}, \widehat{\mathbf{p}})$ solves (3.10) if and only if

$$\widehat{\mathbf{v}} = -L^*\widehat{\mathbf{p}}, \qquad \widehat{\mathbf{p}} = (\partial F_2)_\lambda(L\widehat{\mathbf{v}} + \lambda\widehat{\mathbf{p}}). \tag{4.1}$$

b) Using (4.1), one can introduce a fixed-point iterative algorithm for the computation of $\widehat{\mathbf{v}}$ and $\widehat{\mathbf{p}}$: We first choose $\lambda > 0$ and $\mathbf{p}_0 \in H$; then, for each $n \geq 0$, we set

$$\mathbf{v}_{n+1} = -L^*\mathbf{p}_n, \qquad \mathbf{p}_{n+1} = (\partial F_2)_\lambda(L\mathbf{v}_{n+1} + \lambda\mathbf{p}_n). \tag{4.2}$$

In terms of partial differential problems, (4.2) reads:

$$\begin{cases} -\dfrac{\partial \mathbf{q}_n}{\partial t} - \nu\Delta\mathbf{q}_n + \nabla\Phi_n = 0, & \nabla\cdot\mathbf{q}_n = 0 \quad \text{in } \Omega\times(0,T), \\ \mathbf{q}_n = 0 \quad \text{on } \partial\Omega\times(0,T), & \mathbf{q}(T) = \mathbf{p}_n, \end{cases} \tag{4.3}$$

$$\mathbf{v}_{n+1} = -\mathbf{q}_n \mathrm{X}_\omega \tag{4.4}$$

$$\begin{cases} \dfrac{\partial \mathbf{y}_{n+1}}{\partial t} - \nu\Delta\mathbf{y}_{n+1} + \nabla\Pi_{n+1} = \mathbf{v}_{n+1}\mathrm{X}_\omega, & \nabla\cdot\mathbf{y}_{n+1} = 0 \quad \text{in } \Omega\times(0,T), \\ \mathbf{y}_{n+1} = 0 \quad \text{on } \partial\Omega\times(0,T), & \mathbf{y}(0) = 0, \end{cases} \tag{4.5}$$

$$\left(\lambda + \frac{\alpha}{|\mathbf{p}_{n+1}|}\right)\mathbf{p}_{n+1} = \mathbf{y}_{n+1}(T) - \mathbf{y}_d + \lambda\mathbf{p}_n. \tag{4.6}$$

THEOREM 3 Let $\lambda > 0$ and $\mathbf{p}_0 \in H$ be given and assume $|\mathbf{y}_d| > \alpha$. Then the sequence $\{(\mathbf{v}_n, \mathbf{p}_n)\}$ generated by (4.2) converges strongly in $\mathcal{U} \times H$ to the unique solution to (3.6).

SKETCH OF THE PROOF: Let us introduce the function J^*, with

$$J^*(\mathbf{p}) = F_1^*(-L^*\mathbf{p}) + F_2^*(\mathbf{p}) = \frac{1}{2}\iint_{\omega\times(0,T)} |\mathbf{p}|^2 \;+\; \alpha|\mathbf{p}| \;+\; (\mathbf{y}_d, \mathbf{p}) \qquad (4.7)$$

for all $\mathbf{p} \in H$. Then, as in [2], it is easy to deduce from (4.2) that

$$J^*(\mathbf{p}_n) \geq J^*(\mathbf{p}_{n+1}) + \lambda|\mathbf{p}_n - \mathbf{p}_{n+1}|^2$$

for all n, whence

$$J^*(\mathbf{p}_0) - J^*(\mathbf{p}_{m+1}) \geq \lambda \sum_{n=0}^{m} |\mathbf{p}_n - \mathbf{p}_{n+1}|^2 \qquad \forall m \geq 1 .$$

As in [3], it can be proved that J^* is bounded from below in H and, also, that $\lim_{|\mathbf{p}|\to+\infty} J^*(\mathbf{p}) = +\infty$. This is implied by the unique continuation property for (3.8). Thus,

$$|\mathbf{p}_n| \leq \text{Const.} \qquad \text{and} \qquad \mathbf{p}_n - \mathbf{p}_{n+1} \to 0 \ \text{ in } \ H .$$

Taking into account the fact that L^* is compact and selecting an appropriate subsequence, one finds:

$$\mathbf{p}_\mu \to \mathbf{p} \ \text{ weakly in } \ H, \qquad \mathbf{v}_{\mu+1} \to \mathbf{v} = -L^*\mathbf{p} \ \text{ in } \ \mathcal{U},$$

$$\mathbf{p}_{\mu+1} \to \mathbf{p} \ \text{ weakly in } \ H \qquad \text{and} \qquad L\mathbf{v}_{\mu+1} + \lambda(\mathbf{p}_\mu - \mathbf{p}_{\mu+1}) \to L\mathbf{v} \ \text{ in } \ H .$$

Consequently, $\mathbf{v} = -L^*\mathbf{p}$ and $\mathbf{p} \in \partial F_2(L\mathbf{v})$, i.e. the couple $(\mathbf{v}, \mathbf{p})$ satisfies (3.10). The rest of the proof is easy to achieve.

Penalization. Let $k > 0$ be given. By changing slightly the argument presented in [8], one sees that (3.6) can be approximated by the following penalized problem:

$$\inf_{\mathbf{v}\in\mathcal{U}} \{F_1(\mathbf{v}) + k\, G(L\mathbf{v} - \mathbf{y}_d)\} . \qquad (4.8)$$

Here, $G : H \to \mathbb{R}$ is a continuous, convex and nonnegative function (to be fixed), with $G(0) = 0$. For instance, the choices

$$G(\mathbf{z}) = \frac{1}{2}|\mathbf{z}|^2 \qquad \text{or} \qquad G(\mathbf{z}) = \text{dist}\,(\mathbf{z}, B(0;\alpha))$$

are adequate.

It is clear that (4.8) possesses exactly one solution $(\widetilde{\mathbf{v}}, \widetilde{\mathbf{p}})$, furthermore characterized by the following:

$$\begin{cases} \widetilde{\mathbf{v}} \in \mathcal{U}, & \widetilde{\mathbf{p}} \in H, \\ \widetilde{\mathbf{v}} = -k\, L^*\widetilde{\mathbf{p}}, & \widetilde{\mathbf{p}} \in \partial G(L\widetilde{\mathbf{v}} - \mathbf{y}_d) . \end{cases} \qquad (4.9)$$

It is thus natural to introduce the following iterates: For each $n \geq 1$, we compute the solution of the coupled system

$$\mathbf{v}_n = -n\, L^* \mathbf{p}_n, \qquad \mathbf{p}_n \in \partial G(L\mathbf{v}_n - \mathbf{y}_d)\,. \tag{4.10}$$

A more explicit way to write (4.10) is the following one:

$$\frac{\partial \mathbf{y}_n}{\partial t} - \nu\Delta \mathbf{y}_n + \nabla \Pi_n = \mathbf{v}_n \mathrm{X}_\omega\,, \qquad \nabla \cdot \mathbf{y}_n = 0 \quad \text{in } \Omega \times (0,T)\,, \tag{4.11}$$

$$-\frac{\partial \mathbf{q}_n}{\partial t} - \nu\Delta \mathbf{q}_n + \nabla \Phi_n = 0\,, \qquad \nabla \cdot \mathbf{q}_n = 0 \quad \text{in } \Omega \times (0,T)\,, \tag{4.12}$$

$$\mathbf{y}_n = \mathbf{q}_n = 0 \quad \text{on } \partial\Omega \times (0,T)\,, \qquad \mathbf{y}_n(0) = 0\,, \quad \mathbf{q}_n(T) = \mathbf{p}_n\,, \tag{4.13}$$

$$\mathbf{v}_n = -n\,\mathbf{q}_n \mathrm{X}_\omega\,, \qquad \mathbf{p}_n \in \partial G(\mathbf{y}_n(T) - \mathbf{y}_d)\,. \tag{4.14}$$

For simplicity, let us restrict ourselves to the case

$$G(\mathbf{z}) = \operatorname{dist}(\mathbf{z}, B(0;\alpha)) \qquad \forall \mathbf{z} \in H. \tag{4.15}$$

THEOREM 4 Assume $|\mathbf{y}_d| > \alpha$ and assume that G is given by (4.15). Then for each $n \geq 1$ the system (4.10) possesses exactly one solution $(\mathbf{v}_n, \mathbf{p}_n)$. The sequence $\{(\mathbf{v}_n, \mathbf{p}_n)\}$ converges strongly in $\mathcal{U} \times H$ to the unique solution to (4.8).

SKETCH OF THE PROOF: Let us only present the main ideas in the proof of convergence. In this case, one has

$$\mathbf{p}_n = \frac{\mathbf{y}_n(T) - \mathbf{y}_d}{|\mathbf{y}_n(T) - \mathbf{y}_d|}.$$

Consequently, $|\mathbf{p}_n| = 1$ and

$$|\mathbf{y}_n(T) - \mathbf{y}_d| + (\mathbf{y}_d, \mathbf{p}_n) + n \iint_{\omega\times(0,T)} |\mathbf{q}_n|^2 = 0 \tag{4.16}$$

for all n. Thus, one sees that $\mathbf{y}_n(T)$ is bounded in H and $\mathbf{q}_n \mathrm{X}_\omega$ converges to 0 in $\mathcal{U}$. One also has:

$$\mathbf{q}_n \quad \text{is bounded in } L^\infty(0,T;H) \text{ and } L^2(0,T;V)\,,$$

$$\frac{\partial \mathbf{q}_n}{\partial t} \quad \text{is bounded in } L^2(0,T;V')\,.$$

Hence, at least for a subsequence, one must have

$$\mathbf{p}_\mu = \mathbf{q}_\mu(T) \to \mathbf{q}(T) \text{ weakly in } H$$

where $\mathbf{q}$ solves (3.8) and $\mathbf{q} = 0$ in $\omega \times (0,T)$. Again, from unique continuation we deduce that $\mathbf{q} \equiv 0$ and $\mathbf{p}_\mu$ converges weakly to 0. Using (4.16), we also deduce that $\mathbf{y}_\mu(T)$ converges strongly to $\mathbf{y}_d$. It is now easy to conclude.

Exact regularization and penalization. Other algorithms can be found by combining (exact) regularization and penalization. Among other possibilities, we mention the following, whose convergence properties have not still been analyzed: Choose $\lambda > 0$ and $\mathbf{p}_0 \in H$; then, for each $n \geq 0$, we set

$$\mathbf{v}_{n+1} = -n\,L^*\mathbf{p}_n\,, \qquad \mathbf{p}_{n+1} = (\partial G)_\lambda(L\mathbf{v}_{n+1} - \mathbf{y}_d + \lambda\mathbf{p}_n)\,. \tag{4.17}$$

5 THE CASE OF A NAVIER-STOKES FLOW

In this section, we consider flows governed by the *Navier-Stokes* problem. More precisely, with the same notation used in section 1, we are concerned with the problems

$$\frac{\partial \mathbf{y}}{\partial t} - \nu\Delta\mathbf{y} + (\mathbf{y}\cdot\nabla)\mathbf{y} + \nabla\Pi = \mathbf{v}\mathrm{X}_\omega\,, \qquad \nabla\cdot\mathbf{y} = 0 \qquad \text{in } \Omega\times(0,T)\,, \tag{5.1}$$

$$\mathbf{y} = 0 \quad \text{on } \partial\Omega\times(0,T)\,, \qquad \mathbf{y}(0) = 0\,, \tag{5.2}$$

associated to control functions $\mathbf{v} \in \mathcal{U}$.

THEOREM 5 [9,10] For every $\mathbf{v} \in L^2\left(\omega\times(0,T)\right)^N$, problem (5.1) – (5.2) possesses a weak solution $(\mathbf{y}_\mathbf{v}, \Pi_\mathbf{v})$ such that

$$\begin{cases} \mathbf{y}_\mathbf{v} \in L^2(0,T;V)\cap L^\infty(0,T;H)\cap C^0([0,T];V')\,, \\ \qquad\qquad \dfrac{\partial \mathbf{y}_\mathbf{v}}{\partial t} \in L^\sigma\left(\Omega\times(0,T)\right)^N, \end{cases} \tag{5.3}$$

with $\sigma = 2$ if $N = 2$ and $\sigma = 4/3$ if $N = 3$. Moreover, $\mathbf{y}_\mathbf{v} \in L^\sigma(0,T;W^{2,\sigma}(\Omega)^N)$. When $N = 2$, $\mathbf{y}_\mathbf{v}$ is unique and $\Pi_\mathbf{v}$ is unique up to a constant. In this case, one also has

$$\frac{\partial \mathbf{y}_\mathbf{v}}{\partial t} \in L^2(0,T;V')\,,$$

whence $\mathbf{y}_\mathbf{v} : [0,T] \to H$ is absolutely continuous.

Let us introduce for every $\mathbf{v} \in \mathcal{U}$ the set $\mathbf{Y}_\mathbf{v}(T)$ formed by all final states $\mathbf{y}_\mathbf{v}(T)$ corresponding to weak solutions to (5.1) – (5.2). Again, let $\mathbf{y}_d \in H$ and $\alpha > 0$ be given. Our aim is to solve the following

PROBLEM 3 Find $\mathbf{v} \in \mathcal{U}$ such that $|\mathbf{y}_\mathbf{v}(T) - \mathbf{y}_d| \leq \alpha$ for some $\mathbf{y}_\mathbf{v}(T) \in \mathbf{Y}_\mathbf{v}(T)$.

This is motivated by a conjecture, due to J.-L. Lions (see [11]), which affirms that the set

$$Y = \left(\bigcup_{\mathbf{v}} \mathbf{Y}_\mathbf{v}(T)\right) \cap H$$

is dense in H. Unfortunately, this is unknown. A probably very partial result in this direction is provided by theorem 1.2 in [6], where it is proved that the linear space spanned by Y is dense (see [1] and [5] for other similar results).

Let us present a fixed-point argument which can be regarded as an attempt to solve Problem 3. Let Z be a suitable space of velocity fields $\mathbf{z} = \mathbf{z}(x,t)$. For each $\mathbf{z} \in Z$, we introduce the linear operator $L_{\mathbf{z}}$, which is given by $L_{\mathbf{z}}\mathbf{v} = \mathbf{y}(T)$, with $\mathbf{y} = \mathbf{y}_{\mathbf{v},\mathbf{z}}$ being, together with some Π, the unique solution to:

$$\frac{\partial \mathbf{y}}{\partial t} - \nu\Delta\mathbf{y} + (\mathbf{z}\cdot\nabla)\mathbf{y} + \nabla\Pi = \mathbf{v}\mathrm{X}_{\omega}\,, \qquad \nabla\cdot\mathbf{y} = 0 \qquad \text{in } \Omega\times(0,T)\,, \tag{5.4}$$

$$\mathbf{y} = 0 \quad \text{on } \partial\Omega\times(0,T)\,, \qquad \mathbf{y}|_{t=0} = 0\,. \tag{5.5}$$

For the moment, we assume that Z is such that $L_{\mathbf{z}}$ is well defined and maps continuously $\mathcal{U}$ into H for all $\mathbf{z} \in Z$. Then we introduce the following optimization problems:

$$\inf_{\mathbf{v}\in\mathcal{U}} \{F_1(\mathbf{v}) + F_2(L_{\mathbf{z}}\mathbf{v})\} \tag{5.6}$$

and

$$\inf_{\mathbf{p}\in H} \{F_1^*(-L_{\mathbf{z}}^*\mathbf{p}) + F_2^*(\mathbf{p})\} \tag{5.7}$$

(recall the notation in section 3). Here, $L_{\mathbf{z}}^*$ (the adjoint of $L_{\mathbf{z}}$) is given as follows: $L_{\mathbf{z}}^*\mathbf{p} = \mathbf{q}\mathrm{X}_{\omega}$, with $\mathbf{q}$ being, together with Φ, the unique solution to

$$-\frac{\partial \mathbf{q}}{\partial t} - \nu\Delta\mathbf{q} - \nabla\cdot(\mathbf{z}\mathbf{q}) + \nabla\Phi = 0\,, \qquad \nabla\cdot\mathbf{q} = 0 \qquad \text{in } \Omega\times(0,T)\,, \tag{5.8}$$

$$\mathbf{q} = 0 \quad \text{on } \partial\Omega\times(0,T)\,, \qquad \mathbf{q}|_{t=T} = \mathbf{p}\,. \tag{5.9}$$

Assume that (5.6) and (5.7) possess unique solutions respectively denoted by $\widehat{\mathbf{v}}(\mathbf{z})$ and $\widehat{\mathbf{p}}(\mathbf{z})$ and, also, that these are characterized by the following:

$$\frac{\partial \widehat{\mathbf{y}}}{\partial t} - \nu\Delta\widehat{\mathbf{y}} + (\mathbf{z}\cdot\nabla)\widehat{\mathbf{y}} + \nabla\Pi = \widehat{\mathbf{v}}(\mathbf{z})\mathrm{X}_{\omega}\,, \qquad \nabla\cdot\widehat{\mathbf{y}} = 0 \quad \text{in } \Omega\times(0,T)\,, \tag{5.10}$$

$$-\frac{\partial \widehat{\mathbf{q}}}{\partial t} - \nu\Delta\widehat{\mathbf{q}} - \nabla\cdot(\mathbf{z}\widehat{\mathbf{q}}) + \nabla\Phi = 0\,, \qquad \nabla\cdot\widehat{\mathbf{q}} = 0 \quad \text{in } \Omega\times(0,T)\,, \tag{5.11}$$

$$\widehat{\mathbf{y}} = \widehat{\mathbf{q}} = 0 \qquad \text{on } \partial\Omega\times(0,T)\,, \qquad \widehat{\mathbf{y}}|_{t=0} = 0\,, \quad \widehat{\mathbf{q}}|_{t=T} = \widehat{\mathbf{p}}(\mathbf{z})\,, \tag{5.12}$$

$$\widehat{\mathbf{v}}(\mathbf{z}) = -\widehat{\mathbf{q}}\mathrm{X}_{\omega}\,, \qquad \widehat{\mathbf{y}}|_{t=T} = \mathbf{y}_d + \frac{\alpha}{|\widehat{\mathbf{p}}(\mathbf{z})|}\widehat{\mathbf{p}}(\mathbf{z})\,. \tag{5.13}$$

This will be the case if (5.8) satisfies the unique continuation property for each $\mathbf{z}$. Then, in order to solve Problem 3, it will suffice to prove that a fixed point exists for the mapping $\mathbf{z} \to \widehat{\mathbf{y}}$ (where $\widehat{\mathbf{y}}$ is the one in (5.10) – (5.13)).

There exist Banach spaces Z such that $\mathbf{z} \to \widehat{\mathbf{y}}$ is a well defined, continuous and compact mapping from Z into itself. For instance, one may take

$$Z = L^\infty(0,T;W^{1,\infty}(\Omega)^N)$$

(see [6] for a justification of the unique continuation property). But, contrarily, it does not seem easy to deduce uniform estimates for $\widehat{\mathbf{y}}$ in these spaces. Hence, it is not clear whether Schauder's theorem can be applied in this context. Other similar fixed-point arguments can be thought but all them lead to difficulties of the same kind: The involved space has to be chosen small enough to provide unique continuation and, at the same time, large enough to provide "good" estimates.

To end this section, let us present a second tentative idea. This is inspired by the results in section 4 and, again, is related to exact regularization. Let us choose $\lambda > 0$ and $\mathbf{p}_0 \in H$. Among other possibilities, we introduce the following algorithm:

$$\mathbf{v}_{n+1} = -\mathbf{q}_n \mathrm{X}_\omega \,, \tag{5.14}$$

$$\begin{cases} \dfrac{\partial \mathbf{y}_{n+1}}{\partial t} - \nu\Delta \mathbf{y}_{n+1} + (\mathbf{y}_{n+1}\cdot\nabla)\mathbf{y}_{n+1} + \nabla\Pi_{n+1} = \mathbf{v}_{n+1}\mathrm{X}_\omega \,, \\ -\dfrac{\partial \mathbf{q}_{n+1}}{\partial t} - \nu\Delta \mathbf{q}_{n+1} - (\mathbf{y}_{n+1}\cdot\nabla)\mathbf{q}_{n+1} + \nabla\Phi_{n+1} = 0 \,, \\ \nabla\cdot\mathbf{y}_{n+1} = 0 \,, \qquad \nabla\cdot\mathbf{q}_{n+1} = 0 \quad \text{in } \Omega\times(0,T)\,, \\ \mathbf{y}_{n+1} = \mathbf{q}_{n+1} = 0 \qquad \text{on } \partial\Omega\times(0,T)\,, \\ \mathbf{y}_{n+1}(0) = 0\,, \quad \mathbf{q}_{n+1}(T) = \mathbf{p}_{n+1}\,, \end{cases} \tag{5.15}$$

$$\left(\lambda + \frac{\alpha}{|\mathbf{p}_{n+1}|}\right)\mathbf{p}_{n+1} = \mathbf{y}_{n+1}(T) - \mathbf{y}_d + \lambda\mathbf{p}_n \,. \tag{5.16}$$

It can be proved that (5.15) – (5.16) possesses at least one solution for each n. Thus, it can be interesting to try to prove that $\mathbf{y}_n(T)$ converges towards a field in H at distance $\leq \alpha$ from $\mathbf{y}_d$.

REFERENCES

1. C. Bardos and L. Tartar, Sur l'unicité rétrograde des équations paraboliques et quelques questions voisines, *Arch. Rat. Mech. Anal., 50* (1973), p. 10–25.
2. A. Bermúdez and C. Moreno, Duality methods for solving variational inequalities, *Comput. Math. Appl., v. 7,* 1981, pp. 43–58.
3. C. Fabre, J.-P. Puel and E. Zuazua, Contrôlabilité approchée de l'équation de la chaleur semi-linéaire, *C. R. Acad. Sci. Paris, t. 315,* série I, p. 807–812, 1992.
4. C. Fabre, J.-P. Puel and E. Zuazua, Contrôlabilité approchée de l'équation de la chaleur linéaire avec des contrôles de norme L^∞ minimale, *C. R. Acad. Sci. Paris, t. 316,* série I, p. 679–684, 1993.
5. E. Fernández-Cara and M. González-Burgos, A result concerning controllability for the Navier-Stokes equations, *SIAM J. Control Opt.,* to appear.
6. E. Fernández-Cara and J. Real, On a conjecture due to J.L. Lions, *Nonlinear Anal. T M & A, Vol. 21,* No. 11, pp. 835–847, 1993.
7. E. Fernández-Cara, M. González-Burgos, J.D. Martín and J. Real, *In preparation.*
8. R. Glowinski and J.-L. Lions, Exact and approximate controllability for distributed parameter systems, *to appear.*

9. O.A. Ladyzhenskaya, The Mathematical Theory of Viscous Incompressible Flow, *Gordon and Breach,* New York (1963).

10. J.-L. Lions, Quelques Méthodes de Résolution des Problèmes aux Limites non Linéaires, *Dunod, Gauthiers-Villars,* Paris (1969).

11. J.-L. Lions, Lectures given at the "Collège de France", 1989/90. See also: Remarques sur la contrôlabilité approchée, In *Proceedings of the Jornadas Hispano-Francesas sobre Control de Sistemas Distribuidos,* University of Malaga (Spain), October 1990.

12. S. Mizohata, Unicité du prolongement des solutions pour quelques opérateurs différentiels paraboliques, *Mem. Coll. Sci. Univ. Kyoto, Sér. A 31,* (3) (1958), p. 219–239.

13. J.-C. Saut and B. Scheurer, Unique continuation for some evolution equations, *J. Diff. Equations 66,* (1987), p. 118–139.

On Some Weighting Methods to Improve Output Least Squares Estimation

Jacques Henry INRIA Rocquencourt, Le Chesnay, France

Jean-Pierre Yvon INRIA Rocquencourt, Le Chesnay, France, and Université de Technologie de Compiègne, Compiègne, France

1 Introduction

1.1 Ouput least squares identification framework

Throughout this paper we will deal with the general situation where we want to estimate a set of parameters $\theta \in \Theta_{ad}$ where Θ_{ad} is a subset of the Hilbert space Θ. The observation $z(t;\theta)$ is given by $z(t;\theta) = Cy(t;\theta) \in Z$, Z being the Hilbert space of observations and $y(t;\theta)$ is solution of the state equation :

$$\begin{cases} \dfrac{dy}{dt} + A(\theta)y + f(y) = 0, \\ y(0) = y_0, \end{cases} \tag{1}$$

where A is a bounded or unbounded operator in the Hilbert space H. The output least squares estimation leads to the minimization of the following criterion :

$$J(\theta) = \frac{1}{2}\int_0^T \| z(t;\theta) - z_d(t) \|_Z^2 \, dt,$$

where $z_d(t) \in Z$ is the given observation.

The estimate $\hat{\theta}$ of θ being given by

$$J(\hat{\theta}) = \min_{\theta \in \Theta_{ad}} J(\theta).$$

This problem is usually ill-posed and a classical way to overcome this difficulty consits in regularizing the problem. For instance the Tichonov regularization amounts to introducing a new criterion :

$$J_\varepsilon(\theta) = J(\theta) + \frac{\varepsilon}{2}\left\|\theta - \bar{\theta}\right\|_\Theta^2 \tag{2}$$

where $\bar{\theta}$ is some "reference" value for the parameters.

Remark 1 Throughout this paper we will adopt the following point of view : we have a set of experiments corresponding to several values of the parameters which lie in a neighbourhood of a value $\bar{\theta}$. So, in order to define weightings, we will use a linearization around the point $\bar{\theta}$.

Remark 2 Even if the problem is regularized, the previous problem is difficult to solve : one observes slow convergence of minimization algorithms and classical stopping rules for theses algorithms lead to solutions which can be far from the optimum. □

Remark 3 For the sake of simplicity we will drop the regularizing term $\frac{\varepsilon}{2}\left\|\theta-\bar{\theta}\right\|_{\Theta}^{2}$ in the presentation of the methods, but its introduction leads to straightforward modifications.

1.2 Orientation

Many authors have tried to introduce weighted criterion in order to

- minimize the variance of the estimate in the stochastic framework,
- to improve the efficiency of the minimization process as in [3].

Our approach, which is close to the later idea, has the purpose to improve the conditionning of the criterion.

From now on (section 2.3 excepted) we will assume that there is a finite number of observations and parameters :

$$Z = \mathbb{R}^q, \quad \Theta = \mathbb{R}^p.$$

We will denote by $\mathcal{M}_{q,p}$ the set of $q \times p$ matrices.

Throughout this paper we will introduce modified criterions J_W, in order to

- get a better conditionning of J_W,
- try to orthonormalize the sensitivity functions $S(t)$:

$$S_\theta(t) = \nabla_\theta z(t;\theta) \in \Theta'.$$

There are several ways to introduce weights in the criterion :

- *A weighting local in time*

  Let $W(t)$ be a family of symmetric matrices of order q, $W(t) \geq 0$.The new criterion is given by

$$J_W(\theta) = \frac{1}{2}\int_0^T < z(t;\theta) - z_d(t), W(t)[z(t;\theta) - z_d(t)] >_Z dt.$$

- *A weighting global in time*

  The weighting operator can be introduced by means of a kernel $W(t,\tau)$ which is a symmetric matrix such that

$$W(t,\tau) \geq 0, W(t,\tau) = W(\tau,t)$$

  and the new criterion takes the form :

$$J_W(\theta) = \frac{1}{2}\int_0^T\int_0^T < z(t;\theta) - z_d(t), W(t,\tau)[z(\tau;\theta) - z_d(\tau)] >_Z d\tau\, dt.$$

Remark 4 The Hessian of $J(\theta)$ is given by :

$$\nabla_\theta^2 J(\theta) = \int_0^T S_\theta^*(t) S_\theta(t)\, dt + \int_0^T < z(t;\theta) - z_d(t), \nabla_\theta^2 z(t;\theta) >_Z \, dt$$

and we will denote by G_θ the so-called *Gauss approximation* of the Hessian :

$$G_\theta = \int_0^T S_\theta^*(t) S_\theta(t)\, dt.$$

The previous goals amount to getting G_θ close to I. If the Tichonov regularisation (2) is used the Gauss approximation takes the following form :

$$G_\theta = \int_0^T S_\theta^*(t) S_\theta(t)\, dt + \varepsilon I.$$

□

2 Gauss-Newton weighting

In this section we study the global weighting introduced in the previous section, J_W is then given by

$$J_W(\theta) = \frac{1}{2} \int_0^T \int_0^T [z(t;\theta) - z_d(t)]^T W(t,\tau) [z(\tau;\theta) - z_d(\tau)]\, d\tau\, dt.$$

From now on we will assume that $\bar{\theta}$ is given in Θ and we will use the simplified notation :

$$S(t) := S_{\bar{\theta}}(t) = \nabla_\theta z(t;\bar{\theta}) \in \mathcal{M}_{q,p}$$

2.1 Problem formulation

Definition 1 *We say that the system is* locally identifiable, *around* $\bar{\theta}$, *if*

$$rank \int_0^T S^T(t) S(t)\, dt = p \tag{3}$$

This property will be assumed throuhout this paper.

Problem 1 *One seeks a family of symmetric, positive definite matrices* $W(t,\tau)$ *such that :*

$$W(t,\tau) = W(\tau, t),$$

$$\int_0^T S^T(t) W(t,\tau) S(\tau)\, d\tau\, dt = I, \tag{4}$$

which minimizes

$$\int_0^T \int_0^T \| W(t,\tau) \|_F^2 \, d\tau\, dt \tag{5}$$

where the matrix norm is the Frobenius norm defined by :

$$\| W \|_F^2 = tr\, W^T W = < W, W >_F .$$

Remark 5 The minimization problem (5) is introduced to ensure the uniqueness of W. The choice of the Frobenius norm is somehow arbitrary, its main interest lies in the fact that it allows explicit calculations. □

Formal calculation of W. Let $\mathcal{L}(W,\Lambda)$ be the lagrangian :

$$\mathcal{L}(W,\Lambda)=\frac{1}{2}\int_0^T\int_0^T \| W \|_F^2 \, d\tau\, dt - < \Lambda, \int_0^T S^T(t)W(t,\tau)S(\tau)\, d\tau\, dt - I >_F,$$

where $\Lambda \in \mathcal{M}_{p,p}$ is symmetric. The condition $\dfrac{\partial \mathcal{L}}{\partial W}=0$ leads to the relation :

$$W(t,\tau)=S(t)\Lambda S^T(t). \tag{6}$$

Plugging this result in the constraint (4) gives

$$\int_0^T S^T(t)S(t)\, dt \;\; \Lambda \int_0^T S^T(\tau)S(\tau)\, d\tau = I$$

and Λ is given by

$$\Lambda=\left(\int_0^T S^T(t)S(t)\, dt\right)^{-2}.$$

The matrix W can be factorized and we get

$$W(t,\tau)=B^T(t)B(\tau), \tag{7}$$

$$B(t)=\left(\int_0^T S^T(t)S(t)\, dt\right)^{-1} S^T(t). \tag{8}$$

Actually $B(t)$ is simply the *pseudo inverse* of S :

$$\int_0^T B(t)S(t)\, dt = I,$$

and the estimated parameter $\hat{\theta}_W$ is then given by

$$J_W(\hat{\theta}_W)=\inf_\theta J_W(\theta)=\inf_\theta \frac{1}{2}\left\|\int_0^T B(t)[z_d(t)-z(t;\theta)]\, dt\right\|^2$$

Remark 6 We recall that the previous calculations are based upon the value $\bar{\theta}$ and all the matrices and operators depends on $\bar{\theta}$, (we should have written $W_{\bar{\theta}}(t,\tau), B_{\bar{\theta}}(t)\ \ldots$).

2.2 Properties of $\hat{\theta}_W$

First of all if $\bar{\theta}=\hat{\theta}$ then

$$\hat{\theta}_W=\hat{\theta}.$$

Actually we have

$$\hat{\theta}=\int_0^T B(t)z_d(t)\, dt=\int_0^T B(t)z(t;\hat{\theta})\, dt,$$

which implies that $J_W(\hat{\theta}) = 0$.

Furthermore the Gauss approximation of the Hessian of J_W is the identity matrix at $\hat{\theta}_W$.

A last aspect is the sensitivity analysis. A classical result of sensitivity analysis is :

$$\frac{\| \delta\theta \|}{\| \hat{\theta} \|} \leq \chi^{1/2}(G) \frac{\| \delta z \|}{\| z(\hat{\theta}) \|}$$

where $\chi(G)$ is the condition number of the Gauss approximation of the Hessian. If we introduce the semi-norm :

$$\| z \|_W^2 = \int_0^T \int_0^T z^T(t) W(t,\tau) z(\tau)\, d\tau\, dt = \left\| \int_0^T Bz(t)\, dt \right\|^2 ,$$

we have the following estimation :

$$\frac{\| \delta\hat{\theta}_W \|}{\| \hat{\theta}_W \|} \leq \frac{\| \delta z \|_W}{\| z(\hat{\theta}_W) \|_W} \tag{9}$$

which implies the possibility of having a more efficient stopping rule for the minimization algorithm. On the other hand if we consider a perturbation of the sensitivity operator $S(t)$ we have the classical result :

$$\frac{\| \delta\hat{\theta} \|}{\| \hat{\theta} \|} \leq \chi^{1/2}(G) \frac{\| \delta S \|_{L^2(0,T)}}{\| S \|_{L^2(0,T)}} + \chi(G) \frac{\| z_d - z(\hat{\theta}) \|}{\| z(\hat{\theta}) \|} \frac{\| \delta S \|_{L^2(0,T)}}{\| S \|_{L^2(0,T)}} + O\left(\|\delta S\|^2_{L^2(0,T)} \right) ,$$

then, again, we obtain a better formula :

$$\frac{\| \delta\hat{\theta}_W \|}{\| \hat{\theta}_W \|} \leq \chi^{1/2}(G) \frac{\| \delta S \|_{L^2(0,T)}}{\| S \|_{L^2(0,T)}} + O\left(\|\delta S\|^2_{L^2(0,T)} \right) .$$

which indicates a result less sensitive to perturbation of the operator S. Actually the worst term in $\chi(G)$, which is related to the residual $\{z_d - z(\hat{\theta})\}$, is cancelled thanks to the weighting W. □

Remark 7 The gradient of J_W is given by

$$\nabla_\theta J_W = \left(\int_0^T (z(t;\theta) - z_d(t))^T S^T(t)\, dt \right) \left(\int_0^T S^T(t) S(t)\, dt \right)^{-1} .$$

This is exactly the descent direction for J given by a Gauss-Newton method. This explains our terminology "Gauss-Newton weighting". Neverthelesss a gradient method applied to the minimization of J_W and a Gauss-Newton method applied to the minimization of J are not identical as they aim at minimizing two different functions. □

2.3 Connection with the "sentinel theory" of J.L. Lions

The aim of this section is to show the strong links between the previous approach and the "sentinel theory" by J.L. Lions used in the estimation framework. We start by recalling the basic elements of the theory ([7], [2]). Let us consider a parabolic state equation

$$\begin{cases} \dfrac{\partial y}{\partial t} + A(\theta)y + f(y) = 0, & \text{in } \Omega\times]0,T[, \\ y(x,t) = 0, & (x,t) \in \partial\Omega\times]0,T[, \\ y(x,0) = y_0(x), & x \in \Omega, \end{cases} \tag{10}$$

with the observation

$$z(t;\theta) = y(t;\theta)|_{\mathcal{O}},$$

where $\mathcal{O} \subset \Omega$ is the"observatory" (this means that the observation takes place in a "small" domain $\mathcal{O}$ contained in Ω). In the sequel we will denote :

$$Q = \Omega\times]0,T[,\, \mathcal{Q} = \mathcal{O}\times]0,T[.$$

The linearization about $\bar{y} = y(\bar{\theta})$ can be done by introducing

$$\tilde{y}_i = \frac{\partial y}{\partial \theta_i}(\bar{\theta})$$

which satisfies the following equation :

$$\begin{cases} \dfrac{\partial \tilde{y}_i}{\partial t} + A(\bar{\theta})\tilde{y}_i + f'(\bar{y})\tilde{y}_i = b_i, & \text{in } \Omega\times]0,T[, \\ \tilde{y}_i(x,t) = 0, & (x,t) \in \partial\Omega\times]0,T[, \\ \tilde{y}_i(x,0) = 0, & x \in \Omega, \end{cases} \tag{11}$$

where b_i is defined by

$$b_i = -\frac{\partial A(\bar{\theta})}{\partial \theta_i}\bar{y}.$$

From these definitions we get

$$S(t) = [\ldots, \tilde{y}_i|_{\mathcal{O}}, \ldots].$$

Then one can define a family of sentinels $\{\eta_i\}$ by

$$\eta_i = \int_{\mathcal{Q}} w_i(x,t)y(x,t)\,dx\,dt.$$

Following J.L. Lions, we want to define w_i of minimal norm such that η_i is sensitive to θ_i but insensitive to $\theta_j, j \neq i$.

Problem 2 *Find a family of functions $w_i \in L^2(\mathcal{Q})$ such that :*

$$\frac{\partial \eta_i}{\partial \theta_j} = \int_{\mathcal{Q}} w_i(x,t)\tilde{y}_j(x,t)\,dx\,dt = \delta_{ij},$$

which minimizes :

$$\frac{1}{2}\sum_i \|w_i\|^2_{L^2(\mathcal{Q})}.$$

This problem can be solved by introducing the "adjoint state" q_i :

$$\begin{cases} -\dfrac{\partial q_i}{\partial t} + A^*(\bar{\theta})q_i + f'(\bar{y})q_i = w_i\,\chi_{\mathcal{O}}, & \text{in } \Omega\times]0,T[, \\ q_i(x,t) = 0, & (x,t) \in \partial\Omega\times]0,T[, \\ q_i(x,T) = 0, & x \in \Omega, \end{cases} \tag{12}$$

where $\chi_{\mathcal{O}}$ denotes the characteristic function of the subset $\mathcal{O}$. Then we are led to a family of control problems :

$$\min_w \frac{1}{2}\sum_i \|w_i\|^2_{L^2(Q)}$$

subject to (12) and

$$\int_Q w_i(x,t)\tilde{y}_j(x,t)\,dx\,dt = \int_Q q_i(x,t)b_j(x,t)\,dx\,dt = \delta_{ij}. \tag{13}$$

Corresponding to this new family of control problem there is an other "adjoint state" ρ_i defined by

$$\begin{cases} -\dfrac{\partial \rho_i}{\partial t} + A(\bar{\theta})\rho_i + f'(\bar{y})\rho_i + \displaystyle\sum_{j=1}^{p} p_{ij}b_j\chi_{\mathcal{O}} = 0, & \text{in } \Omega\times]0,T[, \\ \rho_i(x,t) = 0, & (x,t) \in \partial\Omega\times]0,T[, \\ \rho_i(x,0) = 0, & x \in \Omega. \end{cases} \tag{14}$$

where $P = [p_{ij}]$ is the Lagrange multiplier of the family of constraints (13).

The set of optimality conditions is then

$$w_i(t) = -\rho_i(t)\chi_{\mathcal{O}}, \qquad t \in]0,T[.$$

Let Π be the $p^2 \times p^2$ matrix corresponding to the mapping

$$P \stackrel{\Pi}{\longrightarrow} \int_Q q_i(x,t)b_j(x,t)\,dx\,dt,$$

then the condition (13) is equivalent to :

$$\Pi P = I.$$

Remark 8 The matrix Π is positive definite. In fact if we consider, for ρ and $\tilde{\rho}$, the scalar product :

$$\sum_i \int_Q q_i(x,t)\,\tilde{q}_i(x,t)\,dx\,dt,$$

with evident notations, we get

$$\begin{aligned} \sum_i \int_Q q_i(x,t)\tilde{q}_i(x,t)\,dx\,dt &= \sum_i \int_Q q_i(x,t)\left(\sum_j \tilde{p}_{ij}(x,t)b_j(x,t)\right)\,dx\,dt \\ &= <\Pi P, \tilde{P}>, \end{aligned}$$

which implies that P is non negative definite. Furthermore if $<\Pi P, \tilde{P}>= 0$ then we have $\rho_i|_Q \equiv 0$ and

$$w_i \equiv 0 \Rightarrow q_i \equiv 0 \rightarrow P \equiv 0.$$

Then P is uniquely defined by

$$\Pi P = I.$$

This corresponds to the use of the theory of sentinels as a method to control the sensitivity of the estimate with respect to each parameter (see [5]). As a result of the previous analysis we have

$$\begin{aligned} J_w(\theta) &= \frac{1}{2}\sum_i |\eta_i(z_d(x,t)) - \eta_i(y(x,t;\theta))|^2 \\ &= \frac{1}{2}\sum_i \left(\int_Q w_i(x,t)(z_d(x,t) - y(x,t;\theta))\,dx\,dt\right)^2, \end{aligned}$$

where the weights $\{w_i\}$ are calculated at the point $\bar{\theta}$.

Remark 9 The method can be extended in order to render the estimation (locally) insensitive to a family of other parameters which, for instance, are irrelevant. □

3 Weighting local in time

We now consider a family of symmetric weighting matrices $W(t) \in \mathcal{M}_{q,q}$ and the following criterion

$$J_W(\theta) = \frac{1}{2}\int_0^T [z(t;\theta) - z_d(t)]^T W(t)[z(t;\theta) - z_d(t)]dt.$$

The associated problem is

Problem 3 *Find a family of symmetric positive definite matrices $W(t)$ such that*

$$\int_0^T S^T(t)W(t)S(t)\,dt = I$$

which minimizes

$$\frac{1}{2}\int_0^T \|W(t)\|_F^2\,dt.$$

□

Formal calculation of W

We introduce the Lagrangian :

$$\mathcal{L}(W,\Lambda) = \frac{1}{2}\int_0^T \|W(t)\|_F^2\,dt - <\Lambda, \int_0^T S^T(t)W(t)S(t)\,dt - I>.$$

The condition $\dfrac{\partial \mathcal{L}}{\partial W} = 0$ leads to

$$W(t) = S(t)\Lambda S^T(t)$$

and

$$\int_0^T S^T(t)S(t)\Lambda S^T(t)S(t)\,dt = I.$$

Let L be the matrix defined by

$$L = \int_0^T (S^T(t)S(t)) \otimes (S^T(t)S(t))\,dt, \tag{15}$$

where $\otimes$ denotes the Kronecker product. Then we get

$$\begin{cases} L\Lambda = I, \\ W(t) = S(t)\Lambda S^T(t). \end{cases} \tag{16}$$

Existence of Λ results from the *strong identifiability* hypothesis :

Definition 2 *We say that the system is (locally)* strongly identifiable *around* $\bar{\theta}$ *if the mapping* $\mathcal{C} \in \mathcal{L}\left(L^2(0,T;\mathcal{M}_{qq}), \mathcal{M}_{pp}\right)$ *defined by*

$$\mathcal{C}Z = \int_0^T S^T(t)Z(t)S(t)\,dt$$

is onto.

Theorem 1 *[4] The hypothesis of local strong identifiability implies local identifiability.*

Theorem 2 *[4] Under the hypothesis of strong identifiability, the Problem 3 has a unique solution given by (16).*

Remark 10 It is clear that $\Lambda \geq 0 \Rightarrow S(t)\Lambda S^T(t) \geq 0$. Then a crucial problem is the positiveness of Λ if we want to have a well posed problem for the minimization of J_W. Unfortunately this condition is hard to check. Nevertheless we have the following theorem.

Theorem 3 *The solution* Λ *of (16) is positive semidefinite iff the following condition holds : for all matrix* $M \in \mathcal{M}_{qq}$ *we have the implication*

$$\left\{\forall t \in [0,T] : S(t)MS^T(t) \geq 0\right\} \Longrightarrow \{\mathrm{tr}M \geq 0\} \tag{17}$$

Properties of $\hat{\theta}_W$.

The Gauss approximation of $\nabla^2_\theta J_W$ is I. The sensitivity analysis goes along the lines of the previous paragraph, in particular the sensitivity with respect to a perturbation δz of the observation is governed by

$$\frac{\| \delta\hat{\theta}_W \|}{\| \hat{\theta}_W \|} \leq \frac{\| \delta z \|_W}{\| z(\hat{\theta}_W) \|_W} \tag{18}$$

where the norm $\|z\|_W$ is defined by :

$$\|z\|_W^2 = \int_0^T z^T(t)W(t)z(t)\,dt.$$

Finally it must be noticed that if $z_d \neq z(\hat{\theta})$ then $\hat{\theta}_W \neq \hat{\theta}$. □

4 Appendix A : illustration of the strong identifiability condition

We consider the one dimensional parabolic equation

$$\begin{cases} \dfrac{\partial u}{\partial t} - a\dfrac{\partial^2 u}{\partial x^2} - bu = 0, & x \in [0,1], t \geq 0, \\ u(0,t) = u(1,t) = 0, & t \geq 0, \\ u(x,0) = \alpha_1\sqrt{2}\sin \pi x + \alpha_2\sqrt{2}\sin 2\pi x, & x \in [0,1], \end{cases} \tag{19}$$

where a, b are parameter to be estimated and $\theta = [a, b]^T$. The observation is

$$Cu = u(x_0, t)$$

(see [6], [1]). The condition θ identifiable is equivalent to

$$\alpha_1\alpha_2 \sin \pi x_0 \sin 2\pi x_0 \neq 0. \tag{20}$$

The sensitivity function $S(t)$ is given by

$$S(t) = \begin{bmatrix} \dfrac{\partial u}{\partial a}(x_0) \\ \dfrac{\partial u}{\partial b}(x_0) \end{bmatrix} = \begin{bmatrix} -\pi^2 t\sqrt{2}\left(e^{(b-a\pi^2)t}\alpha_1 \sin \pi x_0 + 4e^{(b-4a\pi^2)t}\alpha_2 \sin 2\pi x_0\right) \\ t\sqrt{2}\left(e^{(b-a\pi^2)t}\alpha_1 \sin \pi x_0 + e^{(b-4a\pi^2)t}\alpha_2 \sin 2\pi x_0\right) \end{bmatrix}.$$

The condition of strong identifiability takes the form :

$$\Lambda \in \mathcal{S}_2, S(t)\Lambda S^T(t) = 0 \Longrightarrow \Lambda = 0 \ ?$$

If we denote Λ by

$$\Lambda = \begin{bmatrix} \lambda_1 & \lambda_{12} \\ \lambda_{12} & \lambda_2 \end{bmatrix},$$

the condition $S(t)\Lambda S^T(t) = 0$ takes the form

$$\mathcal{A}\begin{bmatrix} \lambda_1 \\ \lambda_{12} \\ \lambda_2 \end{bmatrix} = \begin{bmatrix} \pi^4 & -2\pi^2 & 1 \\ 4\pi^4 & -5\pi^2 & 1 \\ 16\pi^4 & -8\pi^2 & 1 \end{bmatrix}\begin{bmatrix} \lambda_1 \\ \lambda_{12} \\ \lambda_2 \end{bmatrix} = 0.$$

And the condition of strong identifiability is simply $\det(\mathcal{A}) = 27\pi^6 \neq 0$. Furthermore, by (17), the positiveness of Λ is equivalent to $\alpha_1\alpha_2 < 0$.

5 Appendix B : relaxed version of problem 3

In place of taking into account the constraint

$$\int_0^T S^T(t)W(t)S(t)\,dt = I$$

it is possible to introduce a penalty term :

$$\min_W \left\{ \int_0^T \|(W(t)\|_F^2 \, dt + \frac{1}{\varepsilon} \left\| \int_0^T S^T(t)W(t)S(t)\, dt - I \right\|_F^2 \right\}. \tag{21}$$

The optimality condition satisfied by a solution W of (21) is

$$W(t) = S(t)\Lambda S^T(t),$$
$$(L + \varepsilon I)\Lambda = I,$$

where the matrix L is defined by (15). This problem has a unique solution without the strong identifiability hypothesis.

Finally if the condition of positiveness is not met, Problem 3 can be relaxed to

$$\min_W \int_0^T \|(W(t) - \mu I\|_F^2 \, dt,$$

where μ is a given parameter ($0 \leq \mu \leq 1$) with the constraint

$$\int_0^T S^T(t)W(t)S(t)\, dt = (1-\mu)G + \mu I.$$

This allows to get $W(t) \geq 0$ for μ small enough.

- - - - - - - - - - - - - -

References

[1] Courdesses M., Amouroux M. *Identifiabilité et Identification des Systèmes à parmètres répartis* in proceedings of 3rd Symposium IFAC on Control of DPS, Toulouse, June 1992.

[2] Chavent G. *Generalized Sentinels Defined via Least Squares*, Rapport INRIA, **1932**, (1993)

[3] Goudou X. *Étude numérique et théorique de problèmes inverses et application à la prospection électromagnétique.* Thesis, Paris-IX (1994)

[4] Guyon F., Henry J., Yvon J.P. *On a weighting method improving identifiability of distributed parameter system*, Lecture Notes in Control and Information Sciences, **149**, pp 104-119 (1991)

[5] Kernevez J.P. *Des Sentinelles pour Surveiller l'Environnement*, Book to appear.

[6] Kitamura S., Nakagiri S. *Indentifiabiity of Spatially Varying and Constant Parameters in Distributed Systems of Parabolic Type*, SIAM J. of Cont. Opt. Vol. 15, **5**, pp. 785-802 (1977)

[7] Lions J.L. *Sentinelles pour les Systèmes à Données Incomplètes*, Masson (1992)

Boundary Control of a Schrödinger Equation with Nonconstant Principal Part

Mary Ann Horn[1] and Walter Littman, School of Mathematics, University of Minnesota, Minneapolis, Minnesota 55455

1 INTRODUCTION

In this paper, we obtain a simple proof of exact boundary controllability for the Schrödinger equation with variable principal part by combining a method of W. Littman and S. Taylor with an estimate of B. Vainberg.

In the context of boundary controllability for a Schrödinger equation, Littman and Taylor have introduced a general technique, which, basically stated, says that

$$\textit{local smoothing} + \textit{reversibility} + \textit{uniqueness} \Longrightarrow \textit{exact controllability}$$

(see [4]). Here "uniqueness" is the uniqueness property implying approximate controllability by duality. If control is to be exercised on the whole lateral boundary of a cylindrical domain, $\Omega \times (0, T)$, in space-time, it means that every solution (in an appropriate space) of the homogeneous linear evolution equation in this domain having zero Cauchy data on the lateral boundary must vanish identically. "Reversibility" means that the backward problem (in time) is wellposed. For the system we wish to consider, the difficulty of this approach lies in proving the necessary smoothing properties.

[1]This material is based upon work partially supported under a National Science Foundation Mathematical Sciences Postdoctoral Research Fellowship.

Local smoothing properties of evolution equations are not a newly discovered phenomenon. For example, the propagation of singularities for hyperbolic systems is well known. However, in [4], Littman and Taylor established similar properties for very different classes of equations, i.e., the Schrödinger equation with nonsmooth potentials, and that these smoothing properties can be applied in the method just described.

In our problem, equations exhibiting smoothing properties result in solutions belonging to the class Gevrey-δ, where this Gevrey class is defined with respect to the time variable t. We formulate this more precisely below.

DEFINITION *A function $f(x,t)$ belongs to the class Gevrey-δ with respect to t and uniformly for (x,t) in the compact set K if for all $(x,t) \in K$ and for all $\theta > 0$, there exists a constant $C_{K,\theta} > 0$ such that*

$$|\frac{\partial^n}{\partial t^n} f(x,t)| \leq C_{K,\theta}\theta^n (n!)^\delta.$$

With this definition in mind, we state Littman and Taylor's result on local smoothing properties for a Schrödinger equation with potential in $\mathbb{R}^n$.

THEOREM 1.1 *(See [4].) Consider the initial value problem in $\mathbb{R}^n$:*

$$\begin{aligned} i\tfrac{\partial u}{\partial t} + \Delta u - V(x)u = 0, \quad x \in \mathbb{R}^n, \quad t > 0 \\ u(x,0) = \phi(x) \end{aligned}$$

Assume that V is bounded, measurable and has compact support and that $\phi(x) \in L^2(\mathbb{R}^n)$ and has compact support. Then the solution $u(x,t)$ is of class Gevrey-2 with respect to t uniformly in compact sets of $\mathbb{R}^n \times \{t > 0\}$.

Extensions of the above result to the case of nonconstant coefficients in the principal part of the operator are of interest both mathematically and physically. On the other hand, techniques used in the proof of Theorem 1.1 take advantage of the structure of the Green's function associated with the problem, which will no longer be possible in the more general case. We note that the proof in [4] was only given for $n = 3$ and that Taylor has extended the result to general n.

To begin our study of Schrödinger equations with nonconstant coefficients in the principal part of the differential operator, we consider the following system.

$$\begin{aligned} i\tfrac{\partial u}{\partial t} + \mathcal{A}u = 0, \qquad x \in \mathbb{R}^n, \quad t > 0, \\ u(x,0) = \phi(x), \end{aligned} \tag{1.1}$$

where

$$\mathcal{A} \equiv \sum_{i,j=1}^{n} \frac{\partial}{\partial x_i} a_{ij}(x) \frac{\partial}{\partial x_j} + c(x). \tag{1.2}$$

Note that, by definition, any potential is included in the operator $\mathcal{A}$.

The question we wish to address is that of boundary controllability of the associated system,

$$\begin{aligned} i\tfrac{\partial u}{\partial t} + \mathcal{A}u &= 0, \qquad x \in \Omega, \quad t > 0, \\ u(x,0) &= g(x) \\ u &= h(x,t) \qquad x \in \partial\Omega, t > 0, \end{aligned} \tag{1.3}$$

where $\Omega \subset \mathbb{R}^n$ is a bounded domain. Given initial data in some appropriate space (e.g. $L^2(\mathbb{R}^n)$), can we find a function $h(x,t)$ such that for some $T > 0$, $u(x,t) = 0$ for all $t > T$?

Although various systems of equations have been considered and a variety of techniques developed in the context of boundary control, little has been done specifically for the Schrödinger equation. Hence, the following references are of particular interest. In [2], Lasiecka and Triggiani use the method of multipliers to obtain boundary controllability for the Schrödinger equation with constant coefficients. In [3], Lebeau modifies the results of Bardos, Lebeau and Rauch (see [1]) on exact controllability for hyperbolic equations to apply their techniques to Schrödinger type equations. However, in the statement of his results, Lebeau assumes constant coefficients (i.e., $\mathcal{A}$ replaced by Δ) and an analytic boundary. Although his work appears to extend to the case of nonconstant coefficients, our goal is to avoid the technical difficulties of applying his techniques to (1.3).

To instead take advantage of Littman and Taylor's technique, we prove the following regularity result.

THEOREM 1.2 *Assume the coefficients of $\mathcal{A}$ are real and satisfy the following conditions:*

$$\begin{array}{lll} i.) & a_{ij}(x) = a_{ji}(x) & \text{for all } i,j, \\ ii.) & \sum_{i,j=1}^{n} a_{ij}(x)\xi_i\xi_j \geq \alpha|\xi|^2 & \text{for some } \alpha > 0, \\ iii.) & a_{ij}(x) - \delta_{ij}, c(x) \in C_0^\infty(\mathbb{R}^n), & \end{array} \tag{1.4}$$

where δ_{ij} is the Kronecker delta. Without loss of generality, we assume that the support of these functions is contained in the ball $B_a \equiv \{x : |x| < a\}$.

Assume $\phi(x) \in L^2(\mathbb{R}^n)$ and has compact support. Then the solution $u(x,t)$ to the initial value problem (1.1) is of class Gevrey-2 with respect to t uniformly in compact sets of $\mathbb{R}^n \times \{t > 0\}$.

For the convenience of the reader, in the proof of this theorem, we reproduce some of the arguments and computations from [4]. Extensions to nonsmooth coefficients will appear elsewhere.

It should be noted that using Littman and Taylor's technique to prove exact controllability results in the control function belonging to $C^\infty(\partial\Omega)$ for $t > 0$. As far as Sobolev spaces are concerned, the results are the same as in [4], i.e., for initial data in $H^1(\partial\Omega)$, the Dirichlet controls will also be in $L^2(\Omega)$, indeed, somewhat better.

2 Gevrey Regularity of the Solution: Proof of Theorem 1.2

Step 1: *Laplace Transform.* Assuming the coefficients satisfy (1.4), the operator $\mathcal{A}$ is strongly elliptic and self-adjoint. Additionally, we can apply semigroup theory (see [5]) to find that *$i\mathcal{A}$ is the infinitesimal generator of a strongly continuous group of unitary operators.* Taking the Laplace transform of (1.1) with respect to the time variable, t, we find

$$i\lambda w + \mathcal{A}w = i\phi \tag{2.1}$$

$$\text{or} \quad k^2 w + \mathcal{A}w = i\phi, \tag{2.2}$$

where $\lambda = -ik^2$. Thus, the solution operator mapping $\phi \to w$ is the resolvent operator corresponding to $i\mathcal{A}$, where the resolvent is defined to be

$$R(\lambda; i\mathcal{A}) \equiv (\lambda - i\mathcal{A})^{-1}. \tag{2.3}$$

Since $i\mathcal{A}$ generates a group of unitary operators, (2.1) is solvable for any $\phi \in L^2(\mathbb{R}^n)$ for all λ such that $\Re e\lambda > 0$ and the resulting solution, w, lies in $H^2(\mathbb{R}^n)$.

Step 2: *Estimates for the Modified Resolvent.* To prove the desired estimates for the solution which are needed to show that it is of class Gevrey-2 with respect to time, we must extend the region where (2.1) is solvable, i.e., the region where the resolvent exists and is bounded. However, this cannot be done directly. Instead, we consider a "modified" resolvent operator,

$$R_\chi(\lambda; i\mathcal{A}) \equiv \chi(x) R(\lambda; i\mathcal{A}) \chi(x), \tag{2.4}$$

where $\chi(x) \in C_0^\infty(\mathbb{R}^n)$ and $\chi(x) \equiv 1$ on suppϕ.

Recalling the change of variable, $\lambda = -ik^2$,

$$\begin{aligned} \lambda - i\mathcal{A} &= -i(k^2 + \mathcal{A}) \\ &\Longrightarrow R(\lambda; i\mathcal{A}) = i(k^2 + \mathcal{A})^{-1}. \end{aligned} \tag{2.5}$$

Hence, we will consider the operator,

$$\tilde{R}(k^2; \mathcal{A}) \equiv \chi(x)(k^2 + \mathcal{A})^{-1}\chi(x). \tag{2.6}$$

Notice that the interior term differs from the standard definition of a resolvent operator by only a change of sign. For this operator, Vainberg has shown ([6], Theorem 3) that if

$$U_{\alpha,\beta} \equiv \{k : \alpha|k|^{-1}\exp\{\beta|\Im mk|\} \leq \frac{1}{2}\}, \tag{2.7}$$

then $\tilde{R}(k^2; \mathcal{A})$ has no poles in $U_{\alpha,\beta}$ and

$$\|\tilde{R}(k^2; \mathcal{A})\| \leq c_1|k|^{-1}\exp\{c_2|\Im mk|\}, \qquad k \in U_{\alpha,\beta}. \tag{2.8}$$

In particular, this implies

$$\|\tilde{R}(k^2; \mathcal{A})\| \leq C, \qquad k \in U_{\alpha,\beta}. \tag{2.9}$$

Step 3: *Transformation from k to λ.* Because the results of Step 2 are in terms of the variable k, we need to convert the information derived from Vainberg's work into estimates in terms of λ to obtain corresponding results for $R_\chi(\lambda; i\mathcal{A})$. We begin by transforming the region in which estimate (2.9) is valid. Let

$$S_{a,b,R} \equiv \{k : |k| > R, \Im mk \geq -a\ln|k| + b\}. \tag{2.10}$$

By appropriately choosing a, b, and R, it is a straightforward calculation to show that $S_{a,b,R} \subset U_{\alpha,\beta}$. Restricting the region further, if

$$\mathcal{T}_{a',b'} \equiv \{k : \Im mk \geq -b', |k| \geq a'\}, \tag{2.11}$$

we can choose a' and b' so that $\mathcal{T}_{a',b'} \subset S_{a,b,R}$.

Since the transformation $\lambda = -ik^2$ is two to one on parts of the set $\mathcal{T}_{a',b'}$, we consider the restriction of this set,

$$\chi_{a,b} \equiv \mathcal{T}_{a,b} \cap \{k : -\frac{\pi}{4} < \arg k < \frac{3\pi}{4}\}. \tag{2.12}$$

We now apply the following lemma.

LEMMA 2.1 *([4], Lemma 3) For each $a > 0$ there exists a constant $\rho > 0$ such that the set*

$$\Sigma_{b,\rho} \equiv \{\lambda : \Re e\lambda \geq -b|\Im m\lambda|^{1/2} + \rho\}$$

is contained in $\chi_{a,b}$.

Since the estimate (2.9) is uniform in k, it is also valid for all $\lambda \in \Sigma_{b,\rho}$. Therefore,

$$\|R_{\chi}(\lambda; i\mathcal{A})\| \leq C \qquad \forall \lambda \in \Sigma_{b,\rho}. \tag{2.13}$$

Step 4: *Estimates for the Solution.* Define

$$J(t) \equiv \frac{1}{2\pi i} \int_{\mu - i\infty}^{\mu + i\infty} \frac{\hat{R}(\lambda; i\mathcal{A})}{\lambda^2} e^{\lambda t} d\lambda, \tag{2.14}$$

where μ is chosen sufficiently large to guarantee that the half-plane $\Re e\lambda \geq \mu$ is contained in the domain of analyticity of J.

THEOREM 2.1 *Let K be any compact subset of $\mathbb{R}^n$. Then the mapping $t \to J(t)$ is of class Gevrey-2 with respect to time.*

Proof: *Step 1:* To derive estimates for $J(t)$, we begin by proving that the path of integration can be shifted to the contour Γ, where $\Gamma \equiv \partial\Sigma_{b,\rho}$ with orientation in the direction of increasing $\Im m\lambda$.
Consider the contour $\Gamma_R \equiv \{\lambda = s + iR : a - b|R|^{1/2} \leq s \leq \mu\}$ with orientation in the direction of increasing $\Re e\lambda$. Then

$$\begin{aligned} |\textstyle\int_{\Gamma_R} \frac{\hat{R}(\lambda;i\mathcal{A})}{\lambda^2} e^{\lambda t} d\lambda| &\leq \frac{1}{|R|^2} \int_{a-b|R|^{1/2}}^{\mu} C e^{ts} ds \\ &\leq \frac{C}{|R|^2} \frac{e^{\mu t}}{t} \longrightarrow 0 \quad \text{as} \quad |R| \to \infty \;\; \forall t > 0. \end{aligned} \tag{2.15}$$

Thus, the path of integration can be shifted to the contour Γ.

Step 2: Estimates for Derivatives. As in [4], we parametrize the upper branch of the contour (i.e., $\Im m\lambda \geq 0$) in the following way:

$$\lambda = -\mu + i(\frac{\rho + \mu}{b})^2. \tag{2.16}$$

Then, since the integral is infinitely differentiable, derivatives of $J(t)$ may be bounded in the following way, using (2.9) and the technique of the proof of Theorem 4 in [4]:

$$\begin{aligned} |\frac{\partial^n}{\partial t^n} \textstyle\int_{\Gamma} \frac{\hat{R}(\lambda;i\mathcal{A})}{\lambda^2} e^{\lambda t} d\lambda| &= |\textstyle\int_{\Gamma} \lambda^{n-2} \hat{R}(\lambda; i\mathcal{A}) e^{\lambda t} d\lambda| \\ &\leq C \textstyle\int_0^{\infty} (\mu^2 + (\frac{\rho+\mu}{b})^4)^{(n-2)/2} (1 + 4(\frac{\rho+\mu}{b^2})^2)^{1/2} e^{-\mu t} d\mu \\ &\leq C 2^{(n+1)/2} \textstyle\int_0^{\infty} (\frac{\rho+\mu}{b})^{2n-3} e^{-\mu t} d\mu \qquad \text{since} \;\; \rho > b^2 \\ &< C e^{\rho t} 2^{(n+1)/2} \textstyle\int_0^{\infty} (\frac{\mu}{b})^{2n-3} e^{-\mu t} d\mu \\ &= 2Cbe^{\rho t} (2n-3)! (\frac{2}{b^4 t^4})^{(n-1)/2}. \end{aligned} \tag{2.17}$$

If t is in any compact subset of $\mathbb{R}^+$ and $\theta > 0$, we may choose b so that this expression is bounded by $c\theta^n (n!)^2$, where c is constant. □

THEOREM 2.2 *Assume the coefficients of $\mathcal{A}$ satisfy (1.4). Let K be any compact subset of suppϕ. If $\phi(x) \subset L^2(\mathbb{R}^n)$ has compact support, then the mapping $t \to u(\cdot, t)$ is of class Gevrey-2 with respect to time.*

Proof: By a standard semigroup result (see [5], Corollary 1.7.6),

$$\int_0^t (t-s)S(s)\phi ds = \frac{1}{2\pi i}\int_{\beta-i\infty}^{\beta+i\infty} \frac{R(\lambda; i\mathcal{A})}{\lambda^2}\phi e^{\lambda t} d\lambda, \tag{2.18}$$

where $R(\lambda; i\mathcal{A})$ is the resolvent of the infinitesimal generator of our group $S(t)$, and β is assumed to be sufficiently large. However, since $\chi(x) \equiv 1$ on suppϕ, the above equation implies

$$\chi \int_0^t (t-s)S(s)\chi\phi ds = J(t)\phi. \tag{2.19}$$

Differentiating twice with respect to t then gives

$$\chi(x)u(x,t) = \frac{\partial^2}{\partial t^2} J(t)\phi. \tag{2.20}$$

Hence, since $J(t)$ is of class Gevrey-2 and differentiation with respect to t does not affect Gevrey regularity, $\chi(x)u(x,t)$ is also of class Gevrey-2 with respect to time. □

References

[1] C. Bardos, G. Lebeau, and J. Rauch. Contrôle et stabilisation dans les problèmes hyperboliques. In *Contrôlabilité exacte, perturbations et stabilization de systèmes distribués*, volume 2, Paris, 1989. Masson. Appendix 2.

[2] I. Lasiecka and R. Triggiani. Optimal regularity, exact controllability and uniform stabilization of Schrödinger equations with Dirichlet control. *Differential and Integral Equations*, 5:521–535, 1992.

[3] G. Lebeau. Contrôle de l'equation de Schrödinger. *Journal Math. Pures Appl.*, 71:267–291, 1992.

[4] W. Littman and S. Taylor. Smoothing evolution equations and boundary control theory. *Journal D'Analyse Mathématique*, 59:117–131, 1992.

[5] A. Pazy. *Semigroups of Linear Operators and Applications to Partial Differential Equations.* Springer-Verlag, New York, 1983.

[6] B. R. Vainberg. On a point source in an inhomogeneous medium. *Math. USSR Sbornik*, 23(1):123–148, 1974.

Augmented Lagrangian Formulation of Nonsmooth Convex Optimization in Hilbert Spaces

Kazufumi Ito Department of Mathematics, North Carolina State University, Raleigh, North Carolina

Karl Kunisch Fachbereich Mathematik, Technische Universitat, Berlin, Germany

Abstract. In this paper we discuss the augmented Lagrangian formulation of a class of nonsmooth, convex optimization problems in Hilbert spaces. Based on the augmented Lagrangian formulation, the existence of a Lagrange multiplier and a regularized optimality condition are obtained. The first order augmented Lagrangian method and Uzawa's algorithm are presented and the convergence of the both methods is established and compared. Moreover a Newton-like update using the active set strategy based on the regularized complementarity condition is described.

1 Introduction

In this paper we consider the class of nonsmooth, convex optimization problems in Hilbert spaces X, H of form

$$\text{minimize} \quad f(x) + \varphi(\Lambda x) \quad \text{over } x \in K, \tag{1.1}$$

where K is a closed, convex subset in X, $f : K \to R$ and $\varphi : H \to R$ are convex functions, and $\Lambda : X \to H$ is a bounded affine operator. We assume that f is a continuously differentiable, coercive function and φ is a proper, lower semicontinuous, convex function. Thus, the functional φ represents the nonsmooth component of the cost-functional. Our formulation here is slightly more general than that in [Gl],[ETe], since we allow the additional constraint $x \in K$. For example, if we choose the closed convex subset K as $K = \{x \in X : |x - x_0| \leq \delta > 0\}$, $x_0 \in X$ then (1.1) is minimized locally. For optimal control problems (e.g.,[IK1]) K is a linear submanifold of X defined by linear equality constraints. Problem (1.1) encompasses a wide variety of optimization problems [IK1], including variational inequalities of the first and second kind [Gl]. Specifically, this formulation includes the variational problem for the image restoration problem [ROF] (or Bingham flow problem [GLT]), i.e.,

$$\min \quad \int_\Omega \left(\frac{\mu}{2} |\nabla u|^2 + g\, |\nabla u|\right) dx + \frac{1}{2} \int_\Omega |u - z|^2 \, dx \quad \text{over } u \in H_0^1(\Omega), \tag{1.2}$$

for restoring the original image u from a noisy image $z \in L^2(\Omega)$. Here $\mu > 0$ and $g > 0$ are fixed but can be adjusted to the statistics of noise. If $\mu = 0$ then this problem is equivalent to the image enhancement algorithm in [ROF] based on minimization of the BV semi-norm $\int_\Omega |\nabla u| \, dx$ subject to the standard deviation constraint $\int_\Omega |u - z|^2 \, dx = \sigma^2$. In fact $1/g$ is the Lagrange multiplier associated with the equality constraint $\int_\Omega |u - z|^2 \, dx = \sigma^2$. Here, we choose

$$X = H_0^1(\Omega), \; H = L^2(\Omega) \times L^2(\Omega) \quad \text{and} \quad \Lambda = g \, \text{grad},$$

and define f and φ by

$$f(u) = \frac{1}{2} \int_\Omega (\mu\, |\nabla u|^2 + |u - z|^2) \, dx,$$

and

$$\varphi(v_1, v_2) = \int_\Omega \sqrt{v_1^2 + v_2^2} \, dx.$$

Also, the problem (1.1) includes the inequality type constraint minimization. In this case $\varphi = \chi_C$, where C is a closed convex set in H and χ_S is the characteristic function of a set S:

$$\chi_S(x) = \begin{cases} 0 & \text{if } x \in S \\ \infty & \text{if } x \notin S \end{cases}.$$

For example, for the obstacle problem

$$\text{(1.3)} \qquad \begin{aligned} &\min \int_\Omega \left(\frac{1}{2}|\nabla u|^2 - \tilde{f}\, u\right) dx \quad \text{over } u \in H_0^1(\Omega) \\ &\text{subject to} \quad \phi \le u \le \psi \ \text{ a.e. in } \Omega. \end{aligned}$$

We choose

$$X = H_0^1(\Omega), \ H = L^2(\Omega) \quad \text{and} \quad \Lambda = \text{the natural injection},$$

and define f and φ by

$$f(u) = \int_\Omega \left(\frac{1}{2}|\nabla u|^2 - \tilde{f}\, u\right) dx \quad \text{and} \quad \varphi(v) = \chi_C,$$

where C is a closed convex set defined by $C = \{v \in H : \phi \le v \le \psi \text{ a.e. in } \Omega\}$.

In Section 2 we shall establish existence and a necessary optimality condition for the minimization problem (1.1). The main result of that section will be concerned with the existence of a Lagrange multiplier $\lambda^* \in H$ such that

$$\text{(1.4)} \qquad \begin{aligned} &\langle f'(x^*),\, x - x^* \rangle + (\lambda^*,\, \Lambda(x - x^*))_H \ge 0 \text{ for all } x \in K \\ &\lambda^* = \varphi_c'(\Lambda x^*, \lambda^*), \end{aligned}$$

where the second condition is the regularized version of the complementary condition $\lambda^* \in \partial\varphi(\Lambda x^*)$ and φ_c is the C^1-approximation of φ in terms of the augmented Lagrangian formulation (see (1.9) below). Also, the usage of the term "Lagrange multiplier" is examined in some detail.

We apply a regularization technique based on an augmented Lagrangian formulation in which the non-smooth problem (1.1) is approximated by a sequence of smooth problems. For the finite dimensional case we refer to [Be]. To briefly explain the method, we note that (1.1) is equivalent to

$$\text{(1.5)} \qquad \begin{aligned} &\text{minimize} \quad f(x) + \varphi(\Lambda x - u) \\ &\text{subject to } x \in K \quad \text{and} \quad u = 0 \text{ in } H. \end{aligned}$$

The equality constraint $u = 0$ in (1.5) is the treated by the augmented Lagrangian method. That is, we consider a sequential minimization over $x \in K$ and $u \in H$ of the form

$$\text{(1.6)} \qquad \text{minimize} \quad f(x) + \varphi(\Lambda x - u) + (\lambda,\, u)_H + \frac{c}{2}|u|_H^2$$

where $\lambda \in H$ is a multiplier and c is a positive scalar penalty parameter [Be]. Equivalently, problem (1.6) can be expressed as

$$\text{(1.7)} \qquad \text{minimize} \quad L_c(x, \lambda) = f(x) + \varphi_c(\Lambda x, \lambda) \quad \text{over } x \in K$$

where $\varphi_c(v, \lambda)$ is defined by

$$\text{(1.8)} \qquad \varphi_c(v, \lambda) = \inf_{u \in H} \{\varphi(v - u) + (\lambda, u)_H + \frac{c}{2}|u|_H^2\}$$

for $(c, \lambda) \in R^+ \times H$. Here, $\varphi_c(v, \lambda)$ is C^1 approximation of φ in the following sense. Since φ is a proper, lower semicontinuous, convex function the subdifferential $A = \partial\varphi$ of φ, defined by

$$\partial\varphi(v) = \{v^* \in H : \varphi(u) - \varphi(v) \geq (v^*,\, y - v) \quad \text{for all } y \in H\},$$

is maximal monotone and for $\mu > 0$, the resolvent $J_\mu = (I + \mu\, A)^{-1}$ of A is a contraction defined on all of H. The Yosida approximation A_μ of A is defined by

$$A_\mu v = \mu^{-1}\,(v - J_\mu v).$$

Then $A_\mu v \in AJ_\mu v$ and A_μ is single-valued, monotone, everywhere defined, and Lipschitz continuous (with Lipschitz constant $1/\mu$) [Br],[Pa]. We have the following lemma [IK1].

Lemma 1.1 *For $v,\ \lambda \in H$ the infimum in (1.8) is attained at a unique point $u_c(v, \lambda)$ where $u_c(v, \lambda) = v - J_{1/c}\,(v + c^{-1}\lambda)$. Moreover $\varphi_c(v, \lambda)$ is convex, (Lipschitz) continuously Fréchet differentiable in v and*

$$\varphi_c'(v, \lambda) = \lambda + c\, u_c(v, \lambda) = A_{1/c}\,(v + c^{-1}\lambda).$$

If $\lambda = 0$, then $\varphi_c(v, 0)$ is the proximal approximation of φ. Moreover, if $\lambda \in Av$ then $u_c(v, \lambda) = 0$ and $\varphi_c'(v, \lambda) = \lambda$. The lemma follows from results in [Br],[Pa], observing that (1.8) is equivalent to

$$(1.8)' \qquad \varphi_c(v, \lambda) = \inf_{\tilde{u} \in H}\, \{\varphi(\tilde{u}) + \frac{c}{2}\,|v + c^{-1}\lambda - \tilde{u}|^2\} - \frac{1}{2c}|\lambda|^2$$

where $\tilde{u} = v - u$.

The following dual representation of $\varphi_c(v, \lambda)$ plays an important role in our discussions. Assume that $H^* = H$ and define the convex conjugate function φ^* of φ by

$$\varphi^*(v^*) = \sup_{v \in H}\, \{(v,\, v^*) - \varphi(v)\} \quad \text{for } v^* \in H.$$

Then, we have the following result [IK1].

Lemma 1.2 *For every $x,\ \lambda \in H$*

$$(1.9) \qquad \varphi_c(v, \lambda) = \sup_{y^* \in H}\, \{(x,\, y^*) - \varphi^*(y^*) - \frac{1}{2c}\,|y^* - \lambda|^2\},$$

where the supremum is attained at a unique point $\lambda_c(v, \lambda)$ and we have

$$(1.10) \qquad \lambda_c(v, \lambda) = \lambda + c\, u_c(v, \lambda) = \varphi_c'(v, \lambda) = A_{1/c}\,(v + c^{-1}\lambda).$$

The lemma can be proved using the duality theory in [ETe],[ETu] (see, [IK1]).

In many of the applications including our two examples, the conjugate function φ^* is given by

$$(1.11) \qquad \varphi^*(v) = \chi_{C^*}(v),$$

where C^* is a closed convex set in H. Then, it follows from Lemma 1.2 that for $v,\ \lambda \in H$

$$\varphi_c(v, \lambda) = \sup_{y^* \in C^*}\, \{-\frac{1}{2c}|y^* - (\lambda + c\, v)|_H^2\} + \frac{1}{2c}\,(|\lambda + c\, v|_H^2 - |\lambda|_H^2).$$

Hence, the supremum is attained at

$$(1.12) \qquad \lambda_c(v, \lambda) = Proj_{C^*}(\lambda + c\, v),$$

where $Proj_{C^*}(\phi)$ denotes the projection of $\phi \in H$ onto C^*. For example, if $\varphi(v) = \chi_C(v)$ where C is a closed convex cone in H, then $\varphi^* = \chi_{C^*}$, where $C^* = \{w \in H : (w, v)_H \leq 0 \text{ for all } v \in C\}$ is the dual cone of C. If $\varphi(v) = |v|_H$, then $\varphi^* = \chi_B$, where B is the closed unit ball in H.

The (first-order) augmented Lagrangian method is given as follows.

Algorithm 1

Step 1: Choose a starting value $\lambda_1 \in H$, a positive number c and set $k = 1$.

Step 2: Given $\lambda_k \in H$ find $x_k \in K$ by

$$L_c(x_k, \lambda_k) = \min L_c(x, \lambda_k) \quad \text{over } x \in K.$$

Step 3: Update λ_k by $\lambda_{k+1} = \varphi_c'(\Lambda x_k, \lambda_k)$.

Step 4: If the convergence criterion is not satisfied then set $k = k+1$ and go to Step 2.

It can be proved (see Section 3) that Algorithm 1 converges unconditionally (with respect to $c > 0$) if and only if there exists a Lagrange multiplier $\lambda^* \in \partial\varphi_c(\Lambda x^*)$ such that (1.4) holds, provided that

$$\langle f'(x_1) - f'(x_2), x_1 - x_2 \rangle \geq \sigma\, |x_1 - x_2|_X^2, \tag{1.13}$$

for all x_1, $x_2 \in K$ and some $\sigma > 0$. That is, the non-smooth minimization problem (1.1) is replaced by sequential minimization of the smooth problems (1.7)-(1.8).

Remark 1.3 (1) The update of the Lagrange multiplier in Step 3 is explained as follows. We define the dual functional $d_c : H \to R$ by

$$d_c(\lambda) = L_c(x(\lambda), x(\lambda)) = f(x(\lambda)) + \varphi_c(\Lambda x(\lambda), \lambda) \tag{1.14}$$

for $\lambda \in H$ and $c > 0$, where $x(\lambda) \in K$ is the unique solution to (1.7)-(1.8). It is shown in [IK1, Lemma 4.1] that d_c is continuously Fréchet differentiable with F-derivative

$$d_c'(\lambda) = u(\Lambda x(\lambda), \lambda),$$

where $u_c(x, \lambda)$ is defined in Lemma 1.1. Thus, a steepest ascent step for the dual functional $d_c(\lambda)$ is given by $\lambda_{k+1} = \lambda_k + c\, u(\Lambda x(\lambda_k), \lambda_k)$, which in view of Lemma 1.2 equals $\varphi_c'(x(\lambda_k), \lambda_k)$.

(2) It is shown in [IK1], Section 2 that if a pair $(x^*, \lambda^*) \in K \times H$ satisfies the necessary optimality condition (1.4) for some $c > 0$, then x^* is the minimizer of (1.1) and $\lambda^* \in \partial\varphi(\Lambda x^*)$. Moreover, (1.4) is equivalent to $\lambda^* \in H$ is a fixed point of $\Phi(\lambda) = \varphi_c'(\Lambda \hat{x}(\lambda), \lambda)$ where $\hat{x}(\lambda) \in K$ is the unique solution to

$$\langle f'(\hat{x}(\lambda)), x - \hat{x}(\lambda) \rangle + (\lambda, \Lambda(x - \hat{x}(\lambda))) \geq 0 \quad \text{for all } x \in K.$$

Thus, the update in Step 3 is a fixed point iterate for Φ in the dual variable λ.

(3) We consider the explicit fixed point iterate $\lambda_k = \Phi(\lambda_k)$, i.e.,

$$\lambda_{k+1} = \varphi_c'(\Lambda \hat{x}_k, \lambda_k) \tag{1.15a}$$

where $\hat{x}_k \in K$ is explicitly defined by

$$\langle f'(\hat{x}_k), x - \hat{x}_k \rangle + (\lambda_k, \Lambda(x - \hat{x}_k)) \geq 0 \quad \text{for all } x \in K. \tag{1.15b}$$

If φ satisfies (1.11) then (1.12) implies that (1.15a) can be expressed as $\lambda_{k+1} = Proj_{C^*}(\lambda_k + c\,\Lambda\hat{x}_k)$ and the algorithm (1.15) reduces to the so-called Uzawa's algorithm for the min-max problem

$$\inf_{x \in K} \sup_{u^* \in C^*} f(x) + (\lambda, \Lambda x)_H. \tag{1.16}$$

It will be shown in Section 3 that if there exists a Lagrange multiplier $\lambda \in \partial\varphi(\Lambda x^*)$ satisfying (1.4) and if (1.13) holds, then there exist $0 < \underline{c} < \bar{c} < \infty$ such that for every $c \in [\underline{c}, \bar{c}]$ we have $\hat{x}_k \to x^*$, strongly in X and $|\lambda_k - \lambda^*| \leq |\lambda_0 - \lambda^*|$, where $\{(\hat{x}_k, \lambda_k)\}_{k \geq 0}$ is the sequence generated by Uzawa's algorithm (1.15). In contrast to Uzawa's algorithm, Algorithm 1 is an implicit fixed-point iterate for Φ, i.e., Algorithm 1 is written as

$$\lambda_{k+1} = \varphi_c'(x_k, \lambda_k), \tag{1.17a}$$

where $x_k \in K$ is implicitly defined by

$$\langle f'(x_k), x - x_k \rangle + (\lambda_{k+1}, \Lambda(x - x_k)) \geq 0 \quad \text{for all } x \in K. \tag{1.17b}$$

The above-mentioned unconditional convergence of Algorithm 1 is due to the implicitness of that algorithm.

Both Algorithm 1 and Uzawa's algorithms are of first order. In order to develop a Newton-like (formally, second order) iterate for nonsmooth the minimization problem (1.1) one possibility is to employ the so-called "active set strategy" (e.g., [IK2],[Be]). We describe the method in terms of the image restoration problem (1.2). Let $\mathcal{C}^*$ be the closed convex set in $H = L^2(\Omega) \times L^2(\Omega)$ defined by

$$\mathcal{C}^* = \{\lambda \in L^2(\Omega) \times L^2(\Omega) : |\lambda(x)| \leq 1 \text{ a.e. } x \in \Omega\}. \tag{1.18}$$

It follows from (1.12) that (1.4) is written as

$$\begin{aligned} &-\mu\,\Delta u + u - g\,\text{div}\,\lambda = z \\ &\lambda(x) = \frac{(\lambda + c\,\nabla u)(x)}{\max\,(1, |\lambda + c\,\nabla u|(x))}, \quad \text{a.e. in } \Omega. \end{aligned} \tag{1.19}$$

We note that

$$\sup_{\lambda \in \mathcal{C}^*} \int_\Omega (\nabla u(x), \lambda(x))\, dx = \int_\Omega |\nabla u(x)|\, dx$$

and that (1.19) is the saddle point condition of (1.16). Moreover, we have

$$\lambda(x) = \frac{\nabla u(x)}{|\nabla u(x)|}, \quad x \in A^c$$

where A is the active set defined by

$$A = \{x \in \Omega : |\nabla u(x)| = 0\}.$$

This terminology is motivated by inequality constraints of the type $u \leq 0$, in which case the complementarity condition $(u, \lambda) = 0$ and $\lambda \geq 0$ imply that on the active set $\{x \in \Omega : u(x) = 0\}$ the dual variable $\lambda \geq 0$ is unknown and $\lambda = 0$ otherwise. Here, this active set corresponds to A.

The active set strategy that we propose involves the update of the active set according to the complementarity condition

$$\lambda = \frac{\lambda + c\,\nabla u}{\max\,(1, |\lambda + c\,\nabla u|)},$$

i.e. we set

$$A^k = \{x \in \Omega : |\lambda^k(x) + c\,\nabla u^k(x)| \leq 1\} \tag{1.20}$$

where (u^k, λ^k) is the solution update. The Newton -like update method of (u^k, λ^k) based on the active set strategy is given as follows.

Algorithm 2

Step 1: Choose $c > 0$, u^0 and λ^0 satisfying $|\lambda^0(x)| \le 1$ and set $k = 0$.

Step 2: Set the index sets $A^k = \{x \in \Omega : |\lambda^k(x) + c\,\nabla u^k(x)| \le 1\}$ and $I^k = \{x \in \Omega : |\lambda^k(x) + c\,\nabla u^k(x)| > 1\}$. Let (u, λ) be a solution to

$$-\mu\,\Delta u + u - g\,\mathrm{div}\,\lambda = z, \qquad \begin{cases} \lambda(x) = \dfrac{(\nabla u)(x)}{|(\nabla u(x))|} & \text{for } x \in I^k, \\ (\nabla u)(x) = 0 & \text{for } x \in A^k. \end{cases} \tag{1.21}$$

Step 3: Set $u^{k+1} = u$, $\lambda^{k+1} = \lambda$ and $k = k + 1$. Go to Step 2.

It is not easy to analyze the convergence property of Algorithm 2. However, the finite-step-convergence of the problem discretized by a finite difference scheme and numerical implementation of Algorithm 2 are discussed in [IK2]. In [IK1] the obstacle problem (1.3) with the one-sided inequality constraint $u \le \psi$ a.e. in Ω is studied and the finite-step-convergence is established for the problem approximated by the central difference method.

Let us briefly outline the contents of the rest of this paper. In Section 2 we establish the existence of a Lagrange multiplier and the regularized necessary optimality (1.4) for (1.1). In Section 3 the convergence of Algorithm 1 and Uzawa's method are proved.

Regarding related work, we refer the reader to Bertsekas [Be] who gave a thorough analysis of augmented Lagrangian methods for nonsmooth problems in finite dimensional spaces. In [FG],[Gl] the augmented Lagrangian method was used extensively in the context of optimal control problems governed by partial differential equations and variational inequalities, for elimination of the equation from the set of explicit constraints and for splitting of the equation. In our own previous work we studied augmented Lagrangian methods with first and second order updates for smooth optimization problems.

2 Necessary Optimality Condition and Lagrange Multiplier

In this section we discuss the necessary optimality condition in terms of a Lagrange multiplier rule for the nonsmooth minimization (1.1). Assume that $H = H^*$ and let $\langle \cdot, \cdot \rangle$ denote the dual product of $X^* \times X$. Throughout this and the following section we assume that

$$f \ \text{ and } \ \varphi \ \text{ are bounded below by zero on } \ K \tag{A1}$$

$$\langle f'(x_1) - f'(x_2), x_1 - x_2 \rangle \ge \sigma |x_1 - x_2|_X^2, \quad \text{for some } \sigma > 0 \text{ and all } x_1, x_2 \in K, \text{ and} \tag{A2}$$

$$\varphi(\Lambda x_0) < \infty \quad \text{for some} \quad x_0 \in K. \tag{A3}$$

Since φ is proper there exists an element $v_0 \in D(A) = \{v \in H : \varphi(v) < \infty\}$ and

$$\varphi(\Lambda x) - \varphi(v_0) \ge (v_0^*, \Lambda x - v_0)_H \quad \text{for } v_0^* \in A v_0. \tag{2.1}$$

Hence, $\lim f(x) + \varphi(\Lambda x) \to \infty$ as $|x|_X \to \infty$ and it follows from [ETe] that there exists a unique minimizer $x^* \in K$.

Theorem 2.1 *The necessary and sufficient condition for $x^* \in K$ to be the minimizer of (1.1) is given by*

$$\langle f'(x^*), x - x^* \rangle + \varphi(\Lambda x) - \varphi(\Lambda x^*) \ge 0 \quad \textit{for all } x \in K. \tag{2.2}$$

Remark. Suppose that x^*

$$\partial(\varphi \circ \Lambda + \chi_K)(x^*) = \Lambda^* \partial\varphi(\Lambda x^*) + \partial\chi_K(x^*). \tag{2.3}$$

This holds, for example, if there exists $x \in \text{int}(K)$ such that φ is continuous and finite at Λx (see Propositions 12-13 in §3.2 of [ETu]). It then follows from (2.2) that $-f(x^*) \in \partial(\varphi \circ \Lambda + \chi_K)(x^*)$ and by (2.3)

$$-f(x^*) \in \Lambda^* \partial\varphi(\Lambda x^*) + \partial\chi_K(x^*) = \Lambda^* \partial\varphi(\Lambda x^*) + N_K(x^*),$$

where $N_K(x^*) = \{z \in X^* : \langle z,\, x - x^*\rangle \le 0, \text{ for all } x \in K\}$. This implies that there exists a $\lambda^* \in \partial\varphi(\Lambda x^*)$ such that

$$\langle f'(x^*),\, x - x^*\rangle + (\lambda^*, \Lambda(x - x^*))_H \ge 0 \quad \text{for all } x \in K. \tag{2.4}$$

Conversely, suppose $(x^*, \lambda^*) \in K \times \partial\varphi(\Lambda x^*)$ satisfies (2.4). Then, $\varphi(\Lambda x) - \varphi(\Lambda x^*) \ge (\lambda^*,\, \Lambda(x - x^*))$ for all $x \in K$. Thus, (2.4) implies (2.2). It then follows from Theorem 2.1 that x^* minimizes (1.1).

Next, we consider the regularized minimization problem (1.7), given $c > 0$ and $\lambda \in H$. It follows from Lemma 1.1 that $x \to \varphi_c(\Lambda x, \lambda)$ is convex, continuously differentiable and bounded below by $-\frac{1}{2c}|\lambda|_H^2$. Thus, for $\lambda \in H$ (1.7) has a unique solution $x_c \in K$ and $x_c \in K$ is the solution of (1.7) if and only if x_c satisfies

$$\langle f'(x_c),\, x - x_c\rangle + (\varphi_c'(\Lambda x_c, \lambda),\, \Lambda(x - x_c))_H \ge 0 \quad \text{for all } x \in K. \tag{2.5}$$

Thus, for a fixed $\lambda \in H$ it is very reasonable to conjecture that $(x_c, \lambda_c = \varphi_c'(\Lambda x_c, \lambda))$ converges to (x^*, λ^*) as $c \to \infty$, where $\lambda^* \in \partial\varphi(\Lambda x^*)$ is a Lagrange multiplier such that (2.4) holds, which in turn would guarantee the existence of the Lagrange multiplier. We have the following lemmas regarding this claim [IK1].

Lemma 2.2 *Suppose that Λ is compact. Then, there exists a $\lambda \in H$ such that $\lambda_c = \varphi'(v_c, \lambda)$ is uniformly bounded in $c \ge 1$ if and only if there exists $\lambda^* \in \partial\varphi(\Lambda x^*)$ such that (2.4) holds. Moreover, if one of the above conditions holds, then there exists a subsequence $(x_{\hat{c}}, \lambda_{\hat{c}}) \in X \times H$ such that $x_{\hat{c}} \to x^*$, strongly in X and $\lambda_{\hat{c}} \to \lambda^*$, weakly in H where $(x^*, \lambda^*) \in K \times H$ satisfies (2.4).*

Lemma 2.3 *Assume that dom$(\varphi) = H$. Then, x_c converges strongly to x^* in X as $c \to \infty$ and for each weak cluster point λ^* of $\{\lambda_c\}_{c \ge 1}$ in H, $\lambda^* \in \partial\varphi(\Lambda x^*)$ and (2.4) holds.*

For the image restoration problem (1.2) dom$(\varphi) = H$ and thus Lemma 2.3 implies the existence of Lagrange multiplier. For the two sided obstacle problem (1.3) it is shown in [IK1] that under an appropriate condition on the functions ϕ, ψ which determine the side constraint there exists $\lambda \in H$ such that the condition in Lemma 2.2 is satisfied and thus the existence of Lagrange multiplier is established. The following theorem [IK1] yields the necessary and sufficient optimality condition (1.4) which is equivalent to (2.4) and plays the central role in the development and analysis of our algorithms.

Theorem 2.4 *Assume that there exists a pair $(x^*, \lambda^*) \in K \times H$ that satisfies (2.4). Then the complementarity condition $\varphi(\Lambda x^*) + \varphi^*(\lambda^*) = (\lambda^*,\, \Lambda x^*)$ can be equivalently expressed as*

$$\lambda^* = \varphi_c'(\Lambda x^*, \lambda^*) \tag{2.6}$$

and x^* is the unique solution of

$$\min\ f(x) + \varphi_c(\Lambda x, \lambda^*) \quad \textit{over } x \in K \tag{2.7}$$

for every $c > 0$.

Define the Lagrangian $L : K \times H \to R$ by

$$L(x, \lambda) = f(x) + (\Lambda x, \lambda) - \varphi^*(\lambda).$$

Then, (1.1) is equivalent to the min-max problem

$$\min_{x\in K}\max_{\lambda\in H} L(x,\lambda)$$

and moreover we have the following result.

Theorem 2.5 *The elements $x^* \in K$ and $\lambda^* \in \partial\varphi(\Lambda x^*)$ satisfy (2.4) (equivalently, (1.4)) if and only if (x^*,λ^*) is a saddle point of L, i.e.,*

$$L(x^*,\lambda) \le L(x^*,\lambda^*) \le L(x,\lambda^*) \quad \textit{for all } x\in K,\ \lambda\in H.$$

Proof: The theorem follows from the convexity of f and the fact that $v^* \in \partial\varphi(v)$ if and only if $v \in \partial\varphi^*(v^*)$ [ETu],[ETe]. □

3 Augmented Lagrangian Methods

In this section we analyze the convergence of Algorithm 1 and of Uzawa's algorithm (1.15). The following theorem asserts convergence of Algorithm 1.

Theorem 3.1 *Assume that there exists $\lambda^* \in \partial\varphi(\Lambda x^*)$ such that (2.4) is satisfied. Then the sequence (x_k,λ_k) is well-defined and satisfies*

$$\frac{\sigma}{2}|x_k - x^*|_X^2 + \frac{1}{2c}|\lambda_{k+1}-\lambda^*|_H^2 \le \frac{1}{2c}|\lambda_k-\lambda^*|_H^2 \quad \textit{for all}\quad k=1,2,\ldots, \tag{3.1}$$

and

$$\sum_{k=1}^{\infty}\frac{\sigma}{2}|x_k-x^*|_X^2 \le \frac{1}{2c}|\lambda_1-\lambda^*|_H^2. \tag{3.2}$$

The last inequality implies that $|x_k - x^*|_X \to 0$ as $k\to\infty$.

Proof: From Theorem 2.4 it follows that $\lambda^* = \varphi_c'(v^*,\lambda^*)$ where $v^* = \Lambda x^*$. Next, we establish (3.1). From (2.5) and Step 3

$$\langle f'(x_k),\, x^*-x_k\rangle + (\lambda_{k+1},\, \Lambda(x^*-x_k)) \ge 0$$

and from (2.4)

$$\langle f'(x^*),\, x_k-x^*\rangle + (\lambda^*,\, \Lambda(x_k-x^*)) \ge 0$$

Adding these two inequalities, we obtain

$$\langle f'(x_k)-f'(x^*),\, x_k-x^*\rangle + (\lambda_{k+1}-\lambda^*,\, \Lambda(x_k-x^*)) \le 0. \tag{3.3}$$

From Lemmas 1.1 and 1.2

$$\lambda_{k+1}-\lambda^* = A_{1/c}\,(v_k + c^{-1}\lambda_k) - A_{1/c}\,(v^*+c^{-1}\lambda^*)$$

where $v_k = \Lambda x_k$ and $v^* = \Lambda x^*$. From definition of A_μ we have

$$(A_\mu v - A_\mu\hat v,\, v-\hat v) = \mu\,|A_\mu v - A_\mu \hat v|^2 + (A_\mu v - A_\mu\hat v,\, J_\mu v - J_\mu\hat v) \ge \mu\,|A_\mu v - A_\mu\hat v|^2$$

for $\mu > 0$ and $v,\ \hat v \in H$, since $A_\mu v \in AJ_\mu v$ and A is monotone. Thus,

$$\begin{aligned}(\lambda_{k+1}-\lambda^*,\, v_k - v^*)) &= (\lambda_{k+1}-\lambda^*,\, v_k + c^{-1}\lambda_k - (v^*+c^{-1}\lambda^*))\\ &-c^{-1}(\lambda_{k+1}-\lambda^*,\, \lambda_k-\lambda^*) \ge c^{-1}|\lambda_{k+1}-\lambda^*|^2 - c^{-1}(\lambda_{k+1}-\lambda^*,\, \lambda_k-\lambda^*)\\ &\ge \frac{1}{2c}|\lambda_{k+1}-\lambda^*|^2 - \frac{1}{2c}|\lambda_k-\lambda^*|^2\end{aligned}$$

Hence, (3.1) follows from assumption (1.13) and (3.3). Summing up (3.1) with respect to k we obtain (3.2). □

Next, we prove the convergence of Uzawa's algorithm (1.15).

Theorem 3.2 *If there exists a Lagrange multiplier $\lambda^* \in \partial\varphi(\Lambda x^*)$ satisfying (1.4), then Uzawa's algorithm converges in the following sense. There exist $0 < \underline{c} < \bar{c} < \infty$ such that for $c \in [\underline{c}, \bar{c}]$ $\hat{x}_k \to x^*$, strongly in X and $|\lambda_k - \lambda^*| \le |\lambda_0 - \lambda^*|$.*

Proof: As in the proof of Theorem 3.1, we have

$$\langle f'(\hat{x}_k) - f'(x^*),\, \hat{x}_k - x^*\rangle + (\lambda_k - \lambda^*,\, \Lambda(\hat{x}_k - x^*)) \le 0. \tag{3.4}$$

where $\hat{x}(\lambda_k) \in K$ satisfies (1.15b). It follows from Lemma 1.1 that

$$\varphi_c'(\Lambda\hat{x}_k, \lambda_k) = A_{1/c}(\Lambda\hat{x}_k + c^{-1}\lambda_k).$$

Since the Yosida approximation $A_{1/c}$ is Lipschitz continuous with Lipschitz constant c we have

$$\begin{aligned} |\lambda_{k+1} - \lambda^*|^2 &\le c^2\, |\Lambda(\hat{x}_k - x^*) + c^{-1}\,(\lambda_k - \lambda^*)|^2 \\ &= |\lambda_k - \lambda^*|^2 + 2c\,(\Lambda(\hat{x}_k - x^*), \lambda_k - \lambda^*) + c^2\,|\Lambda(\hat{x}_k - x^*)|^2 \end{aligned} \tag{3.5}$$

It thus follows from (1.13) and (3.4)-(3.5) that

$$|\lambda_{k+1} - \lambda^*|^2 \le |\lambda_k - \lambda^*|^2 - 2c\sigma\,|\hat{x}_k - x^*|^2 + c^2\|\Lambda\|^2\,|\hat{x}_k - x^*|^2.$$

Let us assume that $0 < \underline{c} < \bar{c}$ and α are chosen so that

$$2c\sigma - c^2\,\|\Lambda\|^2 \ge \alpha > 0 \quad \text{if } c \in [\underline{c}, \bar{c}].$$

Then,

$$\alpha\,|\hat{x}_k - x^*|^2 + |\lambda_{k+1} - \lambda^*|^2 \le |\lambda_k - \lambda^*|^2$$

which implies the theorem. □

Theorem 3.2 is proved in [ETe] in the case that φ is defined by (1.11) and C^* is bounded in H.

4 Numerical Example

We give a numerical example for denoising of an image based on Algorithm 2. The unperturbed image is taken from the MATLAB demonstration software and is shown in the upper left corner of Plot 1. In the upper right corner we see the same picture with uniformly distributed random numbers corresponding to 14 percent absolute noise added to the 128×128 pixels that create the image. Let us point out that inside and on the "o" in "God" the maximum and minimum values are .1172 and 0791, respectively, while the noise ranges in $[-.01, .01]$. The discretization of Algorithm 2 is carried out by a one-sided finite difference method and (1.21) is solved by a penalty technique for the constraint $\nabla u = 0$ on A^k. The algorithmic details are given in [IK2]. The parameters in Algorithm 2 are taken to be $c = 50, g = 10^{-4}, \mu = 10^{-12}$. The denoised image after six iterations of Algorithm 2 is given on the bottom left. This result is obtained without pre- or postprocessing of the data.

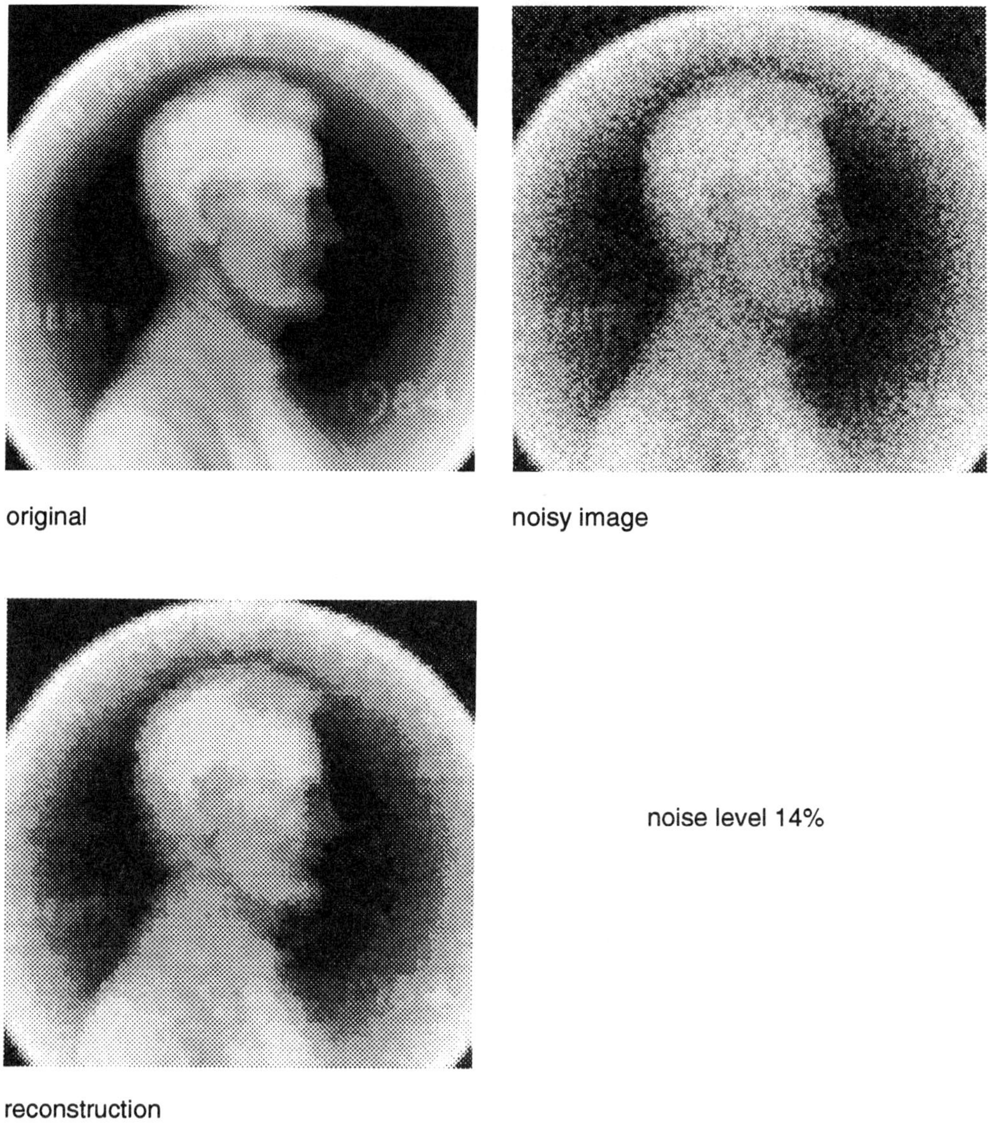

Plot 1

5 References

[Be] D.P.Bertsekas, *Constraint Optimization and Lagrange Multiplier Methods*, Academic Press, Paris (1982).

[Br1] H.Brezis, *Opérateurs Maximaux Monotones et Semi-groupes de Contractions dans les Espaces de Hilbert*, North-Holland, Amsterdam (1973).

[ETe] I.Ekeland and R.Temam, *Convex Analysis and Variational Problems*, North Holland, Amsterdam (1976).

[ETu] I.Ekeland and T.Turnbull, *Infinite Dimensional Optimization and Convexity*, The University of Chicago Press, Chicago (1983).

[FG] M.Fortin and R.Glowinski, *Augmented Lagrangian Methods: Applications to the Numerical Solution of Boundary-Value Problems*, North Holland, Amsterdam, (1983).

[GLT] R.Glowinski, J.L.Lions and R.Tremolieres, *Numerical Analysis of Variational Inequalities*, North Holland, Amsterdam (1981).

[IK1] K.Ito and K.Kunisch, Augmented Lagrangian methods for nonsmooth convex optimization in Hilbert spaces, submitted.

[IK2] K.Ito and K.Kunisch, An active set strategy for image restoration based on the augmented Lagrangian formulation, submitted.

[Pa] A.Pazy, Semi-groups of nonlinear contractions and their asymptotic behaviour, Heriot-Watt Symp. Vol III, R.J.Knops Ed. Pitman Research Notes in Math., 30 (1979), 36-134.

[ROF] L.I.Rudin, S.Osher and E.Fatemi, Nonlinear total variation based noise removal algorithms, Physica D 60 (1992), 259-268.

Analysis of the Control of Wave Generators in a Canal

G. Joly Université de Technologie de Compiègne, Compiègne, France

S. Mottelet Université de Technologie de Compiègne, Compiégne, France

J. P. Yvon INRIA Roquencourt, Le Chesnay, France, and Université de Technologie de Compiègne, Compiègne, France

1 STATEMENT OF THE PROBLEM

The physical system under consideration is a canal where the behaviour of floating bodies is experimental. In such a canal waves are created by mean of plane generators which can be controlled. The paper deals with the mathematical analysis of a two dimensional problem and, more precisely, of various controllability problems related to the control of such a system. For the numerical and experimental aspects of the control of such a system we refer to [Mottelet94], [Joly et al.94].

2 MODEL OF THE CANAL

We consider the rectangular canal on figure 1. This canal is supposed to be wide enough to consider that the waves created by generators located at each end are plane waves. This allows to use a bidimensional model, where the domain Ω is the rectangle $[0, L] \times [0, h]$, represented on figure 1. The frontier $\Gamma = \Gamma_s \cup \Gamma_f \cup \Gamma_1 \cup \Gamma_2$ is represented by

- the free surface $\Gamma_s = \{(x, h) \mid 0 < x < L\}$,
- the bottom of the canal $\Gamma_f = \{(x, 0) \mid 0 < x < L\}$,
- the left end $\Gamma_1 = \{(0, y) \mid 0 < y < h\}$,

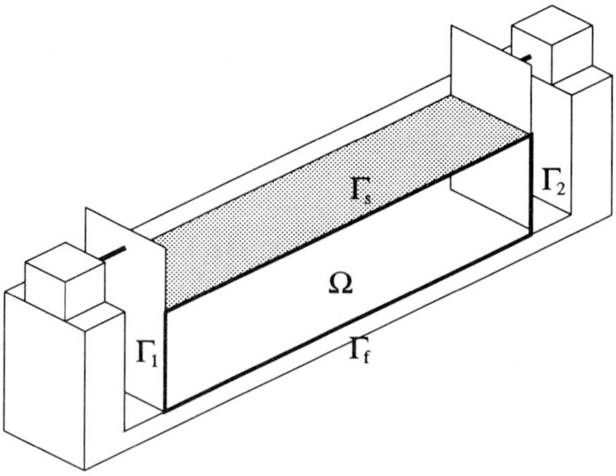

Figure 1: Definition of domain Ω

- the right end $\Gamma_2 = \{(L, y) \mid 0 < y < h\}$.

The fluid is supposed to be perfect, incompressible and irrotational. Let $\vec{V}(x, y, t)$ be the velocity field at time t. From the hypothesis rot $\vec{V} = 0$, there exists a potential ψ defined by

$$\vec{V}(x, y, t) = \vec{\nabla}\, \psi(x, y, t).$$

This together with the incompressibility leads to the classical shallow waters equations

$$\begin{aligned}
&(1) & \Delta\psi &= 0 \text{ in } \Omega \times [0, T],\\
&(2) & \ddot{\psi} + g_0 \frac{\partial \psi}{\partial y} &= 0 \text{ on } \Gamma_s \times [0, T],\\
&(3) & \frac{\partial \psi}{\partial n} &= 0 \text{ on } \Gamma_f \times [0, T],\\
&(4) & \frac{\partial \psi}{\partial n} &= f(y) v_1(t) \text{ on } \Gamma_1 \times [0, T],\\
&(5) & \frac{\partial \psi}{\partial n} &= f(y) v_2(t) \text{ on } \Gamma_2 \times [0, T].
\end{aligned}$$

The dynamic boundary condition (2) is the Bernoulli's condition, linearized under the hypothesis of small perturbations. The boundary conditions on Γ_1 et Γ_2 take into account the action of generators which can produce desired velocities. When the generators have a fixed shape described by the function $f(y)$, the controls $v_1(t)$ and $v_2(t)$ represent the angular velocities (assumed to be small) of these "rigid" generators. When $f(y) = y$ these boundary conditions represent plane generators being able to rotate around an axis located at the bottom of the canal. This is the case on figure 1.

One may consider an hypothetic situation where the velocity of the fluid on Γ_1 and Γ_2 can be arbitrary controlled. In this case we consider the boundary conditions

$$\begin{aligned}
\frac{\partial \psi}{\partial n} &= v_1(y, t), \text{ on } \Gamma_1,\\
\frac{\partial \psi}{\partial n} &= v_2(y, t), \text{ on } \Gamma_2.
\end{aligned}$$

In this case the controls $v_1(y,t)$ and $v_2(y,t)$ are equal to the desired velocity on Γ_1 and Γ_2. The devices to produce these velocities have to be "flexible" generators.

As we will see in the sequel, we only need to specify initial conditions on Γ_s

$$\psi(x,h,0) = \psi_0(x), \tag{6}$$

$$\dot{\psi}(x,h,0) = \psi_1(x). \tag{7}$$

3 MATHEMATICAL ANALYSIS

3.1 Differential equation on Γ_s

The goal of this paragraph is to transform the original problem in a one dimensional problem posed on Γ_s, of the form

$$\begin{cases} \ddot{\varphi} + \mathcal{A}\varphi = \mathcal{B}v & \text{on } \Gamma_s \times [0,T], \\ \varphi(x,0) = \psi_0(x), \\ \dot{\varphi}(x,0) = \psi_1(x), \end{cases} \tag{8}$$

where $\varphi = \psi|_{\Gamma_s}$, and the operators $\mathcal{A}$ and $\mathcal{B}$ are to be defined. For simplicity we will consider only one generator, at the left end of the canal. One can show that $\mathcal{A}$ is the so-called *Dirichlet to Neuman* operator defined in the following way : for φ given on Γ_s one defines

$$\mathcal{A}\varphi = g_0 \left.\frac{\partial \Phi}{\partial y}\right|_{\Gamma_s},$$

where Φ is solution of the problem

$$\begin{cases} \Delta\Phi = 0 & \text{in } \Omega, \\ \dfrac{\partial \Phi}{\partial n} = 0 & \text{on } \Gamma_f \cup \Gamma_1 \cup \Gamma_2, \\ \Phi = \varphi & \text{on } \Gamma_s. \end{cases} \tag{9}$$

When the generator is flexible, the operator $\mathcal{B}$ is defined by $\mathcal{B} : L^2(\Gamma_s) \longrightarrow L^2(\Gamma_s)$, and

$$\mathcal{B}v(.,t) = -g_0 \left.\frac{\partial \varphi_v}{\partial n}\right|_{\Gamma_s},$$

where $\varphi_v(x,y)$ which is solution of the problem

$$\begin{cases} \Delta\varphi_v = 0 & \text{in } \Omega, \\ \varphi_v = 0 & \text{on } \Gamma_s, \\ \dfrac{\partial \varphi_v}{\partial n} = 0 & \text{on } \Gamma_f \cup \Gamma_2, \\ \dfrac{\partial \varphi_v}{\partial n} = v(.,t) & \text{on } \Gamma_1. \end{cases}$$

When the generator is rigid, we have $\mathcal{B} : \mathbb{R} \longrightarrow L^2(\Gamma_s)$, and

$$\mathcal{B}v = \beta(x)v,$$

where we have set $\beta(x) = -g_0 \left.\frac{\partial \varphi_f}{\partial n}\right|_{\Gamma_s}$. The function φ_f is solution of the problem

$$\begin{cases} \Delta\varphi_f &= 0 & \text{in } \Omega, \\ \varphi_f &= 0 & \text{on } \Gamma_s, \\ \dfrac{\partial \varphi_f}{\partial n} &= 0 & \text{on } \Gamma_f \cup \Gamma_2, \\ \dfrac{\partial \varphi_f}{\partial n} &= f(y) & \text{on } \Gamma_1, \end{cases}$$

where $f(y)$ is the function describing the shape of the generator. One can show that if $f \in H^{1/2}(\Gamma_1)$ then $\beta \in H^{1/2}(\Gamma_s)$.

Regularity results

Without using any specific result one can show that $\mathcal{A}$ is a linear and continuous operator from $H^{1/2}(\Gamma_s)$ to $H^{-1/2}(\Gamma_s)$. One can obtain more regularity :

PROPOSITION 3.1 *$\mathcal{A}$ is a linear and continuous operator from $H^1(\Gamma_s)$ to $L^2(\Gamma_s)$.*

PROOF : We first use regularity results of Grisvard [Grisvard85] to show that $\mathcal{A}$ is linear and continuous from $H^{3/2}(\Gamma_s)$ to $H^{1/2}(\Gamma_s)$, and we conclude with results of interpolation theory [Lions et al.68].

□

For some reasons that will appear clearly in the sequel, we will work with zero mean functions. We define the Hilbert space

$$H = \left\{ \varphi \in L^2(\Gamma_s) \;\middle|\; \int_{\Gamma_s} \varphi(x)\, dx = 0 \right\} \equiv L^2(\Gamma_s)/\mathbb{R},$$

and the domain of operator $\mathcal{A}$

$$D(\mathcal{A}) = \left\{ \varphi \in H^1(\Gamma_s) \;\middle|\; \int_{\Gamma_s} \varphi(x)\, dx = 0 \right\} \equiv H^1(\Gamma_s)/\mathbb{R}.$$

One can show that $\mathcal{A}$ is the infinitesimal generator of a strongly semigroup of contractions on H with the domain $D(\mathcal{A})$.

Spectral analysis of $\mathcal{A}$

The eigenvalues and associated eigenfunctions of $\mathcal{A}$, i.e. the functions $w_k(x) \in D(\mathcal{A})$ and the numbers λ_k such that $\mathcal{A} w_k = \lambda_k w_k$ are obtained by solving the problem

$$\begin{cases} \Delta W_k &= 0 & \text{in } \Omega, \\ g_0 \dfrac{\partial W_k}{\partial n} &= \lambda_k W_k & \text{on } \Gamma_s, \\ \dfrac{\partial W_k}{\partial n} &= 0 & \text{on } \Gamma_f \cup \Gamma_1 \cup \Gamma_2. \end{cases}$$

One easily obtains

$$w_k(x) = \sqrt{2/L} \cos \alpha k x, \; k > 0, \; x \in [0, L],$$

and the eigenvalues λ_k are given by

$$\lambda_k = g_0 \alpha k \tanh \alpha k h, \; k > 0.$$

3.2 Formulation as a first order system

The equation (8) can be written in the following way

$$\left\{ \begin{array}{rcl} \dot{\xi} & = & A\xi + Bv, \\ \xi(0) & = & \xi_0, \end{array} \right. \tag{10}$$

where

$$\xi = \begin{pmatrix} \varphi \\ \dot{\varphi} \end{pmatrix}, \quad \xi_0 = \begin{pmatrix} \psi_0 \\ \psi_1 \end{pmatrix}, \quad A = \begin{pmatrix} 0 & I \\ -\mathcal{A} & 0 \end{pmatrix}, \quad B = \begin{pmatrix} 0 \\ \mathcal{B} \end{pmatrix}. \tag{11}$$

If one keeps the natural domain of $\mathcal{A}$, say $H^1(\Gamma_s)$, one can show that the first order system (10) presents some serious pathologies. The first one is that the eigenfunctions of A do not form a basis of the natural state space $D(\mathcal{A}^{1/2}) \times L^2(\Gamma_s)$. The second one is that there is a subset of the state space which does not contribute to the elevation of the surface

$$\eta(x,t) = \frac{1}{g_0}\dot{\varphi}(x,h,t).$$

This means that the system (10) is not observable from η. These two pathologies are the reasons why we redefined the domain of $\mathcal{A}$, and adopted a *minimal realization* of the original system (for more details see [Mottelet94]).

Let us define the system (A, B, C) by

$$\left\{ \begin{array}{rcl} \dfrac{\partial \xi}{\partial t} & = & A\xi + Bv, \\ y & = & C\xi, \\ \xi(0) & = & \xi_0. \end{array} \right. \tag{12}$$

The state space Z is the following :

$$Z = \left\{ \xi = (\xi_1, \xi_2, \xi_3), \xi_1 \in D(\mathcal{A}^{1/2}), \xi_2 \in H, \xi_3 \in \mathbb{R} \right\}.$$

The initial condition ξ_0 is related to the initial conditions (ψ_0, ψ_1) of the original system by

$$\xi_0 = \left(\sum_{k>0} \langle \psi_0 \, , \, w_k \rangle \, w_k \;\; , \;\; \sum_{k>0} \langle \psi_1 \, , \, w_k \rangle \, w_k \;\; , \;\; \langle \psi_1 \, , \, w_0 \rangle \, w_0 \right).$$

The operators A, B and C are defined as follows :

$$A = \begin{pmatrix} \begin{pmatrix} 0 & I \\ -\mathcal{A} & 0 \end{pmatrix} & \\ & 0 \end{pmatrix}, \quad B = \begin{pmatrix} \begin{pmatrix} 0 \\ \widetilde{\mathcal{B}} \end{pmatrix} \\ \mathcal{B}_0 \end{pmatrix}, \tag{13}$$

where the operators $\widetilde{\mathcal{B}}$, $\mathcal{B}_0$ and C are defined by

$$\widetilde{\mathcal{B}}v = \sum_{k>0} \langle \mathcal{B}v \, , \, w_k \rangle \, w_k, \quad \mathcal{B}_0 v = \langle \mathcal{B}v \, , \, w_0 \rangle \, w_0, \quad C\xi = -\frac{1}{g_0}(\xi_2 + \xi_3).$$

The functions w_k, $k > 0$ are the eigenfunctions of $\mathcal{A}$ and $w_0 = \sqrt{1/L}$. The notation $\langle . \, , . \rangle$ denotes the inner product in H, and One can show, by using interpolation results in the quotient spaces [Lions et al.68], that

$$D(\mathcal{A}^{1/2}) = \left\{ \varphi \in H^{1/2}(\Gamma_s) \;\middle|\; \int_{\Gamma_s} \varphi(x)\, dx = 0 \right\} \equiv H^{1/2}(\Gamma_s)/\mathbb{R}.$$

We also have the following result :

PROPOSITION 3.2 *The operator A is the infinitesimal generator of a strongly continuous semigroup on $Z = D(\mathcal{A}^{1/2}) \times H \times \mathbb{R}$, given by the formula*

$$T(t) = \begin{pmatrix} \sum_{k>0} T_k(t) & 0 \\ 0 & 1 \end{pmatrix},$$

where $T_k(t)$ is defined by

$$T_k(t)\begin{pmatrix} \xi_1 \\ \xi_2 \end{pmatrix} = \begin{bmatrix} \cos\omega_k t & \dfrac{\sin\omega_k t}{\omega_k} \\ -\omega_k \sin\omega_k t & \cos\omega_k t \end{bmatrix} \begin{pmatrix} \langle \xi_1 \,,\, w_k\rangle \\ \langle \xi_2 \,,\, w_k\rangle \end{pmatrix} w_k, \tag{14}$$

where $\omega_k^2 = \lambda_k$.

PROOF : We show this by using classical semigroup theory (see [Curtain et al.92]). The explicit representation of the semigroup is obtained by considering the expansion of φ on the eigenfunctions w_k, $k > 0$.

□

From the definition of B we have that $Bv \in Z$, and if ξ_0 is an element of $D(A)$, then

$$\xi(t) = T(t)\xi_0 + \int_0^t T(t-\tau)Bv(\tau)\, d\tau,$$

is the only classical solution of (12) and is such that $\xi \in C^1([0,T]; Z)$. One can show that the eigenpairs (μ_k, ϕ_k) of operator A are given by $\mu_k = \pm j\omega_k$ and $\phi_k = (\frac{1}{\mu_k} w_k, w_k, 0)$ for $k > 0$ and $\mu_0 = 0$, $\phi_0 = (0,0,1)$. We will considerer the family $\{\phi_k, k \in \mathbb{Z}\}$, with $\omega_k = -\omega_{-k}$ for $k < 0$. One can show that $\{\phi_k, k \in \mathbb{Z}\}$ is a Riesz Basis of Z. This means that A is a Riesz-Spectral operator.

4 CONTROLLABILITY PROBLEMS

We focus on the theoretical study of the following problem : is it possible to find a control v such that the surface of the canal is in a given state at a given time ? We will consider the strong version of this question, where the state is to be reached exactly, and the weak version, where we accept that the state is reached approximately. The first version is an exact controllability problem. We will see that exact controllability can eventually be obtained if we consider "flexible" generators. We will see that the case where the generators are "rigid" only allows to eventually obtain approximate controllability.

Let us first recall some definitions. Consider the system

$$\begin{cases} \dot{\xi} &= A\xi + Bv, \\ \xi(0) &= \xi_0, \end{cases} \tag{15}$$

where A is infinitesimal generator of a strongly continuous semigroup $T(t)$ on a Hilbert space Z, B is a bounded operator from U to Z. At a given time t , the control $v(t)$ belongs to a infinite or finite dimensional space U.

DEFINITION 4.1 *The controllability map of system (15) on $[0, \tau]$ (for a finite $\tau > 0$) is the bounded map $\mathcal{H}^\tau : L_2(0,\tau; U) \rightarrow Z$ defined by*

$$\mathcal{H}^\tau v = \int_0^\tau T(\tau - s)Bv(s)\, ds.$$

The natural definition of exact controllability is the following :

DEFINITION 4.2 *The system (15) is exactly controllable on* $[0, \tau]$ *if every element of* Z *can be reached from the origin at time* τ*, say*

$$ran\ \mathcal{H}^\tau = Z.$$

The definition 4.2 does not give a practical tool to show that a given system is exactly controllable. One usually shows this by showing an inverse inequality [Curtain et al.92] :

THEOREM 4.1 *The system (15)is exactly controllable on* $[0, \tau]$ *if there exists* $\gamma > 0$ *such that for every* $z \in Z$*,*

(16)
$$\int_0^\tau \|B^* T^*(s) z\|_U^2 \, ds \geq \gamma \|z\|_Z^2.$$

REMARK 4.1 When U is a finite dimensional space[1], exact controllability cannot hold, because in this case the map $\mathcal{H}^\tau$ is compact ([Curtain et al.92], Theorem 4.1.5). This result seems to be strange, but since we have supposed that B is a bounded operator, we have excluded pointwise or boundary control. It is clear that when one relax this hypothesis, the classical example of the one-dimensional wave equation

(17)
$$\begin{cases} \ddot{\varphi} - \dfrac{\partial^2 \varphi}{\partial x^2} &= 0, \ x \in]0, 1[, \\ \varphi(0, t) &= 0, \\ \varphi(1, t) &= v(t), \end{cases}$$

with pointwise control at $x = 1$, which is well known to be exactly controllable, shows that exact controllability can be obtained with a finite dimensional control, provided that the control operator is not bounded.

□

When one can obtain explicitly the semigroup, for example under the form of an infinite series, theorem 4.1 allows to easily check if exact controllability holds, but this is not a constructive approach to obtain the control. If one wants to show that exact controllability holds, and moreover construct the control, one can use the H.U.M. [2] method introduced by J.L. Lions [Lions88]. In this paper we will use those two approaches.

The concept of exact controllability is often too strong, and sometimes the concept of approximate controllability is more adequate. It is defined in the following way :

DEFINITION 4.3 *The system (15) is approximately controllable on* $[0, \tau]$ *(for a finite* $\tau > 0$*) if for* $\varepsilon > 0$*, it is possible to steer from the origin at a distance* ε *form all elements of* Z *in a finite time* τ*, say*

(18)
$$\overline{ran\ \mathcal{H}^\tau} = Z.$$

To show (18) one usually tries to show a *uniqueness theorem* ([Curtain et al.92], theorem 4.1.7):

(19)
$$B^* T(t)^* \xi = 0 \text{ on } [0, \tau] \Rightarrow \xi = 0.$$

In our application, we are first going to focus on the case where the generators are rigid.

---

[1] It will be the case when we will consider rigid generators

[2] Hilbert Uniqueness Method

4.1 Controllability with a rigid generator

Since we consider a rigid generator, the operator B is defined by

$$Bv = \begin{pmatrix} \begin{pmatrix} 0 \\ \widetilde{\beta} \end{pmatrix} \\ \beta_0 \end{pmatrix} v,$$

where $\widetilde{\beta}$ and β_0 are defined by $\widetilde{\beta} = \sum_{k>0} \langle \beta\,,\, w_k \rangle\, w_k$ and $\beta_0 = \langle \beta\,,\, w_0 \rangle\, w_0$.

Approximate controllability on an interval $[0, \tau]$

In fact our system is not approximately controllable on $[0, \tau]$ for any $\tau > 0$. This result is based on the following lemma :

LEMMA 4.1 *The family $\{e^{jw_k t}\}_{k\in\mathbb{Z}}$ is total and linearly dependent in $L^2(0, \tau)$ for $\tau \leq 2\pi$.*

PROOF : Let us recall that $\omega_k = \sqrt{\lambda_k}$, where $\lambda_k > 0$ are the eigenvalues of $\mathcal{A}$, given by the $\omega_k = \sqrt{\alpha k g_0 \tanh \alpha k h}$, where $\alpha = \pi/L$. For simplicity let us set $L = \pi$ and $g_0 = 1$. Let us define the density of the sequence $\{\omega_k\}_{k\in\mathbb{N}}$ by

$$D = \lim_{k\to\infty} \frac{k}{\omega_k} = \lim_{k\to\infty} \frac{k}{\sqrt{k \tanh kh}} = +\infty.$$

From Levinson [Levinson40], since D is greater that $2A$, $\forall A > 0$, the family $\{e^{jw_k t}\}_{k\in\mathbb{Z}}$ is total in $L^2(-A, +A)$. If one uses a result of Schwartz ([Schwartz59], Th. 4, p. 102) this family is either free and it is a basis, or every vector is dependent of others. We show that we are in the second case by showing that the family $\{e^{jw_k t}\}_{k\in\mathbb{Z}}$ strictly contains a total family in $L^2(0, 2\pi)$. The idea is that for k sufficiently large one has $\omega_k \approx \sqrt{k}$, and that the sequence $\{\sqrt{k}\}_{k\in\mathbb{N}}$ strictly contains $\mathbb{N}$ and that $\left\{e^{jkt}\right\}_{k\in\mathbb{Z}}$ is total in $L^2(0, 2\pi)$. For the detail of the proof see [Mottelet94].

□

We have the following result

THEOREM 4.2 *The system (A, B) is not approximately controllable on $[0, \tau]$, $\forall \tau > 0$.*

PROOF : We show that (19) is not fulfilled. Let us suppose that there exists $\xi \in Z$ such that

$$B^* T^*(t)\xi = 0, \;\; t \in [0, \tau].$$

On can express $B^* T^*(t)\xi$ by using the series expression of the semigroup, and one obtains [Mottelet94]

$$B^* T^*(t)\xi = \sum_{k=-\infty}^{k=\infty} c_k e^{j\omega_k t}.$$

where the nullity of the sequence $\{c_k\}_{k\in\mathbb{Z}}$ is equivalent to the nullity of ξ. Since the family $\{e^{jw_k t}\}_{k\in\mathbb{Z}}$ is not free there is a infinity of sequences $\{c_k\}_{k\in\mathbb{Z}}$, thus an infinity of ξ such that $B^* T^*(t)\xi = 0$. Uniqueness does not hold, so the system is not approximately controllable on $[0, \tau]$.

□

This result is essentially due to the fact that the ω_k gather at infinity, i.e., there is no asymptotic gap between ω_k and ω_{k+1}. In the next paragraph we show that the system is approximately controllable in the sense of another definition.

Generic approximate controllability

One can define a generic concept of approximate controllability where the time τ does not appear explicitly. This is the concept used by R. Curtain and H. Zwart [Curtain et al.92]. Its definition is the following

DEFINITION 4.4 *Let us call $\mathcal{R}$ the reachability subspace*

$$\mathcal{R} = \bigcup_{\tau \in \mathbb{R}^+} Ran\,\mathcal{H}^\tau,$$

The system will be approximately controllable $\overline{\mathcal{R}} = Z$. □

Since we have shown that A is a Riesz-Spectral operator, we can use Theorem 4.2.3 in [Curtain et al.92] which claims that the system will be approximately controllable in the sense of the above definition if

$$B^* \phi_k \neq 0, \ \forall k \in \mathbb{Z},$$

where ϕ_k, $k \in \mathbb{Z}$ are the eigenfunctions of operator A. We have the following result

PROPOSITION 4.1 *If $\langle \beta \, , \, w_k \rangle > 0$, $\forall k$ the system (A, B) is approximately controllable.*

PROOF : This results from the definition of B^*, say

$$B^* \xi = \left\langle \left(\begin{pmatrix} 0 \\ \tilde{\beta} \end{pmatrix} \atop \beta_0 \right), \xi \right\rangle_Z,$$

And one easily shows that $B^* \phi_k = \langle \beta \, , \, w_k \rangle$. The quantity $\langle \beta \, , \, w_k \rangle$ can be obtained by the formulas

$$(20) \qquad \left\{ \begin{array}{rcl} \langle \beta \, , \, w_k \rangle & = & \dfrac{\sqrt{2} g_0}{\sqrt{L} \cosh \alpha k h} \displaystyle\int_0^h \cosh \alpha k y \, f(y) \, dy, \ k > 0, \\ \langle \beta \, , \, w_0 \rangle & = & \dfrac{g_0}{\sqrt{L}} \displaystyle\int_0^h f(y) \, dy. \end{array} \right.$$

If $f(y)$ is a positive function on $[0, h]$, for example $f(y) = y$ in our case, then the condition $\langle \beta \, , \, w_k \rangle \neq 0$ is fulfilled. Hence the system is approximately controllable. □

4.2 Controllability with a flexible generator

Since exact controllability cannot hold when the generator is rigid, we can try to obtain such a result when the generator is flexible. In this case the control $v(y, t)$ represents the normal derivative of the velocity potential on the boundary Γ_1.

We are going to split the original exact controllability problem in two sub-problems of different kind. The first one is a problem concerning an hyperbolic equation with internal control, and the other concerning an elliptic problem :

- First (hyperbolic) problem : does there exist $\widetilde{\beta}(x,t) \in L^2(0,T;H)$ and $\beta_0 \in L^2(0,T)$ such that, $\forall\, T > 0$, $\forall\, \xi_d \in Z$, the solution $\xi(x,t)$ of

$$\begin{cases} \dot{\xi} &= A\xi + \begin{pmatrix} 0 \\ \widetilde{\beta} \\ \beta_0 \end{pmatrix}, \\ \xi(0) &= \xi_0, \end{cases}$$

verifies $\xi(T) = \xi_d$?

- Second (elliptic) problem : for a given $\widetilde{\beta}(x) \in L^2(\Gamma_s)$ and $\beta_0 \in \mathbb{R}$ does there exist $v(y) \in L^2(\Gamma_1)$ such that

$$\begin{cases} \widetilde{\mathcal{B}}v &= \widetilde{\beta}, \\ \mathcal{B}_0 v &= \beta_0 \text{ ?} \end{cases}$$

From the definitions of $\mathcal{B}$, $\widetilde{\mathcal{B}}$ and $\mathcal{B}_0$, these two equalities take the form

$$\mathcal{B}v = \widetilde{\beta} + \beta_0.$$

As far as the hyperbolic is concerned, if we set $\xi = (\varphi,\ \dot{\varphi},\ \varphi_0)$, the system (4.2) splits into the two systems

$$\text{(21)} \qquad \begin{cases} \ddot{\varphi} + \mathcal{A}\varphi &= \widetilde{\beta}(x,t) \quad \text{on } \Gamma_s \times [0,T], \\ \varphi(x,0) &= 0 \\ \dot{\varphi}(x,0) &= 0, \end{cases}$$

$$\text{(22)} \qquad \begin{cases} \dot{\varphi}_0 &= \beta_0(t) \quad t \in [0,T], \\ \varphi_0(0) &= 0, \end{cases}$$

This shows that the determination of $\widetilde{\beta}$ and β_0 are decoupled. The system (22) is finite dimensional hence exactly controllable. So we just have to know whether (21) is exactly controllable or not. To answer this question we propose to use two different techniques. The first one is a classical technique using semigroup theory.

Hyperbolic problem : classical approach

The classical approach consists in using theorem 4.1. The system (21) takes the form

$$\dot{\xi} = A\xi + Bv,$$

with

$$\xi = \begin{pmatrix} \varphi \\ \dot{\varphi} \end{pmatrix}, \quad A = \begin{pmatrix} 0 & I \\ -\mathcal{A} & 0 \end{pmatrix}, \quad v = \widetilde{\beta} \text{ and } B = \begin{pmatrix} 0 \\ I \end{pmatrix}.$$

A is the infinitesimal generator on $Z = D(\mathcal{A}^{1/2}) \times H$ of the semigroup $T(t) = \sum_{k>0} T_k(t)$, where $T_k(t)$ is defined by formula (14). We take v in $L^2(0,T;L^2(\Gamma_s))$ and we apply theorem 4.1. The system will be exactly controllable if there exists $\gamma > 0$ such that

$$\text{(23)} \qquad \int_0^T \|B^*T^*(s)z\|_U^2\, ds \geq \gamma \|z\|_Z^2, \ \forall z \in Z.$$

We here use the same technique as Curtain and Zwart [Curtain et al.92] for the one dimensional wave equation. The calculus of the two members in (23) gives the equivalent inequality

$$\begin{aligned}
\gamma\|z\|^2 &\leq \int_0^\tau \left\| \sum_{k>0} \left(\omega_k \sin \omega_k t \, \langle z_1 \, , \, w_k\rangle + \cos \omega_k t \, \langle z_2 \, , \, w_k\rangle\right) w_k \right\|^2 dt \\
&= \int_0^\tau \sum_{k>0} \left(\omega_k \sin \omega_k t \, \langle z_1 \, , \, w_k\rangle + \cos \omega_k t \, \langle z_2 \, , \, w_k\rangle\right)^2 dt \\
&= \int_0^\tau \sum_{k>0} \Big(\omega_k^2 \sin^2 \omega_k t \, \langle z_1 \, , \, w_k\rangle^2 + 2\omega_k \sin \omega_k t \cos \omega_k t \, \langle z_1 \, , \, w_k\rangle \langle z_2 \, , \, w_k\rangle \\
&\quad + \cos^2 \omega_k t \, \langle z_2 \, , \, w_k\rangle^2\Big) \, dt \\
&= \sum_{k>0} \omega_k^2 \left(\frac{\tau}{2} - \frac{\sin 2\omega_k \tau}{4\omega_k}\right) \langle z_1 \, , \, w_k\rangle^2 + \sin^2 \omega_k \tau \, \langle z_1 \, , \, w_k\rangle \langle z_2 \, , \, w_k\rangle \\
&\quad + \left(\frac{\tau}{2} + \frac{\sin 2\omega_k \tau}{4\omega_k}\right) \langle z_2 \, , \, w_k\rangle^2 .
\end{aligned}$$

We have also

$$\|z\|_Z^2 = \sum_{k>0} \omega_k^2 \langle z_1 \, , \, w_k\rangle^2 + \langle z_2 \, , \, w_k\rangle^2 .$$

One easily shows that (23) is fulfilled with γ given by (see [Curtain et al.92])

$$\frac{1}{\gamma} = \max_{k>0} \frac{2}{\tau - \left|\dfrac{\sin \omega_k \tau}{\omega_k}\right|} .$$

Hence the system is exactly controllable for any $\tau > 0$. □

One the one hand one can see that the result we have obtained does not depend on τ. This is due to the fact that our control is distributed on all the surface Γ_s. On the other hand the technique we have used is rather simple, but it does not give a constructive technique to obtain the control. In the contrary this is possible with the H.U.M. method, as we will see in the next paragraph.

Hyperbolic problem. Approach with the H.U.M. method

We will not describe in details the way the H.U.M. method works, because the case we are interested in is closely related to the case of the wave equation with internal control treated by Lions ([Lions88] pp 397-407).

Let us recall that we have defined $D(\mathcal{A}) = H^1(\Gamma_s)/\mathbb{R}$, $D(\mathcal{A}^{1/2}) = H^{1/2}(\Gamma_s)/\mathbb{R}$ and $H = L^2(\Gamma_s)/\mathbb{R}$. We have the following result :

THEOREM 4.3 *For a given $\tau > 0$ and $(\varphi_0, \varphi_1) \in D(\mathcal{A}^{1/2}) \times H$ there exists a control $\beta \in L^2(\Gamma_s \times [0, T])$ such that the solution of*

$$\text{(24)} \qquad \left\{ \begin{array}{rcll} \ddot{\varphi} + \mathcal{A}\varphi &=& \beta & \text{on } \Gamma_s \times [0, \tau], \\ \varphi(0) &=& \varphi_0, & \\ \dot{\varphi}(0) &=& \varphi_1. & \end{array} \right.$$

verifies $\varphi(\tau) = \dot{\varphi}(\tau) = 0$.

PROOF : We apply the H.U.M. method. We first solve the homogeneous problem

$$(25)\qquad \begin{cases} \ddot{\phi} + \mathcal{A}\phi &= 0 \quad \text{on } \Gamma_s \times [0,\tau], \\ \phi(0) &= \phi_0, \\ \dot{\phi}(0) &= \phi_1. \end{cases}$$

with initial conditions $(\phi_0, \phi_1) \in \mathcal{D}(\Gamma_s) \times \mathcal{D}(\Gamma_s)$. We then define the hilbertian norm

$$\|(\phi_0, \phi_1)\|_{F_1} = \left(\int_{\Gamma_s \times [0,\tau]} \phi^2 \, dx \, dt \right)^{1/2}.$$

Let F_1 be the completed space of $\mathcal{D}(\Gamma_s) \times \mathcal{D}(\Gamma_s)$ with respect to the norm $\| \, . \, \|_{F_1}$. One can show that

$$F_1 = H \times K,$$

where $K = H^{-1/2}(\Gamma_2)/\mathbb{R}$ (this result is the most important part of this proof. See [Mottelet94] for the details). We then consider the problem in reversed time

$$(26)\qquad \begin{cases} \ddot{\psi} + \mathcal{A}\psi &= -\phi \quad \text{on } \Gamma_s \times [0,T], \\ \psi(T) &= 0, \\ \dot{\psi}(T) &= 0. \end{cases}$$

When $\phi \in L^2(0,\tau;H)$ (26) admits a unique solution ψ. We then define the operator

$$\Lambda(\phi_0, \phi_1) = (\dot{\psi}(0), -\psi(0)),$$

which verifies

$$\langle \Lambda(\phi_0, \phi_1), (\phi_0, \phi_1) \rangle = \int_{\Gamma_s \times [0,T]} \phi^2 \, dx \, dt.$$

Hence Λ is an isomorphism form F_1 to is dual F_1' say

Λ *is an isomorphism form* $H \times K$ *to* $H \times D(\mathcal{A}^{1/2})$.

For any initial data $(\varphi_0, \varphi_1) \in D(\mathcal{A}^{1/2}) \times H$ the equation

$$(27)\qquad \Lambda(\phi_0, \phi_1) = (\varphi_1, -\varphi_0)$$

admits a unique solution

$$(\phi_0, \phi_1) \in H \times K,$$

and the control

$$\beta = -\phi \in L(\Gamma_s \times [0,\tau]),$$

where ϕ is the solution of (25) associated to $(\phi_0, \phi_1) \in F_1$ verifying (27), is such that

$$\varphi(\tau; \beta) = \dot{\varphi}(\tau; \beta) = 0.$$

□

Exact controllability of the elliptic problem

Since we know that there exists $\beta(x,t)$ on $[0,\tau]$ such that $\varphi(\tau)=\varphi_d$ and $\dot{\varphi}(\tau)=\psi_d$, we know have to answer the following question : for a given $\beta(x)\in L^2(\Gamma_s)$ does there exist $v(y)\in L^2(\Gamma_1)$ such that $\mathcal{B}v=\beta$, say, do we have $L^2(\Gamma_s)\subset \text{ran } \mathcal{B}$?

The formula (20) allows β to be obtained as a function of $v(y)$, under the form $\beta=\sum_{k\geq 0}\delta_k w_k$ with

$$(28)\qquad \begin{cases} \delta_k=\langle\beta\,,\,w_k\rangle &=& \dfrac{\sqrt{2}g_0}{\sqrt{L}\cosh\alpha kh}\displaystyle\int_0^h \cosh\alpha ky\, v(y)\,dy,\ k>0,\\ \delta_0=\langle\beta\,,\,w_0\rangle &=& \dfrac{g_0}{\sqrt{L}}\displaystyle\int_0^h v(y)\,dy.\end{cases}$$

One can state the inverse problem in the following way : for $\{\delta_k\}_{k\in\mathrm{N}}$ a sequence of $l_2(\mathbb{R})$ does there exist $v(y)$ in $L^2(\Gamma_1)$¿ such that

$$(29)\qquad \int_0^h \cosh\alpha ky\, v(y)\,dy = \delta_k\sqrt{L/2}\frac{\cosh\alpha kh}{g_0},\ k>0$$

$$(30)\qquad \int_0^h v(y)\,dy = \delta_0\sqrt{L}\frac{1}{g_0}\ ?$$

One can show that the answer is no with help of the following lemma :

PROPOSITION 4.2 *The set $\{\cosh\alpha ky\}_{k\in\mathrm{N}}$ is total and linearly dependent in $L^2(0,h)$.*

PROOF : We use a result of Young ([Young80] Th. 15 p. 90) which says that the set $\left\{e^{\alpha ky}\right\}_{k\in\mathrm{N}}$ is total and linearly dependent in $L^2(a,b)\ \forall\ a,b$. □

We have the following result :

THEOREM 4.4 *The elliptic problem is not exactly controllable, say, $L^2(\Gamma_s)\neq \text{ran } \mathcal{B}$.*

PROOF : Let us take a function β in $L^2(\Gamma_s)$ such that only a finite number of δ_k are non zero, say $\delta_k=0$ for $k>K$. We have

$$(31)\qquad \int_0^h \cosh\alpha ky\, v(y)\,dy=0,\ k>K$$

thus for a infinite number of k values. But the set $\{\cosh\alpha ky\}_{k\in\mathrm{N}}$ remains total even if one suppresses a finite number of elements. So we have $v=0$, thus

$$\int_0^h \cosh\alpha ky\, v(y)\,dy=0,\ \forall k\in\mathrm{N},$$

so for $k\leq K$ which contradicts (31). So there are functions in $L^2(\Gamma_s)$ which do not belong to the range of $\mathcal{B}$; this ends the proof. □

Approximate controllability of the elliptic problem

One can show that approximate controllability holds in $L^2(\Gamma_s)$:

THEOREM 4.5 *The elliptic problem is approximately controllable in $L^2(\Gamma_s)$, say*

$$\overline{ran\ \mathcal{B}} = L^2(\Gamma_s).$$

PROOF : Let us take $z \in L^2(\Gamma_s)$. We have to show the uniqueness result $\mathcal{B}^* z = 0 \Rightarrow z = 0$. We define $\mathcal{B}^*$ by taking Ψ solution of

$$(32) \qquad \left\{ \begin{array}{rcl} \Delta\Psi & = & 0 \;\text{ in } \Omega, \\ \Psi & = & z \;\text{ on } \Gamma_s, \\ \dfrac{\partial\Psi}{\partial n} & = & 0 \;\text{ on } \Gamma_f \cup \Gamma_1 \cup \Gamma_2. \end{array} \right.$$

Let us recall that $\mathcal{B}$ is defined by $\mathcal{B}v = -g_0 \left.\dfrac{\partial\Phi}{\partial n}\right|_{\Gamma_s}$, where Φ is solution of

$$(33) \qquad \left\{ \begin{array}{rcll} \Delta\Phi & = & 0 & \text{in } \Omega, \\ \Phi & = & 0 & \text{on } \Gamma_s, \\ \dfrac{\partial\Phi}{\partial n} & = & 0 & \text{on } \Gamma_f \cup \Gamma_2, \\ \dfrac{\partial\Phi}{\partial n} & = & v(y) & \text{on } \Gamma_1. \end{array} \right.$$

We write that $\langle \Delta\Phi\,,\,\Psi \rangle_{L^2(\Omega)} = 0$, and applying Green's formula twice, we obtain that $\mathcal{B}^* z = g_0\, \Psi|_{\Gamma_1}$. So to prove uniqueness we have to show that the following problem

$$(34) \qquad \left\{ \begin{array}{rcl} \Delta\Psi & = & 0 \;\text{ in } \Omega, \\ \Psi & = & 0 \;\text{ on } \Gamma_1, \\ \dfrac{\partial\Psi}{\partial n} & = & 0 \;\text{ on } \Gamma_f \cup \Gamma_1 \cup \Gamma_2, \end{array} \right.$$

has the unique solution $\Psi = 0$. Since Ψ is smooth enough (we have taken z in $L^2(\Gamma_s)$), the Cauchy uniqueness theorem can be applied, say, $\Psi = \frac{\partial\Psi}{\partial n} = 0$ on Γ_1 implies that $\Psi = 0$. So $z = 0$, and approximate controllability holds in $L^2(\Gamma_s)$.

□

This result allows to claim that the original system (hyperbolic+elliptic) is approximately controllable :

THEOREM 4.6 *For any ξ_d in $Z = D(\mathcal{A}^{1/2}) \times H$, for any $\varepsilon > 0$, there exists $v \in L^2(0, T, L^2(\Gamma_1))$ such that the solution of*

$$(35) \qquad \left\{ \begin{array}{rcl} \dot{\xi} & = & A\xi + Bv, \\ \xi(0) & = & \xi_0. \end{array} \right.$$

verifies $\|\xi_d - \xi(T)\|_Z \leq \varepsilon$.

PROOF : One the one hand we know that $\forall\ \xi_d \in Z$ there exists $\beta = \widetilde{\beta} + \beta_0$, belonging to $L^2(0,T;L^2(\Gamma_s))$, where $\widetilde{\beta}$ has zero mean, such that the solution of

$$\left\{\begin{array}{rcl} \dot{\zeta} & = & A\zeta + \begin{pmatrix} 0 \\ \widetilde{\beta} \\ \beta_0 \end{pmatrix}, \\ \zeta(0) & = & \zeta_0. \end{array}\right. \tag{36}$$

verifies $\zeta(T) = \xi_d$. On the other hand, since A is generator of a strongly continuous semi-group, there is continuity of the solution of (36) with respect to the data, say, there exists $C > 0$ such that

$$\|\xi_d - \xi(T)\|_Z \leq C \left\| \begin{pmatrix} 0 \\ \widetilde{\beta} \\ \beta_0 \end{pmatrix} - Bv \right\|_{L^2(0,T;Z)},$$

where ξ is solution of (35). We have, by construction of $\widetilde{\mathcal{B}}$ and $\mathcal{B}_0$,

$$\left\| \begin{pmatrix} 0 \\ \widetilde{\beta} \\ \beta_0 \end{pmatrix} - Bv \right\|_{L^2(0,T;Z)} = \|\beta - \mathcal{B}v\|_{L^2(0,T;L^2(\Gamma_s))}.$$

The approximate controllability result for the elliptic problem allows to claim that $\forall \varepsilon > 0$ there exists $v \in L^2(0,T,L^2(\Gamma_1))$ such that

$$\|\beta - \mathcal{B}v\|_{L^2(0,T;L^2(\Gamma_s))} \leq \varepsilon',$$

or equivalently

$$\|\xi_d - \xi(T)\|_Z \leq \varepsilon,$$

with $\varepsilon = C\varepsilon'$. This ends the proof. □

5 CONCLUSION

One main conclusion of this study id that approximate controllability can be achieved by a rigid generator which, from the point of view of application, is sufficient. The problem with a flexible generator is purely theoretical as its implementation is unrealistic.

This interesting problem can be attacked by various techniques which are related to the semi-group theory, generalized Fourier series and H.U.M. method. This shows the complementarity of the different mathematical tools.

REFERENCES

[Curtain et al.92] Curtain (R.F.) et Zwart (H.J.). – *Lecture notes on distributed parameter systems, preprint.* – Berlin, Springer Verlag, 1992.

[Grisvard85] Grisvard (P.). – *Elliptic problems in nonsmooth domains.* – Pitman, 1985.

[Joly et al.94] Joly (G.), Mottelet (s.) et Yvon (J.P.). – Application of H_∞ control to wave generators in a canal. *In: Proceedings of the SIAM Symposium on Control Problems in Industry, San Diego, 22-23 July 1994, to appear.* – Birkauser.

[Levinson40] Levinson (N.). – *Gap and density theorems.* – AMS Colloqium Publications 26, 1940.

[Lions et al.68] Lions (J.L.) et Magenes (E.). – *Problèmes aux limites non homogènes et applications.* – Paris, Dunod, coll. Travaux et Recherches Mathématiques, 1968.

[Lions88] Lions (J.L.). – *Contrôlabilité exacte, perturbation et stabilisation de systèmes distribués.* – Paris, Masson, Collection Recherches en Mathématiques Appliquées, 1988.

[Mottelet94] Mottelet (S.). – *Quelques aspects théoriques et numériques du contrôle d'un bassin de carnènes.* – PhD thesis, Université de Technologie de Compiègne, 1994.

[Schwartz59] Schwartz (L.). – *Etude des sommes d'exponentielles.* – Herman, 1959.

[Young80] Young (R. M.). – *An introduction to nonharmonic Fourier series.* – Academic Press, 1980.

Boundary Stabilization of Isotropic Elasticity Systems

V. KOMORNIK, Département de mathématique et C.N.R.S., Université Louis Pasteur, 7, rue René Descartes, 67084 Strasbourg Cédex, France

1 INTRODUCTION

This work was motivated by two former papers of Lagnese (1983), (1991). In the first one he studied the boundary stabilization of three-dimensional linear elastodynamic systems under some technical assumptions. His results did not apply to isotropic systems. In his second paper he succeeded to obtain uniform energy decay estimates for *two-dimensional* homogeneous isotropic systems by applying either linear or nonlinear boundary feedbacks. The aim of this paper is to extend his results for the three-dimensional case. Moreover, by applying a carefully chosen feedback and by using an improved version of a method introduced in Komornik and Zuazua (1990), we shall obtain rather precise decay estimates. Applying an elegant elementary method of Cazenave (1990) we will show that these estimates are the best possible if the underlying domain is a three-dimensional ball. Using a general theorem of Russell (1978) we can deduce from our stabilization result an exact controllability theorem of Lions (1988), originally proven by his Hilbert Uniqueness Method.

We remark that the same method has already been used before for the boundary stabilization of the wave equation and for Maxwell's equations in Komornik (1991), (1994a).

Due to space limitations, we only consider here the *linear* stabilization of homogeneous *isotropic* elasticity systems, but our approach can be applied in more general situations, too; cf. also Komornik (1994a) for an adaptation of this method to nonlinear feedbacks for the wave equation and some plate models. A more complete study will be given elsewhere.

Let Ω be a bounded open domain in $\mathbb{R}^n$ having a boundary Γ of class C^2 and denote by $\nu = (\nu_1, \ldots, \nu_n)$ the outward unit normal vector to Γ. Let

---

The author thanks the organisers for their kind invitation to this IFIP Workshop.

$\{\Gamma_0, \Gamma_1\}$ be a partition of the boundary Γ and assume that

$$\overline{\Gamma_0} \cap \overline{\Gamma_1} = \emptyset. \tag{1}$$

Given two positive numbers λ, μ (the Lamé constants in the physical interpretation of the model) and two continuously differentiable functions $A : \Gamma_1 \to \mathbb{R}_+$, $B : \Gamma_1 \to \mathbb{R}_+$ where $\mathbb{R}_+ = [0, +\infty)$, consider the following evolutionary system :

$$u'' - \mu\Delta u - (\lambda + \mu)\text{grad div } u = 0 \quad \text{in} \quad \Omega \times \mathbb{R}_+, \tag{2}$$
$$u = 0 \quad \text{on} \quad \Gamma_0 \times \mathbb{R}_+, \tag{3}$$
$$\mu(\partial_\nu u) + (\lambda + \mu)(\text{div } u)\nu + Au + Bu' = 0 \quad \text{on} \quad \Gamma_1 \times \mathbb{R}_+, \tag{4}$$
$$u(0) = u^0 \quad \text{and} \quad u'(0) = u^1 \quad \text{in} \quad \Omega. \tag{5}$$

Let us denote by $H^1_{\Gamma_0}(\Omega)$ the set of functions $v \in H^1(\Omega)$ satisfying $v = 0$ on Γ_0. Using the standard linear semigroup theory (we omit the details here; cf. e.g. Komornik (1994b) for a similar study of the wave equation), this problem is well-posed : given $u^0 \in H^1_{\Gamma_0}(\Omega)^n$ and $u^1 \in L^2(\Omega)^n$ arbitrarily, there exists a unique *mild* solution

$$u \in C(\mathbb{R}_+; H^1(\Omega)^n) \cap C^1(\mathbb{R}_+; L^2(\Omega)^n),$$

and the linear mapping $(u^0, u^1) \mapsto u$ is continuous with respect to these topologies. Moreover, if $u^0 \in (H^2(\Omega) \cap H^1_{\Gamma_0}(\Omega))^n$, $u^1 \in H^1_{\Gamma_0}(\Omega)^n$ and if the compatibility condition

$$\mu(\partial_\nu u^0) + (\lambda + \mu)(\text{div } u^0)\nu + Au^0 + Bu^1 = 0 \quad \text{on} \quad \Gamma_1$$

is satisfied, then we have the following regularity property :

$$u \in C(\mathbb{R}_+; H^2(\Omega)^n) \cap C^1(\mathbb{R}_+; H^1(\Omega)^n) \cap C^2(\mathbb{R}_+; L^2(\Omega)^n).$$

We say in this case that u is a *strong* solution.

Unlike Lagnese (1983), let us define the energy of the solutions by the formula

$$E = \tfrac{1}{2} \int_\Omega \mu|\nabla u|^2 + (\lambda + \mu)(\text{div } u)^2 + |u'|^2 \, dx + \tfrac{1}{2} \int_{\Gamma_1} Au^2 \, d\Gamma. \tag{6}$$

The choice of this (stronger) energy is essential in our proofs; we note that the same energy function turned out to be very useful in the study of exact controllability; cf. Lions (1988). Then the energy $E : \mathbb{R}_+ \to \mathbb{R}_+$ is a non-increasing function. Indeed, if u is a strong solution, then we have

$$E' = \int_\Omega \mu\nabla u \cdot \nabla u' + (\lambda+\mu)(\text{div } u)(\text{div } u') + u' \cdot u'' \, dx + \int_{\Gamma_1} Au \cdot u' \, d\Gamma$$

$$= \int_\Omega \mu(\nabla u' \cdot \nabla u + u' \cdot \Delta u) + (\lambda+\mu)\big((\text{div } u')(\text{div } u) + u' \cdot \nabla(\text{div } u)\big) \, dx$$

$$+ \int_{\Gamma_1} Au' \cdot u \, d\Gamma$$

$$= \int_\Gamma u' \cdot \big(\mu\partial_\nu u + (\lambda+\mu)(\text{div } u)\nu\big) \, d\Gamma + \int_{\Gamma_1} Au' \cdot u \, d\Gamma$$

$$= \int_{\Gamma_1} u' \cdot \big(\mu\partial_\nu u + (\lambda+\mu)(\text{div } u)\nu + Au\big) \, d\Gamma$$

$$= -\int_{\Gamma_1} B|u'|^2 \, d\Gamma.$$

It follows that

$$E(S) - E(T) = \int_S^T \int_{\Gamma_1} B|u'|^2 \, d\Gamma \, dt \tag{7}$$

for all $0 \le S < T < +\infty$. This identity remains valid for all mild solutions by an easy density argument. Since $B \ge 0$ by assumption, the non-increasingness of E follows.

We are going to obtain exponential decay rate estimates under suitable assumptions. Fix a point $x^0 \in \mathbb{R}^n$ arbitrarily, set

$$m(x) = x - x_0, \quad x \in \mathbb{R}^n,$$

and assume that

$$m \cdot \nu \le 0 \quad \text{on} \quad \Gamma_0, \tag{8}$$

$$m \cdot \nu \ge 0 \quad \text{on} \quad \Gamma_0. \tag{9}$$

Furthermore, set

$$R = \sup_{x \in \Omega} |m(x)| \tag{10}$$

and define the functions A, B by the formulae

$$A = (n-1)\mu(m \cdot \nu)/(2R^2), \quad B = \sqrt{\mu}(m \cdot \nu)/R. \tag{11}$$

Then we have the following

THEOREM 1 Assume that $n \geq 3$. Then every mild solution of (2)–(5) satisfies estimate

$$E(t) \leq E(0)e^{1-(\sqrt{\mu}t/2R)} \tag{12}$$

for all $t \geq 0$.

We note that theorem 1 remains valid without the assumption (1). However, the proof becomes more delicate because the strong solutions are not sufficiently smooth to justify the computations given below. One can overcome this difficulty by applying a domain approximation argument due to Grisvard (1989); cf. also Komornik and Zuazua (1990) for a similar study in the case of the wave equation. Also, we note that analogous results hold in dimensions $n = 1, 2$: one can modify the proofs below as in Komornik (1991) for the wave equation.

If Ω is a three-dimensional ball, then the exponent in (12) is the best possible :

THEOREM 2 Let $n = 3$, $\Omega = B(x^0; R)$, and let T be a positive number such that

$$E(t) \leq E(0)e^{1-(t/T)} \tag{13}$$

for all $t \geq 0$ and for all mild solutions of (2)–(5). Then $T \geq 2R/\sqrt{\mu}$.

Let us note finally that applying a method of Russell (1978) one can deduce easily from theorem 1 the following exact controllability result, analogous to an earlier theorem of Lions (1988), p. 227 :

THEOREM 3 Let $n \geq 3$ and $T > 2R/\sqrt{\mu}$. Given $u^0, u_T^0 \in H^1(\Omega)^n$ and $u^1, u_T^1 \in L^2(\Omega)^n$ arbitrarily, there exists a function $v \in L^2(0, T; L^2(\Gamma_1))$ such that the solution of the problem

$$\begin{gathered}
u'' - \mu\Delta u - (\lambda + \mu)\text{grad div } u = 0 \quad \text{in} \quad \Omega \times (0, T), \\
u = 0 \quad \text{on} \quad \Gamma_0 \times (0, T), \\
\mu(\partial_\nu u) + (\lambda + \mu)(\text{div } u)\nu = v \quad \text{on} \quad \Gamma_1 \times (0, T), \\
u(0) = u^0 \quad \text{and} \quad u'(0) = u^1 \quad \text{in} \quad \Omega
\end{gathered}$$

satisfies the final conditions

$$u(T) = u_T^0 \quad \text{and} \quad u'(T) = u_T^1 \quad \text{in} \quad \Omega.$$

Since the proof of theorem 3 is completely analogous to the proof of a similar result for the wave equation, given in Komornik (1994b), pp. 108-111, we shall only prove theorems 1 and 2 here.

2 PROOF OF THEOREM 1

Turning to the proof of theorem 1, first we note that it is sufficient to prove the estimate (12) for strong solutions : the general case then follows by an easy density argument. We shall henceforth assume that u is a strong solution of (2)–(5). In order to simplify the notation we set

$$Mu = 2m \cdot \nabla u + (n-1)u = 2m_k u_k + (n-1)u \tag{14}$$

(we shall use the summation convention of repeated indexes),

$$a = (n-1)\mu/(2R^2), \quad b = \sqrt{\mu}/R \tag{15}$$

and

$$d\Gamma_m = (m \cdot \nu) d\Gamma. \tag{16}$$

First we prove the following basic identity :

LEMMA 1 For every $0 \le S < T < +\infty$ we have

$$\begin{aligned} 2\int_S^T E\, dt &= \Big[\int_\Omega (Mu)\cdot u'\, dx\Big]_T^S \\ &+ \int_S^T \int_{\Gamma_0} \mu(\partial_\nu u)^2 + (\lambda+\mu)(\operatorname{div} u)^2\, d\Gamma_m\, dt \\ &+ \int_S^T \int_{\Gamma_1} |u'|^2 - (Mu)\cdot(au+bu') \\ &-\mu|\nabla u|^2 - (\lambda+\mu)(\operatorname{div} u)^2 + a|u|^2\, d\Gamma_m\, dt. \end{aligned} \tag{17}$$

Proof We multiply the equation (2) with $2m \cdot \nabla u$ and we integrate by parts in $\Omega \times (S,T)$. We have

$$\begin{aligned} 0 &= \int_S^T \int_\Omega (2m_k\partial_k u_i)(u_i'' - \mu\partial_j^2 u_i - (\lambda+\mu)\partial_i\partial_j u_j)\, dx\, dt \\ &= \Big[\int_\Omega (2m_k\partial_k u_i)u_i'\, dx\Big]_S^T \\ &+ \int_S^T \int_\Omega -m_k\partial_k(u_i')^2 + \mu m_k\partial_k(\partial_j u_i)^2 + 2\mu(\partial_j m_k)(\partial_k u_i)(\partial_j u_i) \\ &+(\lambda+\mu)m_k\partial_k(\partial_j u_j)^2 + 2(\lambda+\mu)(\partial_i m_k)(\partial_k u_i)(\partial_j u_j)\, dx\, dt \\ &+ \int_S^T \int_\Gamma -2\mu\nu_j m_k(\partial_k u_i)(\partial_j u_i) - 2(\lambda+\mu)\nu_i m_k(\partial_k u_i)(\partial_j u_j)\, d\Gamma\, dt \end{aligned}$$

$$= \left[\int_\Omega (2m_k \partial_k u_i) u_i' \, dx \right]_S^T$$

$$+ \int_S^T \int_\Omega (\partial_k m_k)((u_i')^2 - \mu(\partial_j u_i)^2 - (\lambda + \mu)(\partial_j u_j)^2)$$
$$+ 2\mu(\partial_j m_k)(\partial_k u_i)(\partial_j u_i) + 2(\lambda + \mu)(\partial_i m_k)(\partial_k u_i)(\partial_j u_j) \, dx \, dt$$
$$+ \int_S^T \int_\Gamma -2\mu\nu_j m_k(\partial_k u_i)(\partial_j u_i) - 2(\lambda + \mu)\nu_i m_k(\partial_k u_i)(\partial_j u_j)$$
$$+ (m_k \nu_k)(-(u_i')^2 + \mu(\partial_j u_i)^2 + (\lambda + \mu)(\partial_j u_j)^2) \, d\Gamma \, dt.$$

Since $\partial_k m_k = n$, $\partial_i m_k = \delta_{ik}$, $(u_i')^2 = |u'|^2$, $(\partial_j u_i)^2 = |\nabla u|^2$, $\partial_j u_j = \text{div } u$ and $m_k \nu_k = m \cdot \nu$, we can rewrite this identity in the following form :

$$\int_S^T \int_\Omega n|u'|^2 + (2-n)\mu|\nabla u|^2 + (2-n)(\lambda + \mu)(\text{div } u)^2 \, dx \, dt$$
$$= \left[\int_\Omega (2m \cdot \nabla u) \cdot u' \, dx \right]_T^S$$
$$+ \int_S^T \int_\Gamma (m \cdot \nu)(|u'|^2 - \mu|\nabla u|^2 - (\lambda + \mu)(\text{div } u)^2)$$
$$+ (2m_k \partial_k u_i)(\mu \partial_\nu u_i + (\lambda + \mu)\nu_i \text{div } u) \, d\Gamma \, dt.$$

Next we multiply the equation (2) with u and we integrate by parts again in $\Omega \times (S, T)$. We obtain that

$$0 = \int_S^T \int_\Omega u_i(u_i'' - \mu \partial_j^2 u_i - (\lambda + \mu)\partial_i \partial_j u_j) \, dx \, dt$$
$$= \left[\int_\Omega u_i u_i' \, dx \right]_S^T$$
$$+ \int_S^T \int_\Omega -(u_i')^2 + \mu(\partial_j u_i)^2 + (\lambda + \mu)(\partial_i u_i)^2 \, dx \, dt$$
$$+ \int_S^T \int_\Gamma -\mu u_i \partial_\nu u_i - (\lambda + \mu)(\nu_i u_i)\partial_j u_j \, d\Gamma \, dt$$

and hence

$$\int_S^T \int_\Omega -|u'|^2 + \mu|\nabla u|^2 + (\lambda + \mu)(\text{div } u)^2 \, dx \, dt$$
$$= \left[\int_\Omega u \cdot u' \, dx \right]_T^S$$
$$+ \int_S^T \int_\Gamma u_i(\mu \partial_\nu u_i + (\lambda + \mu)\nu_i \text{div } u) \, d\Gamma \, dt.$$

Multiplying this by $(n-1)$ and adding to the preceding identity, we obtain that

$$\int_S^T \int_\Omega |u'|^2 + \mu|\nabla u|^2 + (\lambda+\mu)(\text{div } u)^2 \, dx \, dt$$
$$= \left[\int_\Omega (Mu)\cdot u' \, dx\right]_T^S$$
$$+ \int_S^T \int_\Gamma (m\cdot\nu)\big(|u'|^2 - \mu|\nabla u|^2 - (\lambda+\mu)(\text{div } u)^2\big)$$
$$+(Mu_i)(\mu\partial_\nu u_i + (\lambda+\mu)\nu_i \text{div } u) \, d\Gamma \, dt.$$

Now we simplify the last integral by using the boundary conditions (3) and (4). On Γ_0 we have $u=0$ whence $u'=0$, $Mu_i = 2m\cdot\nabla u_i = 2(m\cdot\nu)\partial_\nu u_i$ and $\text{div } u = (\partial_\nu u)\cdot\nu$. Hence the integrand on Γ_0 is equal to

$$(m\cdot\nu)\big(\mu|\partial_\nu u|^2 + (\lambda+\mu)(\text{div } u)^2\big).$$

Furthermore, on Γ_1 we have

$$\mu\partial_\nu u_i + (\lambda+\mu)\nu_i \text{div } u = -(m\cdot\nu)(au_i + bu_i').$$

Hence

$$\int_S^T \int_\Omega |u'|^2 + \mu|\nabla u|^2 + (\lambda+\mu)(\text{div } u)^2 \, dx \, dt$$
$$= \left[\int_\Omega (Mu)\cdot u' \, dx\right]_T^S$$
$$+ \int_S^T \int_{\Gamma_0} \mu|\nabla u|^2 + (\lambda+\mu)(\text{div } u)^2 \, d\Gamma_m \, dt$$
$$+ \int_S^T \int_{\Gamma_1} |u'|^2 - \mu|\nabla u|^2 - (\lambda+\mu)(\text{div } u)^2 - (Mu_i)(au_i + bu_i') \, d\Gamma_m \, dt.$$

Adding

$$\int_S^T \int_{\Gamma_1} a|u|^2 \, d\Gamma_m \, dt$$

to both sides and using the definition of energy, hence the identity (17) follows. □

In order to prove theorem 1, we shall deduce from the identity (17) an energy inequality. We need two lemmas.

LEMMA 2 The following estimate holds true :

$$\left|\int_\Omega (Mu)\cdot u'\ dx\right| \le (2R/\sqrt{\mu})E. \tag{18}$$

Proof We have

$$\begin{aligned}\int_\Omega &|2m_k\partial_k u_i + (n-1)u_i|^2\ dx - \int_\Omega |2m_k\partial_k u_i|^2\ dx\\ &= \int_\Omega (n-1)^2u_i^2 + 4(n-1)m_ku_i\partial_k u_i\ dx\\ &= \int_\Omega (n-1)^2u_i^2 + 2(n-1)m_k\partial_k(u_i)^2\ dx\\ &= \int_\Gamma (2n-2)|u|^2\ d\Gamma_m + \int_\Omega (1-n^2)|u|^2\ dx\\ &\le (2n-2)\int_\Gamma |u|^2\ d\Gamma_m.\end{aligned}$$

Hence, using (15) and (6) we conclude that

$$\begin{aligned}&\left|\int_\Omega (Mu)\cdot u'\ dx\right|\\ &\le (\sqrt{\mu}/4R)\int_\Omega |2m_k\partial_k u_i + (n-1)u_i|^2\ dx + (R/\sqrt{\mu})\int_\Omega |u'|^2\ dx\\ &\le (\sqrt{\mu}/4R)\int_\Omega |2m_k\partial_k u_i|^2\ dx + (n-1)(\sqrt{\mu}/2R)\int_\Gamma |u|^2\ d\Gamma_m\\ &\qquad +(R/\sqrt{\mu})\int_\Omega |u'|^2\ dx\\ &\le (R/\sqrt{\mu})\Big(\int_\Omega |u'|^2 + \mu|\nabla u|^2\ dx + \int_\Gamma a|u|^2\ d\Gamma_m\Big)\\ &\le (2R/\sqrt{\mu})E. \quad \Box\end{aligned}$$

LEMMA 3 On Γ_1 the following estimate holds true :

$$|u'|^2 - (Mu)\cdot(au+bu') - \mu|\nabla u|^2 - (\lambda+\mu)(\operatorname{div} u)^2 + a|u|^2 \le 2|u'|^2.$$

Proof Using (15) we have

$$\begin{aligned}-(Mu)\cdot(au+bu') &\le 2|m_k\partial_k u||au+bu'| - (n-1)a|u|^2 - (n-1)bu\cdot u'\\ &\le 2R|\nabla u||au+bu'| + (1-n)a|u|^2 + (1-n)bu\cdot u'\\ &\le \mu|\nabla u|^2 + (R^2/\mu)(a^2|u|^2 + b^2|u'|^2 + 2abu\cdot u') + (1-n)a|u|^2 + (1-n)bu\cdot u'\\ &= \mu|\nabla u|^2 + ((1-n)/2)a|u|^2 + |u'|^2.\end{aligned}$$

Therefore

$$\begin{aligned}|u'|^2 - (Mu)\cdot(au+bu') - \mu|\nabla u|^2 - (\lambda+\mu)(\text{div } u)^2 + a|u|^2 \\ \le 2|u'|^2 + ((3-n)/2)a|u|^2 - (\lambda+\mu)(\text{div } u)^2 \\ \le 2|u'|^2.\end{aligned}$$

In the last step we used the assumption $n \ge 3$. ▯

Let us return to the identity (17). Applying the preceding two lemmas and using also the assumptions (8) and (9), we deduce easily from this identity the following inequality for all $0 \le S < T < +\infty$:

$$2\int_S^T E\ dt \le (2R/\sqrt{\mu})E(S) + (2R/\sqrt{\mu})E(T) + 2\int_S^T \int_{\Gamma_1} |u'|^2\ d\Gamma_m\ dt.$$

Thanks to the identity (7) and to the choice (11) of the coefficient B, the last term equals to

$$(2R/\sqrt{\mu})E(S) - (2R/\sqrt{\mu})E(T).$$

Hence

$$\int_S^T E\ dt \le (2R/\sqrt{\mu})E(S),$$

and letting $T \to +\infty$ we conclude that

$$\int_S^\infty E\ dt \le (2R/\sqrt{\mu})E(S), \quad \forall S \ge 0. \tag{19}$$

It is well-known that if a non-increasing function $E : \mathbb{R}_+ \to \mathbb{R}_+$ satisfies for some constant $T > 0$ the estimate

$$\int_S^\infty E\ dt \le TE(S), \quad \forall S \ge 0,$$

then it also satisfies the estimate

$$E(t) \le E(0)e^{1-(t/T)}, \quad \forall t \ge 0;$$

cf. e.g. Komornik (1991) or (1994b) for a simple proof. Since the energy is non-increasing by (7), the estimate (12) of theorem 1 follows from property (19). ▯

3 PROOF OF THEOREM 2

Let $\Omega = B(x^0; R)$, and let T be a positive number such that

$$E(t) \le E(0)e^{1-(t/T)}$$

for all $t \ge 0$ and for all mild solutions of (2)–(5). Then $E(T') < E(0)$ for all $T' > T$ and for all mild solutions of (2)–(5), not identically zero. Therefore it is sufficient to find, for each fixed $0 < T' < 2R/\sqrt{\mu}$, a solution of (2)–(5) satisfying $E(T') = E(0) > 0$.

We follow a method of Cazenave (1990). Fix $0 < T' < 2R/\sqrt{\mu}$ arbitrarily. Set $2\varepsilon = 2R - T'\sqrt{\mu}$, fix $v^1 \in C_c^\infty(\mathbb{R}^3)^3$ such that supp $v^1 \subset B(x^0;\varepsilon)$ and curl $v^1 \not\equiv 0$ and consider the problem

$$v'' - \mu\Delta v = 0 \quad \text{in} \quad \Omega \times \mathbb{R}, \tag{20}$$

$$v = 0 \quad \text{on} \quad \Gamma \times \mathbb{R}, \tag{21}$$

$$v(0) = 0 \quad \text{and} \quad v'(0) = v^1 \quad \text{in} \quad \Omega. \tag{22}$$

We will show below that the solution of this problem vanishes in some neighbourhood of Γ for every $t \in \mathbb{R}$ satisfying $\varepsilon + \sqrt{\mu}|t| < R$. Assuming for the moment this property, set $u(t) = \text{curl}\, v(t - (T'/2))$, $u^0 = v(T'/2)$ and $u^1 = v'(T'/2)$. Then div $u \equiv 0$ and therefore u coincides with the solution of the problem (2)–(5) for $0 \le t \le T'$. Furthermore, since $u(t)$ vanishes in a neighbourhood of Γ for all $0 < t < T'$, we deduce from the equality (7) that the energy of u is constant in $[0, T']$. Since $u'(T'/2) = \text{curl}\, v^1 \ne 0$, it follows that $E(T') = E(0) > 0$.

It remains to prove that the solution v of the above problem vanishes in some neighbourhood of Γ for every $t \in \mathbb{R}$ satisfying $|t| < T'/2 = (R-\varepsilon)/\sqrt{\mu}$. This will be proven if we show that the function $F : (-T'/2, T'/2) \to \mathbb{R}_+$, defined by the formula

$$F(t) = \int_{\varepsilon+\sqrt{\mu}|t|<|x|<R} |v'(t)|^2 + \mu|\nabla v(t)|^2 \, dx,$$

vanishes identically. Indeed, then $v(t)$ is constant in a neighbourhood of Γ for all $t \in (-T'/2, T'/2)$, and this constant must be zero by the boundary condition (21).

Let us first show that F has a global maximum at $t = 0$. Indeed, for $0 < t < T'/2$ we have

$$F'(t) = 2\int_{\varepsilon+\sqrt{\mu}t<|x|<R} v'(t)\cdot v''(t) + \mu\nabla v(t)\cdot\nabla v'(t)\, dx$$
$$-\sqrt{\mu}\int_{\varepsilon+\sqrt{\mu}t=|x|} |v'(t)|^2 + \mu|\nabla v(t)|^2 \, d\Gamma$$

$$= 2\mu \int_{\varepsilon+\sqrt{\mu}t<|x|<R} v'(t)\cdot\Delta v(t) + \nabla v'(t)\cdot\nabla v(t)\, dx$$
$$-\sqrt{\mu}\int_{\varepsilon+\sqrt{\mu}t=|x|} |v'(t)|^2 + \mu|\nabla v(t)|^2\, d\Gamma$$
$$= \sqrt{\mu}\int_{\varepsilon+\sqrt{\mu}t=|x|} 2\sqrt{\mu}v'(t)\cdot\partial_\nu v(t) - |v'(t)|^2 - \mu|\nabla v(t)|^2\, d\Gamma$$
$$\leq \sqrt{\mu}\int_{\varepsilon+\sqrt{\mu}t=|x|} 2\sqrt{\mu}|v'(t)||\nabla v(t)| - |v'(t)|^2 - \mu|\nabla v(t)|^2\, d\Gamma$$
$$= -\sqrt{\mu}\int_{\varepsilon+\sqrt{\mu}t=|x|} \left(|v'(t)| - \sqrt{\mu}|\nabla v(t)|\right)^2 d\Gamma \leq 0.$$

If $-T'/2 < t < 0$, then an analogous computation shows that

$$F'(t) = 2\int_{\varepsilon-\sqrt{\mu}t<|x|<R} v'(t)\cdot v''(t) + \mu\nabla v(t)\cdot\nabla v'(t)\, dx$$
$$+\sqrt{\mu}\int_{\varepsilon-\sqrt{\mu}t=|x|} |v'(t)|^2 + \mu|\nabla v(t)|^2\, d\Gamma$$
$$= \sqrt{\mu}\int_{\varepsilon-\sqrt{\mu}t=|x|} 2\sqrt{\mu}v'(t)\cdot\partial_\nu v(t) + |v'(t)|^2 + \mu|\nabla v(t)|^2\, d\Gamma$$
$$\geq \sqrt{\mu}\int_{\varepsilon+\sqrt{\mu}t=|x|} \left(|v'(t)| - \sqrt{\mu}|\nabla v(t)|\right)^2 d\Gamma \geq 0.$$

Hence $F(t) \leq F(0)$ for all $t \in (-T'/2, T'/2)$. Since $F(t) \geq 0$ for all t and since $F(0) = 0$ by the particular choice of the initial data, we conclude that $F \equiv 0$.

4 REFERENCES

Cazenave, T. (1990), On the propagation of confined waves along the geodesics, J. Math. Anal. Appl. 146(1990), 591-603.

Grisvard, P. (1989), Contrôlabilité exacte des solutions de l'équation des ondes en présence de singularités, J. Math. Pures Appl. 68(1989), 215-259.

Komornik, V. and Zuazua, E. (1990), A direct method for the boundary stabilization of the wave equation, J. Math. Pures Appl. 69(1990), 33-54.

Komornik, V. (1991), Rapid boundary stabilization of the wave equation, SIAM J. Control Opt. 29(1991), 197-208.

Komornik, V. (1994a), Boundary stabilization, observation and control of Maxwell's equations, PanAmerican Math. J. 4(1994), Number 4, 1-15.

Komornik, V. (1994b), Exact Controllability and Stabilization. The Multiplier Method, Masson–John Wiley, Paris, 1994.

Lagnese, J. (1983), Boundary stabilization of linear elastodynamic systems, SIAM J. Control Opt. 21(1983), 968-984.

Lagnese, J. (1991), Uniform asymptotic energy estimates for solutions of the equations of dynamic plane elasticity with nonlinear dissipation at the boundary, Nonlinear Anal. TMA, 16(1991), 35-54.

Lions, J.-L. (1988), Contrôlabilité exacte et stabilisation de systèmes distribués, Vol. 1, Masson, Paris, 1988.

Russell, D. L. (1978), Controllability and stabilizability theory for linear partial differential equations. Recent progress and open questions, SIAM Rev. 20(1978), 639-739.

Time-Minimal Control of Rotating Beams

W. Krabs Fachbereich Mathematik, Technische Hochschule Darmstadt, Darmstadt, Germany

INTRODUCTION: STATEMENT OF THE PROBLEM AND RESULTS.

In [3] we have investigated a problem of minimum norm boundary control of a rotating Euler-Bernoulli beam. For a homogeneous beam this problem is equivalent to the following approximation problem: Find a control $u_T \in L^2(0,T)$ for some $T > 0$ such that

$$\varphi_T(u_T) = \|(k - \frac{1}{3})\, u_T - \int_0^{\cdot} K(\cdot - s)\, u_T(s)\, ds - b\|_{L^2(0,T)} \tag{1.1}$$

is minimal subject to

$$\int_0^T u_T(t)\, z_j(t)\, dt = c_j \quad \text{for} \quad j \in I\!N_0 \tag{1.2}$$

where

$$z_0(t) = 1 \ , \ z_{2k-1}(t) = \sin \sqrt{\lambda_k}\, t \ ,$$
$$z_{2k}(t) = \cos \sqrt{\lambda_k}\, t \ , \ t \in [0,T] \ , \ k \in I\!N \ ,$$

$(c_j)_{j \in I\!N_0}$ is a suitable sequence in ℓ^2, $(\lambda_k)_{k \in I\!N}$ is the sequence of positive solutions of $\cosh \lambda^{\frac{1}{4}} \cos \lambda^{\frac{1}{4}} = -1$, $k > \frac{1}{3}$ is a given constant, and

$$K(t) = \sum_{j=1}^{\infty} k_j \sin \sqrt{\lambda_j}\, t \ ,$$
$$b(t) = \sum_{j=1}^{\infty} (\alpha_j \cos \sqrt{\lambda_j}\, t + \beta_j \sin \sqrt{\lambda_j}\, t) \ , \quad t \in [0,T] \ ,$$

are in $C[0,T]$ for every $T > 0$.
Here we consider the following problem: Given a constant $\beta > 0$ and some time $T_0 \geq 0$, find the smallest possible time $T_\beta \geq T_0$ such that there exists a control $u_\beta \in L^2(0,T_\beta)$ which satisfies (1.2) for $T = T_\beta$ and $u_T = u_\beta$ and $\varphi_{T_\beta}(u_\beta) \leq \beta$ where $\varphi_T(u_\beta)$ is given by (1.1) for $T = T_\beta$ and $u_T = u_\beta$. Every such control $u_\beta \in L^2(0,T_\beta)$ is called time-minimal.
For every $T > 0$ we define

$$U_T = \{u \in L^2(0,T) | \int_0^T u(t)\, z_j(t)\, dt = c_j \ , \ j \in I\!N_0\} \tag{1.3}$$

and

$$U_{T,\beta} = \{u \in U_T | \ \varphi_T(u) \leq \beta\} \ . \tag{1.4}$$

Let $U_{T,\beta}$ be non-empty for some $T \geq T_0$. Then the minimum time

$$T_\beta = \inf\{T \geq T_0 | \ U_{T,\beta} \text{ is non-empty}\} \tag{1.5}$$

is well defined and we can prove the

Existence Theorem: Let $U_{T,\beta}$ be non-empty for some $T \geq T_0$. Then there exists a time-minimal control $u_\beta \in U_{T_\beta,\beta}$.
The proof will be given in Section 2.
Now let $T_0 > 0$ be such that, for some $\delta > 0$,

$$\sqrt{\lambda_j} - \sqrt{\lambda_{j-1}} \geq \frac{2\pi + \delta}{T_0} \quad \text{for all} \quad j \in I\!N \tag{1.6}$$

(where $\lambda_0 = 0$). Such a T_0 exists, since one can show that

$$\lambda_j = ((j - \frac{1}{2})\,\pi + \varepsilon_j)^4 \ , \ j \in I\!N \ , \ \text{with} \ \lim_{j\to\infty} \varepsilon_j = 0 \ .$$

If (1.6) is satisfied for $T_0 > 0$, then also for every $T > T_0$ instead of T_0. By results in [2] it is therefore guaranteed that U_T (1.3) is non-empty for all $T \geq T_0$. Further it follows, for every $T \geq T_0$, that there exists $u_T \in U_T$ such that

$$\varphi_T(u_T) = \inf_{u \in U_T} \varphi_T(u) \tag{1.7}$$

and that u_T is unique (see [3]).
In Section 3 we consider the problem of time-minimal control with (1.2) for $j = 0, \ldots, N$ where $N \in I\!N$. Let T_β^N be the corresponding minimum time (see (3.3)). Then we show under suitable conditions that

$$\lim_{N\to\infty} T_\beta^N = T_\beta \ . \tag{1.8}$$

In Section 4 we show (again under suitable conditions) that for $T = T_\beta^N$ it is necessary that the equation

$$\varphi_T(u_T^N) = \beta \tag{1.9}$$

holds true where u_T^N is the minimizer of (1.1) subject to (1.2) for $j = 0, \ldots, N$.
So in order to determine T_β^N and $u_{T_\beta^N}^N$ one has to solve the equation (1.9) as a necessary condition for $T = T_\beta^N$.
This, together with (1.8), can be used in order to derive an algorithm for approximately determining T_β and a time-minimal control $u_\beta \in U_{T_\beta,\beta}$.

2. EXISTENCE OF TIME-MINIMAL CONTROLS.

For every $T > 0$ we define $S : L^2(0,T) \to L^2(0,T)$ by

$$S(u)(t) = (k - \frac{1}{3})\, u(t) - \int_0^t K(t-s)\, u(s)\, ds\ , \tag{2.1}$$

$t \in [0,T]$, $u \in L^2(0,T)$.
Then $U_{T,\beta}$ defined by (1.4) can also be written in the form

$$U_{T,\beta} = \{u \in U_T |\ \|S(u) - b\|_{L^2(0,T)} \le \beta\}\ . \tag{2.2}$$

Let $T \ge T_0$ (for some $T_0 \ge 0$) be given such that $U_{T,\beta}$ is non-empty. By the definition (1.5) of the minimum time T_β then there exists a sequence $(T_k)_{k\in I\!N}$ of times $T_k \ge T_\beta$ for all $k \in I\!N$, $T_\beta = \lim_{k\to\infty} T_k$, and, for every $k \in I\!N$, some $u_k \in L^2(0,T_k)$ such that

$$\int_0^{T_k} u_k(t)\, z_j(t)\, dt = c_j \quad \text{for all} \quad j \in I\!N_0$$

and

$$\|S(u_k) - b\|_{L^2(0,T_k)} \le \beta\ .$$

The latter inequality implies

$$\|S(u_k) - b\|_{L^2(0,T_\beta)} \le \beta$$

and hence

$$\|S(u_k)\|_{L^2(0,T_\beta)} \le \beta + \|b\|_{L^2(0,T_\beta)} \quad \text{for all} \quad k \in I\!N\ .$$

Since $S : L^2(0,T_\beta) \to L^2(0,T_\beta)$ is the adjoint operator of the operator $S^* : L^2(0,T_\beta) \to L^2(0,T_\beta)$ defined by

$$S_{T_\beta}^*(z)(t) = (k - \frac{1}{3})\, z(t) - \int_t^{T_\beta} K(s-t)\, z(s)\, ds\ , \tag{2.3}$$

$t \in [0,T_\beta]$, $z \in L^2(0,T_\beta)$, which is surjective, it follows by Theorem II.4.4 in [1] that S is invertible on $S(L^2(0,T_\beta))$ and has a bounded inverse S^{-1}, i.e. there is a constant $\gamma > 0$ such that $\|S^{-1}\|_{T_\beta} \le \gamma$. This implies

$$\|u_k\|_{L^2(0,T_\beta)} \le \gamma(\beta + \|b\|_{L^2(0,T_\beta)}) \quad \text{for all} \quad k \in I\!N$$

and in turn the existence of some $u_\beta \in L^2(0,T_\beta)$ and a subsequence $(u_{k_i})_{i\in\mathbb{N}}$ with $u_{k_i} \rightharpoonup u_\beta$ on $[0,T_\beta]$, i.e.

$$\int_0^{T_\beta} u_{k_i}(t)\, z(t)\, dt \to \int_0^{T_\beta} u_\beta(t)\, z(t)\, dt \quad \text{for all} \quad z \in L^2(0,T_\beta)\ .$$

For every $j \in \mathbb{N}_0$ we therefore conclude

$$\begin{aligned}\int_0^{T_\beta} u_\beta(t)\, z_j(t)\, dt &= \int_0^{T_\beta} (u_\beta(t) - u_{k_i}(t))\, z_j(t)\, dt + \int_0^{T_\beta} u_{k_i}(t)\, z_j(t)\, dt \\ &= \underbrace{\int_0^{T_\beta} (u_\beta(t) - u_{k_i}(t))\, z_j(t)\, dt}_{\to 0 \text{ as } i\to\infty} + \underbrace{\int_0^{T_{k_i}} u_{k_i}(t)\, z_j(t)\, dt}_{=c_j \text{ for all } i\in\mathbb{N}} - \underbrace{\int_{T_\beta}^{T_{k_i}} u_{k_i}(t)\, z_j(t)\, dt}_{\to 0 \text{ as } i\to\infty}\ ,\end{aligned}$$

hence,

$$\int_0^{T_\beta} u_\beta(t)\, z_j(t)\, dt = c_j \quad \text{for all} \quad j \in \mathbb{N}_0\ .$$

Since the functional $u \to \|S(u) - b\|_{L^2(0,T_\beta)}$, $u \in L^2(0,T_\beta)$ is sequentially weakly lower semi-continuous, it follows that

$$\|S(u_\beta) - b\|_{L^2(0,T_\beta)} \le \liminf_{i\to\infty} \|S(u_{k_i}) - b\|_{L^2(0,T_\beta)} \le \beta\ ,$$

i.e. $\varphi_{T_\beta}(u_\beta) \le \beta$, which shows that u_β is a time-minimal control.

Addition to the Proof: The assertion

$$\lim_{i\to\infty} \int_{T_\beta}^{T_{k_i}} u_{k_i}(t)\, z_j(t)\, dt = 0 \quad \text{for every} \quad j \in \mathbb{N}_0 \tag{2.4}$$

needs a justification. Choose any $T \ge T_k$ for all $k \in \mathbb{N}$ and define, for all $i \in \mathbb{N}$,

$$\tilde{u}_{k_i}(t) = \begin{cases} u_{k_i}(t) & \text{for} \quad t \in [0, T_{k_i}]\ , \\ 0 & \text{for} \quad t \in (T_{k_i}, T]\ . \end{cases}$$

Then it follows that

$$\|S(\tilde{u}_{k_i}) - b\|_{L^2(0,T)} = \|S(u_{k_i}) - b\|_{L^2(0,T_{k_i})} \le \beta$$

which implies

$$\|u_{k_i}\|_{L^2(0,T_{k_i})} = \|\tilde{u}_{k_i}\|_{L^2(0,T)} \le \|S^{-1}\|_T(\beta + \|b\|_{L^2(0,T)})$$

for all $i \in \mathbb{N}$. Therefore we obtain

$$\begin{aligned}&\left|\int_{T_\beta}^{T_{k_i}} u_{k_i}(t)\, z_j(t)\, dt\right| \le \|u_{k_i}\|_{L^2(0,T_{k_i})} \Big(\int_{T_\beta}^{T_{k_i}} z_j(t)^2\, dt\Big)^{1/2} \\ &\le \|S^{-1}\|(\beta + \|b\|_{L^2(0,T)})(T_{k_i} - T_\beta)^{1/2} \to 0 \quad \text{as} \quad i \to \infty\end{aligned}$$

which proves (2.4).

APPROXIMATION OF TIME-MINIMAL CONTROLS.

Let $T_0 > 0$ be such that (1.6) holds true. Let $\beta > 0$ be chosen such that $\beta < \varphi_{T_0}(u_{T_0})$.
Further let $\hat{T} > T_0$ be such that $\varphi_{\hat{T}}(u_{\hat{T}}) < \beta$.
For every $T \geq T_0$ and every $N \in I\!N$ we define

$$U_T^N = \{u \in L^2(0,T) | \int_0^T u(t)\, z_j(t)\, dt = c_j \;,\; j = 0, \ldots, N\} \tag{3.1}$$

and

$$U_{T,\beta}^N = \{u \in U_T^N \,|\, \varphi_T(u) \leq \beta\} \;. \tag{3.2}$$

Obviously

$$U_{T,\beta} \subseteq U_{T,\beta}^N \quad \text{and} \quad U_{T,\beta}^{N+1} \subseteq U_{T,\beta}^N \quad \text{for all} \quad N \in I\!N$$

and all $T \geq \hat{T}$.
Therefore $U_{T,\beta}^N$ is non-empty for all $N \in I\!N$ and all $T \geq \hat{T}$.
We assume in addition, for every $N \in I\!N$ and every $T \geq \hat{T}$, the existence of some $u \in U_{T,\beta}^N$ with $\int_0^T u(t)\, z_j(t)\, dt = 0$ for all $j \geq N+1$.
If we define, for every $N \in I\!N$,

$$T_\beta^N = \inf\{T \geq T_0 |\; U_{T,\beta}^N \text{ is non-empty}\} \;, \tag{3.3}$$

then it follows that

$$T_\beta^N \leq T_\beta^{N+1} \leq T_\beta \quad \text{for all} \quad N \in I\!N \;. \tag{3.4}$$

It also follows that for every $N \in I\!N$ there is some $u_\beta^N \in U_{T_\beta^N,\beta}^N$.
The proof of this fact is the same as the one given in Section 2 for T_β and $U_{T_\beta,\beta}$ instead of T_β^N and $U_{T_\beta^N,\beta}^N$, respectively.

ASSERTION: $T_\beta = \lim_{N\to\infty} T_\beta^N$,
if $k > \frac{1}{3}$ is such that

$$k - \frac{1}{3} > \hat{T} \int_0^{\hat{T}} K(\tau)^2\, d\tau \;. \tag{3.5}$$

PROOF: From (3.4) we infer the existence of $T^* \in [T_0, T_\beta]$ such that $T^* = \lim_{N\to\infty} T_\beta^N$.
We have to show that $T^* = T_\beta$. For every $N \in I\!N$ we choose $u_\beta^N \in U_{T_\beta^N,\beta}^N$ and define

$$u^N(t) = \begin{cases} u_\beta^N(t) & \text{for} \quad 0 \leq t \leq T_\beta^N \;, \\ 0 & \text{for} \quad T_\beta^N < t \leq T^* \;. \end{cases}$$

Then we have

$$\varphi_{T_\beta^N}(u_\beta^N) = \|S(u_\beta^N) - b\|_{L^2(0,T_\beta^N)} \le \beta \text{ for all } N \in I\!N$$

which implies

$$\|S(u_\beta^N)\|_{L^2(0,T_\beta^N)} \le \|b\|_{L^2(0,T^*)} + \beta \text{ for all } N \in I\!N \ . \tag{3.6}$$

From the assumption (3.5) we deduce that every operator $S : L^2(0,T_\beta^N) \to L^2(0,T_\beta^N)$ has a bounded inverse $S^{-1} : L^2(0,T_\beta^N) \to L^2(0,T_\beta^N)$ whose norm can be estimated in the form

$$\|S^{-1}\| \le \frac{1}{k-\frac{1}{3}-(T_\beta^N(k-\frac{1}{3})\int_0^{T_\beta^N} K(\tau)^2\,d\tau)^{\frac{1}{2}}}$$
$$\le \frac{1}{k-\frac{1}{3}-(\hat{T}(k-\frac{1}{3})\int_0^{\hat{T}} K(\tau)^2\,d\tau)^{\frac{1}{2}}} = \gamma \ .$$

Thus it follows from (3.6) that

$$\|u^N\|_{L^2(0,T_\beta)} = \|u_\beta^N\|_{L^2(0,T_\beta^N)} \le \gamma(\|b\|_{L^2(0,T^*)} + \beta) \text{ for all } N \in I\!N \ .$$

Therefore there exists a subsequence $(u^{N_i})_{i\in I\!N}$ in $L^2(0,T^*)$ and some $u^* \in L^2(0,T^*)$ such that

$$\int_0^{T^*} u^{N_i}(t)\, z(t)\, dt \underset{i\to\infty}{\longrightarrow} \int_0^{T^*} u^*(t)\, z(t)\, dt \text{ for all } z \in L^2(0,T_\beta) \ .$$

For every $j \in I\!N_0$ we then obtain

$$\int_0^{T^*} u^*(t)\, z_j(t)\, dt = \underbrace{\int_0^{T^*} (u^*(t) - u^{N_i}(t))\, z_j(t)\, dt}_{\to 0 \text{ as } i\to\infty}$$
$$+ \int_0^{T_\beta^{N_i}} \underbrace{u^{N_i}(t)\, z_j(t)\, dt}_{=c_j} + \int_{T_\beta^{N_i}}^{T^*} \underbrace{u^{N_i}(t)\, z_j(t)\, dt}_{\to 0 \text{ as } i\to\infty} ,$$

hence

$$\int_0^{T^*} u^*(t)\, z_j(t)\, dt = c_j \ .$$

Since the function $u \to \|S(u) - b\|_{L^2(0,T^*)}$ is weakly lower sequentially continuous on $L^2(0,T^*)$, it follows that

$$\|S(u^*) - b\|_{L^2(0,T^*)} \le \liminf_{i\to\infty} \|S(u^{N_i}) - b\|_{L^2(0,T^*)} \ .$$

For every $i \in I\!N$ we now have

$$\|S(u^{N_i}) - b\|^2_{L^2(0,T^*)} = \|S(u_\beta^{N_i}) - b\|^2_{L^2(0,T_\beta^{N_i})}$$
$$+\|S(u^{N_i}) - b\|^2_{L^2(T_\beta^{N_i},T^*)} \le \beta^2 + \|S(u^{N_i}) - b\|^2_{L^2(T_\beta^{N_i},T^*)}$$

where

$$\|S(u^{N_i}) - b\|^2_{L^2(T_\beta^{N_i},T^*)} \le 2(\|S(u^{N_i})\|^2_{L^2(T_\beta^{N_i},T^*)} + \|b\|^2_{L^2(T_\beta^{N_i},T^*)})$$

$$= 2(\int_{T_\beta^{N_i}}^{T^*} (\int_0^{T_\beta^{N_i}} K(t-s)\, u_\beta^{N_i}(s)\, ds)^2\, dt + \int_{T_\beta^{N_i}}^{T^*} b(t)^2\, dt)$$

$$\le 2(\int_{T_\beta^{N_i}}^{T^*} \int_0^{T_\beta^{N_i}} K(t-s)^2\, ds \int_0^{T_\beta^{N_i}} u_\beta^{N_i}(s)^2\, ds\, dt + \int_{T_\beta^{N_i}}^{T^*} b(t)^2\, dt)$$

$$\le 2([\gamma(\|b\|_{L^2(0,T^*)} + \beta)]^2 \int_{T_\beta^{N_i}}^{T^*} \int_0^{T^*} K(t-s)^2\, ds\, dt + \int_{T_\beta^{N_i}}^{T^*} b(t)^2\, dt \to 0$$

as $i \to \infty$.
Thus we obtain

$$\|S(u^*) - b\|^2_{L^2(0,T^*)} \le \beta^2 + \liminf_{i\to\infty} \|S(u^{N_i}) - b\|_{L^2(T_\beta^{N_i},T^*)} = \beta^2$$

which shows that $u^* \in U_{T^*,\beta}$. Since $T^* \le T_\beta$, it follows that $T^* = T_\beta$ which completes the proof of the Assertion.

HOW TO CALCULATE THE APPROXIMATE TIME-MINIMAL CONTROLS.

Let again $T_0 > 0$ be such that (1.6) holds true which is then also the case for all $T > T_0$. For every such T and every $N \in I\!N$ we at first consider the problem of finding $u_T^N \in U_T^N$ (see (3.1)) such that

$$\varphi_T(u_T^N) \le \varphi_T(u) \quad \text{for all} \quad u \in U_T^N\ . \tag{4.1}$$

In analogy to the result in [3] it follows that this problem has a unique solution u_T^N and that

$$\varphi_T(u_T^N) = \|S(u_T^N) - b\|_{L^2(0,T)} = \|\sum_{j=0}^{N} \lambda_j(T)\, \omega_j^T\|_{L^2(0,T)} \tag{4.2}$$

where

$$\omega_j^T = (S_T^*)^{-1}(z_j) \quad \text{for} \quad j = 0, \cdots, N\ , \tag{4.3}$$

S_T^* is the adjoint operator of $S : L^2(0,T) \to L^2(0,T)$ defined by (2.3) with T instead of T_β, and $(\lambda_0(T), \cdots, \lambda_N(T))^T \in I\!R^{N+1}$ is the unique solution of the linear system

$$\sum_{k=0}^{N} \langle \omega_j^T, \omega_k^T\rangle_{L^2(0,T)}\, \lambda_k(T) = c_j - \langle \nu_b, z_j\rangle_{L^2(0,T)} \quad \text{for} \quad j = 0, \ldots, N \tag{4.4}$$

with

$$\nu_b = S^{-1}(b) \ . \tag{4.5}$$

From (4.2) and (4.4) it follows that

$$\varphi_T(u_T^N)^2 = \sum_{j=0}^{N} (c_j - \langle v_b, z_j \rangle_{L^2(0,T)}) \ \lambda_j(T) \ . \tag{4.6}$$

Now let T_1, $T_2 \geq T_0$ be given arbitrarily. Then we infer from (4.6) that

$$\begin{aligned} |\varphi_{T_1}(u_{T_1}^N)^2 - \varphi_{T_2}(u_{T_2}^N)^2| \leq | \sum_{j=0}^{N} (c_j - \langle \nu_b, z_j \rangle_{L^2(0,T_1)})(\lambda_j(T_1) - \lambda_j(T_2))| \\ + | \sum_{j=0}^{N} (\langle \nu_b, z_j \rangle_{L^2(0,T_2)} - \langle \nu_b, z_j \rangle_{L^2(0,T_1)}) \lambda_j(T_2)| \ . \end{aligned} \tag{4.7}$$

On defining, for every $T \geq T_0$,

$$\begin{aligned} A_N(T) = ((\omega_j^T, \omega_k^T \rangle_{L^2(0,T)})_{j,k=0,\ldots,N} \text{ and} \\ r_N(T) = (c_j - \langle \nu_b, z_j \rangle_{L^2(0,T)})_{j=0,\ldots,N} \end{aligned} \tag{4.8}$$

we obtain from (4.4)

$$A(T_i) \ \lambda(T_i) = r(T_i) \quad \text{for} \quad i = 1, 2 \tag{4.9}$$

where $\lambda(T_i) = (\lambda_0(T_i), \ldots, \lambda_N(T_i))^T$ for $i = 1.2$.
The estimate (4.7) leads to

$$\begin{aligned} |\varphi_{T_1}(u_{T_1}^N)^2 - \varphi_{T_2}(u_{T_2}^N)^2| \leq \|r_N(T_1)\|_2 \ \|\lambda(T_1) - \lambda(T_2)\|_2 \\ + \|r_N(T_2) - r_N(T_1)\|_2 \ \|\lambda(T_2)\|_2 \end{aligned} \tag{4.10}$$

where $\| \cdot \|_2$ denotes the Euclidean norm in $I\!R^{N+1}$.
From the definition (4.8) it follows that

$$\begin{aligned} \|r_N(T_2) - r_N(T_1)\|_2^2 = \sum_{j=0}^{N} (\langle \nu_b, z_j \rangle_{L^2(0,T_2)} - \langle \nu_b, z_j \rangle_{L^2(0,T_1)})^2 \\ = \sum_{j=0}^{N} (\int_{T_1}^{T_2} \nu_b(t) \, z_j(t) \, dt)^2 \leq (N+1) \int_{T_1}^{T_2} \nu_b(t)^2 \, dt (T_2 - T_1)^2 \end{aligned}$$

which implies

$$\lim_{T_1 \to T_2} \|r_N(T_2) - r_N(T_1)\|_2 = 0 \ . \tag{4.11}$$

From (4.9) we deduce

$$\begin{aligned} A_N(T_2)(\lambda(T_2) - \lambda(T_1)) = r_N(T_2) - A_N(T_2) \ \lambda(T_1) \\ = r(T_2) - r(T_1) + (A_N(T_1) - A_N(T_2)) \ \lambda(T_1) \end{aligned}$$

and in turn

$$\begin{aligned}\|\lambda(T_2)-\lambda(T_1)\|_2 &\le \|A_N(T_2)^{-1}\|\,\|r_N(T_2)-r_N(T_1)\|_2\\ &+\ \|A_N(T_2)^{-1}\|\,\|A_N(T_1)-A_N(T_2)\|\,\|\lambda(T_1)\|_2\\ &\le\ \|A_N(T_2)^{-1}\|\,\|r_N(T_2)-r_N(T_1)\|_2\\ &+\ \|A_N(T_2)^{-1}\|\,\|A_N(T_1)-A_N(T_2)\|(\|\lambda(T_1)-\lambda(T_2)\|_2+\|\lambda(T_2)\|_2)\end{aligned} \tag{4.12}$$

where $\|A\|$ denotes the spectral norm of the $(N+1)\times(N+1)-$ matrix A.
Since the functions $T \to \langle\omega_j^T,\omega_k^T\rangle_{L^2(0,T)}$, $j,k=0,\ldots,N$, are continuous on $[T_0,\infty)$, if (3.5) is satisfied, it follows that

$$\lim_{T_1\to T_2}\|A_N(T_1)-A_N(T_2)\|=0\ . \tag{4.13}$$

Thus for T_1 being sufficiently close to T_2 we can assume that

$$\|A_N(T_2)^{-1}\|\,\|A_N(T_1)-A_N(T_2)\|<1\ .$$

Hence the estimate (4.12) leads to

$$\|\lambda(T_2)-\lambda(T_1)\|_2 \le \frac{\|A_N(T_2)^{-1}\|(\|r_N(T_2)-r_N(T_1)\|_2+\|A_N(T_1)-A_N(T_2)\|\,\|\lambda(T_2)\|_2)}{1-\|A_N(T_2)^{-1}\|\,\|A_N(T_1)-A_N(T_2)\|}$$

so that (4.11) and (4.13) imply

$$\lim_{T_1\to T_2}\|\lambda(T_2)-\lambda(T_1)\|_2=0\ . \tag{4.14}$$

As a result we infer from (4.10), (4.11), and (4.14) that

$$\lim_{T_1\to T_2}\varphi_{T_1}(u_{T_1}^N)=\varphi_{T_2}(u_{T_2}^N)\ , \tag{4.15}$$

i.e., the minimum norm function $T\to\varphi_T(u_T^N)$ is continuous on $[T_0,\infty)$.
Now we return to the situation at the beginning of Section 3. Since one can show for every $T\ge T_0$ that

$$\lim_{N\to\infty}\varphi_T(u_T^N)=\varphi_T(u_T)\ ,$$

we can assume for sufficiently large $N\in I\!N$ that $\beta<\varphi_{T_0}(u_{T_0}^N)$. Further it follows that $\varphi_{\hat T}(u_{\hat T}^N)<\beta$ for all $N\in I\!N$. This implies the existence of time-minimal controls $u_{T_\beta^N}^N\in U_{T_\beta^N,\beta}^N$ for every $N\in I\!N$. If $N\in I\!N$ is sufficiently large, then $T_\beta^N\in(T_0,\hat T]$. Further it follows that $\varphi_{T_\beta^N}(u_{T_\beta^N}^N)\le\varphi_{T_\beta^N}(u_\beta^N)\le\beta$. If $\varphi_{T_\beta^N}(u_{T_\beta^N}^N)<\beta$, due to the continuity of the function $T\to\varphi_T(u_T^N)$ on $[T_0,\infty)$, it follows that there exists some $T^N\in(T_0,T_\beta^N)$ such that $\varphi_{T^N}(u_{T^N}^N)=\beta$, i.e., $u_{T^N}^N\in U_{T^N,\beta}^N$. However, this contradicts the definition (3.3) of T_β^N. As a result we obtain

$$\varphi_{T_\beta^N}(u_{T_\beta^N}^N)=\beta \quad \text{for all sufficiently large} \quad N\in I\!N \tag{4.16}$$

which shows that $u_{T_\beta^N}^N$ is time-minimal. Therefore, in order to determine T_β^N and time-minimal controls $u_\beta^N\in U_{T_\beta^N,\beta}^N$ we have to solve the equation (4.16) as a necessary condition. This can, for instance, be done by applying the method of "regula falsi". The starting situation for this could be

$$\varphi_{\hat T}(u_{\hat T}^N)<\beta<\varphi_{T_0}(u_{T_0}^N)$$

which holds true for sufficiently large $N\in I\!N$.

References

[1] Goldberg, S. (1966). Unbounded linear operators. McGraw-Hill Book Company, New York – St. Louis – San Francisco – Toronto – London – Sydney.

[2] Krabs, W. (1992). On moment theory and controllability of one-dimensional vibrating systems and heating processes. Lecture Notes in Control and Information Sciences No 173, Springer-Verlag: Berlin – ... – Budapest.

[3] Krabs, W. (1993). Controllability of a rotating beam. In: Analysis and Optimization of Systems: State and Frequency Domain Approaches for Infinite-Dimensional Systems; edited by R.F. Curtain, A. Bensoussan, and J.L. Lions, Lecture Notes in Control and Information Sciences No 185, Springer-Verlag: Berlin – Heidelberg – New York, pp. 447 – 458.

A Singular Control Approach to Highly Damped Second-Order Abstract Equations and Applications

I. Lasiecka University of Virginia, Charlottesville, Virginia

L. Pandolfi Politecnio di Torino, Torino, Italy

R. Triggiani University of Virginia, Charlottesville, Virginia

1 Introduction, statement of the problem

The class of PDE's dynamics and abstract model. Let Ω be an open, bounded domain of $\mathbf{R}^n$, with smooth boundary Γ, $n \geq 1$. It was pointed out in [L-L-P], [T], that there is a class of second-order (in time) partial differential equations (PDE's) with a 'high' internal damping defined on Ω and subject to a control function u acting on a suitable boundary condition on Γ, that fails to be well posed with u only in $L_2(0,T;L_2(\Gamma))$. Rather, the $L_2(0,T;L_2(\Gamma))$-control function u needs to be supplemented by an additional parameter $u_0 \in L_2(\Gamma)$—which plays the role of a datum at $t = 0$—to yield the set of control data: $\{u_0 \in U;\ u \in L_2(0,T;U);\ U = L_2(\Gamma)\}$. Having fixed the parameter $u_0 \in U$ once and for all, the dynamical class of PDE's in question may then be modelled [L-L-P], [T] by the following input-solution relationship:

$$z_x(t) = e^{At}x + (Lu)(t), \qquad x \in Z; \tag{1.1}$$

$$(Lu)(t) = B_1u(t) + (L_0u)(t); \tag{1.2}$$

$$(L_0u)(t) = \int_0^t e^{A(t-\tau)}B_0u(\tau)d\tau + A\int_0^t e^{A(t-\tau)}B_1u(\tau)d\tau; \tag{1.3}$$

on a Hilbert space Z, which is identified below in (1.6), corresponding to the abstract equation

$$\dot{y} = Ay + B_0u + B_1u_t, \quad \text{in } [\mathcal{D}(A^*)]';\ y(0) = y_0 \in Y, \quad \text{or} \tag{1.4}$$

$$y(t) = y(t;y_0) = e^{At}[y_0 - B_1u(0)] + (Lu)(t), \tag{1.5}$$

at least for $u \in H^1(0,T;U)$, where Y is a Hilbert space and where

$$x = y_0 - B_1u(0) \in Z; \quad Z \equiv [\mathcal{D}((-A^*)^\gamma)]', \qquad \gamma \text{ as in (1.7)}. \tag{1.6}$$

The following abstract assumptions on (1.1) are maintained throughout the present note:

(H.1) $A: \ Y \supset \mathcal{D}(A) \to Y$ is the generator of a strongly continuous semigroup e^{At} on Y, $t \geq 0$, which is analytic on Y for $t > 0$, and which, moreover, is uniformly stable: there exist constants $C \geq 1$ and $\delta > 0$ such that $\|e^{At}\|_{\mathcal{L}(Y)} \leq Ce^{-\delta t}$, $t \geq 0$.

(H.2) B_0 and B_1 are linear operators continuous from the Hilbert space U to $[\mathcal{D}((A^*)^2)]'$ and to $[\mathcal{D}(A^*)]'$, respectively, the dual spaces of $\mathcal{D}((A^*)^2)$, and $\mathcal{D}(A^*)$ with respect to the pivot space Y, where A^* is the Y-adjoint of A; indeed, more precisely, we assume that

$$A^{-(1+\gamma)}B_0 \text{ and } A^{-\gamma}B_1 \in \mathcal{L}(U;Y), \text{ for some constant } 0 \leq \gamma < 1. \tag{1.7}$$

Needless to say, the above abstract setting—to include also the subsequent hypothesis (H.4) = (1.10)—encompasses automatically the class of boundary control problems for second-order (in t) PDE's with high internal damping [L-L-P], [T], which display a parabolic-like dynamical behavior, as reflected by assumption (H.1). Specific examples include the following (i) strongly damped wave equation, and (ii) plate-like equation with "structural" damping:

$$\begin{cases} w_{tt} - \Delta w - \Delta w_t = 0 \\ w(0,\,\cdot\,) = w_0,;\ w_t(0,\,\cdot\,) = w_1 \\ w|_\Sigma = u \end{cases} \quad \text{or} \quad \begin{cases} w_{tt} + \Delta^2 w - \Delta w_t = 0 \text{ in } Q & (1.8a) \\ w(0,\,\cdot\,) = w_0,;\ w_t(0,\,\cdot\,) = w_1 \text{ in } \Omega & (1.8b) \\ w|_\Sigma = u;\ \Delta w|_\Sigma = 0 \text{ in } \Sigma & (1.8c) \end{cases}$$

$Q = (0,\infty) \times \Omega$; $\Sigma = (0,\infty) \times \Gamma$, see [L-L-P], [T] for details.

The optimal control problem. With the z-dynamics in (1.1), we associate the following quadratic cost functional

$$J(u;z_x) \equiv \int_0^\infty \left[\|Rz_x(t)\|_Y^2 + \|u(t)\|_U^2\right] dt, \tag{1.9}$$

where $z_x(t)$ is the solution of (1.1) corresponding to $u \in L_2(0,\infty;U)$ and $x \in Z$. The "observation" operator R is assumed to satisfy the following hypothesis:

(H.3) For some $\bar{\gamma}$, $\gamma < \bar{\gamma}$, we have

$$R(-A)^{\bar{\gamma}} \in \mathcal{L}(Y) \Leftrightarrow R \in \mathcal{L}\left([\mathcal{D}((-A^*)^{\bar{\gamma}})]';Y\right). \tag{1.10}$$

The optimal control problem is then: for $x \in Z$, equivalently, $A^{-\gamma}x \in Y$,

$$\text{minimize } J(u;z_x) \text{ over all } u \in L_2(0,\infty;U). \tag{1.11}$$

This problem admits a unique optimal solution, denoted by $\{u_x^0(t), z_x^0(t)\}$. As with all quadratic cost optimal problems, the main goal is, qualitatively, to provide a feedback representation of the unique optimal pair $\{u_x^0(t), z_x^0(t)\}$, whereby $u_x^0(t)$ is expected to be expressed pointwise in t in terms of $z_x^0(t)$ by means of a suitable Riccati operator, i.e., the (unique) solution of an appropriate algebraic Riccati equation. It is a peculiar feature of the problem under study that the algebraic Riccati equation(s) associated with it is (are), in fact, non-standard.

Literature. For lack of space, we confine ourselves to noticing that the optimal control problem (1.9)–(1.11) for (1.1) has been the object of two contemporaneous studies by the authors [L-L-P] and [T], by means of two different approaches: a change of variable which transforms the original problem into a non-coercive one [L-L-P]; and a direct variational study [T]. Each of these works has lead to a rather complete theory of the optimal control problem (1.9)–(1.11) for (1.1), which, in particular, includes: (i) a non-standard synthesis [here referred to as SYNTH1] of the optimal pair $\{u_x^0(t), z_x^0(t)\}$; (ii) a non-standard algebraic Riccati Equation [here referred to as ARE1], which produces the (unique) Riccati operator entering in to SYNTH1. For a previous study with $u \in H^1(0,\infty;U)$, see [B].

2 Main Results

Orientation. The purpose of this note is to present new results which are obtained by means of a novel approach based on *singular control theory*. More precisely, we shall embed the original optimal control problem (1.9)–(1.11) for (1.1) [at least for suitably smooth x in (1.1), indeed, for $x \in \mathcal{D}(A_F)$, see below (2.2)], into a class of *singular* regulator problems, as an augmented problem having two main features: (i) the control and the observation operators are *bounded* (between the natural spaces); (ii) the functional cost is non-negative. Accordingly, the theory of singular control problems, as presented in [L-W], is applicable to the augmented problem. It eventually leads, through the minimization of the quadratic form appearing in the Dissipativity Inequality, to a new synthesis, SYNTH2, of the optimal pair $\{u_x^0(t), z_x^0(t)\}$, and to a correspondingly new Algebraic Riccati Equation, ARE2, which are different from the SYNTH1 and ARE1 obtained by other approaches [L-L-P], [T], and referred to above. [The singular control approach permits to re-obtain also SYNTH1 and ARE1 [L-P-T], but this is outside the space limitations of the present note.] Only a sketch will be presented here, pinpointing the main steps of the singular control approach while referring to [L-P-T] for all details. We begin with a preliminary result from [L-L-P], [T].

Theorem 2.1. Assume (H.1), (H.2), (H.3). Then:

(i) For each $x \in Z$, there exists a unique optimal pair, denoted by $\{u_x^0(t), z_x^0(t)\}$ satisfying

$$u_x^0 \in C([0,\infty]; U), \quad z_x^0 \in C([0,\infty]; Z). \tag{2.1}$$

Setting

$$\Phi(t)x \equiv z_x^0(t) - B_1 u_x^0(t) \in C([0,\infty]; Z), \tag{2.2}$$

we have that $\Phi(t)$ is an s.c. semigroup on Z, which, moreover, is uniformly stable in $\mathcal{L}(Z)$. Henceforth, we denote by $A_F : Z \supset \mathcal{D}(A_F) \to Z$ the generator of $\Phi(t)$.

(ii) Let now $x \in \mathcal{D}(A_F)$. Then

$$u_x^0 \in C^1([0,\infty]; U). \tag{2.3}$$

(iii) For $x \in Z$, the operator P defined by

$$Px = \int_0^\infty e^{A^*\sigma} R^* R z_x^0(\sigma) d\sigma \tag{2.4}$$

satisfies the following properties:

(i)

$$P : \text{ continuous } [\mathcal{D}((-A^*)^{\bar{\gamma}})]' \to \mathcal{D}((-A^*)^{\bar{\gamma}+\theta}), \quad \theta < 1 \tag{2.5}$$

[*a fortiori*, $Z \to \mathcal{D}((-A^*)^{\bar{\gamma}+\theta})$];

(ii)

$$B_i^* P : \text{ continuous } [\mathcal{D}((-A^*)^{\bar{\gamma}})]' \to U \tag{2.6}$$

[*a fortiori*, $Z \to U$];

(iii) P, viewed as an element of $\mathcal{L}(Y)$, is non-negative, self-adjoint, and

$$(Px, x)_Y = J(u_x^0; z_x^0) = \int_0^\infty \left[\|Rz_x^0(t)\|_Y^2 + \|u_x^0(t)\|_U^2 \right] dt = \min_{u \in L_2(0,\infty;U)} J(u; z_x); \tag{2.7}$$

(iv) $A^*P \in \mathcal{L}(Z;Y)$; $B_1^*A^*P \in \mathcal{L}(Z;U)$; $B_1^*R^*R \in \mathcal{L}(Z;U)$. □ (2.8)

The main results obtained via the singular control approach are given next. To this end, we introduce an additional assumption (also always fulfilled in the PDE's applications that we have in mind, such as those in (1.8)), which complements (H.3):

(H.4) For all $\epsilon > 0$, we have that

$$A^{-(1+\gamma+\epsilon)}B_0,\ A^{-(\gamma+\epsilon)}B_1 \text{ are compact operators in } \mathcal{L}(U;Y). \tag{2.9}$$

Theorem 2.2. Assume hypotheses (H.1), (H.2), (H.3), and (H.4). Then

(i) the operator

$$K \equiv (I - B_0^*PB_1 - B_1^*PB_0) = \{I - [B_0^*PB_1 + (B_0^*PB_1)^*]\} \tag{2.10}$$

is boundedly invertible in $\mathcal{L}(U)$:

$$K^{-1} = [I - B_0^*PB_1 - B_1^*PB_0]^{-1} \in \mathcal{L}(U); \tag{2.11}$$

(ii) the operator P solves also the following (ARE2): $\forall\, y_1, y_2 \in [\mathcal{D}((-A^*)^\gamma)]' \equiv Z$,

$$\begin{aligned}(Ay_1, Py_2)_Y + (Py_1, Ay_2)_Y + (Ry_1, Ry_2)_Y \\ = ([B_0^*P - B_1^*PA]y_1, K^{-1}[B_0^*P - B_1^*PA]y_2)_U\end{aligned} \tag{2.12}$$

(iii) (second feedback synthesis SYNTH2) Let $x \in Z$, then

$$u_x^0(t) = -K^{-1}[B_0^*P - B_1^*PA]z_x^0(t) \in C([0,\infty];U) \cap L_p(0,\infty;U),\ 1 \leq p \tag{2.13}$$

(iv) Moreover, P is the unique solution of (ARE2) within the class of non-negative, self-adjoint operators in $\mathcal{L}(Y)$ satisfying the regularity properties (2.5), (2.6). □

3 Proof of Theorem 2.2 (sketch)

Step 1. Let assume that *the control function $u(t)$ for the original problem* (1.1), (1.9), *be regular, in the sense that $u \in H^1(0,\infty;U)$*, so that $u(0)$ is well defined in U: in the preceding Theorem 2.1(ii), we have seen, in Eqn. (2.3), that the optimal control u_x^0 of problem (1.1), (1.11) is, infact, in $C^1([0,\infty];U)$ if $x \in \mathcal{D}(A_F)$, dense in $Z = [\mathcal{D}((-A^*)^\gamma)]'$. Then, with u regular, the following claim can be directly verified (see [L-L-P], [T, Remark 1.3.2]):

Claim 1. $z_x(t)$ in (1.1) provides a strong solution to the differential equation

$$y_t = Ay + B_0u + B_1u_t,\quad y(0) = y_0,\quad \text{provided } y_0 = x + B_1u(0), \tag{3.1}$$

$$\text{i.e.,}\quad \left.\begin{array}{l} u \in H^1(0,\infty;U) \\ y_0 = x + B_1u(0) \end{array}\right\} \Rightarrow \left\{\begin{array}{l} z_x(t) = y(t;y_0),\ 0 \leq t < \infty \\ z_x(0) = x + B_1u(0) = y(0;y_0) = y_0. \end{array}\right. \tag{3.2}$$

This observation suggests that we consider the following "augmented problem"

$$\left\{\begin{array}{lcll} \zeta_t & = & A\zeta + B_0u + B_1w & \zeta(0) = \zeta_0 \quad (3.3a) \\ u_t & = & w & u(0) = u_0 \quad (3.3b) \end{array}\right.$$

to which we associate the functional cost

$$\mathcal{J}(\zeta_0, u_0; w) = \int_0^\infty \{\|R\zeta(t)\|_Y^2 + \|u(t)\|_U^2\}dt. \tag{3.4}$$

The corresponding control problem is then

$$\inf_{w\in L_2(0,\infty;U)} \mathcal{J}(\zeta_0, u_0; w). \tag{3.5}$$

We note that the quadratic optimal control problem (3.5) for the dynamics (3.3) is *singular*, as the new input w is not penalized in the functional cost (3.4). According to Claim 1, if we compare (3.3) with (3.1), and (3.4) with (1.9), we obtain

Claim 2.

$$\left.\begin{array}{l} u \in H^1(0,\infty;U); \\ \text{[in particular,} \\ u_x^0 \text{ for } x \in \mathcal{D}(A_F)] \\ \\ y_0 = \zeta_0 = x + B_1 u(0) \\ w(t) = u_t(t) \text{ a.e.} \end{array}\right\} \Rightarrow \left\{\begin{array}{rcll} \zeta_{\zeta_0}(t) & = & y(t;y_0) = z_x(t),\ 0 \le t < \infty & (3.6) \\ \zeta_{\zeta_0}(0) & = & y(0;y_0) = z_x(0) = y_0 & \\ & = & x + B_1 u(0) = \zeta_0 & \\ & & & \\ J(u;z_x) & = & \mathcal{J}(x + B_1 u(0), u(0); u_t) & (3.7) \end{array}\right.$$

Step 2. Thus, in view of (3.7), the optimal solution to the original problem (1.1), (1.9), (1.11), and $x \in Z \equiv [\mathcal{D}((-A^*)^\gamma)]'$, should be contained in the solution to the augmented problem (3.3)–(3.5). In fact, see [P.1, Lemma 3],

THEOREM 3.1. Assume (H.1), (H.2), (H.3). With reference to J in (1.9) and to $\mathcal{J}$ in (3.4), we have for any $x \in Z = [\mathcal{D}((-A^*)^\gamma)]'$

$$\min_{u\in L_2(0,\infty;U)} J(u;z_x) = \inf_{u_0\in U} \inf_{w\in L_2(0,\infty;U)} \mathcal{J}(x + B_1 u_0, u_0; w). \quad \Box \tag{3.8}$$

Step 3. Writing the variation of parameter formula for the ζ-equation in (3.3a), or the y-equation in (3.1a), and integrating by parts the integral term involving $B_1 w = B_1 u_t$, we obtain (compare with (1.1)):

$$y(t;y_0) = e^{At}[y_0 - B_1 u(0)] + (Lu)(t); \ \zeta_{\zeta_0}(t) = e^{At}[\zeta_0 - B_1 u(0)] + (Lu)(t). \tag{3.9}$$

Step 4 (the augmented system). With reference to (3.3), we next introduce

$$\sigma(t) = \begin{bmatrix} \zeta(t) \\ u(t) \end{bmatrix}; \ \mathcal{B} = \begin{bmatrix} B_1 \\ I \end{bmatrix} \in \mathcal{L}(U\times U; Z\times U); \ \mathcal{R} = \begin{bmatrix} R & 0 \\ 0 & I \end{bmatrix} \in \mathcal{L}(Y\times U), \tag{3.10}$$

recalling assumptions (H.2) = (1.6) and (H.3) = (1.11), and

$$\mathcal{A} = \begin{bmatrix} A & B_0 \\ 0 & 0 \end{bmatrix} : Z\times U \supset \mathcal{D}(\mathcal{A}) \to Z\times U; \tag{3.11a}$$

$$\mathcal{D}(\mathcal{A}) = \{[z,u] \in Z\times U : \ A[A^{-\gamma}z + A^{-(1+\gamma)}B_0 u] \in Y\}. \tag{3.11b}$$

Thus, we can rewrite the augmented singular problem (3.3) as

$$\sigma_t = \mathcal{A}\sigma + \mathcal{B}w, \quad \sigma(0) = \sigma_0 = [\zeta_0, u_0] = [\zeta(0), u(0)], \tag{3.12}$$

and the corresponding optimal problem is

$$\inf_{w\in L_2(0,\infty;U)} \mathcal{J}(\sigma_0;w); \qquad \mathcal{J}(\sigma_0;w)=\int_0^\infty \|\mathcal{R}\sigma(t)\|^2_{Y\times U}dt, \tag{3.13}$$

PROPOSITION 3.2. (i) Assume (H.1), (H.2). Then, the operator $\mathcal{A}$ in (3.11) generates an s.c. analytic, uniformly bounded semigroup $e^{\mathcal{A}t}$, $t>0$ on $Z\times U$.

(ii) Assume further (H.3). Then, if the operator $\mathcal{F}$ is defined by $\mathcal{F}=[0,-I]$, we have that $\mathcal{A}_{\mathcal{F}}=\mathcal{A}+\mathcal{B}\mathcal{F}$, $\dot{\sigma}=\mathcal{A}_{\mathcal{F}}\sigma$ is the generator of a s.c. analytic semigroup $e^{\mathcal{A}_{\mathcal{F}}t}$, which, moreover, is uniformly stable in $\mathcal{L}(Z)$. In other words, the operator $\mathcal{F}$, through its action $w(t)=\mathcal{F}\sigma(t)=-u(t)$, via (3.10) is a stabilizing feedback operator for the system (3.12), or the pair $\{\mathcal{A},\mathcal{B}\}$. □

The proof is by direct computations based on the resolvent [L-P-T].

Step 5 (The Dissipativity Inequality). Thus, the dynamics (3.12) has the control operator $\mathcal{B}$ bounded, as in (3.10), between the natural spaces, and the operator $\mathcal{R}$ of the non-negative, singular cost (3.13) likewise bounded. Moreover, the dynamics (3.12) is exponentially stabilizable as guaranteed by Proposition 3.2. Thus, we may appeal to the recent treatment [L-W] on singular control theory and obtain

PROPOSITION 3.3. Assume (H.1), (H.2), (H.3). (i) Then, by invoking [L-W, Theorem 3(ii), p. 143], which is legal since the inf of $\mathcal{J}(\sigma_0,w)$ in (3.13) is $>-\infty$, over all $w\in L_2(0,\infty;U)$, $\forall\, \sigma_0\in Z\times U$, we obtain that: there exists a non-negative, self-adjoint, operator $\mathcal{W}\in\mathcal{L}(Z\times U)$ such that

$$\inf_{w\in L_2(0,\infty;U)} \mathcal{J}(\sigma_0;w)=(\mathcal{W}\sigma_0,\sigma_0)_{Z\times U}, \quad \sigma_0\in Z\times U, \tag{3.14}$$

satisfying the so-called "linear operator inequality" or "Dissipativity Inequality" D.I.:

$$2(\mathcal{A}x+\mathcal{B}w,\mathcal{W}x)_{Z\times U}+\|\mathcal{R}x\|^2_{Y\times U}\geq 0, \quad \forall\, x,w\in\mathcal{D}(\mathcal{A})\times U, \tag{3.15}$$

and, indeed, the selected $\mathcal{W}$ is the maximal self-adjoint operator (in the order induced by the inner product on $Z\times U$), among all bounded self-adjoint solutions of the D.I. = (3.15);

(ii) By combining (2.7) with (3.8) and (3.14), we obtain for $x\in Z$, $\sigma_0=[x+B_1u_0,u_0]$ recalling $\mathcal{B}$ from (3.10):

$$\begin{aligned}(Px,x)_Y=\min_{u\in L_2(0,\infty;U)} J(u;z_x) &= \inf_{u_0\in U}\left(\mathcal{W}\begin{bmatrix} x+B_1u_0\\ u_0\end{bmatrix},\begin{bmatrix} x+B_1u_0\\ u_0\end{bmatrix}\right)_{Z\times U}\\ &=\inf_{u_0\in U}\left(\mathcal{W}\left\{\begin{bmatrix} x\\ 0\end{bmatrix}+\mathcal{B}u_0\right\},\begin{bmatrix} x\\ 0\end{bmatrix}+\mathcal{B}u_0\right)_{Z\times U}\end{aligned} \tag{3.16}$$

(iii) Let $x\in\mathcal{D}(A_F)$, and consider the corresponding (unique) optimal pair $\{u_x^0(t)$ and $z_x^0(t)\}$ of the original optimal problem (1.11) for the original dynamics (1.1), satisfying the regularity (2.3) of Theorem 2.1. Then, by invoking Claim 2, in Step 1, Eqn. (3.6) and (3.7), we obtain: the infimum in (3.5) for $\mathcal{J}$ for the ζ-dynamics (3.3) with

$$\zeta_0=x+B_1u_0; \quad u_0=u_x^0(0),\ x\in\mathcal{D}(A_F) \tag{3.17}$$

is achieved by the optimal w for problem (3.5), denoted by $w^0_{\zeta_0}(\cdot)$ and given by

$$w^0_{\zeta_0}(t) = \frac{du^0_x(t)}{dt} \in C([0,\infty];U), \quad \zeta_0 \text{ as in (3.17)}, \tag{3.18}$$

(recall the regularity (2.3)), with corresponding optimal trajectory ζ for problem (3.3), (3.5), denoted by $\zeta^0_{\zeta_0}(t)$, and given by $\zeta^0_{\zeta_0}(t) = z^0_x(t)$, ζ_0 as in (3.17), according to (3.6), so that an optimal $\sigma^0_{\zeta_0}(t)$ defined by (3.10), for the augmented dynamics (3.12), is $\sigma^0_{\zeta_0}(t) = [z^0_x(t),\ u^0_x(t)]$, ζ_0 as in (3.17).

Finally, according to [L-W.1, Lemma 6, p. 149], the D.I. = (3.15) becomes an equality along an optimal pair of problem (3.12), (3.13): for ζ_0 as in (3.17):

$$2(\mathcal{A}\sigma^0_{\zeta_0}(t) + \mathcal{B}w^0_{\zeta_0}(t), \mathcal{W}\sigma^0_{\zeta_0}(t))_{Z\times U} + \|\mathcal{R}\sigma^0_{\zeta_0}(t)\|^2_{Y\times U} \equiv 0; \quad x \in \mathcal{D}(A_F). \quad \square \tag{3.19}$$

Step 6. We next collect a series of consequences of Proposition 3.3, see [L-P-T].

PROPOSITION 3.4. Assume (H.1), (H.2), (H.3). Then

(a) $(\mathcal{B}w, \mathcal{W}x)_{Z\times U} = 0, \quad \forall\, w, x \in U \times Z.$ (3.20)

(b) Setting

$$\mathcal{W} = \begin{bmatrix} P_0 & Q_0 \\ Q_1 & S \end{bmatrix}, \tag{3.21}$$

we have

(i) $(Px, x)_Y = (P_0x, x)_Z, \quad x \in Z;$ (3.22)

(ii)

$$P_0 = (-A)^\gamma[(-A)^{-\gamma}P_0(-A)^\gamma]^*(-A)^{-\gamma}; \quad Q_0 = (-A)^\gamma[Q_1(-A)^\gamma]^*; \tag{3.23}$$

$$Q_1 = [(-A)^{-\gamma}Q_0]^*(-A)^{-\gamma}; \qquad S = S^*, \tag{3.24}$$

where * denotes the adjoints in $\mathcal{L}(Y)$, $\mathcal{L}(Y;U)$; $\mathcal{L}(U;Y)$; $\mathcal{L}(U)$.

(iii)

$$Q_1 = -[(-A)^{-\gamma}B_1]^*(-A)^{-\gamma}P_0; \qquad S = -[(-A)^{-\gamma}B_1]^*(-A)^{-\gamma}Q_0; \tag{3.25}$$

$$P = (-A^*)^{-\gamma}(-A)^{-\gamma}P_0; \tag{3.26}$$

$$Q_0 = -P_0B_1; \quad (-A^*)^{-\gamma}(-A)^{-\gamma}Q_0 = -PB_1. \quad \square \tag{3.27}$$

Step 7. [Simplified version of Dissipativity Inequality] By virtue of the relations of the last proposition, the D.I. = (3.15), and its equality version (3.19), can be simplified.

THEOREM 3.5. Assume (H.1), (H.2), (H.3). Then

(i) the D.I. = (3.15) can be rewritten for all $z \in Z$, $u \in U$ as

$$0 \le 2(z, A^*P[z - B_1u])_Y + 2(u, B_0^*P[z - B_1u])_Y + \|Rz\|^2_Y + \|u\|^2_U \tag{3.28}$$

$$= 2(z, A^*Pz)_Y + \|Rz\|^2_Y + 2([B_0^*P - B_1^*PA]z, u)_U + ([I - (B_0^*PB_1 + B_1^*PB_0)]u, u)_U \tag{3.29a}$$

$$= 2(z, A^*Pz)_Y + \|Rz\|^2_Y + 2(Mz, u)_U + (Ku, u)_U, \tag{3.29b}$$

where we have set for convenience

$$K \equiv \{I - [B_0^*PB_1 + (B_0^*PB_1)^*]\} \in \mathcal{L}(U) \tag{3.30a}$$
$$= \text{bounded, non-negative, self-adjoint on } U \tag{3.30b}$$

$$M \equiv B_0^*P - B_1^*PA \in \mathcal{L}(Z;U), \tag{3.31}$$

where, *a-fortiori* from (2.8), $A^*P \in \mathcal{L}(Z;Y)$ so $PA \in \mathcal{L}(Y;Z)$; $B_0^*P \in \mathcal{L}(Z;U)$ and $B_1 \in \mathcal{L}(U;Z)$ by (1.7). Equivalently, under $z - B_1u = x$, $\forall x \in Z$, $\forall\, u \in U$:

$$2(x + B_1u, A^*Px)_Y + 2(u, B_0^*Px)_Y + \|R(x + B_1u)\|_Y^2 + \|u\|_U^2 \geq 0. \tag{3.32}$$

(ii) Let $x \in \mathcal{D}(A_F)$. Then, for all $t \geq 0$, equality (3.19) [of the D.I. = (3.15)] becomes

$$\begin{aligned} 0 &\equiv 2(z_x^0(t), A^*P[z_x^0(t) - B_1u_x^0(t)])_Y + 2(u_x^0(t), B_0^*P[z_x^0(t) - B_1u_x^0(t)])_Y \\ &\quad + \|Rz_x^0(t)\|_Y^2 + \|u_x^0(t)\|_U^2 & (3.33)\\ &= 2(z_x^0(t), A^*Pz_x^0(t))_Y + \|Rz_x^0(t)\|_Y^2 + 2(u_x^0(t), [B_0^*P - B_1^*PA]z_x^0(t))_U \\ &\quad + (u_x^0(t), [I - B_0^*PB_1 - (B_0^*PB_1)^*]u_x^0(t))_U & (3.34)\\ &\equiv 2(z_x^0(t), A^*Pz_x^0(t))_Y + \|Rz_x^0(t)\|_Y^2 \\ &\quad + 2(Mz_x^0(t), u_x^0(t))_U + (Ku_x^0(t), u_x^0(t))_U. & (3.35)\end{aligned}$$

(iii) Moreover, with $z_x^0(0) - B_1u_x^0(0) = x \in \mathcal{D}(A_F)$, (3.33) for $t = 0$ becomes

$$\begin{aligned} &2(x + B_1u_x^0(0), A^*Px)_Y + 2(u_x^0(0), B_0^*Px)_Y \\ &\qquad + \|R(x + B_1u_x^0(0)\|_Y^2 + \|u_x^0(0)\|_U^2 = 0, \end{aligned} \tag{3.36}$$

where $u_x^0(0) \in U$ by (2.1). □

Proof. (i) By (3.20), (3.21), and (3.11), we compute with $x = [z,u]$,

$$\begin{aligned} (\mathcal{A}x + \mathcal{B}w, \mathcal{W}x)_{Z\times U} &= (\mathcal{A}x, \mathcal{W}x)_{Z\times U} \\ &= ((-A)^{-\gamma}[Az + B_0u], (-A)^{-\gamma}[P_0z + Q_0u])_Y \\ &= (Az, (-A^*)^{-\gamma}(-A)^{-\gamma}P_0z)_Y + (Az, (-A^*)^{-\gamma}(-A)^{-\gamma}Q_0u)_Y \\ &\quad + (B_0u, (-A^*)^{-\gamma}(-A)^{-\gamma}P_0z) + (B_0u, (-A^*)^{-\gamma}(-A)^{-\gamma}Q_0u)_Y & (3.37)\\ &= (Az, Pz)_Y - (Az, PB_1u)_Y + (B_0u, Pz)_Y - (B_0u, PB_1u)_Y, & (3.38)\end{aligned}$$

where in going from (3.37) to (3.38) we have used (3.26) and (3.27) (right). Using (3.38) into the D.I. = (3.15) and recalling $\mathcal{R}$ from (3.10) yields readily (3.28), which is plainly equivalent to (3.32) under $z - B_1u = x \in Z$. The identity between (3.28) and (3.29a) can be verified directly from which (3.29b) follows at once via (3.30) and (3.31). Then, K is plainly self-adjoint by its definition (3.30) and, moreover, is non-negative by setting $z = 0$ in (3.29b) [which is the D.I.].

(ii) If $x \in \mathcal{D}(A_F)$, then the D.I. = (3.15) is satisfied as an equality along an optimal pair $\sigma_{\zeta_0}^0(t) = [z_x^0(t), u_x^0(t)]$ as in (3.19). But, by part (i), the D.I. = (3.15) becomes (3.28) through (3.29b), which specialized as equalities yield (3.33) through (3.35) for $u = u_x^0(t)$ and $z = z_x^0(t)$.

(iii) Setting $t = 0$ in (3.33), i.e., considering there $u_x^0(0)$ and $z_x^0(0) = x + B_1u_x^0(0)$, produces (3.36). □

Step 8. PROPOSITION 3.6. Assume (H.1), (H.2), (H.3). Let $x \in Z \equiv [\mathcal{D}((-A^*)^\gamma)]'$. Then, the optimal pair $\{u_x^0(t), z_x^0(t)\}$ of problem (1.11) for (1.1) satisfies explicitly,

$$Ku_x^0(t) = -Mz_x^0(t) \tag{3.39a}$$

$$[I - B_0^* P B_1 - B_1^* P B_0] u_x^0(t) = -[B_0^* P - B_1^* P A] z_x^0(t), \; t \geq 0, \tag{3.39b}$$

where K and M are defined in (3.30), (3.31).

Proof. The proof is similar to that of identity (3.33) in Theorem 3.5(ii). We return to identity (3.35b) and compare it with the D.I. = (3.29b) evaluated at $z = z_x^0(t)$ fixed. Thus, for fixed $z = z_x^0(t)$, we see that $u_x^0(t)$ minimizes the right-hand side (R.H.S.) of (3.29b), which we rewrite as a function of u (using $((B_0^* P B_1)^* u, u) = (B_0^* P B_1 u, u)$) as:

$$\begin{aligned}&[\text{R.H.S. of (3.29b) for } z = z_x^0(t) \text{ fixed}] \\ &= ([I - 2B_0^* P B_1]u, u)_U + (u, 2[B_0^* P - B_1^* P A] z_x^0(t))_U + z_x^0(t) - \text{term}.\end{aligned} \tag{3.40}$$

The minimum in u of (3.36), i.e., $u_x^0(t)$, satisifies the following identity, as desired,

$$\{[I - 2B_0^* P B_1] + [I - 2B_0^* P B_1]^*\} u_x^0(t) = -2[B_0^* P - B_1^* P A] z_x^0(t). \quad \Box \tag{3.41}$$

Step 9. Identity (3.41) raises the question as to whether the operator on its left-hand side is boundedly invertible in $\mathcal{L}(U)$. The answer is in the affirmative, provided that the compactness assumption (H.4) = (2.9) holds true. Remarkably, to show this result below, we shall need both to invoke [L-P-T, (7.1.6) of Proposition 7.1.2], as well as to follow the strategy of its proof. We begin by recalling an immediate corollary of [L-P-T, Lemma 7.1.1]:

LEMMA 3.7. Assume (H.1), (H.2), (H.3), (H.4). Then, the operators

$$C \equiv B_0^* P B_1 + (B_0^* P B_1)^* \in \mathcal{L}(U); \tag{3.42}$$

$$D \equiv B_1 [I + B_1^* R^* R B_1]^{-1} [B_1^* R^* R + (B_0^* + B_1^* A^*) P] \in \mathcal{L}(Z) \tag{3.43}$$

are compact. $\Box$

We can now establish the desired bounded inversion of $K \equiv [I - C]$ needed in (3.39), as well as the bounded inversion $[I - D]$ needed below in (3.45). Their proofs mimic the ideas of the proof of [L-P-T, Proposition 7.1.2.].

PROPOSITION 3.8. Assume (H.1), (H.2), (H.3), (H.4). With reference to (3.42) and (3.43), we have:

(i) the operator $K = [I - C]$ is boundedly invertible on U; i.e.,

$$K^{-1} = [I - C]^{-1} = [I - (B_0^* P B_1 + B_1^* P B_0)]^{-1} \in \mathcal{L}(U) \tag{3.44a}$$

$$= \text{positive self-adjoint}, \tag{3.44b}$$

thus improving on (3.30b) from non-negative to positive.

(ii) the operator $[I - D]$ is boundedly invertible on Z; i.e.,

$$[I - D]^{-1} = \{I - B_1 (I + B_1^* R^* R B_1)^{-1} [B_1^* R^* R + (B_0^* + B_1^* A^*) P]\}^{-1} \in \mathcal{L}(Z). \quad \Box \tag{3.45}$$

Proof. (i) In view of the compactness of C in Lemma 3.7, it is sufficient (as well as necessary) to prove that $\lambda = 1$ is not an eigenvalue of C. To this end, let

$$Cv = v; \qquad \text{or } [B_0^* P B_1 + B_1^* P B_0]v = v \tag{3.46}$$

for some $v \in U$. The goal is to show that, in fact, $v = 0$. Indeed, select the special point $x_0 = y_0 - B_1 v = -B_1 v$, where $y_0 = 0$ and v as in (3.46). Thus, from (1.2),

$$z_{x_0}^0(0) - B_1 u_{x_0}^0(0) = x_0 = -B_1 v. \tag{3.47}$$

Setting $t = 0$ in (3.39) with $x = x_0$ and replacing $z_{x_0}^0(0) = x_0 + B_1 u_{x_0}^0(0)$ yields, after a cancellation of $B_0^* P B_1 u_{x_0}^0(0)$:

$$[I - B_1^* P(B_0 + AB_1)]u_{x_0}^0(0) = -[B_0^* P - B_1^* PA]x_0. \tag{3.48}$$

Setting $x_0 = -B_1 v$ from (3.47) in (3.48), and invoking (3.46) yields,

$$[I - B_1^* P(B_0 + AB_1)]u_{x_0}^0(0) = [B_0^* P B_1 - B_1^* PAB_1]v = [I - B_1^* P(B_0 + AB_1)]v. \tag{3.49}$$

But the operator $[I - B_1^* P(B_0 + AB_1)]$, being the adjoint of the operator in [L-P-T, Proposition 7.1.2(i)], is boundedly invertible under (H.4). Thus, (3.49) yields, $u_{x_0}^0(0) = u_{x_0=0-B_1 v}^0(0)$ $= v$, recalling x_0. Then, Step 2 of the proof of [L-P-T, Proposition 7.1.2] yields $v = 0$, as desired. Thus, $K^{-1} \in \mathcal{L}(U)$. Hence K is actually positive self-adjoint via [and improving upon] (3.30b), and so is then K^{-1}, establishing (3.40b). A proof of (ii) is similar. □

Step 10. As a corollary of parts (i) and (ii) of Proposition 3.8, we obtain the desired second synthesis of the optimal pair, as well as a critical property that the map $z_x^0(0) \to x$, is an isomorphism on Z.

COROLLARY 3.9. Assume (H.1), (H.2), (H.3), (H.4). Let $x \in Z$. Then:

(i) The optimal synthesis of the optimal pair is also given by (see (3.30), (3.31)):

$$\begin{aligned} u_x^0(t) &= -K^{-1} M z_x^0(t) && (3.50a)\\ &= -[I - (B_0^* P B_1 + B_1^* P B_0)]^{-1}[B_0^* P - B_1^* PA] z_x^0(t) && (3.50b)\\ &\in C([0,\infty]; U) \cap L_p(0,\infty; U), \quad \forall\, p \ge 1 && (3.50c) \end{aligned}$$

(ii)

$$\begin{aligned} z_x^0(0) &= [I - D]x = [I + B_1 G]x && (3.51a)\\ &= \{I - B_1(I + B_1^* R^* R B_1)^{-1}[B_1^* R^* R + (B_0^* + B_1^* A^*)P] && (3.51b) \end{aligned}$$

where $[I - D]$ is an isomorphism Z onto Z. □

Proof. (i) Eqn. (3.50) follows from (3.39) via (3.44) of Proposition 3.8.

(ii) (3.51) is readily obtained after substituting (6.52a) of [L-P-T] for $u_x^0(0) = Gx$ of [L-P-T] into $z_x^0(0) = x + B_1 u_x^0(0)$, and recalling (3.43); then we invoke (3.45) for the critical direction of the isomorphism, since D is bounded by (3.43). □

We finally provide an important identity for the expression (3.29b) of the D.I., which uses $K^{-1} \in \mathcal{L}(U)$, with K^{-1} positive self-adjoint as in (3.44b).

COROLLARY 3.10. Assume (H.1), (H.2), (H.3), (H.4). Then the following identity (3.53) = (3.54) holds true for the right-hand side of (3.29b), which is the expression appearing in the D.I., so that:

$$\begin{aligned} 0 &\leq 2(z, A^*P[z - B_1u])_Y + 2(u, B_0^*P[z - B_1u])_Y + \|Rz\|_Y^2 + \|u\|_U^2 && (3.52)\\ &= 2(z, A^*Pz)_Y + \|Rz\|_Y^2 + 2(Mz, u)_U + (Ku, u)_U && (3.53)\\ &= 2(z, A^*Pz)_Y + \|Rz\|_Y^2 - (Mz, K^{-1}Mz)_U + (Ku + Mz, K^{-1}[Ku + Mz])_U. && (3.54)\end{aligned}$$

Proof. The identity (3.53) = (3.54) is immediately verified with $K^{-1} \in \mathcal{L}(U)$ and positive, self-adjoint by (3.44b). Also, (3.52) is nothing but the D.I. = (3.28). □

Step 11. It remains to show that the operator P in (2.4), satisfies the second Algebraic Riccati Equation (ARE2) = (2.12) [existence], and is, moreover, the unique, non-negative, self-adjoint solution with regularity properties as in (2.5), (2.6) [uniqueness]. This is obtained in the next result, where uniqueness of (ARE2) is reduced to uniqueness of (ARE1), which was already established in [L-L-P.1], [T.1].

PROPOSITION 3.11. (Existence) Assume (H.1), (H.2), (H.3), (H.4). Then, the operator P, satisfies the following (ARE2) = (2.12) for all $y_1, y_2 \in Z$:

$$\begin{aligned}&(Ay_1, Py_2)_Y + (Py_1, Ay_2)_Y + (Ry_1, Ry_2)_Y\\ &\quad = ([B_0^*P - B_1^*PA]y_1, [I - B_0^*PB_1 - B_1^*PB_0]^{-1}[B_0^*P - B_1^*PA]y_2)_U; \qquad (3.55)\end{aligned}$$

or recalling (3.30) and (3.31),

$$(Ay_1, Py_2)_Y + (Py_1, Ay_2)_Y + (Ry_1, Ry_2)_Y = (My_1, K^{-1}My_2). \quad \square \qquad (3.56)$$

Proof. Step 1. Returning to (3.50), we substitute the expression $u_x^0(t) = -K^{-1}Mz_x^0(t)$ into (3.35b), initially for $x \in \mathcal{D}(A_F)$ as required, thus obtaining the following result for all $t \geq 0$ and $x \in \mathcal{D}(A_F)$:

$$\begin{aligned}&2(z_x^0(t), A^*Pz_x^0(t))_Y + \|Rz_x^0(t)\|_Y^2 - 2(Mz_x^0(t), K^{-1}Mz_x^0(t))_Y\\ &\qquad + (KK^{-1}Mz_x^0(t), K^{-1}Mz_x^0(t))_U \equiv 0, \qquad (3.57)\end{aligned}$$

or, explicitly by (3.30), (3.31), after a simplification and setting $t = 0$, still for $x \in \mathcal{D}(A_F)$

$$\begin{aligned}&2(z_x^0(0), A^*Pz_x^0(0))_Y + \|Rz_x^0(0)\|_Y^2\\ &- \; ([I - (B_0^*PB_1 + B_1^*PB_0)]^{-1}[B_0^*P - B_1^*PA]z_x^0(0), [B_0^*P - B_1^*PA]z_x^0(0))_U \equiv 0, \quad (3.58)\end{aligned}$$

since K^{-1} is self-adjoint. Then, (3.58), originally obtained for $x \in \mathcal{D}(A_F)$, can be continuously extended to all $x \in Z$, since all operators in (3.58) are bounded.

We next recall from Corollary 3.9(ii) that the map $x \in Z \to z_x^0(0) \in Z$ is (in particular) surjective, so that we can replace $z_x^0(0)$ in (3.58) with any element $x \in Z$, thus obtaining:

$$\begin{aligned}&2(x, A^*Px)_Y + \|Rx\|_Y^2\\ &\quad -([I - (B_0^*PB_1 + B_1^*PB_0)]^{-1}[B_0^*P - B_1^*PA]x, [B_0^*P - B_1^*PA]x)_U = 0. \quad (3.59)\end{aligned}$$

Finally, by polarization, (3.59) yields the (ARE2) = (3.55) for all $y_1, y_2 \in Z$, as desired. □

PROPOSITION 3.12. Assume (H.1), (H.2), (H.3), (H.4).

(i) Let $\bar{P} = \bar{P}^* \in \mathcal{L}(Y)$ be a non-negative solution of (ARE2) = (3.55), satisfying the regularity properties (2.5), (2.6). Then, $\bar{P}$ solves also the D.I. = (3.28), which is equivalently rewritten in its form (3.29b) as:

$$2(z, A^*\bar{P}z)_Y + \|Rz\|_Y^2 + 2(\bar{M}z, u)_U + (\bar{K}u, u)_U \geq 0, \ \forall\, u \in U, \ z \in Z, \tag{3.60}$$

where $\bar{K}$ and $\bar{M}$ are given by (3.30a), (3.31) with P replaced by $\bar{P}$ now. Hence, $\bar{P}$ is also a solution of the ARE1 in [L-P-T].

(ii) (Uniqueness) Let P be the operator in (2.4) Then, such P is the unique solution of (ARE2) = (3.55) within the class of non-negative, self-adjoint operators in $\mathcal{L}(Y)$ which satisfy the regularity properties (2.5), (2.6). □

Proof. (i) We return to the D.I. in the form expressed by (3.54), which we compare with the (ARE2) = (3.56). Thus, since $\bar{P}$ by assumption satisfies (3.55) = (3.56), then by (3.54),

$$\begin{aligned} 0 &\leq 2(z, A^*\bar{P}z)_Y + \|Rz\|_Y^2 - (\bar{M}z, \bar{K}^{-1}\bar{M}z)_U + (\bar{K}u + \bar{M}u, \bar{K}^{-1}[\bar{K}u + \bar{M}z]) \\ &\text{(by (3.54))} \\ &= 0 + (\bar{K}u + \bar{M}u, \bar{K}^{-1}[\bar{K}u + \bar{M}z]), \end{aligned} \tag{3.61}$$

and so $\bar{P}$ satisfies also the D.I. = (3.29b) with $\bar{K}^{-1} \in \mathcal{L}(U)$ positive self-adjoint. Thus, the arguments of Section 6 [from Theorem 6.5 to Corollary 6.7] of [L-P-T] show that $\bar{P}$ satisfies the (ARE1) as well, as desired.

(ii) By part (i), uniqueness of a non-negative, self-adjoint solution P of the (ARE1) [which is guaranteed by [L-L-P], [T], [L-P-T], where P is the operator (2.4)] possessing the regularity properties (2.5), (2.6) implies uniqueness of a non-negative, self-adjoint solution, i.e., the same operator P in (2.4), of the (ARE2) = (3.55). □

References

[B] F. Bucci, A Dirichlet boundary control boundary problem for the strongly damped wave equation, *SIAM J. Control and Optimization*, 30 (1992), 1092–1100.

[L-L-P] I. Lasiecka, D. Lukes, and L. Pandolfi, Input dynamics and nonstandard Riccati equations with applications to boundary control of damped wave and plate equations, *J. Optim. Theory and Appl.*, to appear.

[L-P-T] I. Lasiecka, L. Pandolfi, R. Triggiani, A singular control approach to highly damped second-order abstract equations (complete manuscript), 1994.

[L-W] J. C. Louis and D. Wexler, The Hilbert space regulator problem and operator Riccati equation under stabilizability, *Annales de la Societe Scientifique de Bruxelles,* T. 105m 4 (1991), 137-165.

[P.1] L. Pandolfi, The standard regulator problem for systems with input delays: An approach through singular control theory, *Appl. Math. and Optim.*, to appear.

[P.2] L. Pandolfi, From singular to regular control systems, *Lecture Notes in Pure and Applied Mathematics*, Marcel Dekker, vol. 165, 1994, 153–166.

[T] R. Triggiani, Optimal quadratic boundary control problems for wave– and plate–like equations with high internal damping: An abstract approach. *Lecture Notes in Pure and Applied Mathematics*, Marcel Dekker, vol. 165, 1994, 215–270.

Uniform Stabilization of a Shallow Spherical Shell with Boundary Dissipation

I. Lasiecka University of Virginia, Charlottesville, Virginia

R. Triggiani University of Virginia, Charlottesville, Virginia

V. Valente Instituto per le Applicazioni del Calcolo, Rome, Italy

1 Introduction

The model. We consider a thin elastic spherical shell whose reference configuration in spherical coordinates (r, θ, ϕ) is the region: $r \in [R-h, R+h], \theta \in [0, \theta_0], \phi \in [0, 2\pi)$, where the half-thickness h of the shell, the middle surface ray R and the opening angle $\theta_0 < \pi$ are given. A spherical shell has two characteristic parameters, *thinness* and *shallowness* defined respectively by: $\alpha = h/R\sin\theta_0$; and $\beta = (R - R\cos\theta_0)/R\sin\theta_0$. We restrict our interest to a *thin* and *shallow* spherical shell, so that for some fixed $\alpha > 0$, we take $a = R\sin\theta_0$ and suppose $\alpha = h/a << 1$ and $\beta << 1$. This latter condition implies θ_0 sufficiently small. Starting from the Koiter linear shell model as in [G-L-V.1], [G-L-V.2], [K.1], we arrive at the shallow shell approximation by introducing the coordinate $\rho = R\theta$ and by replacing $\cot\theta$ by $\frac{1}{\theta}$. Then, the axially symmetric vibrations for the meridional and radial middle surface displacements (u, w) can be written in the following form on $(0, T) \times (0, \rho_0)$:

$$\begin{cases} u_{tt} + \dfrac{e}{R}v_{tt} - L(u) - \dfrac{e}{R}L(v) + \dfrac{(1+\nu)}{R}w' = 0; & \text{(1.1a)} \\ w_{tt} - e\left[v' + \dfrac{v}{\rho}\right]_{tt} + \dfrac{e}{\rho}[L(v)\rho]' - \dfrac{(1+\nu)}{\rho R}(u\rho)' + \dfrac{2(1+\nu)}{R^2}w = 0; & \text{(1.1b)} \end{cases}$$

$$v \equiv \frac{u}{R} + w'; \qquad L(u) \equiv u'' + \frac{u'}{\rho} - \frac{u}{\rho^2}; \qquad \rho_0 \equiv R\theta_0; \quad e \equiv \frac{h^2}{3}, \tag{1.1c}$$

where the prime symbol $'$ denotes differentiation with respect to ρ, along with the initial conditions

$$u(0,\,\cdot\,)=u_0,\ u_t(0,\,\cdot\,)=u_1;\ w(0,\,\cdot\,)=w_0,\ w_t(0,\,\cdot\,)=w_1, \tag{1.1d}$$

and the following homogeneous boundary conditions at $\rho=0$:

$$u=w'=L(v)=0, \qquad \rho=0,\ t>0, \tag{1.1e}$$

and the following dissipative boundary conditions at $\rho=\rho_0$:

$$\begin{cases} u'-\dfrac{(1+\nu)}{R}w+\nu\dfrac{u}{\rho_0} = -u_t-u & \text{(1.1f)} \\ ev' = -v_t \qquad \rho=\rho_0,\ t>0. & \text{(1.1g)} \\ eL(v)-ev_{tt} = w_t & \text{(1.1h)} \end{cases}$$

Well-posedness of system (1.1) will be asserted in Theorem 2.1 below. With system (1.1) we associate the energy functional

$$E(u,w)\equiv E(t)=E_k(t)+E_p(t); \tag{1.2}$$

$$2E_k(t) \equiv \int_0^{\rho_0}[u_t^2+w_t^2+v_t^2]\rho\,d\rho \tag{1.3}$$

$$\begin{aligned} 2E_p(t) \equiv\ & e\int_0^{\rho_0}\left[(v')^2\rho+\frac{v^2}{\rho}\right]d\rho \\ & +(1-\nu)\int_0^{\rho_0}\left[\left(u'-\frac{w}{R}\right)^2\rho+\left(\frac{u}{\rho}-\frac{w}{R}\right)^2\rho\right]d\rho \\ & +\nu\int_0^{\rho_0}\left[\left(u'-\frac{w}{R}\right)\sqrt{\rho}+\left(\frac{u}{\rho}-\frac{w}{R}\right)\sqrt{\rho}\right]^2 d\rho+u^2(\rho_0)\rho_0. \end{aligned} \tag{1.4}$$

The main goal of this note is to announce, and sketch the proof of, the following result in Theorem 2.3: That the dissipation taking place at the boundary of the shell in (1.1f–g–h)—via forces, shears, and moments applied to the edge—causes the energy (1.2) of the (well-posed) system (1.1) to decay exponentially to zero as $t\to\infty$.

Preliminary to the study of the asymptotic behaviour of the dissipative system (1.1) is the need to analyze well-posedness and regularity of its solutions, in order to justify the calculations in the energy method applied to (1.1). The required regularity result, in terms of weighted Sobolev spaces, is given in Theorem 2.2, while a well-posedness result of system (1.1) in the sense of semigroups, is given by Theorem 2.1.

Preliminaries. Stability and regularity results pertaining to the shell model (1.1) require the introduction of the following weighted spaces:

$$L_2^\rho(0,\rho_0)=\{u:\ u\sqrt{\rho}\in L_2(0,\rho_0)\},\ \text{with norm}\ \|u\|_{0,\rho}=\left[\int_0^{\rho_0}u^2\rho\,d\rho\right]^{\frac{1}{2}} \tag{1.5}$$

$$\mathcal{U}^1_\rho(0,\rho_0) = \{u : \frac{u}{\sqrt{\rho}},\ u'\sqrt{\rho} \in L_2(0,\rho_0),\ u(0)=0\} \tag{1.6a}$$

$$\text{with norm } \|u\|_{1,\rho} = \left\{\int_0^{\rho_0}\left[\frac{u^2}{\rho} + (u')^2\rho\right]d\rho\right\}^{\frac{1}{2}}, \tag{1.6b}$$

where we notice that, by Sobolev embedding,

$$\frac{u}{\sqrt{\rho}},\ u'\sqrt{\rho} \in L_2(0,\rho_0) \Rightarrow u \in H^1(\epsilon,\rho_0), \quad 0<\epsilon<\rho_0,\ \epsilon \text{ arbitrary positive,} \tag{1.7}$$

so that the identity

$$u^2(\rho) - u^2(\rho_0) = \int_{\rho_0}^{\rho}\frac{d}{d\rho}(u^2)d\rho = 2\int_{\rho_0}^{\rho}\frac{u}{\sqrt{\rho}}u'\sqrt{\rho}\,d\rho, \quad 0\le\rho\le\rho_0, \tag{1.8}$$

yields $u \in C[0,\rho_0]$, and so the condition $u(0)$ is well defined in (1.6a);

$$\mathcal{W}^2_\rho(0,\rho_0) = \{w : w\sqrt{\rho}\in L_2(0,\rho_0);\ w' \in \mathcal{U}^1_\rho(0,\rho_0);\ w(0)=0\} \tag{1.9a}$$

$$\text{with norm} \quad \|w\|_{2,\rho} = \left\{\int_0^{\rho_0} w^2\rho\,d\rho + \|w'\|^2_{1,\rho}\right\}^{\frac{1}{2}} \tag{1.9b}$$

$$= \left\{\int_0^{\rho_0}\left[w^2\rho + \frac{(w')^2}{\rho} + (w'')^2\rho\right]d\rho\right\}^{\frac{1}{2}}, \tag{1.9c}$$

where, likewise, $w\sqrt{\rho},\ \frac{w'}{\sqrt{\rho}} \in L_2(0,\rho_0) \Rightarrow w \in H^1(\epsilon,\rho_0)$ and the identity

$$w^2(\rho) - w^2(\rho_0) = \int_{\rho_0}^{\rho}\frac{d}{d\rho}(w^2)d\rho = 2\int_{\rho_0}^{\rho}w\sqrt{\rho}\frac{w'}{\sqrt{\rho}}d\rho, \quad 0\le\rho\le\rho_0, \tag{1.10}$$

yields that $w \in C[0,\rho_0]$, so that the condition $w(0)$ is well defined in (1.9a);

$$\mathcal{V}^1_\rho(0,\rho_0) = \{(w,u)\in L^\rho_2(0,\rho_0)\times L^\rho_2(0,\rho_0) : v \equiv \frac{u}{R} + w' \in L^\rho_2(0,\rho_0)\}; \tag{1.11a}$$

$$\text{with norm } \|(w,u)\|^2_{\mathcal{V}^1_\rho} = \|w\|^2_{0,\rho} + \|u\|^2_{0,\rho} + \left\|v = \frac{u}{R} + w'\right\|^2_{0,\rho} \tag{1.11b}$$

$$H^2_{\epsilon,\rho}(0,\rho_0) = \left\{u \in \mathcal{U}^1_\rho : \frac{u}{(\sqrt{\rho})^{3-\epsilon}},\ \frac{u'}{(\sqrt{\rho})^{1-\epsilon}},\ u''(\sqrt{\rho})^{1+\epsilon} \in L^\rho_2(0,\rho_0)\right\}, \tag{1.12}$$

$0<\epsilon<\frac{1}{2}$, with obvious norm. In the sequel we shall drop specification of the interval $(0,\rho_0)$.

Given u, w from (1.1a–b) and recalling $v = \frac{u}{R} + w'$ from (1.1c), we define

$$\begin{aligned} 2\hat{E}_p(u,w) \equiv\ & e\int_0^{\rho_0}\left[(v')^2\rho + \frac{v^2}{\rho}\right]d\rho + u^2(\rho_0)\rho_0 \\ & +(1-\nu)\int_0^{\rho_0}\left[\left(u' - \frac{w}{R}\right)^2\rho + \left(\frac{u}{\rho} - \frac{w}{R}\right)^2\rho\right]d\rho. \end{aligned} \tag{1.13}$$

By comparing (1.13) with (1.4), we see readily that $\hat{E}_p(u,w) \le E_p(t) \le \text{const}_\nu E_p(u,w)$, and thus the energy $\hat{E}_p(u,w)$ is topologically equivalent to the energy $E_p(t)$, so that they may be interchanged in the estimates below. The following lemma provides a topological relation between the energy space and the weighted spaces introduced above.

LEMMA 1.1. $E_p(t)$, equivalently $\hat{E}_p(u,w)$, is well defined, bounded and coercive on the space $\mathcal{U}^1_\rho \times \mathcal{W}^2_\rho$: there exist constant C, $c > 0$ such that for $(u,w) \in \mathcal{U}^1_\rho \times \mathcal{W}^2_\rho$, we have, recalling (1.6b) and (1.9c),

$$c[\|u\|^2_{1,\rho} + \|w\|^2_{2,\rho}] \leq \hat{E}_p(u,w) \sim E_p(t) \leq C[\|u\|^2_{1,\rho} + \|w\|^2_{2,\rho}] \tag{1.14a}$$

$$\|u\|^2_{1,\rho} + \|w\|^2_{2,\rho} = \left\{\int_0^{\rho_0}\left[\frac{u^2}{\rho} + (u')^2\rho\right]d\rho + \int_0^{\rho_0}\left[w^2\rho + \frac{(w')^2}{\rho} + (w'')^2\rho\right]d\rho\right\}. \tag{1.14b}$$

Proof. That $\hat{E}_p(u,w)$ is well defined and bounded for $(u,w) \in \mathcal{U}^1_\rho \times \mathcal{W}^2_\rho$ follows from (1.6), (1.9), (1.13). Next, let $u \in \mathcal{U}^1_\rho$ in (1.6), so that

$$u'\sqrt{\rho} - \frac{u}{\sqrt{\rho}} \equiv f(\rho) \in L_2(0,\rho_0). \tag{1.15}$$

We now multiply (1.15) by $\frac{u}{\sqrt{\rho}}$ and integrate over $[0,\rho_0]$. We then obtain after using $2u'u = \frac{d}{d\rho}(u^2)$ and $u(0) = 0$.

$$\int_0^{\rho_0}\left[u'u - \frac{u^2}{\rho}\right]d\rho = \int_0^{\rho_0} f\frac{u}{\sqrt{\rho}}d\rho = \frac{1}{2}[u^2(\rho_0) - u^2(0)] - \int_0^{\rho_0}\frac{u^2}{\rho}d\rho;$$

$$\int_0^{\rho_0}\frac{u^2}{\rho}d\rho \leq \frac{1}{2}u^2(\rho_0) + \int_0^{\rho_0}|f|\frac{|u|}{\sqrt{\rho}}d\rho \leq \frac{1}{2}\left[u^2(\rho_0) + \int_0^{\rho_0}\left(f^2 + \frac{u^2}{\rho}\right)d\rho\right]; \tag{1.16}$$

$$\int_0^{\rho_0}\frac{u^2}{\rho}d\rho \leq u^2(\rho_0) + \int_0^{\rho_0} f^2 d\rho. \tag{1.17}$$

Next, we estimate from (1.15), after adding and subtracting $w\sqrt{\rho}/R$, with $w \in \mathcal{W}^2_\rho$,

$$\begin{aligned}\int_0^{\rho_0} f^2 d\rho &= \int_0^{\rho_0}\left(u'\sqrt{\rho} - \frac{u}{\sqrt{\rho}}\right)^2 d\rho \\ &\leq 2\int_0^{\rho_0}\left(u'\sqrt{\rho} - \frac{w}{R}\sqrt{\rho}\right)^2 d\rho + 2\int_0^{\rho_0}\left(\frac{u}{\sqrt{\rho}} - \frac{w\sqrt{\rho}}{R}\right)^2 d\rho \qquad (1.18)\\ &\leq 2E_p(t), \qquad (1.19)\end{aligned}$$

where in the last step we have recalled the second integral term in the definition of $E_p(t)$ in (1.4). Substituting (1.19) in (1.17) yields, after recalling the last term in (1.4):

$$\int_0^{\rho_0}\frac{u^2}{\rho}d\rho \leq \frac{1}{\rho_0}u^2(\rho_0)\rho_0 + 2E_p(t) \leq \left(\frac{1}{\rho_0} + 1\right)2E_p(t). \tag{1.20}$$

Recalling (1.18)–(1.20), we obtain

$$\int_0^{\rho_0}(u')^2\rho\, d\rho - 2\int_0^{\rho_0}\left(u'\sqrt{\rho} - \frac{u}{\sqrt{\rho}}\right)^2 d\rho + 2\int_0^{\rho_0}\frac{u^2}{\rho}d\rho \leq \text{ const } E_p(t). \tag{1.21}$$

Then, (1.20) and (1.21) together, say, by (1.6b), that

$$E_p(t) > c_1\|u\|^2_{1,\rho}, \quad c_1 > 0, \tag{1.22}$$

as desired. Moreover, adding and subtracting, we obtain

$$\begin{aligned}\int_0^{\rho_0} w^2\rho\,d\rho &= R^2\int_0^{\rho_0}\left[\frac{w}{R}-\frac{u}{\rho}+\frac{u}{\rho}\right]^2\rho\,d\rho \\ &\le 2R^2\int_0^{\rho_0}\left(\frac{w}{R}-\frac{u}{\rho}\right)^2\rho\,d\rho + 2R^2\int_0^{\rho_0}\frac{u^2}{\rho}d\rho \\ &\le \text{const } E_p(t),\end{aligned} \tag{1.23}$$

where in the last step we have invoked the inequality from the second term on the right-hand side of (1.18) to (1.19), as well as (1.20). Furthermore, still after adding and subtracting

$$\int_0^{\rho_0}\frac{(w')^2}{\rho}d\rho \le 2\int_0^{\rho_0}\left(\frac{u}{R}+w'\right)^2\frac{1}{\rho}d\rho + \frac{2}{R^2}\int_0^{\rho_0}\frac{u^2}{\rho}d\rho \le \text{ const } E_p(t), \tag{1.24}$$

where in the last step we have invoked the first integral term on the right-hand side of (1.4) with $v = \frac{u}{R} + w'$, as well as (1.20). Finally, again after adding and subtracting

$$\int_0^{\rho_0}(w'')^2\rho\,d\rho \le 2\int_0^{\rho_0}\left(\frac{u'}{R}+w''\right)^2\rho\,d\rho + \frac{2}{R^2}\int_0^{\rho_0}(u')^2\rho d\rho \le \text{ const } E_p(t), \tag{1.25}$$

where in the last step we have invoked the first integral term on the right-hand side of (1.4) with $v' = \frac{u'}{R} + w''$, as well as (1.21). Then, (1.23)–(1.25) together, say, by (1.9c), that

$$E_p(t) \ge c_2\|w\|^2_{2,\rho}, \qquad c_2 > 0. \tag{1.26}$$

Finally, (1.22) and (1.26) combined yield the lower bound (coercivity) in (1.14), while the upper bound (boundedness) for $(u, w) \in \mathcal{U}^1_\rho \times \mathcal{W}^2_\rho$ via (1.6), (1.9), (1.13), was already observed at the outset. □

2 Statement of main results

THEOREM 2.1 (Well-posedness of (1.1)) Let

$$u_0 \in \mathcal{U}^1_\rho;\ w_0 \in \mathcal{W}^2_\rho;\ (u_1, w_1) \in \mathcal{V}^1_\rho. \tag{2.1}$$

Then, with reference to problem (1.1), we have

(i) the map $\{u_0, w_0, u_1, w_1\} \to \{u, w, u_t, w_t\}$ defines an s.c. contraction semigroup on the space

$$\mathcal{E} \equiv \left\{[u, w, u_1, w_1] : \ u \in \mathcal{U}^1_\rho;\ w \in \mathcal{W}^2_\rho;\ (u_1, v_1) \in \mathcal{V}^1_\rho\right\}; \tag{2.2a}$$

$$\|\{u, w, u_1, w_1\}\|^2_{\mathcal{E}} = \|u\|^2_{1,\rho} + \|w\|^2_{2,\rho} + \|(u_1, w_1)\|^2_{\mathcal{V}^1_\rho}. \tag{2.2b}$$

(ii) Moreover, the following "trace" regularity holds true for the solutions of (1.1), where $v = \frac{u}{R} + w'$ as in (1.1c):

$$\int_0^\infty \left[u_t^2(t,\rho_0) + w_t^2(t,\rho_0) + ev_t^2(t,\rho_0)\right] dt \leq C\|\{u_0, w_0, u_1, w_1\}\|_{\mathcal{E}}^2. \tag{2.3}$$

THEOREM 2.2 (Regularity) Recalling (1.5)–(1.12), let the initial data be in the domain of the generator of the s.c. semigroup guaranteed by Theorem 2.1(i). This implies that

$$u_0, u_1 \in \mathcal{U}_\rho^1;\ w_0, w_1 \in \mathcal{W}_\rho^2;\ \left(u_0, v_0 = \frac{u_0}{R} + w_0'\right) \in H_{\epsilon,\rho}^2 \times H_{\epsilon,\rho}^2 \tag{2.4}$$

satisfy the first two boundary conditions in (1.1f–g) at $\rho = \rho_0$; i.e.,

$$\begin{cases} u_0' - \dfrac{(1+\nu)}{R} w_0 + \nu \dfrac{u_0}{\rho_0} = -u_1 - u_0 & (2.5a) \\ \\ ev_0' = -v_1 = -\left(\dfrac{u_1}{R} + w_1'\right). & (2.5b) \end{cases}$$

Then the solution of problem (1.1), guaranteed by Theorem 2.1, satisfies the following regularity properties

$$\{u_t, w_t\} \in C([0,\infty];\ \mathcal{U}_\rho^1 \times \mathcal{W}_\rho^2); \tag{2.6}$$

$$\{u_{tt}, w_{tt}\} \in C([0,\infty];\ \mathcal{V}_\rho^1); \tag{2.7}$$

$$\left\{u, v = \frac{u}{R} + w'\right\} \in C\left([0,\infty]; H_{\epsilon,\rho}^2 \times H_{\epsilon,\rho}^2\right), \qquad \forall\, \epsilon > 0. \quad \square \tag{2.8}$$

The main result announced in the present note is

THEOREM 2.3 (Stabilization). Let the initial data $\{u_0, w_0, u_1, w_1\} \in \mathcal{E}$ defined in (2.2). Then, the following energy decay holds true for the solutions of problem (1.1), guaranteed by Theorem 2.1: there exist positive constants M, a, such that

$$E(t) \leq Me^{-at}E(0), \quad t \geq 0. \tag{2.9}$$

Comments on the Literature. (i) Problems related to boundary stabilization of wave-like and plate-like equations have received considerable attention in recent years, and we may only point to the recent book [Lag] and references therein. However, very little is known in the case of the dynamics of shells, particularly in the area of control/stabilization theory. The only exception is the paper [G-L-V], which gives an exact controllability result of a simplified spherical shell, by means of boundary controls whose regularity has not been ascertained. Moreover, [G-L-V] makes the undesirable assumption that R (or θ_0) be suitably small [beyond the requirements of the model].

(ii) It should be noted that the techniques developed in the context of plate/wave-like equations are not adequate to obtain a complete solution of the corresponding control problems of shells. The second equation (1.1b) in w contains the term u''/ρ with variable coefficient, where u'' is the highest term (in u) the first equation (1.1a) in u. Coupling between the u-equation (1.1a) and the w-equation (1.1b) occurs through high-order terms w_{tt}' and w''' in the former, u_{tt}' and u''' in the latter. Unique continuation properties (critical to absorb lower-order terms from the preliminary stabilizability estimate (3.2) below, to obtain the define estimate (3.7)] are poorly understood and developed in the case of systems. Indeed, a new contribution of the present work is a unique continuation result, Theorem 3.3 below, valid for the system of equations (1.1a–b) for shells. Finally, we remark that use of such Theorem 3.3 would also allow one to remove the restrictive assumption of smallness of R [or θ_0] made in [G-L-V] to obtain exact controllability.

3 Sketch of proof of Theorem 2.3

Because of space limitations, we shall confine ourselves to sketch only the proof of the stabilization Theorem 2.3. Details of Theorems 2.1 through 2.3 are given in the full companion paper [L-T-V].

Step 1. A starting point in stabilization theory is, as usual, the derivation of the following energy identity for the (dissipative) problem (1.1), which implies that the energy is non-increasing.

LEMMA 3.1. Let the initial data $\{u_0, w_0, u_1, w_1\}$ be in $\mathcal{E}$ defined by (2.2). Then, the following energy identity holds true for the solution of (1.1) guaranteed by Theorem,

$$E(t) + \rho_0 \int_0^t \left\{ u_t^2(t,\rho_0) + \left[\frac{u_t}{R}(t,\rho_0) + w_t'(t,\rho_0)\right]^2 + w_t^2(t,\rho_0) \right\} dt = E(0). \quad \square \tag{3.1}$$

The above identity is obtained after careful computations based on multiplying the first equation (1.1a) by u_t, the second equation (1.1b) by w_t–for smooth initial data as in the regularity Theorem 2.2 which yields smooth solutions as in (2.6)–(2.8)—and integrating by parts over $[0,\rho_0] \times [0,T]$. The resulting computations are justified for smooth solutions. One first obtains identity (3.1) for smooth initial data and then extend it by density.

Step 2. A first critical step to show stabilization is the establishment of the following preliminary estimate.

LEMMA 3.2. The (semigroup) solutions guaranteed by Theorem 2.1 satisfy the estimate: there is a constant $C_T > 0$ such that

$$\int_0^T E(t)dt + E(T) + E(0)$$
$$\leq C_T \left\{ \int_0^T [u_t^2(t,\rho_0) + w_t^2(t,\rho_0) + v_t^2(t,\rho_0)]dt + l.o.t.(u,w) \right\}, \tag{3.2}$$

where the lower-order terms are

$$l.o.t.(u,w) \leq C \left\{ \int_0^T \int_0^{\rho_0} w^2 \rho \, d\rho + \int_0^T [w^2(t,\rho_0) + u^2(t,\rho_0)]dt \right\}, \tag{3.3}$$

with $T >$ some suitable $T_0 > 0$. $\square$

The proof of Lemma 3.2 is based on the use of suitable multipliers: We again take smooth initial data as in Theorem 2.2, and multiply the first equation (1.1a) by $\left[u'\rho^2 + \frac{u\rho}{2}\right]$

and the second equation (1.1b) by $\left[w'\rho^2 - \frac{w\rho}{2}\right]$, and integrate by parts over $[0, \rho_0] \times [0, T]$. Inequality (3.2) is first obtained for smooth initial data, and then extended.

Step 3. The next step is to dispense, or absorb, the lower-order terms in (3.2). This is done, as usual, by means of a compactness/uniqueness argument. However, in the case of the shell's equations (1.1), the uniqueness part is far from trivial, and is singled out as a result of independent interest.

THEOREM 3.3 (Uniqueness continuation across the boundary). Let $\{u, w\}$ satisfy the original equations (1.1a–b), along with the overdetermined boundary conditions

$$u = w = v = L(v) \equiv 0, \qquad \text{at } \rho = 0,\ 0 \le t \le T_0; \tag{3.4}$$

$$u = u' = w = w' = w'' = w''' \equiv 0, \qquad \text{at } \rho = \rho_0,\ 0 \le t \le T_0, \tag{3.5}$$

for some suitable positive T_0. Then, the solution $\{u, w\}$ must be identically zero: $u \equiv w \equiv 0$ in $[0, T_0] \times [0, \rho_0]$. □

Remark 3.1. The difficulty in establishing the above uniqueness result stems from the fact that we deal with a system (1.1a–b). To obtain Theorem 3.3, we apply Carleman's estimates to each equation. □

Step 4. The result of Theorem 3.3 is critical in absorbing the lower-order terms in (3.2), as established by the following result.

LEMMA 3.4. With reference to the established inequality (3.2), there exists a constant $C_T > 0$ such that

$$l.o.t.(u, w) \le C_T \int_0^T [u_t^2(t, \rho_0) + w_t^2(t, \rho_0) + v_t^2(t, \rho_0)]dt. \quad \square \tag{3.6}$$

Step 5. By combining estimates (3.2) and (3.6), we obtain the definitive inequality,

$$\int_0^T E(t)dt \le C_T \int_0^T [u_t^2(t, \rho_0) + w_t^2(t, \rho_0) + v_t^2(t, \rho_0)]dt, \tag{3.7}$$

which then leads, in a standard way by semigroup methods, to the exponential decay (2.10) of Theorem 2.3.

References

[G-L-V.1] G. Geymonat, P. Loreti, and V. Valente, Exact controllability of a shallow shell model, *International Series of Numerical Mathematics,* Birkhauser Verlag Basel, vol. 107 (1992), 85–97.

[G-L-V.2] G. Geymonat, P. Loreti, and V. Valente, Exact controllability of a thin elstic hemispherical shell via harmonic analysis, *Boundary Value Problems for Partial Differential Equations and Applications,* ed. Masson, 1993.

[K.1] W. T. Koiter, On the foundations of the linear theory of thin elastic shells, *Proc. Kon. Ned. Akad. Wetensch.* B73 (1970), 169–195.

[L.1] J. Lagnese, Boundary stabilization of thin plates SIAM studies in Applied Mathematics, Philadelphia, 1989.

[L-T-V] I. Lasiecka, R. Triggiani, and V. Valente, full manuscript, 1994.

SQP Interior Point Methods for Parabolic Control Problems

F. LEIBFRITZ, E. W. SACHS Fachbereich IV - Mathematik, Universität Trier, D-54286 Trier, Germany

1 INTRODUCTION

We consider optimization problems which occur in the control of heating processes of industrial furnaces. The goal is to heat the metal ingots or ceramic charges such that the interior of them follows a prescribed temperature profile. In addition, certain constraints are imposed such as upper bounds on the temperature during the heating process.

The problem can be formulated as an optimal control problem with a nonlinear partial differential equation which is controlled through the boundary. The objective function is given in a least squares framework. The constraints are pointwise bounds on the states, i.e. the temperature, and are so-called hard constraints in contrast to constraints on the controls. It is well known from control problems with ordinary differential equations that the state constraints require sophisticated numerical methods to solve these problems.

In this paper we consider a finite dimensional approximation of the original problem. We present a method from sequential quadratic programming (SQP) which reduces the nonlinear problem to a sequence of quadratic subproblems. In order to keep the computing times reasonable, each of the quadratic subproblems has to be solved by an efficient algorithm. Traditional methods will not make use of the sparsity of the matrices occurring in the quadratic subproblems. Also the presence of inequality constraints poses a further serious difficulty for these algorithms. In our work we use interior point algorithms to solve the quadratic subproblems. Since these methods are of iterative nature they have two major advantages which we can use in the context of the control problem. First, the sparsity of the matrices can be used in an efficient way and the increase in computing time with a refinement of the time grid is modest. Secondly, the inner iterative method can be stopped during the first iterates of the outer SQP method when one is still far away from the solution at an early stage saving computing time. Then one can adaptively tighten the stopping criterion as the iteration progresses.

One can prove that the superlinear rate of the outer SQP method is retained although the quadratic subproblems are not solved to complete accuracy. Furthermore, the interior point algorithm itself can be shown to be superlinearly convergent. Its subproblems are solved by an iterative equation solver like GMRES, also with an adaptive termination criterion. All the convergence results are supported by numerical results for the control problem mentioned above.

In Section 2 the control problem is presented and discretized to yield a large scale nonlinear programming problem. Note that the underlying problem is still one dimensional in the space variables. The following section contains an inexact SQP method and the precise description of the sparsity pattern. In Section 3 an interior point method is described which is an iterative scheme to solve the quadratic subproblems occurring in the SQP iteration. Then we show how to reduce the number of variables and solve the linear systems during the QP solver iteratively with GMRES. The paper concludes with several numerical examples.

2 PARABOLIC BOUNDARY CONTROL PROBLEMS

As an example for an application of the numerical method we use a parabolic control problem that has already been studied numerically. Burger and Pogu (1991) use Newton's method and a conjugate gradient method whereas Kupfer and Sachs (1992) apply a reduced SQP method, both for the unconstrained case. In Leibfritz and Sachs (1994) a state constrained problem is solved with an exact SQP method combined with an interior point solver for the QP problems and a band LR decomposition for the linear systems.

The infinite dimensional example is for motivation only. The goal is to solve the finite dimensional counterpart in an efficient way with our method. Let $y(x,t)$ denote the temperature at time $t \in [0,T]$ and at the location $x \in [0,1]$. The heating process is governed by a nonlinear heat equation and controlled through the boundary. We use the following notion

C	:	heat capacity,	λ	:	heat conduction,
q	:	source term,	p	:	reference profile,
y	:	state,	u	:	control.

Then the parabolic boundary value problem is described as follows with $Q = (0,1)\times(0,T)$:

$$\begin{aligned}
C(y(x,t))y_t(x,t) - \tfrac{\partial}{\partial x}(\lambda(y(x,t))y_x(x,t)) &= q(x,t) && (x,t) \in Q,\\
\lambda(y(0,t))y_x(0,t) &= c[y(0,t) - u(t)], && t \in (0,T),\\
\lambda(y(1,t))y_x(1,t) &= 0, && t \in (0,T),\\
y(x,0) &= y_0(x), && x \in (0,1),\\
y(x,t) &\leq y_{max}, && (x,t) \in Q.
\end{aligned}$$

It is the goal to follow a desired temperature profile at the uncontrolled boundary $x = 1$.

$$\text{Minimize } \int_0^T [(y(1,t) - p(t))^2 + \gamma u^2(t)]dt.$$

Note that the optimal control problem has inequality constraints on the temperature. Since the temperature plays the role of the state this constitutes a hard constraint in the optimization problem.

In the paper by Burger and Pogu (1991) existence of a solution $y \in L_2[0,T;H^1(0,1)]$ was proven for each control $u \in L_2[0,T]$ under the following assumption:

ASSUMPTION 1 Let $C, \lambda \in C(I\!R)$ and

$$0 < C_1 \le C(t) \le C_2, \qquad 0 < \lambda_1 \le \lambda(t) \le \lambda_2$$

for all $t \in I\!R$ and let $\gamma > 0$.

In order to reformulate the problem in variational form as in Burger and Pogu (1991) define

$$\beta(s) = \int_0^s \lambda(\xi)\, d\xi, \quad \Gamma(s) = \int_0^s C(\xi)\, d\xi\,, \quad s \in I\!R.$$

These functions are sometimes used as vector functions, e.g. $\Gamma_d : I\!R^n \to I\!R^n$

$$\Gamma_d(y) = (\Gamma(y_1), ..., \Gamma(y_n))^T, \quad y \in I\!R^n.$$

The functions β_d, λ_d and C_d are defined in an analogous way.

The finite element discretization in space is done through the use of linear basis functions b_i

$$b_i(x_j) = \delta_{ij}, \qquad i,j = 1, \dots, N{+}1$$

with grid points in $(0,1)$:

$$x_i = (i-1)/N, \qquad i = 1, \dots, N{+}1.$$

The discretization in time is achieved by using step functions with grid points in $(0,T)$:

$$t^j = (j-1)\frac{T}{M}, \qquad j = 1, \dots, M{+}1.$$

This leads finally to the following discrete parabolic control problem with the variables

$$\begin{aligned} y &= ((y^1)^T, \dots, (y^M)^T)^T \in I\!R^{M(N+1)}, \\ u &= (u^1, \dots, u^M)^T \in I\!R^M. \end{aligned}$$

$$\text{Minimize } f(y,u) = \frac{T}{M}\sum_{j=1}^{M}[(y^j_{N+1} - p^j)^2 + \gamma(u^j)^2]$$

subject to

$$\begin{aligned} h^1(y,u) &= A\Gamma(y^1) + D\beta(y^1) + c(y^1_1 - u^1)e^1 - (q^1 + A\Gamma(y^0)) &= 0, \\ h^j(y,u) &= A(\Gamma(y^j) - \Gamma(y^{j-1})) + D\beta(y^j) + c\,(y^j_1 - u^j)e^1 - q^j &= 0, \\ g^k(y,u) &= y^k - y_{max} &\le 0, \end{aligned}$$

for $j = 2, ..., M, \quad k = 1, ..., M$, where

$$\begin{aligned}
(Ay^1)_i &= \frac{M}{T}(y_0, b_i) && i = 1, ..., N+1, \\
A_{ij} &= \frac{M}{T}((b_i, b_j)) && i, j = 1, ..., N+1, \\
D_{ij} &= (\frac{\partial b_i}{\partial x}, \frac{\partial b_j}{\partial x}) && i, j = 1, ..., N+1, \\
q_i^j &= \frac{M}{T}\int_{t^j}^{t^{j+1}} (q(\cdot, t), b_i)\, dt, && i = 1, ..., N+1, \quad j = 1, ..., M, \\
p^{j+1} &= \frac{M}{T}\int_{t^j}^{t^{j+1}} p(t)\, dt, && j = 1, ..., M-1, \\
e^1 &= (1, 0, \dots, 0)^T && \in I\!R^{N+1}.
\end{aligned}$$

In more compact form, this is a finite dimensional minimization problem with variables $z = (y, u) \in I\!R^{M(N+2)}$:

$$\text{Minimize } f(z)$$

subject to

$$\begin{aligned}
h(z) &= 0, \\
g(z) &\leq 0
\end{aligned}$$

where $f : I\!R^{M(N+2)} \to I\!R$, $h : I\!R^{M(N+2)} \to I\!R^{M(N+1)}$ and $g : I\!R^{M(N+2)} \to I\!R^{M(N+1)}$. There are many other optimal control problems with partial and ordinary differential equations which can be phrased in this form. They all exhibit certain sparsity properties which we will exploit in the next sections.

3 INEXACT SQP METHOD

The SQP Newton method is given for the problem with equality and inequality constraints in the following form, see e.g. Gill, Murray and Wright (1984). Define the Lagrangian by

$$L(\pi, \mu, z) = f(z) + (\pi^1, \dots, \pi^M)\, h(z) + (\mu^1, \dots, \mu^M)\, g(z).$$

At each given iterate z_k with Lagrange multipliers π_k, μ_k solve the quadratic problem

$$\begin{aligned}
& & \text{Minimize} \quad & f'(z_k)\zeta + \tfrac{1}{2}\, \zeta^T L_{zz}(\pi_k, \mu_k, z_k)\zeta \\
&(QP) & \text{s.t.} \quad & h'(z_k)\zeta + h(z_k) = 0, \\
& & & g'(z_k)\zeta + g(z_k) \leq 0.
\end{aligned}$$

Let $\delta\pi_k, \delta\mu_k$ be the Lagrange multipliers for the quadratic subproblem (QP). Then the new iterate and new Lagrange multipliers are determined with a step size ρ by

$$(\pi_{k+1}, \mu_{k+1}, z_{k+1}) = (\pi_k, \mu_k, z_k) + \rho(\delta\pi_k, \delta\mu_k, \delta z_k)$$

For the control problem from the previous section we obtain

$$\text{Minimize } 2\frac{T}{M}\sum_{j=1}^{M}(y_{N+1}^j - p^{*j})\delta y_{N+1}^j + 2\gamma\frac{T}{M}\sum_{j=1}^{M} u^j\delta u^j$$
$$+\frac{T}{M}\sum_{j=1}^{M}(\delta y_{N+1}^j)^2 + \gamma\frac{T}{M}\sum_{j=1}^{M}(\delta u^j)^2 + \frac{1}{2}\sum_{j=1}^{M}(\delta y^j)^T LY_j \delta y^j$$

subject to

$$\begin{aligned} G(y^1)\delta y^1 + Q_2\delta u^1 &= -h_1(y,u), \\ H(y^j)\delta y^{j-1} + G(y^j)\delta y^j + Q_j\delta u &= -h_j(y,u), \quad j = 2,\dots,M \\ \delta y^j + y^j - y_{max} &\le 0, \qquad\qquad j = 1,\dots,M. \end{aligned}$$

Here for $j = 1,\dots,M-1$

$$\begin{aligned} LY_j &= 2\frac{T}{M}e_{N+1}e_{N+1}^T + (\pi^j - \pi^{j+1})^T(AC_d'(y^j) + D\lambda_d'(y^j)) \in I\!R^{N+1}\times I\!R^{N+1}, \\ LY_M &= 2\frac{T}{M}\, e_{N+1}\, e_{N+1}^T + (\pi^M)^T(AC_d'(y^M) + D\lambda_d'(y^M)), \end{aligned}$$

and

$$\begin{aligned} G(w) &= AC_d(w) + D\lambda_d(w) + \text{diag}(c, 0, \dots, 0), \\ H(w) &= -AC_d(w), \\ (Q_j)_{ik} &= \begin{cases} -c & \text{if } i = 1 \text{ and } k = j-1 \\ 0 & \text{else} \end{cases}. \end{aligned} \tag{1}$$

From these expressions one can see that the quadratic problem is rather sparse. To underscore this observation note that the Jacobian of the inequality constraints is the identity. For the Jacobian of the equality constraints we have

$$h_y'(y,u) = \begin{pmatrix} G(y^1) & 0 & \cdot & \cdot & \cdot & \cdot & 0 \\ H(y^1) & G(y^2) & 0 & \cdot & \cdot & \cdot & \cdot \\ 0 & \cdot & \cdot & \cdot & & & \cdot \\ \cdot & \cdot & \cdot & \cdot & \cdot & & \cdot \\ \cdot & & \cdot & \cdot & \cdot & \cdot & \cdot \\ \cdot & & & \cdot & \cdot & G(y^{M-1}) & 0 \\ 0 & \cdot & & \cdot & 0 & H(y^{M-1}) & G(y^M) \end{pmatrix} \in I\!R^{M(N+1)\times M(N+1)}.$$

and

$$h_u'(y,u) = (Q_1^T, Q_2^T, \dots, Q_M^T)^T \in I\!R^{M(N+1)\times M}.$$

Additionally, the Hessian of the Lagrangian is also sparse

$$L''(y,u) = \begin{pmatrix} LY_1 & \cdots & \cdots & 0 \\ \vdots & \ddots & & \vdots \\ \vdots & & LY_M & \vdots \\ 0 & \cdots & \cdots & LU \end{pmatrix} \in I\!R^{M(N+2)\times M(N+2)}$$

where

$$LU = L_{uu} = 2\frac{T}{M}\gamma I \in I\!R^{M\times M}.$$

The necessary optimality conditions for the problem (QP) are given by finding $\xi_k \in I\!R^{M(N+2)}$ and Lagrange multipliers $\pi_k \in I\!R^{M(N+1)}$, $\mu_k \in I\!R^{M(N+1)}$ such that

$$\begin{aligned} L_{zz}(\pi_k,\mu_k,z_k)\zeta_k + h'(z_k)^T\pi_k + g'(z_k)^T\mu_k + f'(z_k) &= 0,\\ h'(z_k)\zeta_k + h(z_k) &= 0,\\ g'(z_k)\zeta_k + g(z_k) &\geq 0,\\ \mu_k &\geq 0. \end{aligned} \tag{2}$$

In the next section we will use an iterative scheme to solve (2). We terminate the iteration based on a termination criterion. This depends on a measure of inexactness how good the equations (2) are satisfied:

$$\phi(\pi,\mu,z) = \begin{pmatrix} L_z(\pi,\mu,z) \\ h(z) \\ \min(\mu, -g(z)) \end{pmatrix}.$$

The quantity to compare with is given by

$$\phi_k(\pi,\mu,z) = \begin{pmatrix} L_z(\pi_k,\mu_k,z_k) + L_{zz}(\pi_k,\mu_k,z_k)(z-z_k) \\ h(z_k) + h'(z_k)(z-z_k) \\ \min(\mu, -g(z_k) - g'(z_k)(z-z_k)) \end{pmatrix}.$$

The following statement provides a rule on when to terminate the QP solver and still to retain a locally good rate of convergence for the SQP method. In order to formulate the theorem we need a second order sufficiency condition.

ASSUMPTION 2 Let the vector (π_*,μ_*,z_*) satisfy the necessary optimality conditions $\phi(\pi_*,\mu_*,z_*) = 0$ and the linear independence constraint qualification

$$\nabla h_j(z_*), \nabla g_i(z_*), \quad \forall i,j \text{ s.t. } g_i(z_*) = 0$$

hold. Then for all $\zeta \neq 0$ we have

$$\zeta^T L_{zz}(\pi_*,\mu_*,z_*)\zeta > 0$$

if

$$h'(z_k)\zeta = 0 \quad \nabla g_i(z_*)\zeta = 0 \;\; \forall i \text{ s.t. } (\mu_*)_i > 0.$$

THEOREM 1 Let for π_*,μ_*,z_* the Assumption 2 hold. Then there exists an $\epsilon,\lambda > 0$ such that for all starting data π_0,μ_0,z_0 with

$$\|(\pi_0,\mu_0,z_0) - (\pi_*,\mu_*,z_*)\| \leq \epsilon$$

the SQP iterates satisfying

$$\|\phi_k(\pi_{k+1},\mu_{k+1},z_{k+1})\| \leq \lambda_k\|\phi(\pi_k,\mu_k,z_k)\|$$

converge at a linear rate provided $\lambda_k \in (0, \lambda)$. If in addition $\lim_{k\to\infty} \lambda_k = 0$, then the rate of convergence is superlinear.

For a proof of the theorem we refer to Leibfritz and Sachs (1995). This theorem shows that it is sufficient for the local convergence behavior to solve the quadratic problem only up to a certain accuracy. Whatever rule for termination is chosen, this results in a linear or superlinear rate of convergence.

4 INTERIOR POINT METHOD

We have seen from the previous paragraph that the quadratic subproblem in the SQP method should be solved by an iterative scheme. Since there are inequality constraints involved we use an interior point algorithm. We are aware of only a few approaches to use interior point methods in the context of optimal control problems, e.g. S. Wright (1993) and Boggs, Tolle, Kearsley (1994).

In order to describe the primal-dual method we introduce slack variables ν_k for the linearized inequality constraint in (2), i.e. replace the inequality for g in (2) by:

$$\nu_k + g(z_k)\zeta_k + g(z_k) = 0, \quad \nu_k \geq 0.$$

The resulting nonlinear system in the unknowns $\zeta_k, \pi_k, \mu_k, \nu_k$ is solved by a variant of Newton's method. Let an iterate, now denoted by subindex l, $\zeta_l, \pi_l, \mu_l, \nu_l$ be given such that $\mu_l > 0,\ \nu_l > 0$ satisfy

$$\begin{aligned} L_{zz}\zeta_l + h'(z)^T\pi_l + g'(z)^T\mu_l + f'(z) &= 0, \\ h'(z)\zeta_l + h(z) &= 0, \\ \nu_l + g'(z)\zeta_l + g(z) &= 0. \end{aligned} \tag{3}$$

The positivity of μ and ν is enforced by introducing an additional equation $\mu_i\nu_i = m_l$ for all components i. Equivalently, one can write $N_l\mu_l = e$, where $N_l = diag\ \nu_l$, $M_l = diag\ \mu_l$ and $e = (1, ..., 1)^T$. For fixed l a Newton step looks like

$$\begin{aligned} L_{zz}\delta\zeta + h'(z)^T\delta\pi + g'(z)^T\delta\mu &= 0, \\ h'(z)\delta\zeta &= 0, \\ g'(z)\delta\zeta + \delta\nu &= 0, \\ N_l\delta\mu + M_l\delta\nu &= -M_l\nu_l + m_l e. \end{aligned} \tag{4}$$

Note that m_l will be adjusted for the next iteration. The new iterate is obtained with a step size parameter $\alpha_l > 0$

$$(\zeta_{l+1}, \pi_{l+1}, \nu_{l+1}, \mu_{l+1}) = (\zeta_l, \pi_l, \nu_l, \mu_l) + \alpha_l\ (\delta\zeta, \delta\pi, \delta\nu, \delta\mu)$$

The choice of the step size is done according to Y. Zhang, R. E. Tapia and J. E. Dennis (1992) with $d = M(N+1)$

$$\begin{aligned} \alpha_l &= \frac{-\tau_l}{\min_{i=1,\ldots,d}\{(\mu_l^i)^{-1}\mu_l^i, (\nu_l^i)^{-1}\nu_l^i, -\tau_l\}} \\ \tau_l &= \min\,(\tau, 1 - c_1\,\mu_l^T\nu_l), \quad \tau \in (0,1),\ c_1 > 0. \end{aligned} \tag{5}$$

This rule ensures that the new iterates $(\nu_{l+1}, \mu_{l+1}) = (\nu_l, \mu_l) + \alpha_l\ (\delta\nu, \delta\mu)$ remain positive. The parameter m_l is adjusted according to the following rule which yields this theorem due to Y. Zhang, R. E. Tapia and J. E. Dennis (1992).

THEOREM 2 If the step size is chosen according to (4) and the parameter m_l is adjusted according to

$$\begin{aligned} \sigma_l &= \min(\sigma, c_2\, \mu_l^T \nu_l), \quad \sigma, c_2 > 0, \\ m_l &= \sigma_l \frac{1}{d} \mu_l^T \nu_l, \quad d = M(N+1), \end{aligned}$$

then the duality gap converges locally superlinearly to zero

$$\lim_{l \to \infty} \frac{\mu_{l+1}^T \nu_{l+1}}{\mu_l^T \nu_l} = 0.$$

In our numerical experiments we used the constants

$$c_1 = c_2 = 1, \ \sigma = .005, \ \tau = .99995.$$

5 SPARSITY PATTERNS

In this section we consider the numerical solution of the linear system (4) which has to be solved at each step of the interior point algorithm. Elimination of $\delta\nu$ in the linear system (4) yields

$$\begin{pmatrix} L_{zz} & h'(z)^T & g'(z)^T \\ h'(z) & 0 & 0 \\ g'(z) & 0 & -M_l^{-1}\nu_l \end{pmatrix} \begin{pmatrix} \delta\zeta \\ \delta\pi \\ \delta\mu \end{pmatrix} = \begin{pmatrix} 0 \\ 0 \\ -m_k M_l^{-1} e + N_l e \end{pmatrix}. \tag{6}$$

If we reorder variables for parabolic control problem in the following way

$$\cdots\ \delta y_l^{j-1} \quad \delta\pi_l^j\ \delta\mu_l^j\ \delta\nu_l^j\ \delta y_l^j \quad \delta\pi_l^{j+1}\ \cdots$$

the matrix in (6) exhibits a block diagonal structure

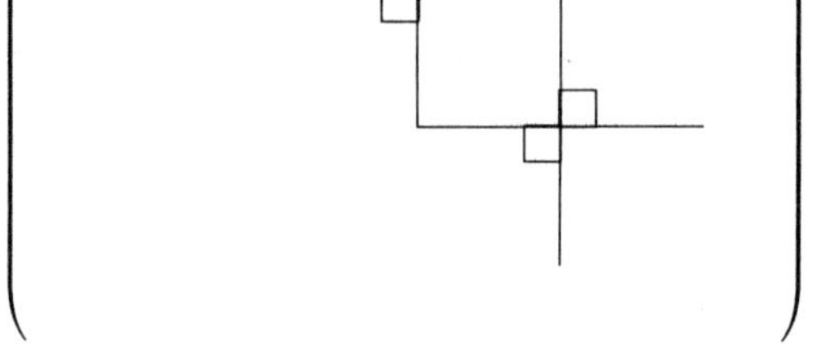

In the case of the control problem we are considering, the structure of each block is also sparse. Set $H_j = H(y^j)$, $G_j = G(y^j)$ from (1).

$$\begin{array}{|c|cccc|c|}
\hline
H_{j-1} & 0 & 0 & -e^1 & G_j & \\
\hline
 & 0 & -M_l^{-1}N_l & 0 & I & \\
 & -e^{1T} & 0 & 2\gamma\, T/M & 0 & \\
\hline
 & G_j & I & 0 & L_{y^j y^j} & H_j \\
\hline
\end{array}$$

If we eliminate for the control problem δy and δu then the dimension is further reduced and the structure of each block is

$$\begin{array}{|c|cc|c|}
\hline
H_{j-1} & -2\gamma\, T/M\ e^1(e^1)^T & G_j M_l^{-1} N_l & \\
\hline
 & G_j & I + L_{y^j y^j} M_l^{-1} N_l & H_j \\
\hline
\end{array}$$

The resulting matrix is of the size $M(2N+2) \times M(2N+2)$ with M blocks. Note that symmetry is lost with the last elimination step. In order to solve the linear system of equations we used an iterative solver, GMRES.

6 NUMERICAL RESULTS

In the following section we present several numerical results for the parabolic control problem introduced in Section 2 with parameters:

$$T = 0.7, \quad y_0 \equiv 0, \quad q \equiv 0, \quad \gamma = 10^{-4}, \quad y_{max} = .45.$$

The nonlinearities in the differential equation are described by

$$C(t) = 4 + t, \quad \lambda(t) = 4 - t.$$

The temperature profile which should be reached is given by

$$p(t) = \begin{cases} 0.7\, t & t \in [0, 0.25] \\ 0.75 & t \in (0.25, 0.45] \\ 0.175 - 0.7\,(t - 0.45) & t \in (0.45, .7] \end{cases}.$$

In order to compare our results we used also a version where the linear systems of the interior point algorithm was solved by a direct method, in this case a band LR decomposition. The following table shows that GMRES(10) is appropriate to solve the linear systems. The number tabulated in column j is the numbers of GMRES iterations necessary to achieve the residual $\|r_j\|$. As a comparison, a band LR solver needs magnitudes of order more in the flop counts.

Table 6.1 Number j of GMRES Iterations

l	j	$\|\|r_j\|\|$ ($N = 18$ $M = 120$)	flops	l	j	$\|\|r_j\|\|$ ($N = 18$ $M = 240$)	flops
1	1	$3.002 \cdot 10^{-01}$	669 122	1	1	$1.858 \cdot 10^{-00}$	1 315 286
2	4	$8.744 \cdot 10^{-14}$	2 751 963	2	4	$1.175 \cdot 10^{-13}$	5 409 387
3	4	$3.838 \cdot 10^{-14}$	2 730 699	3	3	$1.158 \cdot 10^{-09}$	4 026 450
4	3	$2.515 \cdot 10^{-09}$	2 002 686	4	3	$1.156 \cdot 10^{-09}$	3 997 632
5	3	$2.999 \cdot 10^{-10}$	1 982 538	5	3	$1.170 \cdot 10^{-09}$	3 987 066
6	3	$2.125 \cdot 10^{-13}$	1 980 552	6	3	$3.554 \cdot 10^{-10}$	3 986 670
				7	3	$1.366 \cdot 10^{-10}$	3 953 760
				8	3	$1.291 \cdot 10^{-10}$	3 948 618
				9	3	$8.593 \cdot 10^{-13}$	3 947 736
				10	2	$7.808 \cdot 10^{-11}$	2 617 855
Band LR Solver			46 812 960	Band LR Solver			93 625 920

In the following table we illustrate the superlinear convergence of the duality gap during the last few iterations of the interior point algorithm.

Table 6.2 Superlinear Convergence of the Duality Gap

	LR Decomposition			GMRES(10)		
k	l	$d_l = \mu_l{}^T \nu_l$	d_l/d_{l-1}	l	$d_l = \mu_l{}^T \nu_l$	d_l/d_{l-1}
2	10	$4.225 \cdot 10^{-07}$	0.22522	7	$5.137 \cdot 10^{-04}$	0.37320
	11	$2.548 \cdot 10^{-08}$	0.06031	8	$9.657 \cdot 10^{-05}$	0.18799
3	10	$1.408 \cdot 10^{-08}$	0.18341	5	$2.159 \cdot 10^{-07}$	0.43156
	11	$6.407 \cdot 10^{-10}$	0.04549	6	$2.772 \cdot 10^{-09}$	0.01284
4	4	$2.119 \cdot 10^{-08}$	0.84038	4	$1.895 \cdot 10^{-09}$	0.14059
	5	$1.304 \cdot 10^{-08}$	0.61551	5	$1.285 \cdot 10^{-10}$	0.06783
	6	$3.908 \cdot 10^{-09}$	0.29965	6	$4.807 \cdot 10^{-13}$	0.00374
	7	$4.245 \cdot 10^{-10}$	0.10864	7	$9.042 \cdot 10^{-17}$	0.00019
	8	$3.757 \cdot 10^{-12}$	0.00885			
	9	$3.082 \cdot 10^{-16}$	0.00008			

The last table shows the convergence rate of the inexact SQP method which is superlinear or even quadratic for certain examples. The SQP iteration is counted by k and l indicates the number of interior point iterations needed to converge. It shows also that there is no effect by using an inexact iterative solver for the linear systems.

Table 6.3 Convergence Rate of Inexact SQP with LR Decomposition

	$N = 18$ $M = 120$		$N = 18$ $M = 200$		$N = 18$ $M = 240$	
k	l	ε_k	l	ε_k	l	ε_k
1	19	$8.906 \cdot 10^{-01}$	20	$1.819 \cdot 10^{+00}$	22	$2.247 \cdot 10^{+00}$
2	8	$3.688 \cdot 10^{-02}$	11	$9.503 \cdot 10^{-02}$	12	$1.242 \cdot 10^{-01}$
3	5	$1.575 \cdot 10^{-04}$	11	$1.137 \cdot 10^{-04}$	15	$1.020 \cdot 10^{-04}$
4	7	$1.871 \cdot 10^{-08}$	9	$4.974 \cdot 10^{-09}$	9	$1.529 \cdot 10^{-08}$
Convergence Rate of Inexact SQP with GMRES(10)						
1	19	$8.906 \cdot 10^{-01}$	20	$1.819 \cdot 10^{+00}$	22	$2.247 \cdot 10^{+00}$
2	6	$3.442 \cdot 10^{-02}$	8	$9.270 \cdot 10^{-02}$	12	$1.222 \cdot 10^{-01}$
3	7	$1.908 \cdot 10^{-04}$	6	$2.095 \cdot 10^{-04}$	17	$7.932 \cdot 10^{-05}$
4	6	$8.897 \cdot 10^{-07}$	7	$1.709 \cdot 10^{-08}$	10	$1.601 \cdot 10^{-09}$

7 ACKNOWLEDGEMENT

This research was supported by the Bundesministerium für Forschung und Technologie through the research initiative 'Anwendungsorientierte Verbundvorhaben auf dem Gebiete der Mathematik'.

REFERENCES

[1] P. T. Boggs, J. W. Tolle, and A. J. Kearsley. A practical algorithm for general large scale nonlinear optimization problems. Technical Report NISTIR 5407, National Institute of Standards and Technology, Gaithersburg, MD, USA, 1994.

[2] J. Burger and M. Pogu. Functional and numerical solution of a control problem originating from heat transfer. *J. Optim. Theory Appl.*, 68:49 – 73, 1991.

[3] P. E. Gill, W. Murray, M. A. Saunders, and M. H. Wright. Sparse matrix methods in optimization. *SIAM J. Sci. Statis. Comput.*, 5:562–589, 1984.

[4] F.-S. Kupfer and E. W. Sachs. Numerical solution of a nonlinear parabolic control problem by a reduced SQP method. *Computational Optimization and Applications*, 1:113–135, 1992.

[5] F. Leibfritz and E. W. Sachs. Inexact SQP interior point methods and large scale optimal control problems. in preparation.

[6] F. Leibfritz and E. W. Sachs. Numerical solution of parabolic state constrained control problems using SQP- and interior-point-methods. In W. W. Hager, D. W. Hearn, and P. M. Pardalos, editors, *Large Scale Optimization: State of the Art*, pages 251 - 264. Kluwer, 1994.

[7] S. J. Wright. Interior point methods for optimal control of discrete-time systems. *J. Optim. Theory Appl.*, 77:161–174, 1993.

[8] Y. Zhang, R. A. Tapia, and J. E. Dennis. On the superlinear and quadratic convergence of primal-dual interior point linear programming algorithms. *SIAM J. Optimization*, 2:304–323, 1992.

Reverberation Analysis and Control of Networks of Elastic Strings

Günter Leugering, Institut für Mathematik, Universität Bayreuth, D-95440 Bayreuth, Germany

Introduction

We consider a network of uniform strings, which in its reference configuration coincides with a simple planar connected graph G. G contains a set $V = \{v_J | J = 1, \dots, n_v\}$ of vertices and a set $E = \{e_i | j = 1, \dots, n_e\}$ of straight directed edges. To each vertex v_J we assign the set $\mathcal{E}(v_J) = \{i | e_i \leftrightarrow v_J\}$ of indices corresponding to incident edges. The number $\#\mathcal{E}(v_J)$ then gives the edge-degree $d(v_J)$ of v_J. Using $d(v_J)$ we may distinguish multiple vertices (nodes) $\overset{\circ}{V} = \{v_J | d(v_J) > 1\}$ and simple vertices $\partial V = \{v_J | d(v_J) = 1\}$, which we refer to as the boundary of G. The boundary ∂V is devided into those vertices $\partial_D V$ which are clamped and those $\partial_N V = \partial V \backslash \partial_D V$ which are free or under control. We also assign to each $v_J \in V$ a set of signs ϵ_{iJ} according to $\epsilon_{iJ} = 1$ if e_i ends at v_J or -1 if e_i starts at v_J. Let $r : G \to \mathbb{R}^2$ denote the deformation of the graph, and $r_i = r|_{e_i}, r_i(x), x \in [0, l_i]$ its localization. Then r_i splits into longitudinal and vertical deformation according to

$$r_i = u_i e_i + w_i e_i^{\perp} \ .$$

In order to retain connectivity before and after deformation, we require

$$r_i(v_J) = r_j(v_J) \qquad \forall\, i, j \in \mathcal{E}(v_J), v_J \in \overset{\circ}{V} \ . \tag{1}$$

At Dirichlet-nodes we have

$$r_d(v_D) = 0 \qquad \forall\, v_D \in \partial_D V, d \in \mathcal{E}(v_D) \tag{2}$$

(for simplicity we relable v such that at a simple nodes v_J the index j corresponds to the single edge incident at v_J). At Neumann-nodes we require

$$\begin{aligned} r_n'(v_N) &= 0 \quad \forall\, V_N \in \partial_{N_0}(V) \quad n \in \mathcal{E}(V_N) \\ r_n'(v_N) &= f_N \quad \forall\, V_N \in \partial_{N_c}(V) \quad n \in \mathcal{E}(V_N) \ . \end{aligned} \tag{3}$$

Thus, at ∂V_{N_c} we apply controls (shear forces).

We introduce the spaces

$$H = \prod_{i=1}^{n_e} L_2(0, l_i), \ V = \{r \in \prod_{i=1}^{n_e} H^1(0, l_i) | r \text{ satisfies } (1), (2)\} \ ,$$

where, obviously, V is compactly embedded into H.

We define the external work done to the network as

$$W(t) = \sum_{i=1}^{n_e} \int_0^{l_i} r_i \cdot F_i \, dx + \sum_{J=1}^{n_v} r(v_J) \cdot f_J \ , \tag{4}$$

(note that because of (1) $r(v_J) := r_i(v_J) \quad \forall\, i \in \mathcal{E}(v_J)$), where F_i are forces distributed along e_i (which we neglect in the sequel), and f_J are nodal forces, viewed as controls. The kinetic energy is represented as

$$R(t) = \frac{1}{2}\sum_{i=1}^{n_e}\int_0^{l_i} |\dot{r}_i|^2 dx \ , \tag{5}$$

where the dot signifies the time derivative, and the potential energy is taken to be

$$V(t) = \frac{1}{2}\sum_{i=1}^{n_e}\int_0^{l_i} K_i r_i' \cdot r_i'\, dx \ , \tag{6}$$

where $K_i = \operatorname{diag}(p_i^2, q_i^2)$. We further introduce the Lagrangian $L = \int_0^t \{K(s) + W(s) - V(s)\}\, ds$. Applying Hamiltons principle, we arrive at the following system

$$\begin{cases} \ddot{r}_i = K_i r_i'' & i = 1,\dots,n_e,\ x_i \in (0,l_i),\ t \in (0,T) \\ r_d(v_D) = 0 & v_D \in \partial_D V,\ d \in \mathcal{E}(v_D),\ t \in (0,T) \\ r_i(v_J) = r_j(v_J) & \forall\, i,j \in \mathcal{E}(v_J),\ v_J \in \overset{\circ}{V},\ t \in (0,T) \\ \sum_i \epsilon_{iJ} K_i r_i'(v_J) = f_J & v_J \in \overset{\circ}{V} \cup \partial_N V,\ t \in (0,T) \\ r_i(x_i,0) = r_{i0}(x_i), & \dot{r}_i(x_i,0) = r_{i1}(x_i) \qquad x_i \in (0,l_i)\ . \end{cases} \tag{7}$$

It is understood that $r_i(v_J)$ is either $r_i(l_i)$ or $r_i(0)$, depending on whether $\epsilon_{iJ} = 1$ or -1. System (7) is the strong form derived by variations of L with respect to test functions in V. This system is the simplest planar system considered in Lagnese, Leugering and Schmidt [3]. Actually, it is possible to derive, in a similar manner, much more complicated systems consisting of strings, beams, membranes and plates; see [3]. The obvious control problem to consider is that of exact controllability in finite time. In order for this problem to be well-defined, well-posedness of (7) with respect to "energy spaces" is needed. The latter can be studied using the variational form of (7): introduce a form on V

$$(r,\hat{r})_V := \sum_{i=1}^{n_e}\int_0^{l_i} K_i r_i' \cdot \hat{r}_i'\, dx =: < Kr', \hat{r}' >_H \ ,$$

then we find (7) equivalent to

$$< \ddot{r}, \hat{r} >_H + (r,\hat{r})_V = < r, F >_H + \sum_{J=1}^{n_v} \hat{r}(v_J) \cdot f_J \qquad \forall\, \hat{r} \in V \tag{8}$$

with initial data as above.

It is quite clear that there is an operator A associated with the V-form: $(r,\hat{r})_V = < Ar, \hat{r} >_{V^\star,V}$ which is in fact the Riesz isomorphism between V and its dual $V^\star$. It is also quite classical then to establish the following wellposedness result (see [3]).

Theorem 1. [3] $\forall\ (r_0, r_1) \in V \times H, f \in \prod_{J=1}^{n_v} L_2(0,T), F \in H\ \exists | r \in C(0,T,V) \cap C^1(0,T,H) \cap C^2(0,T,V^\star)$ with (8), ((7)). ∎

The problem of exact controllability at time $T > 0$ is then put into the form:

$$(P) \qquad \forall\, (r_0, r_1), (r_{T0}, r_{T1}) \in E := V \times H \quad ?\exists\, f \in U := \prod_{J=1}^{n_v} L_2(0,T)$$

$$\text{such that } r \text{ satisfies } (7),\ ((8)) \text{ and}$$

$$r(T) = r_{T0},\ \dot{r}(T) = r_{T1}\ .$$

This problem has been solved in [3], [6] under the assumption

$$(H)\begin{cases} \text{i)} & \text{G is a tree,} \\ \text{ii)} & \text{the father of all nodes is clamped,} \\ \text{iii)} & \text{the entire youngest generation of nodes is controlled,} \\ \text{iv)} & T \text{ is larger than twice the time a signal needs to travel} \\ & \text{from the youngest nodes to the father of all nodes.} \end{cases}$$

(H) needs some explanation: H i) is obvious, H ii) here we label the tree such that there is exactly one simple $v_D \in \partial_D V$ and corresponding e_d with $r_d(v_D) = r_d(0)$, that v_D is the father of all nodes; H iii) the youngest nodes is the entirety of simple nodes other than the father as in H ii). H iv) is then obvious.

Theorem 2. [3],[6] Under the hypothesis (H) the system (7) is exactly controllable. ∎

Remarks: i) If H ii) fails, it has been shown in [2] that even approximate controllability fails to hold for homogeneous networks. Whether it does hold (or even (P) holds) if the optical-lengths of all edges involved in a circuit are rationally independent is an open (but admittedly academical) question.
ii) If H ii) fails, the coercivity of $(r, \hat{r})_V$ is lost and one can proceed in a quotient space.
iii) If H iii) fails, a variety of situations can occur. Some of the uncontrolled simple nodes (the youngest generation) might be clamped, while others are free. A general result on this problem is also lacking. This note is partly devoted to establishing some results in this direction.
iv) If H iv) fails, that is if T is taken smaller then in H iv) exact controllability of **all** finite energy states is impossible. It remains, however, the intreaging question of characterizing the reachable set. As seen above, theorem 2 leaves us with a lot of open intreaging open problems. ∎

In order to get more insight into special situations, we propose to use the so-called "reverberation analysis", based on the Fourier-(Laplace-) transform, which has been developed in accustics, but which has also been successively used in structural engineering see Howard [1], which we take as a basic reference. We have

$$K_i r_i' = p_i^2 u_i' e_i + q_i^2 w_i' e_i^{\perp}\ .$$

The scattering equation

Let $e_i = (\cos\Theta_i, \sin\Theta_i)$ be the (normalized) edges. $e_i^{\perp} = (-\sin\Theta_i, \cos\Theta_i)$. Let $f_J = \begin{bmatrix} -f_1 \\ -f_2 \end{bmatrix}_J$ be given in global coordinates. We consider a typical node v_J. Let e_i be incident at v_J and also at the neighbouring node v_M ($i = s_{JM} = s_{MJ}$; $s : Ind(v) \times Ind(V) \to Ind(E)$) then we rewrite r_i as r_{MJ} or r_{JM} where $r_{MJ}(v_J) = r_{MJ}(l_i)$, $r_{JM}(v_J) = r_i(0)$, $r_{MJ}(v_M) = r_i(0)$, $r_{JM}(v_M) = r_i(l_i)$. $\tilde{\mathcal{E}}(v_J) := \{M | v_M \leftrightarrow v_J\}$.

$$\begin{cases} \sum_{M\in\tilde{\mathcal{E}}(v_J)} (u'_{JM}\cos\Theta_{JM} - w'_{JM}\sin\Theta_{JM})(0,t) &= -f_{1J}(t) \\ \sum_{M\in\tilde{\mathcal{E}}(v_J)} (u'_{JM}\sin\Theta_{JM} + w'_{JM}\cos\Theta_{JM})(0,t) &= -f_{2J}(t) \end{cases} \tag{9}$$

$$\begin{cases} u_{JM}(0,t) &= U_j(t)\cos\Theta_{JM} + W_J(t)\sin\Theta_{JM} \\ w_{JM}(0,t) &= -U_j(t)\sin\Theta_{JM} + W_J(t)\cos\Theta_{JM}\ , \end{cases} \tag{10}$$

where $(U_J(t), W_J(t))$ is, in global coordinates, the position of node v_J. The idea, pursued e.g. in [1], is to Fourier transform the solutions of the local equations,

$$\begin{cases} \hat{u}_{JM}(x_{JM},\omega) &= a^1_{JM}(\omega)e^{ik^1_{JM}x_{JM}} + d^1_{JM}(\omega)e^{-ik^1_{JM}x_{JM}} \\ \hat{w}_{JM}(x_{JM},\omega) &= a^2_{JM}(\omega)e^{ik^2_{JM}x_{JM}} + d^2_{JM}(\omega)e^{-ik^2_{JM}x_{JM}} \end{cases} \tag{11}$$

where $k^i_{JM} = k^i_{JM}(\omega)$ are the wave numbers corresponding to longitudinal ($i = 1$) and transverse waves ($i = 2$). Upon introduction of $x_{JM} = l_{JM} - x_{MJ}$ we realize that $u_{JM}(x_{JM},\omega) = -u_{MJ}(x_{MJ},\omega)$, w_{JM} analogous. Furthermore we obtain the echo-relations:

$$\begin{cases} \hat{a}^{JM}(\omega) &= -e^{-ik_{JM}l_{JM}}\hat{d}_{MJ} \\ \hat{d}^{JM}(\omega) &= -e^{ik_{JM}l_{JM}}\hat{a}_{MJ} \end{cases} . \tag{12}$$

With this notation we express (10) using $R_\Theta = \begin{pmatrix} \cos\Theta & \sin\Theta \\ -\sin\Theta & \cos\Theta \end{pmatrix}$, as

$$\begin{pmatrix} d^1_{JM} \\ d^2_{JM} \end{pmatrix} = R_{\Theta_{JM}-\Theta_{JK}} \begin{pmatrix} a^1_{JK} + d^1_{JK} \\ a^2_{JK} + d^2_{JK} \end{pmatrix} - \begin{pmatrix} a^1_{JM} \\ a^2_{JM} \end{pmatrix} . \tag{13}$$

We introduce $D_{JM} = \mathrm{diag}((p_{JM})^2 k^1_{JM}, (q_{JM})^2 k^2_{JM})$ and $d_{JM} = (d^1_{JM}, d^2_{JM})^T$, $a_{JM} = (a^1_{JM}, a^2_{JM})^T$. Then (13) reads like

$$d_{JM} = R_{\Theta_{JK}-\Theta_{JM}}(a_{JK} + d_{JK}) - a_{JM}\ . \tag{14}$$

Now (9) can be written as

$$\sum_{M\in\tilde{\mathcal{E}}} R_{-\Theta_{JM}} D_{JM} d_{JM} = \sum_{M\in\tilde{\mathcal{E}}} R_{-\Theta_{JM}} D_{JM} a_{JM} + \frac{1}{i} f_J \tag{15}$$

which, upon using (13), reads like

$$\left(\sum_{M\in\tilde{\mathcal{E}}} R_{-\Theta_{JM}} D_{JM} R_{\Theta_{JM}-\Theta_{JK}}\right) d_{JK} \tag{16}$$

$$= 2\sum_{M} R_{-\Theta_{JM}} D_{JM} a_{JM} - \left(\sum_{M\in\tilde{\mathcal{E}}} R_{-\Theta_{JM}} D_{JM} R_{\Theta_{JM}-\Theta_{JK}}\right) a_{JK} + \frac{1}{i}\hat{f}_J$$

Now $S := \sum_{M\in\tilde{\mathcal{E}}} R_{-\Theta_{JM}} D_{JM} R_{\Theta_{JM}}$. We set $s_{JM} := \sin\Theta_{JM}$, $c_{JM} := \cos\Theta_{JM}$. Then the inverse of S can be computed as

$$S^{-1} = \frac{1}{\Delta} \cdot \tag{17}$$
$$\begin{pmatrix} \sum\limits_{M\in\tilde{\mathcal{E}}} p_{JM}^2 k_{JM}^1 s_{JM}^2 + q_{JM}^2 k_{JM}^2 c_{JM}^2 & -\sum\limits_{M\in\tilde{\mathcal{E}}} (p_{JM}^2 k_{JM}^1 - q_{JM}^2 k_{JM}^2) s_{JM} c_{JM} \\ -\sum\limits_{M\in\tilde{\mathcal{E}}} (p_{JM}^2 k_{JM}^1 - q_{JM}^2 k_{JM}^2) s_{JM} c_{JM} & \sum\limits_{M\in\tilde{\mathcal{E}}} p_{JM}^2 k_{JM}^1 c_{JM}^2 + q_{JM}^2 k_{JM}^2 s_{JM}^2 \end{pmatrix}$$

with

$$\begin{aligned} \Delta \;=\; & \frac{1}{2} \sum_{M\in\tilde{\mathcal{E}}} \sum_{K\in\tilde{\mathcal{E}}} \left\{ (p_{JM}^2 p_{JK}^2 k_{JM}^1 k_{JK}^1 + q_{JM}^2 q_{JK}^2 k_{JM}^2 k_{JK}^2) s_{JMJK}^2 \right. \\ & \left. (p_{JM}^2 p_{JK}^2 k_{JM}^1 k_{JK}^2 + p_{JK}^2 q_{JM}^2 k_{JK}^1 k_{JM}^2) c_{JMJK}^2 \right\} \end{aligned}$$

and $s_{JMJK} := \sin(\Theta_{JM} - \Theta_{JK})$, $c_{JMJK} := \cos(\Theta_{JM} - \Theta_{JK})$.

Equation (16) takes the following form

$$\begin{aligned} d_{JK} \;&=\; 2\sum_{M\in\tilde{\mathcal{E}}} \left(R_{\Theta_{JK}} S^{-1} R_{-\Theta_{JM}} D_{JM} \right) a_{JM} - a_{JK} - \frac{1}{i} U_{\Theta_{JK}} S^{-1} \hat{f}_J \\ &=: \sum_{M\in\tilde{\mathcal{E}}} S_{KM}^J a_{JM} - a_{JK} - B_K^J \hat{f}_J \,. \end{aligned} \tag{18}$$

This equation is essentially a scattering relation between waves arriving at v_J, represented by $a_{JM}, M \in \tilde{\mathcal{E}}$, control inputs f_J at v_J and waves departing into the edge e_{JK} incident at v_J. This is the key to controlling the energy flow in a network. We can now define the total departure at v_J by $d^J := \left(d^{JM_1}, \ldots, d^{JM_{d(v_J)}}\right)^T \in \mathbb{C}^{2d(v_J)}$ and the total arrival at v_J by a^J. Similarily, we define in an obvious way a total scattering matrix $S^J \in \mathbb{C}^{2d(v_J)\times 2d(v_J)}$ and a control input map B^J such that

$$d^J = S^J a^J + B^J f_J \,. \tag{19}$$

Even more, we introduce the scattering matrix $\mathcal{S}$ and control input $\mathcal{B}$ for the entire network by simply taking $\mathcal{S} = \mathrm{diag}(S^{J_1} \ldots S^{J_{n_v}})$, $\mathcal{B} = \mathrm{diag}(B^{J_1} \ldots B^{J_{n_v}})$. Because of $\sum\limits_{J=1}^{n_v} d(v_J) = 2n_e$, we have $\mathcal{S} \in \mathbb{C}^{4n_e \times 4n_e}$. We may write

$$d = \mathcal{S}a + \mathcal{B}\hat{f} \,. \tag{20}$$

Examples: i) If all members of the network are identical i.e. if $l_i = 1$, $p_i = q_i = 1$, hence $k_i(\omega) = \omega \;\forall\, i$, we have with $d(v_J) = m$:

$$\begin{aligned} \begin{pmatrix} d_{JM}^1 \\ d_{JM}^2 \end{pmatrix} \;&=\; \frac{2}{m} \sum_{k\in\tilde{\mathcal{E}}} \begin{pmatrix} c_{JMJK} & s_{JMJK} \\ -s_{JMJK} & c_{JMJK} \end{pmatrix} \begin{pmatrix} a_{JK}^1 \\ a_{JK}^2 \end{pmatrix} \\ & - \begin{pmatrix} a_{JM}^1 \\ a_{JM}^2 \end{pmatrix} + \frac{1}{i\omega m} \begin{pmatrix} c_{JM} & s_{JM} \\ -s_{JM} & c_{JM} \end{pmatrix} \begin{pmatrix} \hat{f}_J^1 \\ \hat{f}_J^2 \end{pmatrix} \end{aligned} \tag{21}$$

ii) We take three uniform strings coupled at v_J. At v_J we have e_{JM}, e_{JN}, e_{JO} as incident edges at the angles $\Theta_{JM} = \pi, \Theta_{JN} = 0, \Theta_{JO} = \pi/2$, respectively. Then the departure into the upright string e_{JO} is given by

$$\begin{pmatrix} d^1_{JO} \\ d^2_{JO} \end{pmatrix} = \frac{1}{3}\begin{pmatrix} -a^1_{JO} - 2a^2_{JM} + 2a^2_{JN} \\ -a^2_{JO} + 2a^1_{JM} - 2a^1_{JN} \end{pmatrix} + \frac{1}{3\omega i}\begin{pmatrix} \hat{f}_2 \\ -\hat{f}_1 \end{pmatrix} ; \tag{22}$$

accordingly, for the other directions we obtain

$$\begin{pmatrix} d^1_{JM} \\ d^2_{JM} \end{pmatrix} = \frac{1}{3}\begin{pmatrix} -a^1_{JM} - 2a^1_{JN} + 2a^2_{JO} \\ -a^2_{JM} - 2a^2_{JN} - 2a^1_{JO} \end{pmatrix} - \frac{1}{3\omega i}\begin{pmatrix} \hat{f}_1 \\ \hat{f}_2 \end{pmatrix} \tag{23}$$

$$\begin{pmatrix} d^1_{JN} \\ d^2_{JN} \end{pmatrix} = \frac{1}{3}\begin{pmatrix} -a^1_{JN} - 2a^1_{JM} - 2a^2_{JO} \\ -a^2_{JN} - 2a^2_{JM} + 2a^1_{JO} \end{pmatrix} - \frac{1}{3\omega i}\begin{pmatrix} \hat{f}_1 \\ \hat{f}_2 \end{pmatrix} . \tag{24}$$

Therefore the total departure at v_J may be written as follows

$$\begin{aligned} d^J &:= \left(d^{JM} d^{JN} d^{JO}\right)^T \in \mathbb{R}^6 \\ S^J &= \frac{1}{3}\begin{pmatrix} -1 & 0 & -2 & 0 & 0 & 2 \\ 0 & -1 & 0 & -2 & -2 & 0 \\ -2 & 0 & -1 & 0 & 0 & -2 \\ 0 & -2 & 0 & -1 & 0 & -2 \\ 0 & -2 & 0 & 2 & -1 & 0 \\ 2 & 0 & -2 & 0 & 0 & -1 \end{pmatrix} \\ B^J &= \frac{1}{3\omega i}\begin{pmatrix} -1 & 0 \\ 0 & -1 \\ 1 & 0 \\ 0 & 1 \\ 0 & 1 \\ -1 & 0 \end{pmatrix} . \end{aligned} \tag{25}$$

To complete the picture, let us assume that we have Dirichlet conditions at v_0 and Neumann conditions at v_M, v_N then the corresponding scattering matrices are simply given by:

$$S^M = \begin{pmatrix} 1 & 0 \\ 0 & 1 \end{pmatrix}, \; S^N = \begin{pmatrix} 1 & 0 \\ 0 & 1 \end{pmatrix}, \; S^0 = \begin{pmatrix} -1 & 0 \\ 0 & -1 \end{pmatrix} . \tag{26}$$

If we control at v_M then $B^M = \begin{pmatrix} 1 & 0 \\ 0 & 1 \end{pmatrix} \frac{1}{\omega i}$. ■

We introduce the vectors $\tilde{d}^J = \left(d^{M_1J} \dots d^{M_{d(v_J)}J}\right)^T$ $\tilde{a}^J = \left(a^{M_1J} \dots a^{M_{d(v_J)}J}\right)^T$ and their global counterparts $\tilde{d} = \left(\tilde{d}^{J_1}, \dots, \tilde{d}^{J_{n_v}}\right)^T$, $\tilde{a} = \left(\tilde{a}^{J_1} \dots \tilde{a}^{J_{n_v}}\right)^T$. We also introduce, as in [1], the diagonal shift matrix

$$\mathcal{P}(x,\omega) := \text{diag}\left(-e^{-i(k^{JM}x^{JM})}\right)_{\substack{J=J_1\dots J_{n_v}\\ M=M_1\dots M_{d(V_J)}}} .$$

For convenience, we put $\mathcal{P}(\mathcal{L},\omega)$ if for each x^{JM} in $P(x,\omega)$ we set l^{JM}. In addition, by the definition of $\tilde{d}$ it is obvious that there is a permutation matrix $\mathcal{U}$ such that $\tilde{d} = \mathcal{U}d$. On the other hand we know that $a = \mathcal{P}(\mathcal{L},\omega)\tilde{d} = \mathcal{P}(\mathcal{L},\omega)\mathcal{U}d$. Using the latter identity we can reformulate the scattering equation as

$$d = \mathcal{S}a + \mathcal{B}f = \mathcal{S}\mathcal{P}(\mathcal{L},\omega)\mathcal{U}d + \mathcal{B}\hat{f}$$

and upon setting $\mathcal{R}(\omega) := \mathcal{S}\mathcal{P}(\mathcal{L},\omega)\mathcal{U}$, we may write this as

$$d = (I - \mathcal{R}(\omega))^{-1}\mathcal{B}\hat{f} \tag{27}$$

as soon as $\det(I - \mathcal{R}(\omega) \neq 0$.

Remark: i) The matrix $\mathcal{R}(\omega)$ is the so-called reverberation matrix.
ii) $\det(I - \mathcal{R}(\omega)) = 0$ is a spectral condition. This condition is precisely the complex analogue of the real spectral condition, derived in [3] based on the work of Nicaise [5] and v. Below [7]. It is for instance possible to check the uncontrolled Example ii) to verify that in the case v_0 is clamped while v_M, v_N are free, strings of eigenvalues occur at

$$a) \qquad \arccos\left(\pm\left(\frac{1}{\sqrt{3}}\right)\right) + k\pi,\ k\pi\ ,$$

whereas in the case where v_0, v_M are clamped and v_N is free, we obtain.

$$b) \qquad \arccos\left(\pm\left(\frac{\sqrt{2}}{\sqrt{3}}\right)\right) + k\pi,\ \left(k+\frac{1}{2}\right)\pi\ .$$

■

Controlling the flow of energy in a network

Once the scattering at a given joint is settled as a above, one might think of various ways to influence the scattering of waves incident at that joint.

1. Serial joints

In a serial joint v_J we have only two adjacent vertices v_M, v_N. We take $\Theta_{JM} = 0,\ \Theta_{JN} = \alpha$ and uniform strings.

$$\begin{pmatrix} d^1_{JM} \\ d^2_{JM} \end{pmatrix} = \begin{pmatrix} \cos\alpha & \sin\alpha \\ \sin\alpha & \cos\alpha \end{pmatrix} \begin{pmatrix} -a^1_{JN} \\ a^2_{JN} \end{pmatrix} + \frac{1}{2i\omega}\begin{pmatrix} \hat{f}_1 \\ \hat{f}_2 \end{pmatrix} .$$

The control objective is to prevent signals from penetrating e_{JM}, that is find controls $\hat{f}_i$ such that $d^1_{JM} \equiv d^2_{JM} \equiv 0$. This is equivalent to

$$\begin{aligned}\hat{f}_1(\omega) &= -2i\omega(\cos\alpha a^1_{JN} - \sin\alpha a^2_{JN})\\ \hat{f}_2(\omega) &= -2i\omega(\sin\alpha a^1_{JN} + \cos\alpha a^2_{JN})\ .\end{aligned}$$

Now $2i\omega a^1_{JN} = i\omega\hat{u}_{JN}(0,\omega)+\hat{u}'_{JN}(0,\omega)$ and $2i\omega a^2_{JN} = i\omega\hat{w}_{JN}(0,\omega)+\hat{w}'_{JN}(0,\omega)$. Therefore

$$\begin{aligned}u'_{JM}(0,t) &= \cos\alpha\,\dot{u}_{JN}(0,t) - \sin\alpha\,\dot{w}_{JN}(0,t)\\ w'_{JM}(0,t) &= \sin\alpha\,\dot{u}_{JN}(0,t) + \cos\alpha\,\dot{w}_{JN}(0,t)\ .\end{aligned}$$

But, using condition (10) — the connectivity condition — we arrive at the simple condition

$$u'_{JM}(0,t) = \dot{u}_{JM}(0,t),\ w'_{JM}(0,t) = \dot{w}_{JM}(0,t) \tag{28}$$

which appears to be independent of the angle α. Equation (28) also follows directly by inspection of the condition $d_{JM} = 0$. These transmission conditions are quite classical, when viewed as transparent boundary conditions at $x = 0$ of the single string e_{JM}.

In a simplified sense, this transmission condition can also be viewed as a collocated active sound control (anti-sound). We dwell on this a bit further.

2. Anti-sound

In the $1-d$-anti-sound problem, a sound wave travels towards positive x. At $x = -l$ measurements of the pressure and its derivative are taken and at $x = 0$ so-called anti-sound is applied in order to prevent the sound from penetrating the region $x > 0$. Mathematically, this problem can readily be reformulated in the spirit of this paper. Indeed, multichannel anti-sound problems fall precisely within this framework, and are, in fact, equivalent to the control problems considered above. Assume that measurements are available at v_N, that the control i) applied at v_J and that the control objective still is to cancel d_{JM}. We use the notation of the preceeding example:

$$\begin{aligned}\frac{1}{2i\omega}\begin{pmatrix}\hat{f}_1\\ \hat{f}^2\end{pmatrix} &= \begin{pmatrix}\cos\alpha & -\sin\alpha\\ \sin\alpha & \cos\alpha\end{pmatrix}\begin{pmatrix}a^1_{JN}\\ a^2_{JN}\end{pmatrix} =\\ &= -e^{-i\omega l}\begin{pmatrix}\cos\alpha & -\sin\alpha\\ \sin\alpha & \cos\alpha\end{pmatrix}\begin{pmatrix}d^1_{NJ}\\ d^2_{NJ}\end{pmatrix}\end{aligned}$$

(l = length of string e_{JN}). Now $-2i\omega d^1_{NJ} = -i\omega\hat{u}_{NJ}(0,\omega) + \hat{u}'_{NJ}(0,\omega)$; $-2i\omega d^2_{NJ} = -i\omega\hat{w}_{NJ}(0,\omega)+w'_{NJ}(0,\omega)$. Also note that $\hat{u}_{NJ}(0,\omega) = -\hat{u}_{JN}(l,\omega), \hat{u}'_{NJ}(0,\omega) = u'_{JN}(l,\omega)$ etc.

We are lead to the condition

$$\begin{aligned}f_1(t) &= \cos\alpha\,(\dot{u}_{JN}(l,t-l) + u'_{JN}(l,t-l))\\ &\quad - \sin\alpha\,(\dot{w}_{JN}(l,t-l) + w'_{JN}(l,t-l))\\ f_2(t) &= \sin\alpha\,(\dot{u}_{JN}(l,t-l) + u'_{JN}(l,t-l))\\ &\quad + \cos\alpha\,(\dot{w}_{JN}(l,t-l) + w'_{JN}(l,t-l))\end{aligned}$$

In contrast to the previous case, the corresponding transmission condition does depend on the angle of incidence! If $\alpha = \pi$, this is the collinear case, the transmission condition above reduces to

$$\begin{cases} u'_{JM}(0,t) - u'_{JN}(0,t) &= \dot{u}_{JN}(l,t-l) + u'_{JN}(l,t-l) \\ w'_{JM}(0,t) - w'_{JN}(0,t) &= \dot{w}'_{JN}(l,t-l) + w'_{JN}(l,t-l)\ . \end{cases} \tag{29}$$

If we change the direction of e_{JN} such that $u = u_{JM}\ x > 0,\ u(x) = u_{JN}(-x)\ x < 0$, then (29) reads like

$$\begin{cases} u'(0_+,t) - u'(0_-,t) &= \dot{u}(-l,t-l) - u'(-l,t-l) \\ w'(0_+,t) - w'(0_-,t) &= \dot{w}(-l,t-l) - w'(-l,t-l)\ . \end{cases} \tag{30}$$

In (30) (as in (29)) vertical and longitudinal motion is decoupled as expected. It interesting to note that letting $l \to 0$ in (30), the latter reduces to the classical condition (28).

Remark: i) If one assumes that measurement of $\dot{u}(x,t)$ is available only and this measurement can be taken at $x = -l,\ x : -2l$ in the serial arrangement so that "anti-sound" is again applied at $x = 0$ then we find (in the notation of (30))

$$\begin{cases} f_1(t) &= -2(\dot{u}(-2l,t-2l) - \dot{u}(-l,t-3l)) + f_1(t-2l) \\ f_2(t) &= -2(\dot{w}(-2l,t-2l) - \dot{w}(-l,t-3l)) + f_2(t-2l)\ . \end{cases} \tag{31}$$

Note that (31) is a finite difference relation for f_1, f_2. Again it can be seen that (31) reduces to (28) when $l \to 0$ in (31).
ii) A similar analysis can be performed also for Euler-Bernoulli and Timoshenko beams. However, in those cases it is not possible to arrive at local transmission conditions, and therefore approximations have to be considered. We do not dwell on this here and refer to a future paper. See however, Mace [4].
iii) It is obvious from the preceeding discussion that the problem of finding absorbing or transparent boundary/node conditions is easily settled using the scattering approach. As in ii) this approach has been used to derive transparent boundary conditions for Euler-Bernoulli and Timoshenko beams. Again those conditions are non-local in time. Again, we leave a more detailed exposition of this issue to a forthcoming paper and just mention that for the Euler Bernoulli-beam the following transparent boundary conditions, say at $x = 1$, hold.

$$\begin{cases} w''(1,t) &= -\dot{w}(1,t) - \sqrt{2}\dfrac{1}{\Gamma(\frac{1}{2})}\displaystyle\int_0^t \frac{1}{(t-s)^{1/2}}\dot{w}'(1,t-s)\,ds \\ w''(1,t) &= \dot{w}'(1,t) + \sqrt{2}\dfrac{1}{\Gamma(\frac{1}{2})}\dfrac{d}{dt}\displaystyle\int_0^t \frac{1}{(t-s)^{1/2}}\dot{w}(1,t-s)\,ds\ . \end{cases} \tag{32}$$

(w is taken to satisfy the Euler-Bernoulli equation $\ddot{w}+w^{(4)} = 0$ and, for example, clamped boundary conditions $w(0,t) = 0 = w'(0,t)$). ■

3. Multiple joints

We consider the configuration as in example ii) (the prototype of a divining rod). The scattering matrices are given again by (22)–(24). If, for instance, we want prevent the signals from penetrating the upright string (e_{JO}), we have to set $(d^1_{JO}, d^2_{JO})^T \equiv 0$. By the same analysis as in 1.) or 2.), we find the following transmission conditions

$$w'_{JO}(0,t) = \dot{w}_{JO}(0,t), \quad u'_{JO}(0,t) = \dot{u}_{JO}(0,t) \; . \tag{33}$$

It is interesting to observe that under this condition the departure in the strings e_{JM}, e_{JN} is given by

$$\begin{pmatrix} d^1_{JM} \\ d^2_{JM} \end{pmatrix} = \begin{pmatrix} -a^1_{JM} & +a^2_{JO} \\ -a^2_{JM} & -a^1_{JO} \end{pmatrix}, \begin{pmatrix} d^1_{JN} \\ d^2_{JN} \end{pmatrix} = \begin{pmatrix} -a^1_{JN} & -a^2_{JO} \\ -a^2_{JN} & +a^1_{JO} \end{pmatrix}. \tag{34}$$

This shows very clearly that condition (33) dynamically separates the strings e_{JM} and e_{JN} (which are, of course, still connected!). Energy is flowing from the upright string into the others, but not back. There is no transmission between e_{JM}, e_{JN}, and (33), (34) is clearly not dissipative.

In general, at a vertex v_J with $d(v_J) = m$ we set (say) $d_{JM} = 0$ compute the corresponding control f and then look at the resulting departure into the remaining incident edges. The result is

$$d_{JK} = -a_{JK} + R_{\theta_{JK}-\theta_{JM}} a_{JM}, \quad \forall K \in \tilde{\mathcal{E}}(v_{JM}).$$

But this relation involves only the vertices v_{JM}, v_{JK} bounding the edge e_{JK}. Therefore the phenomenon, observed for the tripod is completely general and the control (33) *strippes off* the energy from the link represented by e_{JO} into *all* those edges which are incident at v_J. Again, this control is not dissipative. Nevertheless, the following result obtains:

Theorem 3. Let G be a tree. Given controls (33) at all vertices v_J (with the exception of the father of all nodes, which in turn is clamped) directed towards the next generation branching off from v_J into e_{JO}, $\forall O \in \tilde{\mathcal{E}}(v_J)$, all solutions will *disappear* in time $T = diameter(G) + length(e_{FM})$, where e_{FM} is the root of the tree. ∎

4. Minimal departure

The last observation concerning (33), (34) shows that it is only possible to cancel reflections into *one* direction using multiple-joint controls. One might, therefore, want to consider such controls that minimize the total departure at a given joint:

$$\min_{f \in \mathbb{C}^2} \|S^J a^J + B^J f\| \; .$$

This will give a feedback $f(a^J)$ which can be transformed back to the time domain. This concept is rather general, in that one can single out a region within the given network and ask for minimal departure from that region (a stealthy region!), or one asks for maximal total departure if one wants to save that region. In our tripod example, we obtain

$$\begin{aligned} f_1(t) &= \frac{1}{2}\{\dot{u}_{JM}(0,t) + u'_{JM}(0,t) - \dot{u}_{JN}(0,t) - u'_{JN}(0,t) \\ &\qquad -\dot{w}_{JO}(0,t) - w'_{JO}(0,t)\,\} \\ f_2(t) &= \frac{1}{2}\{\dot{w}_{JM}(0,t) + w'_{JM}(0,t) - \dot{w}_{JN}(0,t) - w'_{JN}(0,t) \\ &\qquad +\dot{u}_{JO}(0,t) + u'_{JO}(0,t)\,\} \end{aligned}$$

which leads to the transmission condition

$$\begin{cases} -u'_{JM}(0,t) + u'_{JN}(0,t) - 3w'_{JO}(0,t) &= -\dot{u}_{JM}(0,t) \\ -w'_{JM}(0,t) + w'_{JN}(0,t) - 3u'_{JO}(0,t) &= -\dot{w}_{JM}(0,t) \ . \end{cases} \tag{35}$$

5. Optimal location
In a large network only a limited number of nodes can be equipped with controls as described above. The problem then arises that given a total maximum number of controls and, say, given a region (a set of nodes and edges) to save, find the optimal location of those controls which satisfy an overall performance criterium. This problem, for flexible networks, appears to be completely open.

Representing the motion of a network by reverberation

We go back to (27). We note that $a = \mathcal{P}(\mathcal{L},\omega)\mathcal{U}d$ where d is given by (27). Therefore, the Fouriertransform (11) of the solution to (7) can be constructed. To d and a there corresponds the vector $\underline{\hat{r}}(x,\omega) \in \mathbb{C}^{4n_e}$. Each member $(u_{JM}, w_{JM})^T$ is also stored as $(u_{MJ}, w_{MJ})^T$ according to the definition of d, a. We have

$$\begin{aligned} \underline{\hat{r}}(x,\omega) &= -\mathcal{P}(-x,\omega)a(\omega) - \mathcal{P}(x,\omega)d(\omega) \\ &= (\mathcal{P}(\mathcal{L}-x,\omega)\mathcal{U} - \mathcal{P}(x,\omega))\, d(\omega) \\ &= (\mathcal{P}(\mathcal{L}-x,\omega)\mathcal{U} - \mathcal{P}(x,\omega))\, (I - \mathcal{R}(\omega))^{-1}\mathcal{B}(\omega)\hat{f}(\omega) \ . \end{aligned} \tag{36}$$

Hence, taking the inverse Fourier transform

$$\underline{r}(x,t) = \int_{-\infty}^{\infty} (\mathcal{P}(\mathcal{L}-x,\omega)\mathcal{U} - \mathcal{P}(x,\omega))(I - \mathcal{R}(\omega))^{-1}\mathcal{B}(\omega)\hat{f}(\omega)\cdot e^{i\omega t}\, d\omega \ . \tag{37}$$

This relation has been derived in [1] for axial (longitudinal) waves. We note that (36) is also the complex counterpart to the representation of solution of a scalar version of (7) (out-of-the-plane displacements) by J. v. Below [7].

Remark: The importance of (37) lies in the fact that
i) $(I - \mathcal{R}(\omega))^{-1}$ can be approximated by its Neumann series: $(I - \mathcal{R}(\omega)^{-1} = I + \mathcal{R}(\omega) + \mathcal{R}^2(\omega) + \ldots$. Therby, taking into account higher powers of $\mathcal{R}(\omega)$, the reverberations with respect to more distant nodes (from a given node) are accounted for.
ii) Such an approximation has obvious advantages for the numerical simulation of such networks. Investigations in this direction are on their way.
iii) For small times and simple networks (37) and its approximation can serve for an explicit (analytical) representation of solutions. This turns out to be important in order to answer some open question about exact controllability.

On exact controllability

We are no going to use the explicit representation (37) to show exact controllability in an examplaric situation, where "energy-methods" have failed so far. We take the divining rod of Example ii), where the scattering matrices are given by (25), (26). As pointed out in the remark iii) following Theorem 1, the question as to whether exact controllability holds in the case that not all of the "youngest nodes" are under control, has not been settled by Theorem 1. The simplest example of this kind is the divining rod (tripod) with one root clamped (the father) and just one of the remaining two simple nodes (the youngest generation) is under control. There are two situations of interest: i) the remaining uncontrolled node is clamped or ii) free.

Using the spectrum, which could be extracted from [3] explicitly, together with the corresponding eigenelements, it was shown by author earlier (this, however, hasn't been published) that in case i) not even approximate controllability holds, whereas in case ii) spectral controllability obtains. It is known, however, in the context of two collinear identical strings that midspan-controls can uniformly stabilic the system if the endpoints satisfy different boundary conditions. This suggests the that the divining rod (string) should be *exactly* controllable in case ii). We are going to confirm this conjecture using reverberation analysis. We have $\mathcal{S} = \operatorname{diag}(S^J, S^M, S^N, S^0) =: \operatorname{diag}(S^J, S^{MN0})$ with the 6×6 matrix $S^{MN0} = \operatorname{diag}(S^M, S^N, S^0)$. We control the system at the vertex v_M, hence $B^J = 0_{6\times 6}$, $B^N = B^0 = 0_{2\times 2}$, $B^M = \begin{pmatrix} 1 & 0 \\ 0 & 1 \end{pmatrix} \frac{1}{\omega i}$. In this situation the permutation matrix $\mathcal{U}$ is simply $\mathcal{U} = \begin{pmatrix} 0 & I \\ I & 0 \end{pmatrix}$, $I =: d_{6\times 6}$. Using this the reverberation matrix $\mathcal{R}(\omega)$ is established as follows

$$\begin{aligned} \mathcal{R}(\omega) &= \mathcal{S}\mathcal{P}(\mathcal{L},\omega)\mathcal{U} = \\ &= -e^{i\omega l} \begin{pmatrix} 0 & S^J \\ S^{MN0} & 0 \end{pmatrix}. \end{aligned}$$

Hence with $\mu = \exp(-i\omega l)$, $R(\mu, A) := (\mu I - A)^{-1}$

$$(I - \mathcal{R}(\omega))^{-1} = \begin{pmatrix} \mu^2 R(\mu^2, S^J S^{MN0}) & -\mu R(\mu^2, S^J S^{MN0}) S^J \\ -\mu R(\mu^2, S^{MN0} S^J) S^{MN0} & \mu^2 R(\mu^2, S^{MN0} S^J) \end{pmatrix}.$$

On the other hand

$$\begin{aligned} \mathcal{P}(\mathcal{L} - x, \omega)\mathcal{U} - \mathcal{P}(x,\omega) &= \begin{pmatrix} e^{-i\omega x} I & -e^{-i\omega(l-x)} I \\ -e^{-i\omega(l-x)} I & e^{-i\omega x} I \end{pmatrix} \\ \mathcal{B}(\omega) &= \left(0_{6\times 2},\, B^M, 0_{2\times 2},\, 0_{2\times 2},\, 0_{2\times 2}\right)^T \\ &=: (0, B^{MN0})^T . \end{aligned}$$

Therefore, the integral in (37) can be expressed as

$$\underline{r}(x,t) =$$

$$\int_{-\infty}^{\infty} \mu^2 \begin{pmatrix} -e^{i\omega(t-(x+l))}R(e^{2i\omega l}, S^J S^{MN0})S^J - \\ e^{i\omega(t-(2l-x))}R(e^{2i\omega l}, S^J S^{MN0})S^J - \\ -e^{i\omega(t-(l-x))}R(e^{2i\omega l}, S^{MN0})S^J) \\ -e^{i\omega(t-x)}R(e^{2i\omega l}, S^{MN0})S^J) \end{pmatrix} \cdot B^{MN0}\hat{f}(\omega)\, d\omega\ . \tag{38}$$

Since the eigenvalues of the regular matrices $S^J S^{MN0}(S^{MN0}S^J)$ have absolute value 3, $R(e^{2i\omega l}, S^J S^{MN0})$, $(R(e^{2i\omega l}, S^{MN0}S^J))$ can easily be developed into series. It is then obvious from (37) that given a final time t and given that the controls vanish for $t < 0$, we only need to take into account a finite number of terms in that series expansion, that is to say, we only need to account for a finite number of reverberations. In order to study exact controllability it is reasonable to stipulate $t = 6l$ as control time. This implies that only terms quadratic in $S^J S^{MN0}$, $S^{MN0}S$ in the expansion of $R(e^{2i\omega l}, S^J S^{MN0})$ etc. are involved. Also note that $\underline{r}(6l, x) \in \mathbb{R}^{12}$, as each element, with to degrees of freedom each, appears twice according to the 2 different orientations. Therefore, as it is also readily seen from (38), the second "half" of $\underline{r}$ is the mirror image of the first. We look at $\underline{r}_1 = (\underline{r})_{i=1}^6$. Set $F(t) = (f_1(t), f_2(t), 0, 0, 0, 0)^T$ and note that B^{MN0} contains a factor $(\omega i)^{-1}$. We find

$$\begin{cases} \underline{r}_1'(6l, x) = S^J F(5l - x) - F(5l + x) + S^J S^{MN0} S^J F(3l - x) \\ \quad -S^{MN0}S^J F(3l + x) + (S^J S^{MN0})^2 S^J F(l - x) + (S^{MN0}S^J)^2 F(l + x) \\ \dot{\underline{r}}_1(6l, x) = -S^J F(5l - x) - F(5l + x) - S^J S^{MN0} S^J F(3l - x) \\ \quad -S^{MN0}S^J F(3l + x) - (S^J S^{MN0})^2 S^J F(l - x) - (S^J S^{MN0})^2 F(l + x) \\ \quad \forall\, x \in (0, l)\ . \end{cases} \tag{39}$$

Formula immediately reveals the regularity stated in Theorem 1. The equations in (39) constitute a system of 12 equations. But we have also 12 functions to choose from: $f_i(l - x)$, $f_i(l + x)$, $f_i(3l - x)$, $f_i(3l + x)$, $f_i(5l - x)$, $f_i(5l + x)$ $\quad i = 1, 2$. Therefore, we have for each $x \in [0, l]$ a 12×12 system of equations to solve. Exact controllability, thus, reduces to the question as to whether the coefficient matrix has full rank ($= 12$). It turns out that matrix has indeed full rank if v_N is free and has rank $= 8$ if v_N is clamped. In the latter case we know already that even approximate controllability fails to hold.

Theorem 4. The divining rod with one simple node clamped, one simple node free and the third simple node controlled by functions $f_1, f_2 \in L_2(0, 6l)$ as exactly controllable in $V \times H$ and time $T = 6l$. ∎

Remark: i) Other configurations are currently under investigation.
ii) We conjecture similar results for Timoshenko-systems.
iii) We note that viscoelastic networks of the Boltzmann-type can be treated in much the same way.
iv.) If G contains cycles, one strategy to pursue is to consider a suitable spanning tree, arrange for controls as in Theorem 3 at all nodes, other than the root, such that at each vertex, where a cutting edge branches off, the control accounts for the additional edge-degree. Then the entire motion will ultimately end up in all the cutting edges.

Now, if one chooses *strategic points* in all of those cutting edges, and apply classical open-loop controls there, null-controllability - with this mixed strategy - obtains. As the corresponding closed-loop controls (33) are not dissipative, this strategy might only be of theoretical interest. ■

References

[1] S. M. Howard, *Transient stress waves in trusses and frames*, Ph.D. thesis, 1990.

[2] J. E. Lagnese, G. Leugering, and E. J. P. G. Schmidt, *Controllability of planar networks of timoshenko beams*, SIAM J Control. and Optim. **31** (1993), 780–811.

[3] ———, *Modelling, analysis and control of multi-link flexible structures*, Birkhäuser Verlag, 1994.

[4] B.R. Mace, *Active control of flexural vibrations*, J. of Sound and Vibration **114** (1987), 253–270.

[5] S. Nicase, *Some results on spectral theory over networks applied to nerve impulse transmission*, , Lecture Notes in Mathematics, Springer-Verlag, 1985, pp. 533–541.

[6] E. J. P. G. Schmidt, *On the modelling and exact controllability of networks of vibrating strings*, SIAM J. of Control and Optimization **30** (1992), 229–245.

[7] J. von Below, *A characteristic equation associated to an eigenvalue problem on c^2-networks*, Lin. Alg. and its Appl. **71** (1985), 309–325.

Remarks on the Control of Sensitive Distributed Systems

J. L. Lions Collège de France, Paris, France

1. Introduction.

Let us consider a distributed system whose state, denoted by $y = y(t; v)$, is given by the solution of the Partial Differential Equation

$$\frac{\partial y}{\partial t} + A(y) = v \,, \tag{1.1}$$

where y is subject to appropriate boundary conditions (which are not made explicit for the time being) and to the initial condition

$$y|_{t=0} = b \,. \tag{1.2}$$

In (1.1) (1.2), A denotes a Partial Differential Operator (linear or not), v denotes the control function and b is given in the state space.

To be more precise, we assume that there is (at least) one solution in a real Hilbert space H of (1.1) (1.2), where b is given in H, and where

$$v \in F \tag{1.3}$$

where F is a (closed) subspace of $L^2(0, T; H)$, where $T > 0$ is given.

Given v and b, we assume there is at least one solution $y(t; v, b)$ which is continuous from $t \in [0, T] \to H$.

□

Approximate controllability.

We are given an element $y^T \in H$ and a number $\beta > 0$ arbitrarily small.

We say that the system **is approximately controllable** if we can find $v \in F$ such that there is a solution $y(t; v, b)$ of (1.1) (1.2) (and the appropriate boundary) such that

$$y(T; v, b) \in y^T + \beta B \tag{1.4}$$

where

$$B = \text{unit ball in } H. \tag{1.5}$$

(In other words, there is approximate controllability when one can drive the system in the given time T from an arbitrary state b to a state which is arbitrary close to another arbitrary state y^T).

□

If we assume that the system is approximately controllable, then one can drive the system into the ball (1.4) in infinitely many ways. It is then reasonable to consider the problem to find

$$inf. \frac{1}{2} \| v \|_F^2 \tag{1.6}$$

when v is subject to (1.4).

□

Let us now consider a system which is **very sensitive to Initial conditions.** In other words, the (or a) solution of (1.1) (1.2) changes very much under small variations of b. In such a case we should allow b to span a **neighborhood** of a given element of H and to consider **the worst** possible situation.

In order to simplify the **notations**, we restrict the question to the case where b spans a neighborhood of the **origin** (it is without importance for linear problems but it is a restriction in the non linear situations).

Then we replace (1.6) by

$$\left| \begin{array}{l} \sup\limits_{\|b\|_H \leq \alpha} \inf\limits_{v \in F} \frac{1}{2} \| v \|_F^2 , \\ y(T; v, b) \in y^T + \beta B. \end{array} \right. \tag{1.7}$$

We define in this way a function

$$\phi(\alpha) = \text{formula in } (1.7). \tag{1.8}$$

A priori this function **can** be infinite for α large enough, and this can indeed be the case, as we show in Section 2 below.

We study more closely the function $\phi(\alpha)$ in Section 3, in particular situations. We derive from this study some conjectures for the general situation.

□

Before we proceed with technical details, a few remarks are in order.

Remark 1.1.

What we have in mind in the first place is the Navier Stokes System.

If we assume the control to be distributed, the problem becomes

$$\left| \begin{array}{ll} \frac{\partial y}{\partial t} + y\nabla y - \mu\Delta y = v\chi_{\mathcal{O}} - \nabla p & \text{in}\,\Omega\times(0,T)\ ,\ \mu>0\ , \\ \text{div}\ y = 0 & \text{in}\,\Omega\times(0,T) \end{array} \right. \tag{1.9}$$

where

$\chi_{\mathcal{O}}$ = characteristic function of a set $\mathcal{O}\subset\Omega$,

$\Omega\subset\mathbf{R}^d$, $d=2$ or $d=3$,

$v=\{v_1,v_2,v_3\}\in L^2(\mathcal{O}\times(0,T))^3$ (if d = 3).

We assume that

$$y=0 \quad \text{on} \quad \Gamma=\partial\Omega\ ,\ t\in(0,T) \tag{1.10}$$

and we take (if $d=3$) (1.2) with $b\in H$, where

$$H=\{\varphi|\varphi\in(L^2(\Omega))^3 \quad , \quad \text{div}\ \varphi=0\ ,\varphi n=0 \quad \text{on} \quad \Gamma\}$$

where n denotes the normal vector to Γ (oriented towards the exterior of Ω).

We have conjectured in J.L. LIONS [1] [2] that in such a case we have approximate controllability. A very interesting result of J.M. CORON [1] proves this in the case $\mu=0$, $d=2$ (Euler equations in 2 dimensions). It would be interesting to study the function $\alpha\to\phi(\alpha)$ in these situations (without forgetting that this problem makes sense with β arbitrary small only if one has approximate controllability).

□

Remark 1.2.

For the Stokes system (i.e. without the non linear term in (1.9)) Approximate controllability is known and **all** what we are going to say in the next sections can be extended. We wish to mention here in this connection a class of problems. We consider the system

$$\left| \begin{array}{l} \frac{\partial y}{\partial t} - \mu \Delta y = v \chi_{\mathcal{O}} - \nabla p \, , \\ \text{div } y = 0 \end{array} \right. \tag{1.11}$$

There is still approximate controllability (cf. J.L. LIONS [3], I.I. DIAZ-A.V. FURSIKOV [1]) (and all what is presented in Sections 2 and 3 can be extended) if

$$v = \{0, v_2, v_3\} \, , \tag{1.12}$$

i.e. the first component of the control is zero!

One can go one step further. We prove in E. ZUAZUA and the author (J.L. LIONS, E. ZUAZUA [1]) that approximate controllability is still true **in a generic sense with respect to the domain** Ω (of a particular geometrical structure) if

$$v = \{0, 0, v_3\}. \tag{1.13}$$

In such a case, one can still extended what we present below in Section 3 , but the explicit formula are not given here.

□

Remark 1.3.

Systems very sensitive to (among other things) initial conditions abound in weather prediction. One can then work with **"ensemble forecasting"** (cf. R.A. KERR [1]). If one wants to understand the (possible) consequences of human actions (thought of as "controls"), one is led to problems similar to the above. We shall return to this point on another occasion (J.L. LIONS [4]).

□

Remark 1.4.

Instead of taking "ensemble averaging", as alluded to in Remark 1.3. above, one can take **time averages** around T, i.e. replace condition (1.4) by

$$\frac{1}{h} \int_{T-h}^{T} y(t; v, b) dt \in y^T + \beta B, \tag{1.14}$$

other things being uncharged (one can even consider **several** time averages. cf. J.L. LIONS [5] [6]). The formula of Section 3 can be extended to these situations. Technical details are not given here.

□

Remark 1.5.

The results presented in Section 3 are technically related to **the measure of controllability** as introduced in J.L. LIONS [7] (for other goals).

□

2. Simple properties of the function $\phi(\alpha)$.

It is obvious that the function $\phi(\alpha)$ as defined by (1.7) (1.8) is always an increasing function of α.

Let us give a very simple example where $\phi(\alpha) \to +\infty$ as $\alpha \to \infty$.

We consider in $\Omega \subset \mathbf{R}^3$ (to fix the ideas) the state equation

$$\frac{\partial y}{\partial t} - \Delta y = v\chi_{\mathcal{O}} \qquad \text{in } \Omega \times (0,T) \tag{2.1}$$

where

$$y = 0 \qquad \text{on } \Gamma \times (0,T), \qquad \Gamma = \partial\Omega , \tag{2.2}$$

and

$$y(x,0) = y(0) = b(x), b \text{ given in } L^2(\Omega). \tag{2.3}$$

We assume that $v \in L^2(\mathcal{O} \times (0,T))$.

Assume by contradiction that $\phi(\alpha)$ remains bounded as $\alpha \to +\infty$.

Let us choose in (2.3)

$$b(x) = \alpha\theta(x), \tag{2.4}$$

where θ is an eigen function

$$-\Delta\theta = \lambda\theta , \quad \theta = 0 \quad \text{on} \quad \Gamma , \quad \|\theta\|_{L^2(\Omega)} = 1 \tag{2.5}$$

If $\phi(\alpha)$ is bounded, we can find a family v_α of controls such that as $\alpha \to \infty$

$$\|v_\alpha\|_{L^2(\mathcal{O}\times(0,T))} \leq c \,(= \text{constant}) \tag{2.6}$$

and

$$y_\alpha(T) \in y^T + \beta B \tag{2.7}$$

where $y_\alpha = y(t; v_\alpha, \alpha\theta)$. If we introduce φ_α as the solution of

$$\left| \begin{array}{l} \dfrac{\partial \varphi_\alpha}{\partial t} - \Delta \varphi_\alpha = v_\alpha \chi_{\mathcal{O}}, \\ \varphi_\alpha = 0 \text{ on } \Gamma \times (0,T),\ \varphi_\alpha(0) = 0 \end{array} \right. \tag{2.8}$$

then

$$y_\alpha = \varphi_\alpha + \alpha\theta \quad \exp(-\lambda t). \tag{2.9}$$

By virtue of (2.6), $\|\varphi_\alpha(T)\|_{L^2(\Omega)} \leq c_1$ (suitable constant) so that (2.7) (2.9) would imply $\|\alpha\theta exp(-\lambda T)\|_{L^2(\Omega)} \leq c_2$, i.e. α exp $(-\lambda T) \leq c_2$ as $\alpha \to \infty$, which is absurd.

□

Remark 2.1.

In fact we have proven more, namely

$$\phi(\alpha) = +\infty \quad \text{for } \alpha \text{ large enough.} \tag{2.10}$$

□

Remark 2.2.

The result we have just given is trivial because of the linearity of the system. Assuming approximate controllability, do we have a similar result for, say, the Navier Stokes system, as presented in Remark 1.1. above? This seems very likely but seems also to be an open question.

□

3. An identity for $\phi(\alpha)$ with some applications.

We now restrict the problem to the following linear situation :

$$\frac{\partial y}{\partial t} + Ay = v\chi_{\mathcal{O}} \quad \text{in} \quad \Omega \times (0,T), \quad \Omega \subset \mathbf{R}^3 , \ \mathcal{O} \subset \Omega \tag{3.1}$$

where

$$A = \text{second order elliptic operator,} \tag{3.2}$$

and where y is subject to

$$y = 0 \quad \text{on } \Gamma \times (0,T) \tag{3.3}$$

$$y(0) = b \quad , \quad b \in L^2(\Omega), \tag{3.4}$$

and where in (3.1) we have

$$v \in L^2(\mathcal{O} \times (0,T)). \tag{3.5}$$

□

Remark 3.1.

The boundary condition (3.3) can be replaced by any boundary condition such that the problem is well set.

□

Remark 3.2.

What we are going to present easily extends to higher order operators. For instance one can consider the Kuramoto-Sivashinski model

$$\frac{\partial y}{\partial t} + \Delta^2 y + \gamma \Delta y = v\chi_{\mathcal{O}} \quad , \quad \gamma > 0 \tag{3.6}$$

subject to

$$y = \Delta y = 0 \qquad \text{on } \Gamma \times (0,T)\,, \tag{3.7}$$

other things being unchanged.

□

We return to (3.1)...(3.5). We have

$$\phi(\alpha) = \sup_{b \in \alpha B} (\inf_v \frac{1}{2} \iint_{\mathcal{O}\times(0,T)} v^2 dx dt) \tag{3.8}$$

where b, v are subject to

$$y(t; v, b)(\text{solution of } (3.1)\ (3.3)\ (3.4)) \in y^T + \beta B. \tag{3.9}$$

Let us set

$$J(b) = \inf_v \frac{1}{2} \iint_{\mathcal{O}\times(0,T)} v^2 dx dt \quad , \quad \text{with } (3.9). \tag{3.10}$$

Of course

$$\phi(\alpha) = \sup_{b \in \alpha B} \quad J(b). \tag{3.11}$$

We now transform $J(b)$ using duality arguments. The final formula we are going to obtain in this fashion is as follows.

Given $f \in L^2(\Omega)$, let ψ be the solution of the backward equation

$$\left| \begin{array}{lll} -\dfrac{\partial \psi}{\partial t} + A^* \psi = 0 & \text{in} & \Omega \times (0,T)\,, \\ \psi(T) = f & \text{in} & \Omega\,, \\ \psi = 0 & \text{on} & \Gamma \times (0,T) \end{array} \right. \tag{3.12}$$

where A^* stands for the adjoint of A. Then

$$\phi(\alpha) = - \inf_{f \in L^2(\Omega)} [\frac{1}{2} \iint_{\mathcal{O}\times(0,T)} \psi^2 \, dx \, dt + \beta \| f \| - (f, y^T) - \alpha \| \psi(0) \|] \tag{3.13}$$

where $\| f \| = \| f \|_{L^2(\Omega)}$, $(f, y^T) = \int_\Omega f \, y^T \, dx$.

The proof uses the duality theorem of FENCHEL and T.R. ROCKAFELLAR [1] (cf also a proof in I. EKELAND and R. TEMAM [1]). Let us define the following convex proper functionals :

$$F_1(v) = \frac{1}{2} \iint_{\mathcal{O}\times(0,T)} v^2 \, dx \, dt \qquad \text{on } L^2(\mathcal{O} \times (0,T)) , \tag{3.14}$$

$$F_2(f) = \begin{cases} 0 & \text{iff } \in y^T - z(T; b) + \beta B \\ +\infty & \text{otherwise} \end{cases} \tag{3.15}$$

where $z(t; b)$ is defined by

$$\left| \begin{array}{l} \dfrac{\partial z}{\partial t} + Az = 0 , \quad z = 0 \quad \text{on } \Gamma \times (0,T) , \\ z(0) = b. \end{array} \right. \tag{3.16}$$

We have

$$y(t; v, b) = \varphi(t; v) + z(t; b) \tag{3.17}$$

where

$$\frac{\partial \varphi}{\partial t} + A\varphi = v\chi_{\mathcal{O}} , \quad \varphi = 0 \quad \text{on} \quad \Gamma \times (0,T), \quad \varphi(0) = 0. \tag{3.18}$$

Let us define

$$L \in \mathcal{L}(L^2(\mathcal{O} \times (0,T)) \, ; \, L^2(\Omega)) \tag{3.19}$$

by

$$Lv = \varphi(T; b). \tag{3.20}$$

Then $y(T; v, b) = Lv + z(T; b)$ so that we can write

$$J(b) = \inf_{v \in L^2(\mathcal{O}\times(0,T))} [F_1(v) + F_2(Lv)]. \tag{3.21}$$

Using FENCHEL-ROCKAFELLAR formula, we obtain

$$J(b) = \inf_{f \in L^2(\Omega)} [F_1^*(L^* f) + F_2^*(-f)] \tag{3.22}$$

where F_i^* denotes the conjugate function of F_i and L^* the adjoint of L.

One has

$$F_1^*(v) = F_1(v)$$

$$F_2^*(f) = \sup_g (f,g) - F_2(g) = (f, y^T - z(T;b)) + \beta\|f\|.$$

The adjoint L^* of L is given by

$$L^* f = \psi\chi_{\mathcal{O}}$$

where ψ is the solution of (3.12). Then (3.22) gives

$$J(b) = -\inf_f [\, \frac{1}{2} \iint_{\mathcal{O}\times(0,T)} \psi^2 \, dx \, dt + \beta \, \| \, f \, \| \; - (f, y^T) \; + (f, z(T;b))]. \tag{3.23}$$

But taking the scalar product of (3.12) with $z(t;b)$ gives (after using Green's formula)

$$(f, z(T;b)) = (\psi(0), b). \tag{3.24}$$

Using (3.24) into (3.23), (3.11) becomes

$$\phi(\alpha) = -\inf_{b\in\alpha B} \inf_f [\, \frac{1}{2} \iint_{\mathcal{O}\times(0,T)} \psi^2 \, dx \, dt + \beta \, \| \, f \, \| \; - (f, y^T) \; + (\psi(0), b) \,]$$

hence (3.13) follows.

□

Remark 3.3.

Formula (3.13) is very useful for the numerical approximative of the problem cf. R. GLOWINSKI and J.L. LIONS [1] for the case $\alpha \neq 0$. The case $\alpha = 0$ will be studied (from the numerical point of view) in the near future.

□

Remark 3.4.

The method used above is restricted to **linear** state equations. **For non linear** systems, it can only suggest some heuristic arguments, used in numerical computations (R. GLOWINSKI - J.L. LIONS [2]).

□

Remark 3.5.

Another approach to this kind of question is to restrict oneself to controls v which, in a sense, make the state at time T "insensitive" or "less sensitive" to variations of the initial condition b. cf. J.L. LIONS [8] [9].

□

Bibliography

[1] J.M. CORON, On the controllability of 2-D incompressible perfect fluid. JMPA 1995 (to appear).

[1] J.I. DIAZ and A.V. FURSIKOV, Approximate controllability of the Stokes system on cylinders by external unidirectional forces. JMPA. (to appear).

[1] I. EKELAND and R. TEMAM, *Analyse convexe et problèmes variationnels.* Dunod, Gauthier Villars. Paris 1974.

[1] R. GLOWINSKI and J.L. LIONS, Exact and approximate controllability for distributed parameter systems. Acta Numerica. 1994, p. 269-378 and 1995, p.

[2] Book in preparation.

[1] R.A. KERR, Taking the long view of weather. Science. Vol. 266, 23 December 1994, p. 1940.

[1] J.L. LIONS, Exact Controllability for distributed systems. Some trends and some problems, in *Applied and Industrial Mathematics*, R. SPIGLER (ed.), Kluwer, 1991, p. 59-84.

[2] Are there connections between turbulence and controllability ?. Springer Verlag Lecture Notes in Control and Information Sciences 144 (1990), A. BENSOUSSAN and J.L. LIONS, (ed.) Proceeding Int. Conf. INRIA, Antibes).

[3] Lectures in the College de France, 1991-1992.

[4] Mathematical questions in climatology. ICIAM 95.

[5] Quelques remarques sur la contrôlabilité en liaison avec des questions d'environnement, in *"Les grands systèmes des Sciences et de la Technologie"*, dedicated to R. DAUTRAY, Masson. J. HOROWITZ et J.L. LIONS, ed., 1994, p. 525-537.

[6] Remarks on approximate controllability for parabolic systems, in *"The finite element method in the 1990's"*, dedicated to O. ZIENKIEWICZ, Ed E. ONATE, J. PERIAUX, A. SAMUELSON, Springer Verlag, 1991, p.612-620.

[7] Measure of controllability. Proc. Georgian Acad. of Sciences, 1 (1993), p. 53-59.

[8] Insensibilization of functionals connected with distributed systems. 21e Colloque National d'Analyse Numérique. Rambouillet, June 1989. D. CIORANESCU, ed.

[9] Insensitive controls, in *Computational Mathematics and Applications*, CNR Istituto di Analisi Numerica, E. MAGENES ed., 1989, p. 285-299.

[1] J.L. LIONS and E. ZUAZUA, To appear.

[1] T.R. ROCKAFELLAR, Duality and stability in extremum problems involving convex functions. Pac. J. Math. 21 (1967), p. 167-187.

Stabilization and Periodic Solutions of a Hybrid System Arising in the Control of Noise

SORIN MICU Facultatea de Matematica-Informatica, Universitatea din Craiova, 1100, Romania and Universidad Carlos III de Madrid, 28913 Leganés, Spain.

ENRIQUE ZUAZUA Departamento de Matemática Aplicada, Facultad de Ciencias Matemáticas, Universidad Complutense, 28040 Madrid, Spain.

1 INTRODUCTION

In the last years, the construction of new fuel efficient turboprop engines has motivated the development of an active control methodology for interior pressure field chambers. Generally, the active control of noise consists in the generation of an appropriate secondary pressure wave which optimally interferes with the primary one.

In this paper we study a simplified model for this problem, introduced in Banks et al. (1990), consisting of a two-dimensional interior cavity with a flexible boundary. This can be considered as a hybrid system in which the acoustic vibrations of an interior fluid are coupled with the mechanical vibrations of a string. For other examples of hybrid systems, such as those coupling strings or beams with rigid bodies, see Littman and Markus (1988 b) or Hansen and Zuazua (1992). For simplicity we modify slightly the model introduced in Banks et al. (1990) considering on the boundary an active string instead of an Euler-Bernoulli beam.

In Sections 2 and 3 we present the mathematical formulation of the problem and we give a result of existence, uniqueness and stability of solutions.

The aymptotic properties are studied in Sections 4 and 5. We prove the convergence to zero of the energy of the system for every initial data and we note one of the most important mathematical features of the model: the very weak dissipation leads to a non uniform decay rate of the energy. This is a relevant mathematical property of a hybrid system.

Finally, in Section 6, we study the non-homogeneous problem with a second term acting on the flexible part of the boundary. This part transmits noise or vibrations from the exterior field to the interior cavity via fluid-structure interaction. It is natural to consider that the non-homogeneous term is periodic in time. Under this assumption, we study the existence of time-periodic solutions which is equivalent to the fact that all trajectories are bounded. The non uniform decay rate does not allow to apply a standard fix point argument. To overcome this difficulty we use a Fourier decomposition argument.

2 THE MATHEMATICAL MODEL

We consider the two-dimensional cavity $\Omega = (0,1) \times (0,1)$ in which the interior acoustic vibrations are coupled with the mechanical vibration of a string located in the subset $\Gamma_0 = \{(x,0) : x \in (0,1)\}$ of the boundary of Ω.

To describe the acoustic wave motion let $\vec{v}$ be the velocity, p the pressure and ρ the density of the fluid in our domain. Also, we consider that, at rest, the pressure p_0 and the density ρ_0 are constant. The linearized equations for the propagation of sound in an inviscid, elastic and compressible fluid, describing small disturbances, are:

$$\begin{cases} \rho' + \rho_0 \, div\, \vec{v} = 0 & \text{in} \quad \Omega \times (0,\infty) \\ \rho_0 \vec{v}\,' + \nabla p = 0 & \text{in} \quad \Omega \times (0,\infty). \end{cases} \tag{1}$$

We denote by $'$ the time derivative.

Let W be the transversal displacement (in the plane of Ω) of the string which is assumed to be dissipative and with Neumann boundary conditions. On it is acting the interior pressure p of the fluid:

$$\begin{cases} W'' - W_{xx} - W'_{xx} = p - p_0 & \text{on} \quad \Gamma_0 \times (0,\infty) \\ W_x(0) = W_x(1) = 0 & \text{for} \quad t \in (0,\infty). \end{cases} \tag{2}$$

On Γ_0 we impose the condition of continuity of velocity which results from the assumption that the string is impenetrable to the fluid. The part $\Gamma_1 = \partial\Omega \setminus \Gamma_0$ of the boundary of Ω is rigid and impenetrable, thus leading to zero normal velocity. We obtain the following boundary conditions:

$$\begin{cases} \vec{v} \cdot \nu = 0 & \text{on} \quad \Gamma_1 \times (0,\infty) \\ \vec{v} \cdot \nu = W' & \text{on} \quad \Gamma_0 \times (0,\infty). \end{cases} \tag{3}$$

In studying sound waves, it is usual to assume that $p = f(\rho)$. In the case of small perturbations, we can consider that the relation between p and ρ is linear:

$$p - p_0 = c_0^2(\rho - \rho_0) \tag{4}$$

where c_0 is the speed of sound in our fluid.

We introduce now a velocity potential Φ, such that $\nabla\Phi = \vec{v}$. The second equation of (1) implies that $p - p_0 = -\rho_0\Phi'$ and consequently, from (4) we get that $\rho' = -\frac{1}{c_0^2}\rho_0\Phi''$.

Considering that the speed of sound, c_0, and the density at rest, ρ_0, are 1, we obtain the following system in Φ and W:

$$\begin{cases} \Phi'' - \Delta\Phi = 0 & \text{in } \Omega \times (0,\infty) \\ \dfrac{\partial\Phi}{\partial\nu} = 0 & \text{on } \Gamma_1 \times (0,\infty) \\ \dfrac{\partial\Phi}{\partial y} = -W' & \text{on } \Gamma_0 \times (0,\infty) \\ W'' - W_{xx} - W'_{xx} + \Phi' = 0 & \text{on } \Gamma_0 \times (0,\infty) \\ W_x(0,t) = W_x(1,t) = 0 & \text{for } t \in (0,\infty) \\ \Phi(0) = \Phi^0, \quad \Phi'(0) = \Phi^1 & \text{in } \Omega \\ W(0) = W^0, \quad W'(0) = W^1 & \text{on } \Gamma_0. \end{cases} \tag{5}$$

In Micu and Zuazua (1994) we have studied a similar system in which the dissipative term was W' instead of $-W'_{xx}$.

We define the energy associated with this system by:

$$E(t) = \frac{1}{2}\int_\Omega (|\nabla\Phi|^2 + (\Phi')^2) + \frac{1}{2}\int_{\Gamma_0} ((W_x)^2 + (W')^2). \tag{6}$$

The system has a dissipative nature. Indeed, multiplying in (5) the first equation by Φ', the fourth equation by W' and integrating by parts, we get, formally, that:

$$dE(t)/dt = -\int_{\Gamma_0} (W'_x)^2 \le 0.$$

The initial data $(\Phi^0, \Phi^1, W^0, W^1)$ is considered in the space of finite energy:

$$X = H^1(\Omega) \times L^2(\Omega) \times H^1(\Gamma_0) \times L^2(\Gamma_0). \tag{7}$$

For $f^i = (f_1^i, f_2^i, f_3^i, f_4^i) \in X$ with $i \in \{1,2\}$, we define the inner product in X by:

$$\langle f^1, f^2 \rangle = \int_\Omega (\nabla f_1^1 \nabla f_1^2 + f_1^1 f_1^2) + \int_\Omega f_2^1 f_2^2 + \int_{\Gamma_0} ((f_3^1)_x (f_3^2)_x + f_3^1 f_3^2) + \int_{\Gamma_0} f_4^1 f_4^2.$$

Define now the following space:

$$\mathcal{D}(\mathcal{A}) = \{(\Phi^0, \Phi^1, W^0, W^1) \in H^2(\Omega) \times H^1(\Omega) \times H^1(\Gamma_0) \times H^1(\Gamma_0) : (W^0 + W^1) \in H^2(\Gamma_0),$$

$$\frac{\partial\Phi^0}{\partial\nu} = 0 \text{ on } \Gamma_1, \ \frac{\partial\Phi^0}{\partial y} = -W^1 \text{ on } \Gamma_0, \ (W^0 + W^1)_x(0) = (W^0 + W^1)_x(1) = 0\}$$

and the operator $\mathcal{A}$ defined in $\mathcal{D}(\mathcal{A})$ as follows:

$$\mathcal{A}(\Phi, \Psi, W, V) = (-\Psi, -\Delta\Phi, -V, -(W+V)_{xx} + \Psi).$$

We can consider now the following abstract Cauchy problem:

$$\begin{cases} U' + \mathcal{A}U = 0 \\ U(0) = U_0 \\ U(t) = (\Phi, \Phi', W, W')(t) \in \mathcal{D}(\mathcal{A}). \end{cases} \tag{8}$$

At the end of the next section we will discuss the relation between the original system (5) and (8).

3 EXISTENCE AND UNIQUENESS OF SOLUTIONS

First, we have a classical result of existence, uniqueness and stability for the system (8):

THEOREM 1 *i) Strong solutions: If $(\Phi^0, \Phi^1, W^0, W^1) \in \mathcal{D}(\mathcal{A})$ then there exists a unique strong solution of (8) $(\Phi, \Phi', W, W') \in \mathcal{C}([0,\infty), \mathcal{D}(\mathcal{A})) \cap \mathcal{C}^1([0,\infty), X)$.*

Moreover, for any strong solution, the associated energy (6) satisfies:

$$\frac{dE}{dt}(t) = -\int_{\Gamma_0} (W_x')^2. \tag{9}$$

ii) Weak solutions: If $(\Phi^0, \Phi^1, W^0, W^1) \in X$ then there exists a unique solution of (8) with the property: $(\Phi, \Phi', W, W') \in \mathcal{C}([0,\infty), X)$.

For two weak solutions (Φ, W) and $(\bar{\Phi}, \bar{W})$ we have the following stability property:

$$\begin{aligned} &\int_\Omega (|\nabla(\Phi - \bar{\Phi})|^2 + |(\Phi - \bar{\Phi})'|^2) + \int_{\Gamma_0} (|(W - \bar{W})_x|^2 + |(W - \bar{W})'|^2) \leq \\ &\leq \int_\Omega (|\nabla(\Phi^0 - \bar{\Phi}^0)|^2 + |\Phi^1 - \bar{\Phi}^1|^2) + \int_{\Gamma_0} (|(W^0 - \bar{W}^0)_x|^2 + |W^1 - \bar{W}^1|^2). \end{aligned} \tag{10}$$

Proof: We prove that the operator $\mathcal{A} + \mathcal{I}$ is maximal monotone in X and we apply the Hille-Yosida theory (see Cazenave and Haraux (1990), Theorem 3.1.1, p.37).

Indeed, if $U = (\Phi, \Psi, W, V) \in \mathcal{D}(\mathcal{A})$ then $\langle(\mathcal{A} + \mathcal{I})U, U\rangle \geq \int_{\Gamma_0} (V_x)^2 \geq 0$, which means that $\mathcal{A} + \mathcal{I}$ is monotone.

On the other hand, for all $F = (f_1, f_2, f_3, f_4) \in X$ we can find a unique solution $U = (\Phi, \Psi, W, V) \in \mathcal{D}(\mathcal{A})$ for the equation $(\mathcal{A} + \mathcal{I})U = F$. This is equivalent to solve the following system:

$$\begin{cases} -\Psi + \Phi = f_1 \\ -\Delta\Phi + \Psi = f_2, \ \dfrac{\partial \Phi}{\partial \nu} = 0 \text{ on } \Gamma_1 \text{ and } \dfrac{\partial \Phi}{\partial \nu} = V \text{ on } \Gamma_0 \\ -V + W = f_3 \\ -(W + V)_{xx} + \Psi + V = f_4 \text{ and } (W + V)_x(0) = (W + V)_x(1) = 0 \end{cases} \tag{11}$$

First, we consider the variational equation corresponding to (11), which consists in finding (Φ, W) in $H^1(\Omega) \times H^1(\Gamma_0)$ such that, for all $(\varphi, u) \in H^1(\Omega) \times H^1(\Gamma_0)$:

$$\begin{aligned} &\int_\Omega \nabla\Phi\nabla\varphi + \int_\Omega \Phi\varphi - \int_{\Gamma_0} W\varphi + 2\int_{\Gamma_0} W_x u_x + \int_{\Gamma_0} \Phi u + \int_{\Gamma_0} Wu = \\ &\int_\Omega (f_1 + f_2)\varphi - \int_{\Gamma_0} f_3\varphi + \int_{\Gamma_0} (f_3)_x u_x + \int_{\Gamma_0} (f_1 + f_3 + f_4)u. \end{aligned} \tag{12}$$

The left side of the equation (12) defines a continuous and coercive bilinear form in $(H^1(\Omega)\times H^1(\Gamma_0))^2$ while the right side defines a continuous linear form in $H^1(\Omega)\times H^1(\Gamma_0)$.

Applying Lax-Milgram's Lemma it results that (12) has a unique solution (Φ, W) in $H^1(\Omega) \times H^1(\Gamma_0)$. Finally, in view of the classical regularity results for the Laplace's operator, this implies that $\mathcal{A} + \mathcal{I}$ is maximal.

■

Define now the space $\tilde{X} = \tilde{H}^1(\Omega) \times L^2(\Omega) \times \tilde{H}^1(\Gamma_0) \times L^2(\Gamma_0)$ where $\tilde{H}^1$ is the quotient space $\tilde{H}^1 = H^1/P_0$ and P_0 is the set of all constant functions and $\tilde{\mathcal{D}}(\mathcal{A})$ the subspace of $\tilde{X}$ defined in the same way as $\mathcal{D}(\mathcal{A})$:

$$\tilde{\mathcal{D}}(\mathcal{A}) = \{(\Phi^0, \Phi^1, W^0, W^1) \in \tilde{H}^2(\Omega) \times H^1(\Omega) \times \tilde{H}^1(\Gamma_0) \times H^1(\Gamma_0) : (W^0 + W^1) \in H^2(\Gamma_0),$$

$$\frac{\partial \Phi^0}{\partial \nu} = 0 \text{ on } \Gamma_1, \ \frac{\partial \Phi^0}{\partial y} = -W^1 \text{ on } \Gamma_0, \ (W^0 + W^1)_x(0) = (W^0 + W^1)_x(1) = 0\}.$$

Observing that if (Φ, Φ', W, W') is the solution of (8) with initial data $(\Phi^0, \Phi^1, W^0, W^1)$ then $(\Phi + c_1, \Phi_t, W + c_2, W_t)$ is the solution with initial data $(\Phi^0 + c_1, \Phi^1, W^0 + c_2, W^1)$ we can obtain a similar theorem of existence uniqueness and stability in $\tilde{X}$ but in addition, in view of the coercivity of the energy in this space, we have that for all strong solutions:

$$\begin{aligned} &\| (\Phi, \Phi', W, W')(t) \|_{\tilde{X}} \leq \| (\Phi^0, \Phi^1, W^0, W^1) \|_{\tilde{X}}, \quad \forall t \geq 0, \\ &\| \mathcal{A}(\Phi, \Phi', W, W')(t) \|_{\tilde{X}} \leq \| \mathcal{A}(\Phi^0, \Phi^1, W^0, W^1) \|_{\tilde{X}}, \quad \forall t \geq 0 \end{aligned} \tag{13}$$

which means that the trajectories are bounded in $\tilde{\mathcal{D}}(\mathcal{A})$.

We want to see now when the boundary conditions $W_x(0) = W_x(1) = 0$ make sense and for which initial data the solution is relatively compact in $\tilde{X}$. To do this we have to ask for more regularity of W^0.

Property 1 *If the initial data $(\Phi^0, \Phi^1, W^0, W^1)$ belongs to the space:*

$$\tilde{\mathcal{D}} = \{(\Phi^0, \Phi^1, W^0, W^1) \in \tilde{\mathcal{D}}(\mathcal{A}) : W^0 \in \tilde{H}^2(\Gamma_0) \textit{ and } W^0_x(0) = W^0_x(1) = 0\}$$

the corresponding solution of (8), (Φ, Φ', W, W'), is in $\mathcal{C}_b([0,\infty), \tilde{\mathcal{D}})$, where $\mathcal{C}_b([0,\infty), \tilde{\mathcal{D}}) = \mathcal{C}([0,\infty), \tilde{\mathcal{D}}) \cap L^\infty(0,\infty; \tilde{\mathcal{D}})$. Moreover, $\tilde{\mathcal{D}}$ is dense and compact in $\tilde{X}$.

Proof: We have from (13) that $W'(t) + W(t) = V(t) \in \mathcal{C}_b([0,\infty), H^2(\Gamma_0))$. Thus $W(t) = e^{-t}W_0 + \int_0^t e^{s-t}V(s)ds$ and, as soon as W^0 is considered in $H^2(\Gamma_0)$, we deduce that $W \in \mathcal{C}_b([0,\infty), H^2(\Gamma_0))$. The same argument shows that the boundary conditions $W_x(0) = W_x(1) = 0$ are fulfilled for all $t \geq 0$ if $W^0_x(0) = W^0_x(1) = 0$.

■

Remark: Under the hypothesis that the initial data of the problem (8) belongs to $\tilde{\mathcal{D}}$, this problem is equivalent to the original system (5).

4 STRONG STABILIZATION

Concerning the asymptotic behavior of solutions we prove first the strong stabilization in the energy space.

THEOREM 2 *For each initial data in X the corresponding weak solution of (8) tends asymptotically towards zero in $\tilde{X}$, that is,* $\lim_{t\to\infty} E(t) = 0$.

Proof: The main tools of our analysis is an extension of the well known Invariance Principle of La Salle and Holmgren's Uniqueness Theorem.

Observe first that it is sufficient to consider the case of initial data $(\Phi^0, \Phi^1, W^0, W^1)$ in $\tilde{\mathcal{D}}$. A standard density argument and the property of stability (10) enable us to complete the proof.

In this case Theorem 1 and Property 1 give an unique strong solution $U = (\Phi, \Phi', W, W')$ for the equation (8), with $\{U(t)\}_{t\geq 0}$ bounded in $\tilde{\mathcal{D}}$. Since $\tilde{\mathcal{D}} \subseteq \tilde{X}$ with compact inclusion, we have that $\{U(t)\}_{t\geq 0}$ is relatively compact in $\tilde{X}$. So, all we have to prove is that the unique limit point in $\tilde{X}$ of the trajectory $\{U(t)\}_{t\geq 0}$ as $t \to \infty$ is $(0,0,0,0)$.

For this take $(t_n)_{n\geq 0}$ with $U(t_n) \to (z^0, z^1, v^0, v^1)$ in $\tilde{X}$ when $n \to \infty$ and show that

$$z^0 = z^1 = 0, v^0 = v^1 = 0 \text{ in } \tilde{X} \tag{14}$$

i.e. there exist two constants c_1 and c_2 such that $z^0 = c_1$, $z^1 = 0$, $v^0 = c_2$ and $v^1 = 0$.

For some $T > 0$, that will be fixed later on, define, for all $n \geq 0$, the family of translated solutions,

$$\Phi_n(t) = \Phi(t + t_n), \quad W_n(t) = W(t + t_n), \quad t \in [0, T].$$

By the Theorem of Ascoli Arzela the sequence $(\Phi_n, \Phi'_n, W_n, W'_n)_{n\geq 0}$ is relatively compact in $\mathcal{C}([0,T]; \tilde{X})$ and hence there is a subsequence, also denoted by $(\Phi_n, \Phi'_n, W_n, W'_n)_{n\geq 0}$ and $(z, z', v, v') \in \mathcal{C}([0,T]; \tilde{X})$, such that $(\Phi_n, \Phi'_n, W_n, W'_n) \to (z, z', v, v')$ in $\mathcal{C}([0,T]; \tilde{X})$.

Observe that $z(0) = z^0, v(0) = v^0, z'(0) = z^1, v'(0) = v^1$ and so if we prove that z and v are constant in space and time we obtain the stated result.

Since E is a non increasing and non negative function it has a finite limit when t tends to infinite. On the other hand since

$$E(t_n) - E(t_n + T) = \int_{t_n}^{t_n+T} \int_{\Gamma_0} (W'_x)^2 = \int_0^T \int_{\Gamma_0} ((W_n)'_x)^2$$

we can pass to limit and obtain that

$$\int_0^T \int_{\Gamma_0} (v'_x)^2 = 0 \tag{15}$$

It results that (z, z', v, v') satisfies:

$$\begin{cases} z'' - \Delta z = 0 & \text{in } \Omega \times (0,T) \\ \frac{\partial z}{\partial \nu} = 0 & \text{on } \Gamma_1 \times (0,T) \\ \frac{\partial z}{\partial y} = -v' & \text{on } \Gamma_0 \times (0,T) \\ v'' - v_{xx} + z' = 0 & \text{on } \Gamma_0 \times (0,T) \\ v_x(0,t) = v_x(1,t) = 0 & \text{for } t \in (0,T). \end{cases} \tag{16}$$

Consider now the functions $Y = z_x$ and $\varphi = v_x$ which satisfy:

$$\begin{cases} Y'' - \Delta Y = 0 & \text{in } \ \Omega \times (0,T) \\ Y = 0 & \text{for } \ x \in \{0,1\}, (y,t) \in (0,1) \times (0,T) \\ Y_y = 0 & \text{for } \ y \in \{0,1\}, (x,t) \in (0,1) \times (0,T) \\ \varphi_{xx} = Y' & \text{on } \ \Gamma_0 \times (0,T) \\ \varphi(0,t) = \varphi(1,t) = 0 & \text{for } \ t \in (0,T). \end{cases} \tag{17}$$

Finally define the function $\eta = Y''$ which satisfies:

$$\begin{cases} \eta'' - \Delta\eta = 0 & \text{in } \ \Omega \times (0,T) \\ \eta_y = 0 & \text{on } \ \Gamma_0 \times (0,T) \\ \eta = 0 & \text{on } \ \Gamma_0 \times (0,T). \end{cases} \tag{18}$$

We can apply now Holmgren's Uniqueness Theorem (see Hörmander (1990), Theorem 8.6.5, p.309 and Lions (1988), Theorem 8.1, p.88) which implies that $\eta = 0$ in $\Omega \times (1, T-1)$ if $T > 2$ and so

$$Y(t,x,y) = Y_1(x,y)t + Y_2(x,y) \text{ in } \Omega \times (1, T-1).$$

We deduce that, for $i \in \{1,2\}$,

$$\begin{cases} \Delta Y_i = 0 & \text{in } \Omega \\ \dfrac{\partial Y_i}{\partial n} = 0 & \text{for } y \in \{0,1\}, x \in (0,1) \\ Y_i = 0 & \text{for } x \in \{0,1\}, y \in (0,1) \end{cases} \tag{19}$$

which implies that $Y = 0$ in $\Omega \times (1, T-1)$ and $\varphi_{xx} = 0$ on $\Gamma_0 \times (1, T-1)$.

It follows that $z_x = 0$ and $v_{xxx} = 0$ implying that z and v are constant in time and space. Hence $z = 0$ and $v = 0$ in $\tilde{X}$ which concludes the proof.

■

Remark: We have proved the strong stabilization of the system in $\tilde{X}$ but not in X because the energy is not coercive in X. We have that $\nabla\Phi, \Phi_t, W_x$ and W_t tend to zero in the L^2 norm when t goes to infinity but not necessarily Φ and W. Observe that $\Phi = 1$ and $W = 1$ is a constant solution in time of our system. However, from a physical point of view, only the convergence in $\tilde{X}$ is relevant, since it represents the fact that the pressure and velocity of the interior fluid tend to its rest values.

5 THE LACK OF UNIFORM DECAY

In this paragraph we observe that the rate of decay is not uniform. Results like this are typical for linear hybrid systems in which the dissipation is very weak: it can force the strong stabilization but it cannot assure the uniform decay.

THEOREM 3 *The rate of decay of the energy is not exponential in $\tilde{X}$, i.e. there is no $C > 0$ and $\omega > 0$ such that $E(t) \leq CE(0)e^{-\omega t}$ for all weak solution.*

Proof: It consists in constructing solutions with decay rates slower than any preassigned exponential function. To do this we look for a sequence of solutions $\{(\Phi_n, W_n)\}_{n\geq 1}$ for the equation (5) of the type $(\Phi_n, W_n) = e^{\lambda_n t}(\varphi_n, v_n)$ where $\varphi_n = \varphi_n(y)$ and $v_n \in \mathbb{R}$ and $\mathcal{R}e\,\lambda_n \to 0$ when $n \to \infty$.

We can see that the equation (5) has solutions of this form if (φ_n, v_n) satisfies

$$\begin{cases} \lambda^2\varphi - \varphi_{yy} = 0 \text{ for } y \in (0,1) \\ \varphi_y(1) = 0, \ \varphi_y(0) = -\lambda v \\ \lambda^2 v + \lambda v + \lambda\varphi(0) = 0. \end{cases} \tag{20}$$

It results that $\varphi(y) = cosh\,(\lambda(y-1))$ and $v = sinh\,\lambda$ is solution of (20) if λ is solution of the algebric equation: $e^{2\lambda} - 1 = -\dfrac{2}{\lambda+2}$.

To study this equation we consider, for $n \in \mathbb{N}$, the squares γ_n with centers $\alpha_n = n\pi\, i$ and sides $\frac{3}{n\pi} < \varepsilon_n < \frac{4}{n\pi}$.

We can apply now Rouché' s Theorem and we find that, for n large enough, there is a solution of this equation in each square γ_n. This concludes the proof.

■

Remark: Taking $(\Phi_{k,n}, W_{k,n}) = e^{\lambda_{k,n}t}(\varphi_{k,n}, v_{k,n})$ where $\varphi_{k,n}(x,y) = \cos(k\pi x)\xi_{k,n}(y)$ and $v_{k,n}(x) = \cos(k\pi x)u_{k,n}$ with $u_{k,n} \in \mathbb{R}$, we can prove that, for each $k \in \mathbb{N}$, there is a sequence of eigenvalues $\lambda_{n,k}$ with $\mathcal{R}e\,\lambda_{k,n} \to 0$ as n tends to infinity.

Remark: It is well known that the lack of exponential decay implies the lack of uniform decay. In this case the norm of the semigroup $S(t)$ in $\mathcal{L}(\tilde{X}, \tilde{X})$ is one for all $t > 0$. Littman and Markus (1988 a) showed that, in this situation, the rate of decay of a particular trajectory may be slower than any preassigned function that decreases to zero as $t \to \infty$.

6 EXISTENCE OF PERIODIC SOLUTIONS

In this section we consider a non-homogeneous time-periodic term and we study the existence of periodic solutions for the system:

$$\begin{cases} \Phi'' - \Delta\Phi = 0 & \text{in } \ \Omega \times (0,\infty) \\ \frac{\partial\Phi}{\partial\nu} = 0 & \text{on } \ \Gamma_1 \times (0,\infty) \\ \frac{\partial\Phi}{\partial y} = -W' & \text{on } \ \Gamma_0 \times (0,\infty) \\ W'' - W_{xx} - W'_{xx} + \Phi' = f & \text{on } \ \Gamma_0 \times (0,\infty) \\ W_x(0,t) = W_x(1,t) = 0 & \text{for } \ t \in (0,\infty). \end{cases} \tag{21}$$

The non-homogeneous periodic term f represents a perturbing force on the active part of the boundary due to an exterior noise source. We give a positive answer to this problem in the case of a smooth function f.

THEOREM 4 *Assume that $f \in W^{1,2}(0,T; L^2(0,1))$ is periodic with respect to t, of period T. If $f(t) = \sum_{n=0}^{\infty} f_n(t)cos(n\pi x)$ and:*

$$\begin{cases} \sum_{n=0}^{\infty} \| (f_n)' \|_{L^2(0,T)}\, exp(\frac{3\pi}{4}n) < \infty, \\ \sum_{n=0}^{\infty} (n+1) \| f_n \|_{L^2(0,T)}\, exp(\frac{3\pi}{4}n) < \infty \end{cases} \tag{22}$$

then there exists a unique weak solution of (21) periodic in $\tilde{X}$ with period T.

Proof: First we observe that since the decay rate of the energy is not uniform it is not possible to use directly a fix point method for the Poincaré's map. To avoid this difficulty we reduce the system to an infinite sequence of one-dimensional problems. To treat the one-dimensional problem we first introduce an artificial dissipation which produces the uniform decay of the energy. Then by classical arguments we obtain a periodic solution. Then we let the perturbative term go to zero. In this way it is proved that each one-dimensional problem has a unique time-periodic solution.

To acomplish this we write $f(t,x) = \sum_{n=0}^{\infty} f_n(t)cos(n\pi x)$ and study, for each n, the problem of the existence of time-periodic solutions for the one-dimensional system:

$$\begin{cases} \psi'' - \psi_{yy} + n^2\pi^2\psi = 0 & \text{for } y \in (0,1), t \in (0,\infty) \\ \frac{\partial \psi}{\partial y}(1) = 0 & \text{for } t \in (0,\infty) \\ \frac{\partial \psi}{\partial y}(0) = -v' & \text{for } t \in (0,\infty) \\ v'' + n^2\pi^2 v' + n^2\pi^2 v + \psi'(0) = f_n & \text{for } t \in (0,\infty). \end{cases} \tag{23}$$

We prove the existence of a unique time-periodic solution of (23) in three steps.

Step 1. For ε sufficiently small we consider the problem

$$\begin{cases} \psi'' - \psi_{yy} + \varepsilon\psi' + n^2\pi^2\psi = 0 & \text{for } y \in (0,1), t \in (0,\infty) \\ \frac{\partial \psi}{\partial y}(1) = 0 & \text{for } t \in (0,\infty) \\ \frac{\partial \psi}{\partial y}(0) = -v' & \text{for } t \in (0,\infty) \\ v'' + n^2\pi^2 v' + n^2\pi^2 v + \psi'(0) = f_n & \text{for } t \in (0,\infty). \end{cases} \tag{24}$$

It is easy to see that the term $\varepsilon\psi'$ produces an uniform decay rate for the energy of the homogeneous system:

$$F(t) = \frac{1}{2}\int_0^1 (|\, \psi_y\,|^2 + |\, \psi'\,|^2 + n^2\pi^2 \,|\, \psi\,|^2) + \frac{1}{2}((v')^2 + n^2\pi^2 v^2).$$

Indeed, if we consider the following perturbation of F:

$$G(t) = F(t) + \delta(\int_0^1 (\psi_t\psi) + w_t w + \psi(0)w), \text{ for all } t \geq 0$$

we deduce that, for δ small enough,

$$\frac{1}{2}F(t) < G(t) < \frac{3}{2}F(t) \text{ for all } t \geq 0 \text{ and}$$

$$G'(t) \leq -\beta F(t) \text{ for all } t \geq 0 \text{ and for } \beta < \varepsilon.$$

The last two inequalities imply the exponential decay of the energy $F(t)$.

In this case for $k = k(\varepsilon) \in I\!N$ large enough, we can apply the Banach's Fix Point Theorem for the contractive operator:

$$\mathcal{J}^k_{\varepsilon,n} : X \to X,\ \mathcal{J}^k_{\varepsilon,n}(\psi^0, \psi^1, v^0, v^1) = (\psi(kT), \psi'(kT), v(kT), v'(kT)).$$

Since $\mathcal{J}^k_{\varepsilon,n}$ is contractive it has a unique fix point u_0. It results that

$$\mathcal{J}^1_{\varepsilon,n} u_0 = \mathcal{J}^1_{\varepsilon,n}(\mathcal{J}^k_{\varepsilon,n} u_0) = \mathcal{J}^k_{\varepsilon,n}(\mathcal{J}^1_{\varepsilon,n} u_o).$$

Since $\mathcal{J}^k_{\varepsilon,n}$ has a unique fix point we get that u_0 is the unique fix point for $\mathcal{J}^1_{\varepsilon,n}$. This fix point gives us a unique periodic solution $(\psi_{\varepsilon,n}, v_{\varepsilon,n})$.

Step 2. We obtain some estimates for the periodic solutions found on the first step which are independent of ε .

Integrating in time the derivative of the energy of the system and the derivative of the energy of the system obtained by taking a time-derivative in (24), we get that:

$$\begin{aligned} n^2\pi^2 \int_0^T (v'_{\varepsilon,n})^2\, dt \ &\le \int_0^T f_n^2\, dt, \\ n^2\pi^2 \int_0^T (v''_{\varepsilon,n})^2\, dt \ &\le \int_0^T (f'_n)^2\, dt \end{aligned} \tag{25}$$

The equation on Γ_0 of (24) implies that $n^2\pi^2 \int_0^T v_{\varepsilon,n} = \int_0^T f$. Applying now Poincaré's inequality we deduce that:

$$\int_0^T v^2_{\varepsilon,n} \le \frac{C}{n^2\pi^2} \int_0^T f^2. \tag{26}$$

We go back to the equation for v in (24) and we obtain:

$$\int_0^T (\psi'_{\varepsilon,n})^2(t,0)\, dt \ \le C((n^2\pi^2+1) \int_0^T f_n^2\, dt \ + \frac{1}{n^2\pi^2} \int_0^T (f'_n)^2\, dt\ ). \tag{27}$$

Since $\int_0^T \psi_{\varepsilon,n}(t,y) dt \ = 0$, for all $y \in (0,1)$ we can apply Poincaré's inequality

$$\int_0^T (\psi_{\varepsilon,n})^2(t,0)\, dt \ \le C \int_0^T (\psi'_{\varepsilon,n})^2(t,0)\, dt. \tag{28}$$

We can apply now multiplier techniques (see Lions (1988), Lemma 1.3, p.139). Multiplying the first equation of (24) by $(1-y)\Psi_y$, integrating by parts and applying Gronwall's Lemma we obtain that

$$\int_0^T (\psi_t^2 + \psi_y^2 + n^2\pi^2\psi^2)\, dt\ (y) \le \int_0^T (\psi_t^2 + \psi_y^2 + \alpha\psi^2)\, dt\ (0)\ exp((\frac{3\pi n}{2} + \frac{\varepsilon}{2})y). \tag{29}$$

With the estimates (25), (27) and (28) for $\psi_{\varepsilon,n}(t,0)$, $(\psi_{\varepsilon,n})_y(t,0)$ and $\psi'_{\varepsilon,n}(t,0)$ in $L^2(0,T)$, we deduce that

$$\begin{aligned} &\int_0^T ((\psi'_{\varepsilon,n})^2 + ((\psi_{\varepsilon,n})_y)^2 + n^2\pi^2\psi^2_{\varepsilon,n})(t,y)\, dt \ \le \\ &\le c' \int_0^T [(f'_n)^2 + (1+n^2)(f_n)^2]\, dt\ exp(\frac{3\pi n}{2} y), \ \text{ for all } y \in (0,1). \end{aligned} \tag{30}$$

Step 3. The last estimates show that the set of periodic solutions corresponding to each ε, $\{(\psi_{\varepsilon,n}, \psi'_{\varepsilon,n}, v_{\varepsilon,n}, v'_{\varepsilon,n})\}_{\varepsilon>0}$, is bounded in $L^2(0,T;Y)$ with $Y = H^1(0,1)\times L^2(0,1)\times \mathbb{R} \times \mathbb{R}$. Since our problem is linear we can pass to the limit when $\varepsilon \to 0$ and we obtain for each $n \in \mathbb{N}$ a weak periodic solution (ψ_n, v_n) of equation (23).

We turn now to our two dimensional problem and consider the series:

$$\Phi = \sum_{n=0}^{\infty} \psi_n(t,y)cos(n\pi x) \text{ and } W = \sum_{n=0}^{\infty} v_n(t)cos(n\pi x)$$

where, for each $n \in \mathbb{N}, (\psi_n, v_n)$ is the periodic solution found on step 3. The two series converge in $L^2(0,T;H^1(\Omega)) \cap H^1(0,T;L^2(\Omega)$ and in $H^1(0,T;H^1(\Gamma_0))$ respectively, if f verifies (22). On the other hand (Φ, W) is the periodic solution we were looking for. The uniqueness follows from Theorem 2 since the difference of two solutions of the non-homogeneous problem is a solution of the homogeneous problem. ■

Remark: Condition (22) holds for functions f with the property that $f(t)$ belongs to the space of Gevrey's functions of exponent one defined by: $\partial_x^j f(t,x)$ are continuous on $0 \le t < \infty$, $x \in [0,1]$ and, in addition, for each compact set $K \subset [0,\infty) \times [0,1]$ and each constant $\theta > 0$ there is a constant C_θ such that $| \partial_x^j f(t,x) | \le C_\theta \theta^j j^j$ for all $j = 1, 2, ..$ when $(t,x) \in K$. This is a class of analytic functions (see Hörmander (1990), p. 281).

7 COMMENTS

In Banks et al. (1990) a two-dimensional model is presented in which, on the subset Γ_0 of the boundary, a fixed end Euler-Bernoulli beam is considered. The methods developed in this paper can be adapted to this type of problems.

It is possible to give results like Theorem 1 and Theorem 2 (with very similar proofs) for the case of a string located in Γ_0 but with Dirichlet boundary conditions. Nevertheless, the methods used for proving Theorem 3 and 4, based on the separation of variables, cannot be applied in this case.

The results of Sections 2 and 3 can be generalized for any domain. If Ω is a bounded open set in $\mathbb{R}^2$ with smooth boundary and Γ_0 is a part of the boundary of the domain it is sufficient to change in (1) the equation on Γ_0 by

$$W'' - \frac{d^2W}{d\tau^2} + \frac{d^2W'}{d\tau^2} + \Phi' = 0 \quad \text{on} \quad \Gamma_0 \times (0,\infty)$$

where $\frac{d}{d\tau}$ is the derivative in the tangential direction. The rest of the results of this paper may be extended to some particular geometries, for instance when Ω is a ball of $\mathbb{R}^2$.

The last result given in Theorem 4 requests a lot of regularity for the non-homogeneous term f. This phenomena may be expected since the dissipation is very weak and is located in a relatively small part of the boundary. The assumptions of Theorem 4 are very probably sharp. Indeed, we have assumed that f and f' belong to $L^2(0,T;L^2(\Gamma_0))$. However, when dealing with classical damped wave equations like, for instance,

$$\left\{ \begin{array}{ll} u'' - \Delta u = 0 & \text{in } \Omega \times (0,\infty) \\ \dfrac{\partial u}{\partial \nu} + u' = f & \text{on } \partial\Omega \times (0,\infty) \end{array} \right. \tag{31}$$

the fact that $f \in L^2(0,T;L^2(\partial\Omega))$ suffices to guarantee that the energy of the trajectories remains bounded. In Theorem 4 we have imposed an extra assumption on f'. This is due to the hybrid boundary conditions we are dealing with, which are roughly of second order. We also assumed that $f(t)$ belongs to some Gevrey class. This is reasonable in view of the structure of the dissipation region which implies a very weak decay rate of high frecuencies. Indeed, as Ralston (1969) observed, (and later on Bardos, Lebeau and Rauch (1988) applied in the context of the control and stabilization of the wave equation in bounded domains) if one characteristic ray escapes to the dissipative region we can not expect an uniform decay. In this case every segment $\{(x, y_0) : x \in (0,1)\}$, for any $y_0 \in (0,1)$, constitutes a ray with such a property.

Acknowledgments: The first author was supported by the Project MATAROU TEMPUS JEP 2797 during his stay at the Departamento Matemática Aplicada, Universidad Complutense de Madrid, when part of this work was carried out. The work of the second author has been supported by grants PB93-1203 of the DGICYT (Spain) and SC1-CT91-0732 and CHRX-CT94- 0471 of the European Community.

REFERENCES:

1. Banks, H.T., Fang, W., Silcox, R.J. and Smith, R.C. (1990) Approximation Methods for Control of Acustic/Structure Models with Piezoceramic Actuators, *Journal of Intelligent Material Systems and Structures*, to appear.
2. Bardos, C., Lebeau, G. and Rauch, J. (1988) Contrôle et stabilisation dans les problèmes hyperboliques, Appendix in Lions (1988), pp. 492-537.
3. Cazenave, T. and Haraux, A. (1990) *Introduction aux problèmes d'évolution semilinéaires*, Mathématiques et Applications, 1, Ellipses, Paris.
4. Grisvard, P. (1985) *Elliptic Problems in Non-smooth Domains*, Pitman.
5. Hansen, S. and Zuazua, E. (1992) Exact Controllability and Stabilization of a Vibrating String with an Interior Point Mass, *SIAM J. Cont. Optim.*, to appear.
6. Hörmander, L. (1990) *The Analysis of Linear Partial Differential Operators I*, Springer-Verlag.
7. Lions, J.L. (1988) *Contrôlabilité exacte, perturbations et stabilisation de systèmes distribués, Tome 1: Contrôlabilité exacte*, Masson, RMA, Paris.
8. Littman, W. and Marcus, L. (1988 a) Some Recent Results on Control and Stabilization of Flexible Structures, Univ. Minn., Mathematics Report 87-139.
9. Littman, W. and Marcus, L. (1988 b) Exact Boundary Controllability of a Hybrid System of Elasticity, *Arch. Rat. Mech. Anal.*, 103, 3, pp. 193-236.
10. Micu, S.D. and Zuazua, E. (1994) Propriétés qualitatives d'un modèle hybride bi-dimensionnel intervenant dans le contrôle du bruit , *C.R. Acad. Sci. Paris*, to appear.
11. Ralston, J (1969) Solutions of the wave equation with localized energy, *Comm. Pure Appl. Math.* 22, pp. 807-823.

Differentiability of Projection and Applications

Michel Pierre
Universite de Nancy I
Departement de Mathematiques, B.P. 239
U.R.A-C.N.R.S. 750, Projet NUMATH INRIA Lorraine
54506 Vandoeuvre–Les–Nancy Cedex, France

Jan Sokołowski
Universite de Nancy I
Departement de Mathematiques, B.P. 239
U.R.A-C.N.R.S. 750, Projet NUMATH INRIA Lorraine
54506 Vandoeuvre–Les–Nancy Cedex, France
and Systems Research Institute, Polish Academy of Sciences
01–447 Warsaw, ul. Newelska 6, Poland

1 Introduction

The differential stability of metric projection onto convex sets in Sobolev spaces is considered in [2], [3]. In particular the form of the tangent cone and the conical differential of the projection is obtained in the case of $K(\Omega) \subset H^2(\Omega) \cap H_0^1(\Omega)$ defined below, under assumption that the so–called coincidence set is compact. In the present paper the assumption is relaxed using the form of nonnegative functional on the cone of nonnegative functions in the Sobolev space $H^2(\Omega) \cap H_0^1(\Omega)$.

In Section 5 an optimal control problem with the state constraints is considered. The optimality system for the control problem is established and the sensitivity of optimal solutions with respect to the perturbations of the cost functional is performed.

In Section 6 dealing with the variational inequality for the Kirchhoff plate we study the differentiability of solutions with respect to deformations of the domain of integration. The variational inequality is of the form

$$w \in K(\Omega) \ : \ \int_\Omega \Delta w \Delta(\varphi - w)dx \geq \int_\Omega f(\varphi - w)dx \quad \forall \, \varphi \in K(\Omega)$$

and the set $K(\Omega)$ of admissible displacements is given as

$$K(\Omega) = \{\varphi \in H^2(\Omega) \cap H_0^1(\Omega) | \varphi(x) \geq \psi(x) \text{ in } \Omega\} \ .$$

The so-called shape derivative $w'(\Omega;V)$ of solutions to this inequality is given as a solution to the following variational inequality

$$w' \in S: \quad \int_\Omega \Delta w' \Delta(\varphi - w')dx \geq 0 \ \ \forall \varphi \in S \, .$$

In proving results like this we use the results on the conical differentiability of the metric projection onto the set K. In particular an example is constructed in Section 3 showing that in the case of variational inequalities of the fourth order we cannot in general expect to obtain the conical differentiability of solution to the variational inequality with respect to the perturbations of the right–hand side as it is the case for the second order elliptic variational inequalities as shown in [1].

2 Tangent cone

Let us consider the space $H = H^2(\Omega) \cap H_0^1(\Omega)$ with the scalar product

$$(y, z)_H = \int_\Omega \Delta y \Delta z dx, \quad \forall y, z \in H \, . \tag{2.1}$$

Denote by K the following convex cone

$$K = \{\varphi \in H \mid \varphi \geq \psi \quad \text{in } \Omega\} \, . \tag{2.2}$$

We assume that $\psi \in H^2(\Omega)$, $\psi(x) \leq 0$ on $\partial\Omega$, therefore set K is nonempty.

Let $u \in K$ be a given element, denote $\Xi = \{x \in \Omega \mid u(x) = \psi(x)\}$.

The following variational inequality represents an obstacle problem for a simply supported Kirchhoff plate

$$u \in K: \int_\Omega \Delta u \Delta(\phi - v) - \int_\Omega f(\phi - v) \geq 0 \ \forall \phi \in K \tag{2.3}$$

The metric projection onto the set K, $z = P_K y \in K$, $y \in H$, is given by

$$\int_\Omega \Delta z(x) \Delta(\varphi - z)(x)dx \geq \int_\Omega \Delta y(x) \Delta(\varphi - z)(x)dx \quad \forall \varphi \in K \, . \tag{2.4}$$

Therefore

$$u = P_K(F) \text{ with } F \in H \ : \int_\Omega \Delta F \Delta \phi - \int_\Omega f\phi = 0 \quad \forall \phi \in H \tag{2.5}$$

We denote

$$C_K(z) = \{\varphi \in H | \exists t > 0 \text{ such that } z + t\varphi \in K\} \tag{2.6}$$

The form of tangent cone

$$T_K(z) = \ \text{cl}(C_K(z)),$$

cl stands for closure, is derived for any element z in convex set K.

Theorem 1 *For any $u \in K$ the tangent cone is given in the following form*

$$T_K(u) = \{\varphi \in H^2(\Omega) \cap H_0^1(\Omega) | \tilde{\varphi} \geq 0 \; C_2\text{-q.e. on } \{\tilde{u} = 0\} \subset \Omega \quad \text{and } \widetilde{\frac{\partial \varphi}{\partial n}} \leq 0 \; C_{\frac{1}{2}}\text{-q.e. on } \{\widetilde{\frac{\partial u}{\partial n}} = 0\} \subset \Gamma\} \tag{2.7}$$

here C_2-q.e. (resp. $C_{\frac{1}{2}}$-q.e.) means quasi everywhere with respect to the $H^2(\Omega)$ (resp. $H^{\frac{1}{2}}(\Omega)$) capacity, and the notation $\tilde{}$ refers to the corresponding quasi-continuous representation (we refer the reader e.g. to [5] for the details on the capacity).

The following notation is used for $u = P_K(F)$

$$S_K(u) = T_K(u) \cap [F - u]^\perp \tag{2.8}$$

Let us consider the following functional

$$L(\phi) = \int_\Omega \Delta u \Delta \phi - \int_\Omega f\phi \tag{2.9}$$

where, in general $f \in H'$ and $u \in H$ is a given element.

Theorem 2 *Assume that*

$$L(\phi) \geq 0 \quad \forall \phi \in K \tag{2.10}$$

then there exist Radon measures $\mu \geq 0$, $\nu \geq 0$ supported in Ω (resp. $\partial\Omega$) which does not charge the sets of null $H^2(\Omega)$-capacity (resp. $H^{\frac{1}{2}}(\partial\Omega)$-capacity), such that

$$L(\phi) = \int_\Omega \tilde{\phi} d\mu - \int_{\partial\Omega} \widetilde{\frac{\partial \phi}{\partial n}} d\nu \quad \forall \phi \in H^2(\Omega) \cap H_0^1(\Omega) \tag{2.11}$$

Hence

$$S_K(u) = \{\varphi \in T_K(u) | \int_\Omega \tilde{\varphi} d\mu - \int_{\partial\Omega} \widetilde{\frac{\partial \varphi}{\partial n}} d\nu = 0\} \tag{2.12}$$

It is also useful to characterize the following cone $\mathfrak{S}_K(u) \subset S_K(u)$,

$$\mathfrak{S}_K(u) = \text{cl}(C_K(f) \cap [f - g]^\perp) = \{\varphi \in [F - u]^\perp \cap H | \exists \{\varphi_m\} \in H, \int_\Omega \widetilde{\varphi_m} d\mu - \int_{\partial\Omega} \widetilde{\frac{\partial \varphi_m}{\partial n}} d\nu = 0 \quad t_m > 0,\; u + t_m \varphi_m \geq 0,\; \varphi_m \to \varphi \text{ in } H^2(\Omega)\} \tag{2.13}$$

3 Counterexample

An example is constructed such that $\mathfrak{S}_K(u) \neq S_K(u)$.

Let us consider $I =]-1, 1[$ and $H \equiv H_0^1(I) \cap H^2(I)$ with the scalar product

$$(u, v) = \int_I u'' v'' dx \tag{3.1}$$

Consider the function $f(\cdot) \in H$ of the form

$$f(x) = \frac{1}{2}x^2\eta(x) - \frac{1}{6}1^+(x)x^3 + \frac{1}{12}x^3 + \frac{1}{4}x^2 + \frac{1}{6}x + \frac{1}{3} \tag{3.2}$$

where $\eta(\cdot) \in C^\infty(I)$, $0 \le \eta(x) \le 1$, $\eta(-1) = \eta(1) = 0$, $\eta' \neq 0$ on ∂I and $\eta(x) = 1$ for $x \in (-\frac{1}{2}, \frac{1}{2})$.

The metric projection $u = P_K f$ onto the set

$$K = \{v \in H | v(x) \ge 0\} \tag{3.3}$$

is given as

$$u(x) = \frac{1}{2}x^2\eta(x) \tag{3.4}$$

if the variational inequality is satisfied

$$\int_I (u - f)''(v - u)'' dx \ge 0 \quad \forall v \in K \tag{3.5}$$

We have

$$\begin{aligned} S_K(u) &= \{\varphi \in H | \varphi(0) = 0\} \\ \mathfrak{S}_K(u) &= \{\varphi \in H | \varphi(0) = \varphi'(0) = 0\} \end{aligned} \tag{3.6}$$

for

$$\text{supp}\eta = \overline{I} \tag{3.7}$$

4 Conical differentiability

For a given element $g \in H$, such that $f = P_K(g)$ let us define the following convex cone in the space H

$$S = T_K(f) \cap [f - g]^\perp . \tag{4.1}$$

Definition 1 The set $K \subset H$ is polyhedric at $f \in K$, if for any $g \in H$ such that $f = P_K g$ it follows

$$T_K(f) \cap [f - g]^\perp = \text{cl}(C_K(f) \cap [f - g]^\perp) . \tag{4.2}$$

Remark. Let us recall [1] that if the latter condition is satisfied for given elements $(f, g) \in H \times H$, $f = P_K(g)$ then for all $h \in H$ and for $t > 0$, t small enough

$$P_K(g + th) = P_K g + tP_S h + o(t) \ . \tag{4.3}$$

We return to the example constructed in Section 3. In particular the conical derivative of metric projection at f given by (3.2) does not exists for the direction

$$h(x) = \frac{1}{2}(x^+)^2 - \frac{1}{12}x^3 - \frac{1}{4}x^2 - \frac{1}{6}x \tag{4.4}$$

The proof is divided into few lemmae.

First, it is easy to see that the latter direction $h \in S_K(u)$ is such that $h \notin \mathfrak{S}_K(u)$.

Proposition 1 *The following equality*

$$P_K(f + th) = u + th + o(t), \ \ in \ H^2(I) \tag{4.5}$$

cannot hold for $t > 0$, t sufficiently small

Lemma 1 *Let us denote*

$$P_K(f + th) = u + t(h + w_t) \ . \tag{4.6}$$

If we have $w_t \to 0$ in $H^2(I)$, then

$$w_t(0) \geqq \frac{t}{2}h'(0)^2 + o(t) \tag{4.7}$$

Lemma 2 *The element w_t defined above enjoys the following property*

$$-w_t(0) + \int_I w_t''(v - u)'' - t\int_I w_t''(h + w_t)'' \geqq 0 \tag{4.8}$$

Proof of Proposition 1. We use lemmae. For $v = u$ in (4.8) it follows that

$$w_t(0) \leqq t\|w_t''\|_{L^2(I)}\|h'' + w_t''\|_{L^2(I)} \ . \tag{4.9}$$

If (4.5) is valid, $w_t \to 0$ in $H^2(I)$ with $t \downarrow 0$ and in view of (4.9) we have

$$w_t(0) \leqq t\varepsilon(t) \tag{4.10}$$

hence a contradiction with (4.7) since h is seclected in such a way that $h'(0) = -\frac{1}{6} \neq 0$.

Proof of Lemma 1. We denote $u_t = P_K(f + th)$, and by x_t a point with the following properties

(i) x_t is a local minimum for $u_t(\cdot)$,
(ii) $x_t \to 0$ with $t \downarrow 0$.

Since $u(x) = \frac{1}{2}x^2$ for $|x| < \delta$, with $\delta > 0$, we have

$$0 = u_t'(x_t) = u'(x_t) + t(h'(x_t) + w_t'(x_t)) \tag{4.11}$$

or

$$0 = x_t + t(h'(0) + \varepsilon(t)) \tag{4.12}$$

using the continuity of $h'(\cdot)$ at $x = 0$ and taking into account that

$$\text{if } w_t \to 0 \text{ in } H^2(I) \text{ then } w_t' \to 0 \text{ in } C(I) \text{ and } w_t'(x_t) \to 0 \ .$$

Therefore

$$x_t = -th'(0) + o(t) \tag{4.13}$$

and from (4.6) it follows that

$$0 \leqq u_t(x_t) = u(x_t) + t(h'(x_t) + w_t'(x_t))$$

thus we obtain

$$0 \leqq \frac{1}{2}x_t^2 + t[x_t h'(0) + w_t(0) + x_t\varepsilon(t)] \tag{4.14}$$

using the differentiability of $h(\cdot)$ at $x = 0$ and the following inequality

$$|w_t(x_t) - w_t(0)| \leqq |x_t| \|w_t'\|_{L^\infty(I)} \text{ along with } \|w_t'\|_{L^\infty(I)} \to 0 \text{ as } t \downarrow 0$$

In view of (4.12), (4.14) takes the following form

$$0 \leqq \frac{t^2}{2}h'(0)^2 + t[-th'(0)^2 + w_t(0)] + t^2\varepsilon(t)$$

hence

$$w_t(0) \geqq \frac{t}{2}h'(0)^2 + t\varepsilon(t)$$

which completes the proof of Lemma 1.

Proof of Lemma 2. The element u_t satisfies variational inequality

$$\int_I (v - u_t)''(f + th - u_t)'' \leqq 0$$

for all $0 \leqq v \ni (H^2 \cap H_0^1)(I)$, using (4.6)

$$\int_I (v - (u + th + tw_t))''(f - u - tw_t)'' \leqq 0 \ .$$

Hence

$$0 \geqq \int_I (f - u)^{(4)}(v - (u + th + tw_t)) - t\int_I (v - u)''w_t'' + t^2 \int_I w_t''(h + w_t)''$$
$$+ \Big[(f - u)''(v - (u + th + tw_t))' - (f - u)^{(3)}(v - (u + th + tw_t))\Big]_{-1}^{+1}$$

Since we have $(f-u)^{(4)} = -\delta(0)$ in $\mathcal{D}'(I)$ and $\Big[(f-u)''\Big]_{-1}^{+1} = 0$, it follows that

$$0 \geqq tw_t(0) - v(0) - t\int_I (v-u)''w_t'' + t^2\int_I w_t''(h+w_t)''$$

Taking into account that $v(0) = 0$ the latter inequality implies (4.8)

5 An application to control problem

It is clear that Theorems 1, 2 hold in the case of the space $H^2(\Omega)\cap H_0^1(\Omega)$ with the scalar product

$$(y,z) = \int_\Omega yz + \Delta y \Delta z\ .$$

Let us consider an optimal control problem with the following state equation, cost functional and state constraints.

State equation:

$$y \in H_0^1(\Omega)\ :\ -\Delta y = u \text{ in } \Omega \quad u \in L^2(\Omega)$$

Cost functional:

$$J(u) = \frac{1}{2}\int_\Omega (y-z_d)^2 dx + \frac{1}{2}\int_\Omega u^2 dx$$

State constraints:

$$y(x) \geq \Psi(x)$$

The optimal state $y^\star = y^\star(z_d)$ is given by

$$\frac{1}{2}\int_\Omega [(y^\star - z_d)^2 + (\Delta y^\star)^2]dx = \min_{H\ni v\geq\Psi} \frac{1}{2}\int_\Omega [(v-z_d)^2 + (\Delta v)^2]dx$$

We shall consider the differentiability of the mapping

$$L^2(\Omega) \ni z_g \mapsto y^\star(z_d) \in H^2(\Omega)\cap H_0^1(\Omega)\ .$$

We assume for simplicity, that $\Psi(x) = 0$. In the case $\Psi(x) \neq 0$ we can obtain the same result, assuming that $\Psi(\cdot) \in H^2(\Omega)$ and $\Psi(x) \leqq 0$ on $\partial\Omega$.

By the application of Theorem 1 we have the following optimality system for the control problem under consideration.

Proposition 2 *If $\{u^\star, y^\star\}$ is a unique solution to the control problem then there exixt measures $\mu \geqq 0, \nu \geqq 0$ supported in Ω and on $\partial\Omega$, respectively, such that the following system is satisfied*

State equation:

$$y \in H_0^1(\Omega) \; : \; -\Delta y^\star = u^\star \; in \; \Omega$$

Adjoint state equation:

$$\int_\Omega p^\star \Delta\varphi = -\int_\Omega (y^\star - z_d)\varphi + \int \widetilde{\varphi} d\mu - \int \widetilde{\frac{\partial\varphi}{\partial n}} d\nu$$
$$p^\star \in L^2(\Omega), \quad \forall\varphi \in H^2(\Omega) \cap H_0^1(\Omega)$$

Optimality conditions:

$$u^\star = p^\star \; a.e. \; in \; \Omega$$

$$\int (\widetilde{\varphi} - \widetilde{y^\star}) d\mu - \int \left[\widetilde{\frac{\partial\varphi}{\partial n}} - \widetilde{\frac{\partial y^\star}{\partial n}}\right] d\nu \leqq 0 \quad \forall 0 \leqq \varphi \in H^2(\Omega) \cap H_0^1(\Omega)$$

Remark. We denote by $\widetilde{\varphi}$ the quasicontinuous representative of $\varphi \in H^2(\Omega)$ and by $\widetilde{\frac{\partial\varphi}{\partial n}}$ the quasicontinuous representative of $\frac{\partial\varphi}{\partial n} \in H^{\frac{1}{2}}(\partial\Omega)$. The above optimality system is obtained without the standard assumption that $H^2(\Omega) \cap H_0^1(\Omega) \subset C(\overline{\Omega})$ which holds by the Sobolev embedding theorem for $\Omega \subset \mathbb{R}^N$, $N \leqq 3$.

Let us consider the stability of solutions to the control problem with respect to perturbations of the element z_d in the direction ψ.

We denote $f, h \in H^2(\Omega) \cap H_0^1(\Omega)$ solutions to the linear equations

$$\int_\Omega [\Delta f \Delta\varphi + f\varphi] dx = \int_\Omega z_d \varphi$$
$$\int_\Omega [\Delta h \Delta\varphi + h\varphi] dx = \int_\Omega \psi\varphi \quad \forall\varphi \in H^2(\Omega) \cap H_0^1(\Omega)$$

It is easy to see that the optimal state $y_t^\star$ to the above control problem with the element z_d replaced by $z_d + t\psi$ is given by the metric projection onto the cone K of nonnegative elements in the Sobolev space $H^2(\Omega) \cap H_0^1(\Omega)$, i.e.

$$y_t^\star \equiv y^\star(z_d + t\psi) = P_K(f + th)$$

Proposition 3 *Assume that condition (4.1) is satisfied, then*

$$y_t^\star = y^\star + tw^\star + o(t) \; in \; H^2(\Omega)$$
$$u_t^\star = u^\star + tv^\star + o(t) \; in \; L^2(\Omega)$$

where $\{v^\star, w^\star\} \in L^2(\Omega) \times H^2(\Omega) \cap H_0^1(\Omega)$ *is an optimal solution control–state for the following control problem:*

State equation:

$$w \in H_0^1(\Omega) \; : \; -\Delta w = v \; in \; \Omega \quad v \in L^2(\Omega)$$

Cost functional:

$$I(v) = \frac{1}{2}\int_{\Omega}(w-\psi)^2 dx + \frac{1}{2}\int_{\Omega} v^2 dx$$

State constraints:

$$w(x) \geq 0 \;\; C_2\text{-}q.e. \; on \; \{y^\star = 0\} \subset \Omega$$

$$\frac{\widetilde{\partial w}}{\partial n} \leq 0 \;\; C_{\frac{1}{2}}\text{-}q.e. \; on \; \{\frac{\widetilde{\partial y^\star}}{\partial n} = 0\} \subset \Gamma$$

$$\int_{\Omega} \tilde{w} d\mu - \int_{\partial\Omega} \frac{\widetilde{\partial w}}{\partial n} d\nu = 0$$

6 Shape sensitivity analysis of obstacle problem

Let $\Omega \subset \mathbb{R}^2$ be a given domain with smooth boundary $\Gamma = \partial\Omega$. We derive the form of the shape derivative $w'(\Omega; V)$ of the solution $w(\Omega)$ to obstacle problem (2.3) for the Kirchhoff plate in the direction of a vector field $V(.,.)$. We define a family $\{\Omega_t\} \subset \mathbb{R}^2, t \in [0, \delta)$, of domains as follows, $\Omega_t = T_t(V)(\Omega)$, we refer the reader to [4] for details.

For $t > 0, t$ small enough the set

$$K(\Omega_t) = \{\varphi \in H^2(\Omega_t) \cap H_0^1(\Omega_t) \mid \varphi(x) \geqq 0, \text{ in } \Omega\} \tag{6.1}$$

is a non-empty, closed convex subset of the Sobolew space $H^2(\Omega_t) \cap H_0^1(\Omega_t)$.

Let there be given an element $f \in H^2(\mathbb{R}^2)$. We denote by $w_t \in H^2(\Omega_t)$, $t \in [0, \delta)$, the unique solution of the following variational inequality

$$\begin{aligned} &w_t \in K(\Omega_t) \; : \\ &\int_{\Omega_t} \Delta w_t \Delta(\varphi - w_t) dx \geqq \int_{\Omega_t} f(\varphi - w_t) dx, \quad \forall \varphi \in K(\Omega_t) \; . \end{aligned} \tag{6.2}$$

For $t = 0$, we denote $w(x) = w_0(x)$, $x \in \Omega$, and let $\mu \geqq 0, \nu \geqq 0$ be the Radon measures defined by

$$\int_{\Omega} \varphi d\mu - \int_{\partial\Omega} \frac{\widetilde{\partial\varphi}}{\partial n} d\nu = \int_{\Omega} \{\Delta w \Delta\varphi - \varphi f\} dx \; .$$

Furthermore we denote by $\widetilde{w_t}(x)$, $x \in \mathbb{R}^2$, an extension of the element $w_t(x)$, $x \in \Omega_t$, $t \in [0, \delta)$.

It can be shown that the element $\widetilde{w_t}_{|\Omega} \in H^2(\Omega)$ is right-differentiable with respect to t, at $t = 0$. We denote $V(0) = V(0, \cdot)$.

Theorem 3 *Assume that condition*

$$T_K(w) \cap [w - F]^{\perp} = \ cl(C_K(w) \cap [w - F]^{\perp}) \ . \tag{6.3}$$

is satisfied, $w = P_K(F)$, F *defined in Eq. (2.5)*
Then for $t > 0, t$ *small enough*

$$\widetilde{w_t}_{|\Omega} = w + tw' + o(t), \quad in \ H^2(\Omega) \tag{6.4}$$

where $\|o(t)\|_{H^2(\Omega)}/t \to 0$ *with* $t \downarrow 0$.
The shape derivative $w' \in H^2(\Omega)$ *is given by a unique solution of the following variational inequality*

$$w' \in S_v = \{\varphi \in H^2(\Omega) \mid \varphi = -v\frac{\partial w}{\partial n} \ on \ \partial\Omega, \int_{\Omega} \varphi d\mu - \int_{\partial\Omega} \frac{\widetilde{\partial\varphi}}{\partial n} d\nu = 0 \tag{6.5}$$

$$\varphi(x) \geqq 0 \ on \ \{w = 0\} \subset \Omega, \ \frac{\widetilde{\partial\varphi}}{\partial n} \leq 0 \ C_{\frac{1}{2}}\text{-}q.e. \ on \ \{\frac{\widetilde{\partial w}}{\partial n} = 0\} \subset \Gamma\}$$

$$\int_{\Omega} \Delta w' \Delta(\varphi - w') dx \geqq 0, \quad \forall \varphi \in S_v(\Omega)$$

here we denote $v(x) = \langle V(0,x), n(x) \rangle_{\mathbb{R}^2}$, $x \in \partial\Omega$; $n(x)$, $x \in \partial\Omega$ *is the unit outward normal vector on* $\partial\Omega$ *and*

References

[1] F. Mignot, Controle dans les inequations variationelles elliptiques. Journal of Functional Analysis, **22** (1976) 25-39.

[2] M. Rao and J. Sokołowski, Differential Stability of Solutions to Parametric Optimization Problems. Mathematical Methods in Applied Sciences, 14 (1991) 281-294.

[3] M. Rao and J. Sokołowski, Polyhedricity of convex sets in Sobolev space $H_0^2(\Omega)$. Nagoya Math. J., **130** (1993) 101–110.

[4] J. Sokołowski and J.P. Zolesio, Introduction to Shape Optimization. Shape sensitivity analysis. Springer Series in Computational Mathematics Vol. 16, New York, 1992.

[5] P.W. Ziemer, Weakly Differentiable Functions. Springer Verlag, New York, 1989.

Applications of Exact Controllability to Some Inverse Problems for the Wave Equations

Jean-Pierre Puel Université de Versailles Saint-Quentin et Centre de Mathématiques Appliquées, Ecole Polytechnique, Palaiseau, France

Masahiro Yamamoto Department of Mathematical Sciences, University of Tokyo, Tokyo, Japan

Abstract: We consider two inverse problems for wave equations, namely identification of sources and identification of potentials. In the first case, we give a constructive result of uniqueness and stability showing the equivalence with an exact controllability problem. We study the second problem in its linearized version in a first step. We prove a uniqueness and stability result for a generic set of datas, using again an auxiliary exact controllability problem. We then apply these results for proving a local result of uniqueness and stability for the full problem. These results have been announced in [PY1].

1.Identification of sources

The problem of identification of sources has already been considered in [Ya1], [Ya2] and [ESY] in an analogous way. Nevertheless we present here in a simplifed and constructive version. A related question has been studied by J.-L. Lions using an optimal control approach (see [L2])

Let Ω be an open regular bounded set of R^N , let Γ be the boundary of Ω and let Γ_0 a non empty subset of Γ. For every $T > 0$, we denote by $\Sigma = \Gamma \times (0,T)$ and $\Sigma_0 = \Gamma_0 \times (0,T)$. Let f be a function defined on Ω with values in R (f only depends on the x variable). We consider the following problem (writing ' for the time derivative)

$$(1.1) \qquad \begin{cases} y'' - \Delta y = f \text{ in } \Omega \times (0,T), \\ y_{/\Sigma} = 0, \\ y(0) = y_0 \ ; \ y'(0) = y_1 \text{ in } \Omega, \end{cases}$$

where y_0 and y_1 are initial datas defined on Ω with values in R. We suppose that the source f is unknown (depending only on the space variable), but that we have an additional information, namely that $\partial y/\partial \nu_{/\Sigma_0}$ is given (for example it could be measured). The problem is then, knowing $\partial y/\partial \nu_{/\Sigma_0}$, to "identify" f and this gives raise to three questions:

(i) If (with non ambiguous notations $\frac{\partial y}{\partial \nu}_{/\Sigma_0} = \frac{\partial \hat{y}}{\partial \nu}_{/\Sigma_0}$, do we have $f = \hat{f}$? (Uniqueness).

(ii) Can we obtain an estimate like

$$||f||_{X_1} \leq M||\frac{\partial y}{\partial \nu}_{/\Sigma_0}||_{X_2}$$

with appropriate X_1 and X_2 ? (Stability).

(iii) Can we recover f from $\frac{\partial y}{\partial \nu}_{/\Sigma_0}$ by a constructive method ?

It turns out that these questions are here completely and simply related to exact controllability and observability.

Let us introduce the following equation : for $\Phi_0 \in L^2(\Omega)$ and $v \in L^2(\Sigma_0)$ there exists a unique $\Phi \in C([0,T];L^2(\Omega)) \cap C^1([0,T];H^{-1}(\Omega))$ such that

$$\text{(1.2)} \qquad \begin{cases} \Phi'' - \Delta\Phi = 0, \\ \Phi = v_0 \text{ on } \Sigma_0, \\ \Phi = 0 \text{ on } \Sigma - \Sigma_0, \\ \Phi(0) = \Phi_0 \ ; \ \Phi'(0) = 0, \end{cases}$$

We know from classical results ([L1] ; [BLR]) that if Γ_0 and T satisfy a geometric condition (G) given below, there exists $v = v_0 \in L^2(\Sigma_0)$ such that Φ satisfies

$$\Phi(T) = 0 \text{ and } \Phi'(T) = 0.$$

Moreover, we can choose $v_0 = C(\Phi_0)$ (by Lions'H.U.M.) such that C is a continuous linear map from $L^2(\Omega)$ to $L^2(\Sigma_0)$. Condition (G) which has been given in [BLR] can be expressed as follows

(G) For every $x \in \Omega$ and every optical ray starting from point x at time $t = 0$, this ray hits Γ_0 before time T at a non diffractive point.

We can now state the main result of this section

Theorem 1.1

Let us assume that Γ_0 and T satisfy condition (G). Then there exists a positive constant M such that if $y_0 \in H^2(\Omega) \cap H_0^1(\Omega)$ and $y_1 \in H_0^1(\Omega)$,

$$\text{(1.3)} \qquad (||y_1||^2_{H_0^1(\Omega)} + ||\Delta y_0 + f||^2_{L^2(\Omega)})^{\frac{1}{2}} \leq M||\frac{\partial y'}{\partial \nu}||_{L^2(\Sigma_0)}.$$

Moreover we have

$$\text{(1.4)} \qquad \forall \Phi_0 \in L^2(\Omega), \int_\Omega f\Phi_0 dx = -\int_{\Sigma_0} C(\Phi_0)\frac{\partial y'}{\partial \nu}d\Sigma - \int_\Omega \Delta y_0 \Phi_0 dx.$$

Remarks

1) We then have

$$||f||_{L^2(\Omega)} \le M||\frac{\partial y'}{\partial \nu}||_{L^2(\Sigma_0)} + M'(||y_1||^2_{H^1_0(\Omega)} + ||\Delta y_0||^2_{L^2(\Omega)})^{\frac{1}{2}}.$$

In particular, if $y_0 = 0$ and $y_1 = 0$,

$$||f||_{L^2(\Omega)} \le M||\frac{\partial y'}{\partial \nu}||_{L^2(\Sigma_0)},$$

and this estimate is optimal.

2) Relation (1.4) gives the possibility of determining f practically by choosing for Φ_0 elements of a basis of a subspace of $L^2(\Omega)$, for example a basis of finite elements. We must notice that for each choice of Φ_0 we have to solve an exact controllability problem !

3) It is possible to recover f only on a subset ω of Ω by choosing suitable Φ_0.

4) If (G) is not satified but T is large enough ($T > 2diam(\Omega)$) we still have the uniqueness property without stability.

Proof of Theorem 1.1 : It is very simple as if $z = y'$ we have

$$\text{(1.5)} \qquad \begin{cases} z'' - \Delta z = 0 \text{ in } \Omega \times (0,T), \\ z_{/\Sigma} = 0, \\ z(0) = y_1 \ ; \ z'(0) = \Delta y_0 + f \text{ in } \Omega, \end{cases}$$

and we know from classical observability results that condition (G) implies (1.3). Moreover, multiplying (1.5) by Φ we obtain (1.4).

2. Identification of potentials

We now consider the following problem

$$\text{(2.1)} \qquad \begin{cases} u'' - \Delta u + pu = 0, \\ u_{/\Sigma} = h, \\ u(0) = a \ ; \ u'(0) = 0, \end{cases}$$

where h is independent of t, $h \in H^{1/2}(\Gamma)$, $a \in H^1(\Omega)$, $a_{/\Gamma} = h$, and where p is independent of t with $p \in L^\infty(\Omega)$. Here the inverse problem consists in recovering the potential p (which is supposed to be unknown) with the knowledge of $\partial u/\partial \nu_{/\Sigma_0}$ (which can be measured for example). Of course we are interested in answering the same three questions (i), (ii), and (iii) as in section 1.

In a first step, we consider the formal linearization of problem (2.1) in the neighborhood of a given potential p and the corresponding solution u of (2.1). Writing $\delta p = q$ and $\delta u = y$ we then obtain

$$(2.2)\qquad \begin{cases} y'' - \Delta y + py + qu = 0, \\ y_{/\Sigma} = 0, \\ y(0) = 0 \ ; \ y'(0) = 0, \end{cases}$$

We want to identify q (again independent of t) knowing $\partial y / \partial \nu_{/\Sigma_0}$.

Remark

It may happen that u is independent of time (u is then solution of an elliptic equation). Nevertheless, it is important that the (formal) linearization is made on the evolution equation.

Taking the derivative with respect to time in (2.2) and writing $z = y'$ we obtain

$$(2.3)\qquad \begin{cases} z'' - \Delta z + pz + qu' = 0, \\ z_{/\Sigma} = 0, \\ z(0) = 0 \ ; \ z'(0) = -qa, \end{cases}$$

We now introduce an auxiliary exact controllability problem. If $\Phi_0 \in L^2(\Omega)$ we consider the following control problem

$$(2.4)\qquad \begin{cases} \Phi'' - \Delta \Phi + p\Phi = 0, \\ \Phi_{/\Sigma_0} = v, \\ \Phi_{/\Sigma - \Sigma_0} = 0, \\ \Phi(0) = \Phi_0 \ ; \ \Phi'(0) = 0, \end{cases}$$

The following result (in a slightly weaker form) is proved by Komornik in [Ko]

Theorem 2.1

If $p \in L^\infty(\Omega)$ and if Γ_0 and T satisfy condition (G), then there exists $v_0 \in L^2(\Sigma_0)$ such that

$$(2.5)\qquad \Phi(T) = 0 \text{ and } \Phi'(T) = 0.$$

Moreover by choosing v_0 given by Lions'H.U.M., the mapping $\Phi_0 \to v_0 = C(\Phi_0)$ is linear continuous from $L^2(\Omega)$ to $L^2(\Sigma_0)$.

Considering now the solution Φ of (2.4) corresponding to $v = v_0 = C(\Phi_0)$ and multiplying (2.3) by Φ we obtain

$$\forall \Phi_0 \in L^2(\Omega), \int_\Omega q(x)(a(x)\Phi_0(x) + \int_0^T u'(x,t)\Phi(x,t)dt)dx = \int_{\Sigma_0} \frac{\partial z}{\partial \nu} C(\Phi_0) d\Sigma.$$

So that, if we define

$$S(a;\Phi_0) = \int_0^T u'(x,t)\Phi(x,t)dt \tag{2.6}$$

$$T(a;\Phi_0) = a\Phi_0 + S(a;\Phi_0) \tag{2.7}$$

we have

$$\forall \Phi_0 \in L^2(\Omega),\ \int_\Omega q(x)T(a;\Phi_0)(x)dx = \int_{\Sigma_0} \frac{\partial z}{\partial \nu} C(\Phi_0)d\Sigma. \tag{2.8}$$

Suppose that we know that the mapping $\Phi_0 \to T(a;\Phi_0)$ is invertible (with continuous inverse) from $L^2(\Omega)$ to itself. Then for every $\Psi_0 \in L^2(\Omega)$, there exists $\Phi_0 \in L^2(\Omega)$ such that $T(a;\Phi_0) = \Psi_0$ and $|\Phi_0|_{L^2(\Omega)} \leq M_0\ |\Psi_0|_{L^2(\Omega)}$. Then

$$\forall \Psi_0 \in L^2(\Omega),\ \int_\Omega q(x)\Psi_0(x)dx = \int_{\Sigma_0} \frac{\partial z}{\partial \nu} C(\Phi_0)d\Sigma$$

and therefore,

$$\exists M > 0 \text{ such that } |q|_{L^2(\Omega)} \leq M|\frac{\partial y'}{\partial \nu}|_{L^2(\Sigma_0)}.$$

So it is an important problem to study the invertibility of $T(a;.)$.

Remark

In the case where u is independent of t, then $S(a;\Phi_0) = 0$ and $T(a;\Phi_0) = a\Phi_0$. In this case the invertibility of $T(a;.)$ is given by the condition $\frac{1}{a} \in L^\infty(\Omega)$.

Lemma 2.2

If $u'' \in L^1(0,T;L^\infty(\Omega))$, the mapping $\Phi_0 \to S(a;\Phi_0)$ is compact from $L^2(\Omega)$ to $L^2(\Omega)$.

The proof of Lemma 2.2 is technical and the details are given in [PY2].

¿From this lemma and the Fredholm alternative, we easily deduce

Theorem 2.3

Let us assume that Γ_0 and T satisfy condition (G), that $p \in L^\infty(\Omega)$, $u'' \in L^1(0,T;L^\infty(\Omega))$ and that

$$\exists \alpha > 0\ ,\ a(x) \geq \alpha \text{ a.e. in } \Omega, \tag{2.9}$$

then we have the following alternative

(i) either the mapping $\Phi_0 \to T(a;\Phi_0)$ is invertible and

$$\exists M > 0 \text{ such that } |q|_{L^2(\Omega)} \leq M|\frac{\partial y'}{\partial \nu}|_{L^2(\Sigma_0)};$$

(ii) or there exists an integer N_0 and a finite dimensional subspace W_{N_0} of $L^2(\Omega)$ of dimension N_0 , depending only on a, p and h, such that if $\Pi_{N_0}^{\perp}$ is the projection on $W_{N_0}^{\perp}$ we have

$$\exists M > 0 \text{ such that } |\Pi_{N_0}^{\perp}(aq)|_{L^2(\Omega)} \leq M|\frac{\partial y'}{\partial \nu}|_{L^2(\Sigma_0)}.$$

Remark

1) In case (ii) of the alternative, aq (therefore q) can be determined up to a finite number of components.

2) From the Fredholm alternative, we know that case (i) of the alternative occurs if and only if the kernel of $T^*(a;.) = aI + S^*(a;.)$ is reduced to $\{0\}$. We can show that $\Psi_0 \in KerT^*(a;.)$ if and only if there exists $\theta_0 \in H_0^1(\Omega)$ such that if θ and X are solutions of

$$\begin{cases} \theta'' - \Delta\theta + p\theta = u'\Psi_0, \\ \theta_{/\Sigma} = 0, \\ \theta(0) = \theta_0 \ ; \ \theta'(0) = \Psi_0, \end{cases}$$

$$\begin{cases} X'' - \Delta X + pX = 0, \\ X_{/\Sigma_0} = \dfrac{\partial\theta}{\partial\nu}, \\ X_{/\Sigma-\Sigma_0} = 0, \\ X(0) = 0 \ ; \ X'(0) = 0, \end{cases}$$

then $X(T) = 0$ and $X'(T) = 0$.

It would be interesting to find cases of existence of non zero Ψ_0 or to show that in some situations, $KerT^*(a;.) = \{0\}$.

In addition to the hypotheses of Theorem 2.3, let us assume now that

$$p(x) \geq 0 \text{ a.e. in } \Omega, \tag{2.10}$$

$$h(x) \geq \alpha \text{ a.e. in } \Omega. \tag{2.11}$$

Then we can define a_1 as the solution of

$$\begin{cases} -\Delta a_1 + pa_1 = 0 \text{ in } \Omega, \\ a_1 = h \text{ on } \Gamma, \end{cases}$$

and we have

$$a_1(x) \geq \alpha \text{ a.e. in } \Omega.$$

let us define for $r \in R$ or $r \in C$

(2.12) $$a_r = ra_1 + (1-r)a.$$

We see that

(2.13) $\exists\beta > 0$, $\exists\epsilon_0 > 0$ such that $\forall\epsilon$, $0 < \epsilon < \epsilon_0$, $\forall r \in [-\epsilon, 1+\epsilon]$, $a_r(x) \geq \beta$ a.e. in Ω.

Let u^r be solution of

(2.14) $$\begin{cases} u^{r\prime\prime} - \Delta u^r + pu^r = 0, \\ u^r_{/\Sigma} = h, \\ u^r(0) = a_r \ ; \ u^{r\prime}(0) = 0, \end{cases}$$

and define $S_r(\Phi_0) = S(a_r; \Phi_0)$; $T_r(\Phi_0) = T(a_r; \Phi_0)$.

We have $u^0 = u$; $u^1 = a_1$ and $u^r = ru^1 + (1-r)u^0$.

It is easy to show that :

- $r \to S_r(.)$ can be extended as a function defined on $D_0 \subset C$ with $[-\epsilon, 1+\epsilon] \in D_0$ with values in $\mathcal{L}(L^2(\Omega))$.

-$r \to S_r(.)$ is holomorphic.

-$\forall r \in D_0$, $\Phi_0 \to S_r(\Phi_0)$ is compact.

Then using a result on holomorphic perturbations by Kato [Ka] we have the following alternative :

(i) either $KerT_r^* \neq \{0\} \forall r \in D_0$,

(ii) or there exists a finite set $\Lambda \subset D_0$ such that $\forall r \in D_0 - \Lambda$, $KerT_r^* = \{0\}$ and T_r is invertible. Here, $r = 1 \in D_0$ and $S_1 = 0$, therefore case (i) is impossible. Then there exists at most a finite set Λ such that $\forall r \in D_0 - \Lambda$, T_r is invertible. In particular, we deduce the following generic result

Theorem 2.4

Let us assume that Γ_0 and T satisfy condition (G), that $u'' \in L^1(0,T; L^\infty(\Omega))$ and that (2.9),(2.10),(2.11) are satisfied. Then for every $\epsilon > 0$, there exists $b \in L^\infty(\Omega)$, $|b - a|_{L^\infty(\Omega)} \leq \epsilon$ such that $\Phi_0 \to T(b; \Phi_0)$ is invertible and such that if u_b is solution of (2.1) with $u_b(0) = b$ and y_b is solution of (2.2) associated with $u = u_b$, then

(2.15) $$\exists M(b) > 0 \text{ such that } |q|_{L^2(\Omega)} \leq M(b) \left|\frac{\partial y_b'}{\partial \nu}\right|_{L^2(\Sigma_0)}.$$

Now, assuming that a, p, h are such that we have the stability property for the linearized problem (2.2), we can show a local result of stability for the nonlinear problem (2.1) provided that a and h are regular enough.

Let u be solution of (2.1), y be solution of (2.2) and v be solution of

$$(2.16) \qquad \begin{cases} v'' - \Delta v + (p+q)v = 0, \\ v_{/\Sigma} = h, \\ v(0) = a \ ; \ v'(0) = 0, \end{cases}$$

If $w = v - u - y$, we obtain, writing M for various constants,

$$(2.17) \qquad |q|_{L^2(\Omega)} \le M|\frac{\partial y'}{\partial \nu}|_{L^2(\Sigma_0)} \le M|\frac{\partial (v-u)'}{\partial \nu}|_{L^2(\Sigma_0)} + M|\frac{\partial w'}{\partial \nu}|_{L^2(\Sigma_0)}.$$

But from standard regularity result we have

$$|\frac{\partial w'}{\partial \nu}|_{L^2(\Sigma_0)} \le M|q(v-u)'|_{L^1(0,T;L^2(\Omega))}$$

and if $H^1(\Omega) \in L^s(\Omega)$ ($s > 2$) and r=2(s/2)' ($r > 2$),

$$\begin{aligned} |\frac{\partial w'}{\partial \nu}|_{L^2(\Sigma_0)} &\le M|q|_{L^r(\Omega)}|(v-u)'|_{L^1(0,T;H^1(\Omega))} \\ &\le M|q|^2_{L^r(\Omega)}[|v'|_{L^1(0,T;H^1(\Omega))} + |a|_{H^1(\Omega)}] \\ &\le M|q|_{L^r(\Omega)}|q|_{L^\infty(\Omega)}[|v'|_{L^1(0,T;H^1(\Omega))} + |a|_{H^1(\Omega)}]. \end{aligned}$$

Therefore, from (2.17) we see that if $|q|_{L^\infty(\Omega)}$ is small enough we have

$$(2.18) \qquad \exists M > 0 \text{ such that } |q|_{L^2(\Omega)} \le M|\frac{\partial (v-u)'}{\partial \nu}|_{L^2(\Sigma_0)}$$

which is a local stability result for (2.1).

Remark

We obtain a stability result (2.18) in the $L^2(\Omega)-norm$ assuming that q is small enough in the $L^\infty(\Omega) - norm$. It would be interesting to fill this gap which is due to technical difficulties.

References

[BLR] C. Bardos, G. Lebeau et J. Rauch. Sharp sufficient conditions for the observation, control and stabilization of waves from the boundary. SIAM J. on Control and Optimization, 30 (1992), 1024-1065.

[ESY] H.W. Engl, O. Scherzer et M. Yamamoto. Uniqueness of forcing terms in linear partial differential equations with overspecified boundary data. Inverse Problems, à paraître.

[Ka] T. Kato. *Perturbation Theory for Linear Operators.* Springer-Verlag, 1976, Berlin.

[Ko] V. Komornik. A new method of exact controllability in short time and applications. Annales de la Faculté des Sciences de Toulouse, 10, (1989), pp. 415-464.

[L1] J.-L. Lions. *Contrôlabilté exacte, Perturbations et Stabilisation de Systèmes Distribués.* Tome 1, Masson, 1988, Paris.

[L2] J.-L. Lions. Problèmes aux limites non classiques pour équations d'évolution. Revista Matematica de la Universidad Complutense de Madrid, Vol. 1, 1-2-3, 1988.

[PY1] J.-P. Puel, M. Yamamoto. Applications de la contrôlabilité exacte à quelques problèmes inverses hyperboliques. Note aux C.R.A.S. Paris Série I, 1995, to appear.

[PY1] J.-P. Puel, M. Yamamoto. Generic well posedness in a multidimensional hyperbolic inverse problem. To appear

[Ya1] M. Yamamoto. Well-posedness of some inverse hyperbolic problem by the Hilbert Uniqueness Method. Journal of Inverse and Ill-posed Problems, à paraître.

[Ya2] M. Yamamoto. Stability, reconstruction formula and regularization for an inverse source hyperbolic problem by a control method, à paraître.

* J.-P. P. : Université de Versailles Saint-Quentin et Centre de Mathématiques Appliquées, Ecole Polytechnique, 91128 Palaiseau Cedex, France.

** M. Y. : Department of Mathematical Sciences, the University of Tokyo, 3-8-1 Komaba Meguro Tokyo 153, Japan.

Approximation of the Optimal Control of Vibrations

John Read
FB Mathematik - AG 10
TH-Darmstadt
Schlossgartenstrasse 7
64289 Darmstadt, Germany
E-mail: read@mathematik.th-darmstadt.de

0 Introduction

In this paper we apply certain geometrical arguments to the problem of approximating the norm and the time minimal distributed control of a vibrating process in a Hilbert space governed by an abstract linear wave equation. The dynamics involved here are well understood if the eigenelements are known. One thinks of the vibrating string of unit length where H is $L^2(0,1)$, the eigenfunctions are $\varphi_j(x) = \sqrt{2}\sin j\pi x$ and the corresponding eigenvalues are $\lambda_j = (j\pi)^2$ for $j \in \mathbb{N}$. For fixed $T > 0$, the time in which control should occur, the norm minimal control can be written explicitly. One accomplishes this by solving a convex optimization problem, the so called T-**dual problem**. The T-dual functional χ_T depends on $T > 0$ and the p-norm to be minimized. If the eigenelements are unknown, a standard approach is to approximate them by using the finite element method, which provides us with a method of approximating the norm (time) minimal control.

We extend two earlier papers. In [3] the finite element method is employed to approximate (1.9) for $p = 2$. The geometry of the T-dual problem is not discussed and no convergence rate is derived. We place these arguments in a geometrical setting which allows the approximation problem to be understood properly. Exploiting the geometry involved we are able to not only give a more general sufficient condition for convergence than that given in [3], but we are also able to show that this condition is also necessary not only for $p = 2$ but for all $p \in [2, \infty]$. For $p \geq 2$ we show that the T-dual functional is uniformly convex, allowing stronger convergence results than we get for $p = \infty$. The case $p = \infty$ is treated in [4], where calculations are based on truncation and knowledge of the eigenelements. Thus, the finite dimensional subspace used in [4] is $V_M = \text{span}\{\varphi_j\}_{j=1}^M$, which is in practice not viable. Convergence results are weak star, which we extend to convergence in measure. Using Egoroff's Theorem we obtain a subsequence which converges almost uniformly. Our geometrical approach then allows us to take this one step further and apply Asplund space theory in order to show that for "almost" all initial data we can define a (point) separating family of semi-norms on $L^\infty((0,T),\ell^2)$, such that with respect to any semi-norm of this family the convergence of the whole sequence of approximate control functions is uniform. Thus, we are able to weaken significantly the assumptions made (exact knowledge of the eigenelements), while at the same time strengthening the convergence obtained (norm or semi-norm convergence instead of weak*). Using these geometrical ideas we go on to consider the problem of approximating time minimal control. All proofs omitted here can be found in the companion work [5] which provides an indepth analysis of this problem.

1 Basic Results

This section outlines some of the more classical aspects of linear theory, which we will feel free to employ. We remind the reader that the statement $p \geq 2$ rules out the case $p = \infty$. The problem of distributed control of a vibrating process can be placed in a Hilbert space in which the differential equation is expressed by an abstract wave equation of the form

$$\ddot{\psi}(t) + A\psi(t) = \omega(t) \text{ for } t \in (0, \infty) \tag{1.1}$$

where H is a separable Hilbert space, $\psi : (0, \infty) \to H$ and $\ddot{\psi}$ denotes the second derivative with respect to time t [1]. The linear operator A maps a dense subspace $D(A)$ into H and is self adjoint and positive definite with a unique self adjoint square root operator $A^{1/2}$ defined on $D(A^{1/2})$ where $D(A) \subset D(A^{1/2}) \subset H$. In addition we assume A has a complete orthonormal sequence of eigenfunctions $\{\varphi_j\} \subset D(A)$ for $j \in \mathbb{N}$ with a corresponding monotonically increasing sequence of positive eigenvalues $\{\lambda_j\} \subset \mathbb{R}$ satisfying $\lim \lambda_j = \infty$. One can then derive a representation of the operators A and $A^{1/2}$ and their domains in terms of the eigenelements [1]. We call $D(A^{1/2})$ the "energy space". If we choose as the initial position and velocity of our system

$$\begin{aligned} \psi(0) &= \psi_0 \in D(A^{1/2}), \\ \dot{\psi}(0) &= \dot{\psi}_0 \in H, \end{aligned} \tag{1.2}$$

then for each ω in the Bochner 2-space $L^2((0,\infty), H)$ denoted by W^2_∞, there exists a unique "weak solution" of (1.1) and (1.2) in the sense defined in [1]. We define the family of closed subspaces of W^2_∞

$$W^2_T := \{\omega \in W^2_\infty : \omega(t) = 0 \text{ for } t > T \text{ a.e.}\}.$$

The problem of T-null-controllability is: for a given $T > 0$ find a function $\omega \in W^2_T$ so that the corresponding weak solution ψ of (1.1) and (1.2) satisfies the end conditions

$$\begin{aligned} \psi(T) &= 0, \\ \dot{\psi}(T) &= 0. \end{aligned} \tag{1.3}$$

Definition: A **T-control function** is any $\omega \in W^2_T$ such that (1.1), (1.2) and (1.3) hold.

As in [2], it is easy to see that W^2_∞ can be identified, via the natural isomorphism between H and ℓ^2, with the Bochner 2-space $X^2_\infty := L^2((0,\infty), \ell^2)$, which is a Hilbert space when endowed with the usual norm. We define the family of closed subspaces $X^2_T \subset X^2_\infty$ in a manner entirely analogous to the definition of W^2_T and see that each X^2_T can be identified with W^2_T via the aforementioned isomorphism. We will denote the norm on X^2_∞ by a different symbol than that used for the norm on W^2_∞. A second Hilbert space which will be of fundamental interest to us is $Y := \ell^2(\mathbb{R}^2)$ with norm induced by the inner product on Y given by $\langle x, y\rangle = \Sigma(x^1_j y^1_j + x^2_j y^2_j)$ where $x = (x^1_j, x^2_j)_{j\in\mathbb{N}}$. Here both $\{x^1_j\}$ and $\{x^2_j\}$ are elements of ℓ^2. A similar representation holds for $y \in Y$.

Thus, upon carrying out this isomorphism, using a "weak version" of the variation of constants formula and imposing the end conditions (1.3), the problem of T-null-controllability is reduced to solving the linear Hilbert space equation

$$S(f) = c \neq 0, \tag{1.4$_2$}$$

where $S : X_T^2 \to Y$. The value of $c \in Y$ is given by $c = (c_j^1, c_j^2)_{j \in \mathbb{N}}$, where for each $j \in \mathbb{N}$

$$\begin{aligned} c_j^1 &:= \sqrt{\lambda_j}\langle \psi_0, \varphi_j \rangle_H, \\ c_j^2 &:= -\langle \dot{\psi}_0, \varphi_j \rangle_H. \end{aligned} \tag{1.5}$$

The operator S is defined by $S = (S_j)_{j \in \mathbb{N}}$, where for each $j \in \mathbb{N}$

$$S_j(f) := \begin{pmatrix} \int_0^T f_j(t) \sin \sqrt{\lambda_j} t \, dt \\ \int_0^T f_j(t) \cos \sqrt{\lambda_j} t \, dt \end{pmatrix} \quad \text{for} \quad f \in X_T^2.$$

Clearly, S is linear, and by using Hölder's inequality it is easy to show that S is bounded with $\|S\| \leq \sqrt{T}$. We denote the Hilbert adjoint of S by S^*, so that by definition we get that $\langle S(f), x \rangle_Y = \langle f, S^* x \rangle_{X_\infty^2}$ for all $f \in X_T^2$ and all $x \in Y$. It is not difficult to show that $S^* : Y \to X_T^2$ is given by $S^* = (S_j^*)_{j \in \mathbb{N}}$ where the operator S_j^* acts on x to give

$$(S_j^* x)(t) := \begin{cases} x_j^1 \sin \sqrt{\lambda_j} t + x_j^2 \cos \sqrt{\lambda_j} t & \text{for} \quad t \in [0, T] \\ 0 & \text{for} \quad t > T. \end{cases} \tag{1.6}$$

We define the real vector space $Z_T := S^*(Y) \subset X_T^2$. In [5] we show that $Y = S(Z_T)$. Hence, S is a surjective bounded linear operator on X_T^2. This implies the solution set of (1.4)$_2$ is some closed, affine subspace $\Omega_T^2 \subset X_T^2$. Moreover, there is a one to one correspondence between the set Ω_T^2 and the set of T-control functions in the sense that if $f \in \Omega_T^2$, then $\omega = \Sigma f_j \varphi_j$ is a T-control function, and if ω is a T-control function, then $(f_j)_{j \in \mathbb{N}} \in \Omega_T^2$ where $f_j = \langle \omega, \varphi_j \rangle_H$. This is simply the isomorphism between H and ℓ^2. As in [5], we define W_∞^p and X_∞^p for $p \in [1, \infty]$ as the Bochner p-spaces with their usual norms $|||\cdot|||_p$ and $\|\cdot\|_p$, as well as and the families of closed subspaces $W_T^p \subset W_\infty^p$ and $X_T^p \subset X_\infty^p$.

Given $p \in [2, \infty]$ **the problem of norm minimal T-controllability** is: for a given $T > 0$ find (if possible) a (the unique) control function which minimizes $|||\cdot|||_p$. This problem is equivalent, via the aforementioned isomorphism, to finding for a given $T > 0$ a (the unique) element of $\Omega_T^p := X_T^p \cap \Omega_T^2$, which minimizes the norm $\|\cdot\|_p$. That is, for fixed $T > 0$ we wish to find (if possible) a (the unique) p-norm minimal element of the solution set of the linear Banach space equation (1.4)$_p$ which is the restriction of (1.4)$_2$ to $X_T^p \subset X_T^2$. Using Hölder's inequality we get $\|S_p\| \leq T^{1/q}$ where $p^{-1} + q^{-1} = 1$. Since $Z_T \subset X_T^p$, the surjectivity of S_p follows. Thus, Ω_T^p is a nonempty affine subspace of X_T^p.

We will denote the vector space Z_T equipped with the norm $\|\cdot\|_q$ for $q \subset [1, \infty]$ by Z_T^q. Our next result is fundamental to understanding the approach we will take. It shows that for all $q \in [1,\infty]$ the topology induced by $\|S^* \cdot\|_q$ on Y is equivalent to the topology induced by the norm on Y. Thus, Z_T^q has topolgical properties similar to those of Y. We start by defining $\varepsilon(T) := \frac{1}{2}\inf\{1 - \sin\alpha/\alpha : \alpha \geq \sqrt{\lambda_1}T\}$, which clearly satisfies $\varepsilon(T) \in (0,1]$ for any $T > 0$ and is $\mathcal{O}(T^2)$ as $T \to 0^+$. With this notation we can consider

Lemma 1.1: For $T_1 > T_0 \geq 0$ and for all $q \geq 1$

$$(\varepsilon(T_1 - T_0))^q \cdot (T_1 - T_0) \cdot \|x\|^q \leq (\chi_{T_1}(x))^q - (\chi_{T_0}(x))^q \leq (T_1 - T_0) \cdot \|x\|^q, \tag{1.7}$$

which holds for all $x \in Y$.

Proof: We set $T := T_1 - T_0$. Then, by using the polar form $x_j^1 = \rho_j \cos\theta_j$ and $x_j^2 = \rho_j \sin\theta_j$, integrating, using Lebesgue's monotone convergence theorem, noting that $\|(S^*x)(t)\| \leq \|x\|$ for all $t \in [0, T_1]$ and using Hölder's inequality we have

$$\begin{aligned}
\varepsilon(T)\cdot T\cdot\|x\|^2 &\leq \frac{1}{2}\sum_{j=1}^{\infty}\rho_j^2 T\left[\left(\sqrt{\lambda_j}T - \cos(\sqrt{\lambda_j}T + 2\theta_j)\sin\sqrt{\lambda_j}T\right)/\sqrt{\lambda_j}T\right] \\
&= \int_{T_0}^{T_1}\|(S^*x)(t)\|^2\,dt \\
&\leq \|x\|\cdot T^{1/p}\cdot\left(\int_{T_0}^{T_1}\|(S^*x)(t)\|^q\,dt\right)^{1/q}.
\end{aligned}$$

We get the first inequality of (1.7) by dividing by $T^{1/p}$ and taking the q-th power. The second follows from the estimate $\|(S^*x)(t)\| \leq \|x\|$ for all $t \in [0, T_1]$. □

For all $q \in [1,\infty]$ we get $\varepsilon(T)\cdot T^{1/q}\|x\| \leq \|S^*x\|_q \leq T^{1/q}\|x\|$. This implies that each space Z_T^q is a closed subspace of X_T^q and all q-norms for $q \in [1,\infty]$ generate the same topology on Z_T. We remark that for $p \geq 2$ it is easy to see that the spaces $(X_T^p)'$ and X_T^q are isomorphic. For $p = \infty$ this is no longer true, but even here we still get that the adjoint operator of S_p denoted by S_p' satisfies $S_p' : Y' \to X_T^q$. Moreover, it has the form

$$\langle (S_p'x) \mid f\rangle = \int_0^T \langle (S^*x)(t), f(t)\rangle dt,$$

so that by Hölder's inequality and the proper choice of $f \in X_T^p$ we obtain $\|S_p'x\| = \|S^*x\|_q$. The next theorem gives us a necessary and sufficient condition for $\Omega_T^p \cap \bar{\mathcal{B}}_\beta \neq \emptyset$ [5]. We adopt the notation

$$\mathcal{U}_T^q := \{x \in Y : \|S^*x\|_q \leq 1\}, \tag{1.8}$$

which we use in proving

Lemma 1.2: There exists $f \in \Omega_T^p \cap \bar{\mathcal{B}}_\beta$ if and only if $\langle c, x\rangle \leq \beta\cdot\|S^*x\|_q$ for all $x \in Y$.

Lemma 1.3: There exists $\tilde{x} \in \mathcal{U}_T^q$ such that $\langle c, \tilde{x}\rangle = \beta(T, c) := \sup\{\langle c, x\rangle : x \in \mathcal{U}_T^q\}$.

Theorem 1.4: For each $p \in [2, \infty]$ there exists a unique p-norm minimal $f \in \Omega_T^p \cap \bar{\mathcal{B}}_\beta$, where β is given by $\beta(T, c) := \sup\{\langle c, x\rangle : x \in \mathcal{U}_T^q\}$.

Formally, we write $\{f\} = \Omega_T^p \cap \bar{\mathcal{B}}_\beta$ where $\beta = \beta(T, c)$.

Using the Cauchy-Schwarz and Hölder inequalities we see that any (the unique) p-norm minimal element of Ω_T^p has the form

$$f(t) := \begin{cases} \langle c, \tilde{x}\rangle \|(S^*\tilde{x})(t)\|^{q-2}\,(S^*\tilde{x})(t) & \text{for} \quad t \in [0, T] \\ 0 & \text{for} \quad t > T, \end{cases} \tag{1.9}$$

which for $p = \infty$ is called a bang-bang control. Here $\tilde{x}$ is any solution of the T-**dual problem** in Y, which is to maximize $\psi_c(x) := \langle c, x\rangle$ subject to $\chi_T(x) := \|S^* x\|_q \leq 1$.

By Lemma 1.3 the set of solutions to the T-dual problem, the T-**dual solution set**, is nonempty. The derivation of formula (1.9) is similar to that of the finite dimensional case which can be found in [2]. One can easily see that the set of $t \in [0, T]$ where $\|(S^*\tilde{x})(t)\| = 0$ is a finite set (maybe empty). This means that, in as much as we can define $S^*\tilde{x} \in X_T^2$ everywhere on $[0, T]$, our definition of f given by (1.9) is legitimate, if we agree on the value of f at those points where $\|(S^*\tilde{x})(t)\| = 0$. On this set we let f be zero. Using $\varepsilon(T)$ as in Lemma 1.1, we can estimate the p-norm of f in (1.9) by

$$T^{-1/q}\|c\| \;\leq\; \|f\|_p \;\leq\; (\varepsilon(T) \cdot T^{1/q})^{-1}\|c\|. \tag{1.10}$$

The unique p-norm minimal control for (1.1) and (1.2) is regained from (1.9) by writing

$$\omega(t) := \begin{cases} \sum\limits_{j=1}^{\infty} f_j(t)\varphi_j & \text{for} \quad t \in [0, T] \\ 0 & \text{for} \quad t > T. \end{cases} \tag{1.11}$$

The formulation of the T-dual problem is of course dependent on $T > 0$ via the definition of S^*, and thus, the T-dual solution is likewise dependent on $T > 0$. Naturally, f in (1.9) does not remain unaffected by our choice of $T > 0$. We will, at times, emphasize this dependency by writing f_T instead of f in (1.9).

Now the T-dual functional χ_T defines an equivalent norm on Y, which is uniformly convex for $p \geq 2$. Further, for $p \geq 2$ this norm is uniformly Fréchet differentiable on open sets whose norm closure does not contain the origin. For $p = \infty$ it is uniformly Fréchet differentiable on open sets whose weak closure does not contain the origin. This gradient is continuous, and we get the estimates $\varepsilon T^{1/q} \leq \|\nabla\chi_T(x)\| \leq T^{1/q}$ where $\varepsilon \geq 0$ is defined in Lemma 1.1. Further, if $\tilde{x}$ is the T-dual solution, then $c = \lambda\nabla\chi_T(\tilde{x})$, where $\lambda = \langle c, \tilde{x}\rangle = \|f\|_p$.

The T-dual solution set is nonempty, bounded and contained in a hyperplane orthogonal to $c \in Y$. Further, if $\tilde{x}$ is any T-dual solution, then $\chi_T(\tilde{x}) = 1$. For $p \geq 2$ the T-dual solution is unique; as it is for $p = \infty$ if there are at least two indices $j \neq k \in \mathbb{N}$ such that $(c_i^1)^2 + (c_i^2)^2 \neq 0$ for $i = j, k$. If $\tilde{x}$ is any T-dual solution, then $(\tilde{x}_k^1)^2 + (\tilde{x}_k^2)^2 = 0$ if and only if $(c_k^1)^2 + (c_k^2)^2 - 0$.

2 The Finite Element Method

As in [5] we introduce an M-dimensional subspace $V_M \subset D(A^{1/2})$, where we generate the M approximate eigenvalues $\{\lambda_j^M\}_{j=1}^M$ with corresponing orthonormal eigenfunctions $\{\varphi_j^M\}_{j=1}^M$ We define the error $e(\lambda^M, L) := \max_{j\leq L} |\lambda_j^M - \lambda_j|$, and we assume we have chosen the sequence of subspaces V_M such that there exists a strictly monotonically increasing sequence $M : \mathbb{N} \to \mathbb{N}$ with the property

$$\lim_{L\to\infty} e(\lambda^{M(L)}, L) = 0. \tag{2.1}$$

We will see that this condition together with $\lim \|c_L - c\| = 0$ play the crucial role in all our convergence results. Next, we define the sequence of linear Hilbert space equations (for L sufficiently large) by

$$S^L(f) = c_L \neq 0, \qquad (1.4)_2^L$$

where $S^L : X_T^2 \to Y$ is the bounded linear operator given by $S^L = (S_j^L)_{j\in\mathbb{N}}$. Here, S_j^L is derived from S_j by replacing λ_i with λ_i^M for $i = 1, ..., L$. The components of $c_L \in Y$ are defined by (1.5) for $j = 1, ..., L$ but using the approximate eigenelements. For $j > L$ we define $(c_L)_j^1 = (c_L)_j^2 = 0$. For each definition made in Section 1 we now define the L-th approximate expression by replacing $(1.4)_2$ with $(1.4)_2^L$. Doing this, we get $(1.4)_p^L$, $(1.6)^L$, χ_T^L and the L-th approximate T-dual problem with unique solution $\tilde{x}_L$.

Lemma 2.1: If (2.1) holds, then the sequence of L-th approximate T-dual functionals converges uniformly to χ_T on bounded sets.

Lemma 2.2: Let χ_T be uniformly Fréchet differentiable on $\mathcal{V}$. Then we can show that $\lim \|\nabla\chi_T^L(x) - \nabla\chi_T(x)\| = 0$ and that this limit is uniform in $x \in \mathcal{V}$.

Lemma 2.3: If $\lim \|c_L - c\| = 0$, and if (2.1) holds, then $\lim \|\nabla\chi_T^L(\tilde{x}_L) - \nabla\chi_T(\tilde{x}_L)\| = 0$.

Theorem 2.4: Suppose (2.1) holds. Let $\tilde{x}_L$ and $\tilde{x}$ be the L-th approximate and exact T-dual solution respectively, with which we generate f^L and f by formula $(1.9)^L$ and (1.9). We give the following equivalencies:

For $p \geq 2$:	For $p = \infty$:
(1) $\lim \|c_L - c\| = 0$	(1) $\lim \|c_L - c\| = 0$
(2) $\lim\langle c_L, \tilde{x}_L\rangle = \langle c, \tilde{x}\rangle$ and $\lim \|\tilde{x}_L - \tilde{x}\| = 0$	(2) $\lim\langle c_L, \tilde{x}_L\rangle = \langle c, \tilde{x}\rangle$, $c_L \rightharpoonup c$ and $\tilde{x}_L \rightharpoonup \tilde{x}$
(3) $f^L \rightharpoonup f$ and $\|f^L\|_p \to \|f\|_p$	(3) $f^L \overset{\star}{\rightharpoonup} f$ and $\lim \|f^L\|_\infty = \|f\|_\infty$
(4) $\lim \|f^L - f\|_p = 0$	(4) $\lim \|f^L - f\|_1 = 0$

It is of interest to consider convergence in the ∞-norm, which provides us with a very clear picture of how well f^L is approximating f on $[0, T]$.

Theorem 2.5: Under the assumptions of Theorem 2.4 if $\lim \|c_L - c\| = 0$, then we obtain $\|f^L(t) - f(t)\| \to 0$ uniformly on $[0, T]$ for $p \geq 2$; whereas for $p = \infty$ there is some subsequence of $\{f^L\}$ which converges almost uniformly to f on $[0, T]$.

3 Time Minimal Controls and Approximation

In order to better understand the relationship between the problems of norm and time minimal controls, we restate the former and give the latter, each in the equivalent version obtained via the isomorphism between H and ℓ^2.

For a fixed $p \in [2, \infty]$ **the problem of norm minimal T-controllability** is: for a given $c \in Y$ and $T > 0$ define $\gamma_0 := \inf\{\gamma > 0 : \Omega_T^p \cap \bar{\mathcal{B}}_\gamma \neq \emptyset\}$, and find (if possible) a (the unique) $f \in \Omega_T^p \cap \bar{\mathcal{B}}_{\gamma_0}$.

For a fixed $p \in [2, \infty]$ **the problem of time minimal γ-controllability** is: for a given $c \in Y$ and $\gamma > 0$ define $T_0 := \inf\{T > 0 : \Omega_T^p \cap \bar{\mathcal{B}}_\gamma \neq \emptyset\}$, and find (if possible) a (the unique) $f \in \Omega_{T_0}^p \cap \bar{\mathcal{B}}_\gamma$.

We will abuse the definition of S^* somewhat. This operator is inextricably linked with $T > 0$, as a cursory examination of its definition will reveal. We will, however, take the liberty of letting T vary without warning. It should be obvious from the definition why this is permissible. Now, the behavior of β near zero and infinity is governed by (1.10). In Lemma 1.3 we defined the function $\beta(\cdot, c)$, which relates to any $T > 0$ the smallest p-norm any element of Ω_T^p can have. This function is clearly key to understanding the relationship between norm and time minimal controls. In order to simplify notation, we will suppress the explicit dependence on c and write instead $\beta(T) = \beta(T, c)$ for all $T > 0$. Our study of this function will be based on Lemma 1.1. For each $q \in [1, 2]$ and $0 < \gamma < \delta$ we recall the two inequalities

$$(\delta^{1/q} - \gamma^{1/q}) \leq (\gamma^{(1-q)/q}/q) \cdot (\delta - \gamma) \quad \text{and} \tag{3.1}$$

$$(\delta^{1-q}/q) \cdot (\delta^q - \gamma^q) \leq (\delta - \gamma). \tag{3.2}$$

With Lemma 1.1 as well as (1.10), (3.1) and (3.2) we can derive

Lemma 3.1: For fixed $p \in [2, \infty]$ the function β defined in Lemma 1.3 is a strictly monotonically decreasing homeomorpism of $(0, \infty)$ onto $(0, \infty)$. Further, it is uniformly Lipschitz continuous on any closed interval not containing the origin and uniformly decrescent on any bounded interval.

Proof: We let $0 < T_0 < T_1$, and we let $\tilde{x}_i$ be the unique T_i-dual solution for the problem of T_i-norm minimal control for $i = 0, 1$. Clearly, we have that $\chi_{T_i}(\tilde{x}_i) = 1$, which we use in (1.7) to get $1 < [1 - \varepsilon(T_1 - T_0)^{1/q} \cdot (T_1 - T_0)\|\tilde{x}_1\|^q]^{-1/q} \leq (\chi_{T_0}(\tilde{x}_1))^{-1}$. From this, we derive strict monotonicity by observing that $\tilde{x}_1/\chi_{T_0}(\tilde{x}_1) \in \mathcal{U}_{T_0}^q$, which implies by the definition of the T-dual solution that $\langle c, \tilde{x}_1\rangle/\chi_{T_0}(\tilde{x}_1) \leq \langle c, \tilde{x}_0\rangle$. Thus, we get

$$\begin{aligned} \beta(T_1) &< [1 - \varepsilon(T_1 - T_0)^{1/q} \cdot (T_1 - T_0)\|\tilde{x}_1\|^q]^{-1/q}\beta(T_1) \\ &\leq \langle c, \tilde{x}_1\rangle/\chi_{T_0}(\tilde{x}_1) \leq \beta(T_0). \end{aligned}$$

To show the Lipschitz continuity we let $\bar{T} > 0$, and without loss of generality we assume that $\varepsilon(\bar{T}) \leq \varepsilon(T)$ for all $T \geq \bar{T}$. Then chose $T_1 > T_0 \geq \bar{T}$ with corresponding T_i-dual solutions $\tilde{x}_i$ for $i = 0, 1$. Rearranging the second inequality of (1.7) with $x = \tilde{x}_0$ yields

$$\chi_{T_1}(\tilde{x}_0) \leq [1 + (T_1 - T_0) \cdot \|\tilde{x}_0\|^q]^{1/q}.$$

Further, since $\tilde{x}_0/\chi_{T_1}(\tilde{x}_0) \in \mathcal{U}^q_{T_1}$, we get by the definition of $\tilde{x}_1$ that $\langle c, \tilde{x}_0\rangle/\chi_{T_1}(\tilde{x}_0) \leq \langle c, \tilde{x}_1\rangle$. Using these facts, the monotonicity and the definition of β we are justified in writing

$$\begin{aligned}\beta(T_0) &\leq \langle c, \tilde{x}_0\rangle \cdot \left([1+(T_1-T_0)\|\tilde{x}_0\|^q]^{1/q}/\chi_{T_1}(\tilde{x}_0)\right)\\ &\leq [1+(T_1-T_0)\|\tilde{x}_0\|^q]^{1/q}\cdot\beta(T_1). \end{aligned}\tag{3.3}$$

If we take the q-th power of (3.3), rearrange things a bit and use (3.1) with $\delta = \beta^q(T_0)$ and $\gamma = \beta^q(T_1)$, we get

$$\begin{aligned}0 \leq \beta(T_0) - \beta(T_1) &\leq (\beta^{1-q}(T_1)/q)\cdot(\beta^q(T_0) - \beta^q(T_1))\\ &\leq (\beta(T_1)/q)\cdot\|\tilde{x}_0\|^q\cdot(T_1 - T_0)\\ &\leq (\beta(\bar{T})/q)\cdot(\varepsilon^q(\bar{T})\bar{T})^{-1}\cdot(T_1-T_0),\end{aligned}$$

which is what we want to show.

To show that β is uniformly decrescent we let $\hat{T} > 0$ and choose $0 < T_0 < T_1 \leq \hat{T}$ with corresponding T_i-dual solutions $\tilde{x}_i$ for $i = 0, 1$. Rearranging the first inequality of (1.7) with $x = \tilde{x}_1$ yields

$$\chi_{T_0}(\tilde{x}_1) \leq [1 - \varepsilon^q(T_1-T_0)\cdot(T_1-T_0)\cdot\|\tilde{x}_1\|^q]^{1/q}. \tag{3.4}$$

Further, since $\tilde{x}_1/\chi_{T_0}(\tilde{x}_1) \in \mathcal{U}^q_{T_0}$, we get by the definition of $\tilde{x}_0$ that $\langle c, \tilde{x}_1\rangle/\chi_{T_0}(\tilde{x}_1) \leq \langle c, \tilde{x}_0\rangle$. Using this fact, (3.4) and definition of β we are justified in writing

$$\begin{aligned}\beta(T_1) &\leq \beta(T_1)\cdot[1-\varepsilon^q(T_1-T_0)\cdot(T_1-T_0)\cdot\|\tilde{x}_1\|^q]^{1/q}/\chi_{T_0}(\tilde{x}_1)\\ &\leq \beta(T_0)\cdot[1-\varepsilon^q(T_1-T_0)\cdot(T_1-T_0)\cdot\|\tilde{x}_1\|^q]^{1/q}\end{aligned}\tag{3.5}$$

If we take the q-th power of (3.5), rearrange things a bit and use (3.2) with $\delta = \beta(T_0)$ and $\gamma = \beta(T_1)$, we get the inequality

$$\begin{aligned}&(\beta(\hat{T})\cdot\hat{T}^{-1}/q)\cdot\varepsilon^q(T_1-T_0)\cdot(T_1-T_0)\\ &\leq (\beta(T_0)\cdot\|\tilde{x}_1\|^q/q)\cdot\varepsilon^q(T_1-T_0)\cdot(T_1-T_0)\\ &\leq (\beta^{1-q}(T_0)/q)\cdot[\varepsilon^q(T_1-T_0)\cdot(T_1-T_0)\cdot\|\tilde{x}_1\|^q]\cdot\beta^q(T_0)\\ &\leq (\beta^{1-q}(T_0)/q)\cdot(\beta^q(T_0)-\beta^q(T_1))\\ &\leq \beta(T_0)-\beta(T_1)\end{aligned}$$

which is what we want to show.

These arguments together with (1.10) imply that β is a (Lipschitz) continuous, strictly monotone mapping from $(0,\infty)$ onto $(0,\infty)$. Thus, it defines a homeomorphism. □

Now for fixed $p \in [2,\infty]$ and for any $c \in Y$ and $\gamma > 0$ we define

$$T_0 := \inf\{T > 0 : \text{ there exists a control function } \omega \in W^p_T \cap \bar{\mathcal{B}}_\gamma\}.$$

Then **the problem of time minimal γ-controllability** is to show that there exists a (unique) T_0-control function ω with $|||\omega|||_p \leq \gamma$. This is of course equivalent (modulo the homomorphism between H and ℓ^2) to the formulation at the beginning of this section, which by applying Lemma 3.1 present us with a method of solving this problem. The theory of Section 1 shows that for each $T > 0$ the function f_T defined by (1.9) satisfies $\{f_T\} = \Omega_T^p \cap \bar{\mathcal{B}}_{\beta(T)}$. Thus, for each $\alpha > 0$ we have

$$\{f_{\beta^{-1}(\alpha)}\} = \Omega^p_{\beta^{-1}(\alpha)} \cap \bar{\mathcal{B}}_{\beta(\beta^{-1}(\alpha))} = \Omega^p_{\beta^{-1}(\alpha)} \cap \bar{\mathcal{B}}_{\alpha}. \tag{3.6}$$

We rewrite (3.6) with $\alpha = \gamma$ and $T_0 = \beta^{-1}(\gamma)$, to get

$$\{f_{T_0}\} = \Omega^p_{T_0} \cap \bar{\mathcal{B}}_{\gamma},$$

where f_{T_0} is defined by (1.9). Further, for any $T < T_0$ we know that $\beta(T)$ is the smallest p-norm any element of Ω_T^p may have. Since $\beta(T) > \beta(T_0) = \gamma$, we get that $\Omega_T^p \cap \bar{\mathcal{B}}_\gamma = \emptyset$ for all $T < T_0$. Thus, for $\gamma > 0$ the unique time minimal γ-control is given by (1.10), which we define in terms of f_{T_0}, as defined in (1.9) with $T_0 = \beta^{-1}(\gamma)$.

We begin by showing that, once again, the conditions (2.1) and $\lim \|c_L - c\| = 0$ will play a crucial role in the approximation of time minimal controls. For each $L \geq L_0$ we define the mapping $\beta^L : (0,\infty) \to (0,\infty)$, where

$$\beta^L(T) := \langle c_L, \tilde{x}_L\rangle. \tag{3.7}$$

As one might expect, by $\tilde{x}_L$ we mean the L-th approximate T-dual solution, as discussed in Section 2. An immediate result of this definition and Theorem 2.4 is that if (2.1) holds and $\lim \|c_L - c\| = 0$ then $\beta^L(T) \to \beta(T)$ for all $T \in (0,\infty)$. We will strenghten this to uniform convergence on $[\bar{T},\infty)$ for any $\bar{T} > 0$. To these ends we consider

Lemma 3.2: We fix $p \in [2,\infty]$ and use the notation of Corollary 2.2. If (2.1) holds, then we have $\lim \|\chi_T^L(x) - \chi_T(x)\| = 0$ and this limit is uniform for $x \in \mathcal{V}$ and for T on compact sets (not containing the origin).

Proof: Using Lemma 2.1 and some easy manipulations [5], we can find a continuous, strictly monotonically increasing function E_T with $E_T(0) = 0$ such that for all $\alpha > 0$

$$\|\nabla\chi_T^L(x) - \nabla\chi_T(x)\| \leq T^{1+1/q} \cdot e(\lambda^{M(L)}, L) \cdot (\|x\| + 1)/\sqrt{\lambda_1}\alpha + E_T(|\alpha|). \tag{3.8}$$

In [5] we see that E_T can be estimated above, independently of $T > 0$ on bounded subsets of $\mathbb{R}^+$, by some continuous, strictly monotonically increasing function E with $E(0) = 0$. Thus, on bounded sets, E_T can be made independent of T. If we replace E_T with E and set $\alpha = \alpha_L := e(\lambda^{M(L)}, L)^{1/2}$, then the right hand side of (3.8) is a zero sequence and this is uniform in x and T on bounded sets. □

Theorem 3.3: If (2.1) holds and $\lim \|c_L - c\| = 0$, then for each $\varepsilon > 0$ and $\bar{T} > 0$, there exists $L_\varepsilon \in \mathbb{N}$ such that for all $L \geq L_\varepsilon$ and all $T \geq \bar{T}$ we have

$$|\beta^L(T) - \beta(T)| < \varepsilon.$$

That is, $\beta^L \to \beta$ uniformly on closed sets not containing the origin.

Proof: Let $\varepsilon > 0$ be given. By (1.10) there is some $T_\varepsilon > 0$ such that $\beta(T_\varepsilon) < \varepsilon/3$. By Theorem 3.4 there is some $L_1 \in \mathbb{N}$ such that if $L \geq L_1$, then $\beta^L(T_\varepsilon) < 2\varepsilon/3$. For all $T \geq T_\varepsilon$ monotonicity lets us write

$$\begin{aligned} |\beta^L(T) - \beta(T)| &\leq \beta^L(T) + \beta(T) \\ &\leq \beta^L(T_\varepsilon) + \beta(T_\varepsilon) \\ &< \varepsilon. \end{aligned}$$

From the proof of Theorem 3.4 as given in [5] we see that for all $T \in I_\varepsilon := [\bar{T}, T_\varepsilon]$ we get

$$|\beta^L(T) - \beta(T)| \leq \|c_L - c\| \cdot \|\tilde{x}_L\| + \|c\| \cdot \|\hat{x}_L\| \cdot |\chi_T(\tilde{x}_L) - 1| + \|c\| \cdot \gamma^L(\hat{x}_L). \tag{3.9}$$

Using (1.7) we get an estimate on the value of $\|\tilde{x}_L\|$, which is uniform in $T \in I_\varepsilon$. Lemma 2.1 and (1.7) let us estimate the second term of the right hand side of (3.9) uniformly on I_ε. From the definition of γ^L in [5] we see that we can estimate the third term uniformly on I_ε. Thus, there is some $L_2 \in \mathbb{N}$ such that if $L \geq L_2$ and $T \in I_\varepsilon$, then (3.9) is less than ε. □

The next result will put us in a position to do for the approximation of time minimal that which we did in Theorems 3.4 and 3.5 for norm minimal controls. What we now show is that the sequence of minimal times for the L-th approximate problem of time minimal γ_L-controllability converges to the minimal time of the problem of time minimal γ_0- controllability if $\gamma_L \to \gamma_0$. We are relying on the reader's imagination to define this L-th approximate problem analogously to the exact time minimal problem given at the outset of this section.

Corollary 3.4: Suppose the hypotheses of Theorem 3.3 hold. Assume $0 < \bar{\alpha} < \gamma_0 < \hat{\alpha}$. If there exists a sequence $\{\gamma_L\} \subset [\bar{\alpha}, \hat{\alpha}]$, chosen so that $\gamma_L \to \gamma_0$, then $T_L \to T_0$, where $T_0 := \beta^{-1}(\gamma_0)$ and $T_L := \beta_L^{-1}(\gamma_L)$.

Proof: By (1.10) it is easy to see that there exist $0 < \bar{T} < \hat{T}$ such that $T_L \in [\bar{T}, \hat{T}]$ for all $L \in \mathbb{N}$. Then by Theorem 3.3 we get

$$|\beta(T_L) - \gamma_0| \leq |\beta(T_L) - \beta_L(T_L)| + |\beta_L(T_L) - \gamma_0| \to 0,$$

which together with the fact that β is decrescent implies

$$(\beta(\hat{T}) \cdot \hat{T}^{-1}/q) \cdot |\varepsilon^q(T_L - T_0) \cdot (T_L - T_0)| \leq |\beta(T_L) - \beta(T_0)| \to 0. \square$$

Theorem 3.5: Using the notation of Corollary 3.4 let $\tilde{x}_L$ be the L-th approximate T_L-dual solution, with which we generate $f_{T_L}^L$ by formula $(1.9)^L$, and let $\tilde{x}$ be the T_0-dual solution, with which we generate f_{T_0} by formula (1.9). Assume that (2.1) holds and that $\gamma_L \to \gamma_0$. Then the following four conditions are equivalent for $p \geq 2$:

(i) $\lim \|c_L - c\| = 0$

(ii) $\lim \langle c_L, \tilde{x}_L\rangle = \langle c, \tilde{x}\rangle$, $\lim \|\tilde{x}_L - \tilde{x}\| = 0$ and $T_L \to T_0$

(iii) $\lim \|f_{T_L}^L\|_p = \|f_{T_0}\|_p$, $f_{T_L}^L \rightharpoonup f_{T_0}$ and $T_L \to T_0$

(iv) $\lim \|f_{T_L}^L - f_{T_0}\|_p = 0$ and $T_L \to T_0$

Proof: If we assume (i) is true, then Theorem 3.3 holds, and $T_L \to T_0$ by Corollary 3.4. The definitions of β and β_L give the first statement of (ii). We note that $T_L \to T_0$ and (2.1) together imply that $\chi^L_{T_L}$ converges uniformly to χ_{T_0} on bounded sets, allowing us to derive the remaining statement in (ii) with arguments similar to those of Theorem 2.4.

Together, these assumptions imply that $\|S^{L*}\tilde{x}_L - S^*\tilde{x}\|_q \to 0$, so that (ii)$\Rightarrow$(iv) can be demonstrated as in Theorem 3.4.

Likewise, the proof of (iv)$\Rightarrow$(i) carries over if we define S^L_p and S_p in terms of the upper bound $T := \sup\{T_L\}$, but we leave the definitions of S^{L*} and S^* as they are in Theorem 3.4.

Of course, (iii)$\Leftrightarrow$(iv) is still a result of uniform convexity. □

Theorem 3.5 (cont.): For $p = \infty$ the assumptions given above ensure the equivalence of:

(i) $\lim \|c_L - c\| = 0$

(ii) $\lim\langle c_L, \tilde{x}_L\rangle = \langle c, \tilde{x}\rangle, \quad c_L \rightharpoonup c, \quad \tilde{x}_L \rightharpoonup \tilde{x}$ and $T_L \to T_0$

(iii) $\lim \|f^L_{T_L}\|_\infty = \|f_{T_0}\|_\infty, \quad f^L_{T_L} \overset{*}{\rightharpoonup} f$ and $T_L \to T_0$

Just as in Theorem 2.5 we may consider convergence in the norm on X^∞_∞ for any $p \geq 2$.

Theorem 3.6: Let $p \geq 2$. Under the assumptions of Theorem 5.7 if $\lim \|c_L - c\| = 0$, then for each $\varepsilon > 0$ there exists an $L_\varepsilon \in \mathbb{N}$ such that if $L \geq L_\varepsilon$ we obtain $\|f^L_{T_L}(t) - f_{T_0}(t)\| < \varepsilon$ for all $t \in I^L := [0, T_0] \cap [0, T_L]$.

For $p = \infty$ we can say more than we did in Theorem 3.5.

Theorem 3.6 (cont.): Under the assumptions of Theorem 3.5 if $\lim \|c_L - c\| = 0$, then there is some subsequence of $\{f^L_{T_L}\}$ which converges almost uniformly to f_{T_0} on $[0, T_0]$.

We can extend Theorems 3.5 and 3.6 to get the convergence of ω^L to ω given in (1.11). We define the error in the approximating eigenfunctions as $e(\varphi^M, L) := \max_{j\leq L} \|\varphi^M_j - \varphi_j\|_H$.

Theorem 3.7: If $\lim \|c_L - c\| = 0$, if $\lim \sqrt{L}\, e(\varphi^{M(L)}, L) = 0$ and if (2.1) holds, then for any $p \geq 2$ and any $\delta > 0$ we get $\lim \|\omega^L(t) - \omega(t)\|_H = 0$ uniformly on $[0, T_0 - \delta]$, and hence by the uniform boundedness of $\|\omega^L(t) - \omega(t)\|_H$ (in t and in L) we get $|||\omega^L - \omega|||_p \to 0$, whereas for $p = \infty$ we get $\omega^L \overset{*}{\rightharpoonup} \omega$, and we get the existence of a subsequence of $\{\omega^L\}$ which converges to ω almost uniformly on $[0, T]$.

Now, let us define the mapping $\chi^* : Y \to \mathbb{R}$ by $\chi^*(c) := \langle c, \tilde{x}_c\rangle$, where $\tilde{x}_c$ is the T-dual solution for $c \neq 0$ and $\chi^*(0) = 0$. It turns out that χ^* is the dual norm for the Banach space Y with norm χ_T [5] and as such is a continuous, convex function on the Asplund space Y. Thus, it is Fréchet differentiable on a dense G_δ subset of Y, which we denote by $\mathcal{X}$.

Let $c \in \mathcal{X}$ with T-dual solution $\tilde{x}_c$. Let $\alpha_c := \max\{\|(S^*\tilde{x}_c)(t)\| : t \in [0, T]\}$. For each $\delta \in (0, \alpha_c)$ we let $I_\delta := \{t \in [0, T] : \|(S^*\tilde{x}_c)(t)\| \geq \delta\}$. Note that $0 \notin \mathcal{X}$, so $\alpha_c > 0$. Relative to this family of sets we define the family of semi-norms on X^∞_T by

$$\|f\|_\delta := \operatorname*{ess\,sup}_{t\in I_\delta} \|f(t)\|_{\ell^2} \quad \text{for each } \delta \in (0, \alpha_c).$$

Clearly, this family of semi-norms separates points on X_T^∞.

Theorem 3.5 (revised): For $p = \infty$ and $c \in \mathcal{X}$ we have the equivalence of the following three conditions:

(i) $\lim \|c_L - c\| = 0$

(ii) $\lim\langle c_L, \tilde{x}_L\rangle = \langle c, \tilde{x}\rangle$, $\tilde{x}_L \to \tilde{x}$ and $T_L \to T_0$

(iii) $\lim \|f_{T_L}^L\|_\infty = \|f_{T_0}\|_\infty$, $\lim \|f^L(t) - f(t)\|_\delta = 0$ for all $\delta \in (0, \alpha_c)$ and all $t \in I_\delta$ and $T_L \to T_0$.

It is of interest to note that the set of T-dual solutions such that $I_\delta = [0, T]$ for some $\delta > 0$ is itself a weakly open and dense subset of the set of all T-dual solutions. If we can determine that our dual solution is indeed such a creature, then we have an even better convergence result, namely convergence in the ∞-norm of the sequence of approximate norm minimal controls to the actual norm minimal control.

References

[1] Fattorini H.: The Optimal Problem for Distributed Control of Systems Described by the Wave Equation. In:Azis et.al. (eds.):*Control Theory of Systems Governed by Partial Differential Equations*, Academic Press, N.Y., 1977.

[2] Krabs W.: *On Moment Theory and Controllability of One-Dimensional Vibrating Systems and Heating Processes*, Springer Verlag, Berlin 1992.

[3] Krabs W. and Lamp U.: On the Approximate Calculation of Time-Minimal Distributed Controls of Vibrations by Finite Element Approximations, *Applied Mathematics and Optimization*, 27 (1993), 179 - 190.

[4] Krabs W. and Yan Z.: Numerical Solution of Minimum Norm Problems of Distributed Control of Vibrations, *Applied Mathematics and Optimization*, 21 (1990), 243 - 264.

[5] Read J.: The Approximation of the Optimal Control of Vibrations - A Geometrical Approach, *Mathematical Methods in the Applied Sciences*, submitted

A Static Formation Theory for Active Elastic Materials

David L. Russell
Department of Mathematics
Virginia Tech, Blacksburg, VA 24061

1. Introduction We use the term *formation theory* to describe a certain class of optimal design and control problems for active elastic systems. The subject matter concerns the deliberate alteration of the equilibrium configuration of an elastic structure by means of attached or embedded actuators responding to external control signals. To the extent that the controlled configuration is maintained constant over an extended time interval, or varies slowly enough so that inertial effects can be ignored, we have what we designate as *static* formation theory.

The name *formation theory* is suggested by a comparison of the viewpoint appropriate to this theory with that commonly evident in classical elasticity studies. In elasticity the term *deformation* generally refers to a displacement away from some nominal equilibrium configuration of the system. In the *control theory* of elastic systems such deformations are usually viewed as undesirable and controls are designed to suppress them. The emphasis is on restoration of the equilibrium state. By contrast, formation theory takes as its point of departure the realization that in the age of active, or "smart", materials it may be both desirable and feasible to achieve a variety of different geometric configurations for the same elastic structure at different times in response to varying operational requirements, such configurational transformations to be realized via control signals applied to actuators having specific distribution patterns within the structure.

The tasks inherent in formation theory include *model reformulation* to include various types of attached or embedded actuators, determination of feasible and/or optimal actuator spatial distributions, the *design* problem, and determination of feasible and/or optimal signals to induce actuation, the *control* problem. The last two tasks also involve the formulation and use of *optimality criteria* involving best "fit to target" notions and actuator cost considerations.

To begin the informal part of our mathematical discussion let us consider a general, though over-simplified, formulation in the context of linear elasticity. We suppose that we have a body consisting of a matrix material along with embedded or attached actuator elements. The attachment/embedding process results in both systems being constrained to a *common geometry*, which we suppose parametrized by a vector (or vector function) w in an appropriate inner product space, the inner product being conveniently represented by $< u, v >= v^*u$. Different w correspond to different configurations of the material. In the static problem, which is what we consider here, the combined matrix/actuator structure has an associated *potential energy* form which, in the simplest situation, can be written as

$$V = \frac{1}{2}(w^*C^*ECw + (Dw + Bu)^*K(Dw + Bu)). \qquad (1.01)$$

Here $Cw \to ECw$ is the stress/strain relationship of the matrix material while $Dw + Bu \to K(Dw + Bu)$, involving the external control signal u, which in application alters the equilibrium configuration of the actuator structure, is the corresponding stress/strain relationship for that structure. Minimization of V leads to the formal equilibrium equation:

$$(C^*EC + D^*KD)w + D^*KBu = 0. \qquad (1.02)$$

The stiffness operators E and K of the respective matrix and actuator structures,

$$E = E(F, a, V, b, ...), K = K(F, a, V, b, ...),$$

depend on, among other things, the spatial densities $F, V, ...$ of the various types of actuators present, and corresponding orientation parameters $a, b, ...$ which describe the manner in which the actuators are installed in the matrix. The operators E and K are, in general, non-negative, with K generally increasing as $F, V, ...$ increase while the operator E generally decreases due to replacement of matrix material by actuator material - what we call the *actuator stiffening effect.*

2. A General Two-Dimensional Formation Problem; Equilibrium Conditions

The most elementary aspects of formation theory may be explored in the context of beams, plates, etc. In [ ] we have discussed problems of this sort in the context of the Timoshenko beam model. Additional work of this sort will appear in [ ]. In the present article we explore the form which problems of formation theory assume in the context of two-dimensional linear elasticity (cf.[ ],[ ]). To this end, then, we suppose that the matrix material constitutes an isotropic cylindrical elastic body whose cross section, orthogonal to the axis of the cylinder, occupies a region $\mathcal{R}_0 \subset \mathbf{R}^2$ with smooth boundary Σ_0 consisting of finitely many simple closed curves, as shown in Figure 7 at the end of this article. We denote by $\sigma = [\sigma_x \, \sigma_y]^t$ the unit outward normal vector to Σ_0. With $X = [x \, y]^t$ we have $\sigma = \sigma(X), X \in \Sigma_0$. It is important to note that the equations appearing here are specific to this *cylindrical* configuration; the equations describing in-plane deformations of thin plates, e.g., are somewhat different due to transverse thinning or thickening in response to in-plane expansion or compression, respectively.

We will suppose that the matrix material occupying $\mathcal{R}_0$ has associated *Lame' constants* λ and ν; the corresponding values of *Young's modulus* and the *shear modulus* are $E = \frac{\lambda(1+\nu)(1-2\nu)}{\nu}$ and $G = \frac{E}{2(1+\nu)}$, respectively. In a formation operation it is supposed that the configuration of the material changes so that it occupies a new region $\mathcal{R}$ with boundary Σ; the point $X \in \mathcal{R}$ is mapped to the new point $X + \Xi(X), \Xi(X) = [\xi(x,y) \, \eta(x,y)]^t$ so that $\mathcal{R} = \{X + \Xi(X) | X \in \mathcal{R}_0\}$. We will suppose the displacement functions $\xi(x,y), \eta(x,y)$ to be twice continuously differentiable functions of x, y for the purposes of this article. Assuming that the volume of actuator material is quite small compared with matrix material volume, so that replacement effects are negligible, we may express the *potential energy* for the matrix material, in terms of the displacement functions, in the form, with $\nabla \equiv [\frac{\partial}{\partial x} \, \frac{\partial}{\partial y}], \nabla \cdot \Xi \equiv div\Xi$,

$$\mathcal{V}_0 = \frac{1}{2} \int \int_{\mathcal{R}_0} [\nu \|\nabla \xi\|^2 + \nu \|\nabla \eta\|^2 + (\lambda + \nu)(\nabla \cdot \Xi)^2] dx \, dy. \qquad (2.01)$$

We describe an actuator structure, which we will call an *actuator screen*, which we may imagine as composed of, e.g., piezoelectric filaments, in $\mathcal{R}_0$ in the following way. We suppose that $F_1(X), F_2(X), \ldots, F_m(X)$ are smooth vector fields defined on $\mathcal{R}_0 \cup \Sigma_0$; $F_k(X) = [f_k(x,y) g_k(x,y)]^t$. The field $F_k(X)$ describes both the volume density, $D_k(X) \equiv \|F_k(X)\|$ of the k-th actuator material and the direction (assuming $F_k(X) \neq 0$), $\Phi_k(X) = \frac{F_k(X)}{\|F_k(X)\|}$ of the k-th actuator filament structure at the point X. It will also be convenient to define the related 2×2 projection matrix $P_k = \Phi_k \Phi_k^t$ for later use. It will be supposed that the k-th actuator material has Young's modulus E_k; we assume that the Poisson effect for the actuator filament structures is negligible.

Associated with each actuator family $F_k(X)$ is a corresponding *control signal* function $u_k(X)$; the control action is assumed to be such that an actuator filament of length l when $u_k = 0$ has length $(1-u_k)l$ when the control signal u_k is applied to it. In order to develop the potential energy form for the actuator structure we consider an actuator filament of length l located near $X \in \mathcal{R}_0$ with direction $\Phi_k(X)$. Supposing the filament embedded in the matrix material so that no slippage takes place it is easy to see that under deformation as described earlier in terms of $\Xi(X)$ the new length is approximately equal to $l\|(I + \frac{\partial \Xi}{\partial X}(X))\Phi_k(X)\|$, corresponding to a *change in length* from the controlled equilibrium length $(1 - u_k(X)l$

$$\delta l \simeq l(\Phi_k(X)^t \frac{\partial \Xi}{\partial X}(X)\Phi_k(X) + u_k(X)).$$

Taking into account the assumed Young's modulus and supposing the cross sectional area of the filament to be δA, familiar formulae from mechanics show the potential energy increment to be $\frac{1}{2} E_k \delta A l(\Phi_k^t \frac{\partial \Xi}{\partial X} \Phi_k + u_k)^2$. Integrating we obtain the potential energy form for the k-th actuator family:

$$\mathcal{V}_k = \frac{1}{2} \int\int_{\mathcal{R}_0} E_k D_k(X)(\Phi_k(X)^t \frac{\partial \Xi}{\partial X}(X)\Phi_k(X) + u_k(X))^2 dX. \qquad (2.02)$$

The total potential energy (matrix plus all actuators) is then clearly

$$\mathcal{V} = \mathcal{V}_0 + \mathcal{V}_1 + \mathcal{V}_2 \ldots + \mathcal{V}_m. \qquad (2.03)$$

The vector function $\Xi(X)$ yielding the modified configuration due to the action of the control functions $u_k(X)$ through the actuator families $F_k(X)$ minimizes $\mathcal{V}$. Since the $u_k(X)^2$ terms in (2.03) play no role in this minimization operation we can equivalently use the overall energy expression (here using $\Phi_k(X)\Phi_k(X)^t \equiv P_k(X)$ in (2.02))

$$\tilde{\mathcal{V}} = \frac{1}{2} \int\int_{\mathcal{R}_0} [\nu\|\nabla \xi\|^2 + \nu\|\nabla \eta\|^2 + (\lambda + \nu)(\nabla \cdot \Xi)^2$$

$$+\sum_{k=1}^{m} E_k D_k(X)\Phi_k(X)^t \frac{\partial \Xi}{\partial X}(X)P_k(X)\frac{\partial \Xi}{\partial X}(X)\Phi_k(X)$$

$$+2\sum_{k=1}^{m} E_k u_k(X)F_k(X)^t \frac{\partial \Xi}{\partial X}(X)\Phi_k(X)]dX. \qquad (2.04)$$

The first sum in (2.04) corresponds to the actuator stiffening effect described in the preceding section while the second sum may be described as the *control potential.*

In order to develop the partial differential equations and boundary conditions appropriate to our theory it will be convenient to introduce some notation. Since the gradient of a scalar function ϕ is traditionally denoted $\nabla\phi$ and the divergence of a vector field F is denoted by $\nabla \cdot \phi$, let us use the symbol ∇M, for a given differentiable 2×2 matrix $M(x, y)$, to denote the row vector obtained from M by differentiating its first row with respect to x, its second row with respect to y, and adding the two. We denote by $\nabla^* M$ the column vector obtained by applying the corresponding operations to the columns of M.

We may then apply the familiar Hamilton's principle to the potential energy (2.04) to obtain the equilibrium equations. As far as the first line of (2.04) is concerned this operation is entirely standard. In order to treat the variation of the second line it is necessary to note, using the notation of the preceding paragraph, that

$$\nabla \cdot [E_k D_k(X)\Phi_k(X)^t \frac{\partial \Xi}{\partial X}(X)P_k(X)\delta\Xi(X)\Phi_k(X)]$$

$$= \nabla^*[E_k D_k(X)P_k(X)\frac{\partial \Xi}{\partial X}(X)P_k(X)]\delta\Xi(X)$$

$$+E_k D_k(X)\Phi_k(X)^t \frac{\partial \Xi}{\partial X}(X)P_k(X)\frac{\partial \delta\Xi}{\partial X}(X)\Phi_k(X). \qquad (2.05)$$

That observation having been made it is a simple matter to obtain the equilibrium equations in the form of a system of two second order partial differential equations

$$\nu\nabla^2\Xi + (\lambda+\nu)\nabla(\nabla\cdot\Xi)$$

$$+\sum_{k=1}^{m} \nabla^*[E_k D_k P_k \frac{\partial \Xi}{\partial X} P_k + E_k D_k u_k P_k] = 0. \qquad (2.06)$$

The corresponding boundary conditions are

$$\nu \frac{\partial \Xi}{\partial \sigma} + (\lambda + \nu)(\nabla \cdot \Xi)\sigma + \sum_{k=1}^{m} E_k D_k u_k P_k \sigma = 0. \qquad (2.07)$$

Here it is significant to note that the controls u_k appear linearly in the boundary conditions (2.07) while their *derivatives* appear linearly in the partial differential equations of the system (2.06).

3. Some Problems in the Formation Theory of Two-Dimensional Elastic Structures

Given the actuator screen $F_k(X), k = 1, 2, ..., m$, in the region $\mathcal{R}_0$ with boundary Σ_0, we can ask which regions $\mathcal{R}$, with boundaries Σ, can be achieved as modified configurations of the elastic body with the use of various control functions $u_k(X), k = 1, 2..., m$. Since we are assuming the actuator structure to be given, we are restricting attention here to just the control aspect of the overall formation problem as described in Section 1. It turns out that this control problem can be posed in a number of different ways.

i) "Mapping Control" This is the question of which displacement functions $\Xi(X) = [\xi(x, y)\ \eta(x, y)]^t$ can be achieved as solutions of the system (2.06), (2.07) by appropriate choice of the control functions $u_k(X), k = 1, 2, ..., m$. Clearly, more than the mere "shape" of the modified region R is involved here; we are also concerned with the behavior of $\Xi(X)$ in the interior of the region $\mathcal{R}_0$. In many cases it will turn out that this is an unnecessarily over-ambitious objective since the boundary Σ, subject to some mild conditions to avoid pathologies, determines the region R and hence the transformed configuration of the body.

ii) "Full Boundary" Control This is a less ambitious objective than the one just outlined; one specifies $\Xi(X), X \in \Sigma_0$, thereby specifying Σ and its interior R, provided the mapping $X \to X + \Xi(X)$ is one-to-one with Jacobian everywhere positive.

iii) "Boundary Image" Control Here we do not specify $\Xi(X)$ point by point on Σ_0 but rather we specify only that $\Xi(\Sigma_0) = \Sigma, \Sigma$ being appropriately specified. One way in which this might be done is to suppose that

$$\Sigma = \{\Xi = [\xi\ \eta]^t | \phi(\Xi) = \phi(\xi, \eta) = c\}$$

and then impose the conditions

$$\phi(\Xi(X)) \equiv c,\ X \in \Sigma_0,\ \Xi(\Sigma_0) = \Sigma.$$

As in Case ii) one needs to impose additional one-to-one and Jacobian requirements. However, if we are not careful, some physically inadmissible situations can arise, as shown in Figure 8 where an annulus, with boundary curves directed as indicated, is mapped "inside out" onto another annulus, the outer boundary of the original annulus being mapped onto the inner boundary of the image annulus and vice versa. To avoid situations like this it seems likely that some type of *homotopy* requirement might need to be imposed whereby it is required that there should be a continuum of displacement functions $\Xi(X, s), s \in [0, 1]$, with $X + \Xi(X, 0) \equiv X, X + \Xi(X, 1) \equiv X + \Xi(X)$, it being further required that the Jacobian of $X + \Xi(X, s)$ should remain uniformly positive as s varies in $[0, 1]$. The family of displacements $\Xi(X, s)$ will, in many cases, be easy to realize through use of controls $s\, u_k(X), k = 1, 2, ..., m$; the problem is to enforce the one-to-oneness and the Jacobian condition.

Throughout the remainder of this article we will consider Case ii); the case of full boundary control. It turns out that this problem has some interesting features which can be studied with the aid of a duality relationship which we now proceed to develop. To this end, we introduce an *adjoint* or *dual* state $\Omega(X) = [\omega(x, y)\, \zeta(x, y)]^t$ and, with $\Xi(X) = [\xi(x, y)\, \eta(x, y)]^t$ the physical, or *primal*, state already introduced for our elastic system, we consider the bilinear form

$$\int\int_{\mathcal{R}_0} \nabla \cdot \Big[\nu[\omega\frac{\partial \xi}{\partial x}\ \zeta\frac{\partial \eta}{\partial y}]^t - \nu[\xi\frac{\partial \omega}{\partial x}\ \eta\frac{\partial \zeta}{\partial y}]^t$$

$$+(\lambda+\nu)[\omega(\frac{\partial \xi}{\partial x} + \frac{\partial \eta}{\partial y})\ \zeta(\frac{\partial \xi}{\partial x} + \frac{\partial \eta}{\partial y})]^t - (\lambda+\nu)[\xi(\frac{\partial \omega}{\partial x} + \frac{\partial \zeta}{\partial y})\ \eta(\frac{\partial \omega}{\partial x} + \frac{\partial \zeta}{\partial y})]^t$$

$$+\sum_{k=1}^{m} E_k D_k P_k \frac{\partial \Xi}{\partial X} P_k \Omega - \sum_{k=1}^{m} E_k D_k P_k \frac{\partial \Omega}{\partial X} P_k \Xi\Big]\, dX. \qquad (3.01)$$

We form the indicated divergence and see, algebraically, that with

$$\mathcal{P}(\Omega) \equiv \nu\nabla^2\Omega + (\lambda+\nu)\nabla(\nabla\cdot\Omega) + \sum_{k=1}^{m} E_k D_k \nabla^*(P_k \frac{\partial \Omega}{\partial X} P_k), \qquad (3.02)$$

$$\mathcal{B}(\Omega) \equiv \nu \frac{\partial \Omega}{\partial \sigma} + (\lambda + \nu)(\nabla \cdot \Omega)\sigma, \tag{3.03}$$

(3.01) becomes

$$\int\int_{\mathcal{R}_0} [\Omega^t \mathcal{P}(\Xi) - \Xi^t \mathcal{P}(\Omega)] dX. \tag{3.04}$$

Then applying Gauss' theorem to (3.01) (equivalently (3.04)) we obtain

$$\int\int_{\mathcal{R}_0} [\Omega^t \mathcal{P}(\Xi) - \Xi^t \mathcal{P}(\Omega)] dX$$

$$= \int_{\Sigma_0} [\Omega^t \mathcal{B}(\Xi) - \Xi^t \mathcal{B}(\Omega)] ds. \tag{3.05}$$

We require now that Ω should satisfy the *adjoint system* consisting of the partial differential equations and boundary conditions (cf. (3.02),(3.03))

$$\mathcal{P}(\Omega) = 0; \;\; \mathcal{B}(\Omega) = \Omega_0, \tag{3.06}$$

where Ω_0 is a two-dimensional vector valued function defined on Σ_0; its component normal to Σ_0 is a *normal stress* and its component tangent to Σ_0 is a *traction.* Using (3.06) in (3.05) we now have

$$\int\int_{\mathcal{R}_0} \Omega^t \mathcal{P}(\Xi) dX = \int_{\Sigma_0} \left[\Omega^t \mathcal{B}(\Xi) - \Xi^t \Omega_0\right] ds. \tag{3.07}$$

Using the divergence theorem again we can modify (3.07) to

$$\int\int_{\mathcal{R}_0} \left[\Omega^t \mathcal{P}(\Xi) + \nabla \cdot \left[\sum_{k=1}^{m} (E_k D_k u_k P_k \Omega)\right]\right] dX$$

$$= \int_{\Sigma_0} \left[\Omega^t \left[\mathcal{B}(\Xi) + E_k D_k u_k P_k \sigma\right] - \Xi^t \Omega_0\right] ds. \tag{3.08}$$

Now the boundary condition (2.07) is just

$$\mathcal{B}(\Xi) + E_k D_k u_k P_k \sigma = 0.$$

Using this in the second line of (3.08) and expanding the divergence in the first line of (3.08) we arrive at

$$\int_{\Sigma_0} {\Omega_0}^t \Xi ds$$

$$= -\int\int_{\mathcal{R}_0} \Omega^t \Big[\mathcal{P}(\Xi) + \nabla^*\Big(\sum_{k=1}^{m}(E_k D_k u_k P_k)\Big)\Big] dX$$

$$-\int\int_{\mathcal{R}_0} \sum_{k=1}^{m}\Big(E_k D_k u_k {\Phi_k}^t \frac{\partial \Omega}{\partial X}\Phi_k\Big) dX. \qquad (3.09)$$

But the system of partial differential equations (2.06) is just

$$\mathcal{P}(\Xi) + \nabla^*\Big(\sum_{k=1}^{m}(E_k D_k u_k P_k)\Big) = 0$$

so (3.09) finally yields the fundamental equation

$$\int_{\Sigma_0} {\Omega_0}^t \Xi ds + \int\int_{\mathcal{R}_0} \sum_{k=1}^{m}\Big(E_k D_k u_k {\Phi_k}^t \frac{\partial \Omega}{\partial X}\Phi_k\Big) dX = 0. \qquad (3.10)$$

We see then that if the boundary stress vector Ω_0 in the second equation of (3.06) is orthogonal to all boundary displacements $\Xi|_{\Sigma_0}$, so that the first integral in (3.10) is zero, then the second integral there must be zero for all controls $u_k, k = 1, 2, ..., m$. This clearly implies

$${\Phi_k}^t \frac{\partial \Omega}{\partial X}\Phi_k \equiv 0, x \in supp(D_k) \subset \mathcal{R}_0, k = 1, 2..., m. \qquad (3.11)$$

As a consequence it can be seen that we have *approximate* full boundary control, with the indicated actuator screen and controls, in the sense that the achievable boundary displacements must be dense relative to the norm in ${L^2}_{\Sigma_0}$, just in case (3.11), which is a statement about nonextension/noncontraction of the adjoint state Ω determined by Ω_0 parallel to the filaments of the actuator screen $F_k, k = 1, 2, ..., m$, implies that, in fact, $\Omega_0 \equiv 0$ on Σ_0. This will be recognized as an *observability condition* on the adjoint system.

Let us conclude the work here with a specific application of the duality theory. Let us suppose that the region $\mathcal{R}_0$ consists of a rectangle $|x| \leq a, |y| \leq b$ with $x = a$ identified with $x = -a$ to form a periodic domain in the x direction. We consider a single actuator family $F_1(x, y) \equiv [d\ 0]^t$, corresponding to filaments oriented in the horizontal direction with a uniform volume density d and modulus of elasticity $E_1 \equiv e$. Then $D_1(x, y) \equiv d, \Phi_1(x, y) \equiv [1\ \ 0]^t$, $P_1(x, y) = \begin{pmatrix} 1 & 0 \\ 0 & 0 \end{pmatrix}$. Here we find that (cf.(2.06))

$$\nabla^*\left(E_k D_k P_k \frac{\partial \Xi}{\partial X} P_k\right) = ed[\frac{\partial^2 \xi}{\partial x^2}\ \ 0]^t$$

and, with $u_1 \equiv u$ here,

$$\nabla^* \left(E_k D_k u_k P_k \right) = ed[0 \ \frac{\partial u}{\partial x}].$$

It follows then that the "primal" system (2.06) takes the form

$$(\nu + ed)\frac{\partial^2 \xi}{\partial x^2} + \nu\frac{\partial^2 \xi}{\partial y^2} + (\lambda + \nu)\frac{\partial}{\partial x}(\nabla \cdot \Xi) + ed\,\frac{\partial u}{\partial x} = 0, \tag{3.12}$$

$$\nu\frac{\partial^2 \eta}{\partial x^2} + \nu\frac{\partial^2 \eta}{\partial y^2} + (\lambda + \nu)\frac{\partial}{\partial y}(\nabla \cdot \Xi) = 0, \tag{3.13}$$

while the primal boundary conditions become, since $\sigma = [0\,1]^t$ on $y = b$ and $\sigma = [0\ -1]^t$ on $y = -b$,

$$\pm\nu\frac{\partial \xi}{\partial y} + ed\,u = 0, \tag{3.14}$$

$$\pm\nu\frac{\partial \eta}{\partial y} \pm (\lambda + \nu)(\nabla \cdot \Xi) = 0, \tag{3.15}$$

with the "+" sign applying on $y = b$ and the "-" sign on $y = -b$. On the other hand the dual system in $\Omega = [\omega\ \zeta]^t$ is

$$(\nu + ed)\frac{\partial^2 \omega}{\partial x^2} + \nu\frac{\partial^2 \omega}{\partial y^2} + (\lambda + \nu)\frac{\partial}{\partial x}(\nabla \cdot \Omega) = 0, \tag{3.16}$$

$$\nu\frac{\partial^2 \zeta}{\partial x^2} + \nu\frac{\partial^2 \zeta}{\partial y^2} + (\lambda + \nu)\frac{\partial}{\partial y}(\nabla \cdot \Omega) = 0, \tag{3.17}$$

while the dual boundary conditions are

$$\pm\nu\frac{\partial \omega}{\partial y} = \omega_0, \tag{3.18}$$

$$\pm\nu\frac{\partial \zeta}{\partial y} \pm (\lambda + \nu)(\nabla \cdot \Omega) = \zeta_0, \tag{3.19}$$

with the same sign conventions as in (3.15). Here the duality relationship reads

$$\int_{\{y=\pm b\}} (\omega_0 \xi + \zeta_0 \eta)\, dx + \int\!\!\int_{\mathcal{R}_0} edu\frac{\partial \omega}{\partial x} dx dy = 0, \tag{3.20}$$

Thus if $[\omega_0\ \zeta_0]^t$ is orthogonal in $L^2\{y = \pm b\}$ to the achievable boundary displacements $[\xi\ \eta]^t$, we conclude that the dual state $[\omega\ \zeta]$ determined by ω_0 and ζ_0 via the boundary conditions (3.14), (3.15) must satisfy

$$\frac{\partial \omega}{\partial x} \equiv 0. \tag{3.21}$$

Let us suppose, then, that $\Omega = (\omega,\, \zeta)^t$ is a solution of (3.16), (3.17) for which (3.21) is true. Using (3.21) in (3.18), with the assumption of smoothness, we may differentiate that equation with respect to x to obtain

$$\nu \frac{\partial^3 \omega}{\partial x^1 \partial y^2} = (\lambda + \nu) \frac{\partial^3 \zeta}{\partial y^1 \partial x^2}. \tag{3.22}$$

Then we may apply (3.21) again to show that the left hand side of (3.22) is identically zero. Then the right hand side must also be zero, which, together with the assumed x-periodicity of $\zeta(x, y)$, shows that $\frac{\partial^2 \zeta}{\partial x^2} \equiv \rho(x)$, a function of x alone. Again taking the x-periodicity of $\zeta(x, y)$ into account, we see further see that we must have

$$\zeta(x, y) \equiv r(x) + t(y). \tag{3.23}$$

Using (3.21) and (3.23) in (3.17) we now have

$$\nu \frac{\partial^2 \zeta}{\partial x^2} + (\lambda + 2\nu) \frac{\partial^2 \zeta}{\partial y^2} = \nu \frac{d^2 r}{dx^2} + (\lambda + 2\nu) \frac{d^2 t}{dy^2} = 0,$$

which shows that both $\frac{d^2 r}{dx^2}$ and $\frac{d^2 t}{dy^2}$ are constant, say equal to c. Then we clearly have

$$\frac{\partial^2 \zeta}{\partial x^2} \equiv -\frac{\lambda + 2\nu}{\nu} \frac{\partial^2 \zeta}{\partial y^2} \equiv c \tag{3.24}$$

and we conclude that $\zeta(x, y)$ must have the form

$$\zeta(x, y) = \zeta_{00} + \zeta_{10}\, x + \zeta_{01}\, y + \zeta_{11}\, x\, y + \frac{c}{2}\, x^2 - \frac{\nu c}{2(\lambda + 2\nu)}\, y^2. \tag{3.25}$$

The x-periodicity condition forces ζ_{10}, ζ_{11} and c to all be zero. Then we have

$$\zeta(x, y) = \zeta_{00} + \zeta_{01}\, y. \tag{3.26}$$

Then (3.16) shows that

$$\nu \frac{\partial^2 \omega}{\partial y^2} \equiv 0 \tag{3.27}$$

and this together with (3.21) shows that $\omega(x, y)$ also has the form

$$\omega(x, y) = \omega_{00} + \omega_{01}\, y. \tag{3.28}$$

From (3.26) and (3.28) we infer that

$$\omega(x, y) \equiv c_0 \, + \, c_1\, y, \;\; \zeta(x, y) = d_0\, , + d_1\, y. \tag{3.29}$$

The boundary conditions (3.18) and (3.19) then give

$$\omega_{01} = \frac{\omega_0}{\nu} \quad \text{on} \quad y = b, = -\frac{\omega_0}{\nu} \quad \text{on} \quad y = -b, \tag{3.30}$$

$$\zeta_{01} = \frac{\zeta_0}{\nu + (\lambda + \nu)} \quad \text{on} \quad y = b, = -\frac{\zeta_0}{\nu + (\lambda + \nu)} \quad \text{on} \quad y = -b. \tag{3.31}$$

These equations show that ω_0 and ζ_0 have arbitrary, independent, values of opposite sign and equal absolute value on $y = b$ and $y = -b$. It follows from the duality principle developed above that all displacements $\xi(x, \pm 1)$, $\eta(x, \pm 1)$ achievable with the actuator/control structure indicated necessarily satisfy

$$\int_{-a}^{a} \xi(x,b)dx - \int_{-a}^{a} \xi(x,-b)dx = 0, \int_{-a}^{a} \eta(x,b)dx - \int_{-a}^{a} \zeta(x,-b)dx = 0. \tag{3.32}$$

Thus, e.g., it is not possible, using the actuator structure indicated, to produce non-identically-zero vertical or horizontal displacements having a single sign on $y = b$ and the opposite sign on $y = -b$. In particular, one cannot produce a constant shear deformation or a uniform vertical expansion or a uniform vertical contraction with the horizontal actuators assumed in this example. The second of these conclusions might seem counter-intuitive since the Poisson effect would be expected to produce a uniform vertical expansion if the actuators are used to produce a uniform horizontal contraction. What prevents this from being true is the assumed periodicity in the x direction which precludes uniform horizontal contraction.

References

[1] D. L. Russell: *An introduction to the formation theory of active materials*, to appear in the Proceedings of the CODAC Conference on Control and Active Materials, Blacksburg, VA, April, 1994

[2] D. Y. Gao & D. L. Russell: *An extended Timoshenko beam model with applications to active materials*, submitted to J. Appl. Math. & Opt.

[3] R. L. Bisplinghoff & H. Ashley: *Principles of Aeroelasticity*, John Wiley & Sons, Inc., New York, 1962; Reprinted by Dover Publications, Inc., New

York, 1975

[4] P. C. Chou & N. J. Pagano: *Elasticity; Tensor, Dyadic and Engineering Approaches*, D. Van Nostrand & Co., Princeton, NJ, 1967; Reprinted by Dover Publications, Inc., New York, 1992

Optimal Control of Free Boundaries

Srdjan Stojanovic
Department of Mathematical Sciences
University of Cincinnati
Cincinnati, OH 45221-0025
U.S.A.

Abstract

New proof of the perturbation formula for regular free boundaries in the obstacle problem is given. The approach introduced here enables study of parabolic problems.

1 Introduction

In the last twenty years or so there have been lots of works on so called Optimal control theory of variational inequalities (see [14, 9, 1, 10, 11, 8]). One could say that the problem of distributed control, i.e., the problem where the objective is to control a solution of the variational inequality (more precisely, solution of the obstacle problem) is well understood. Also, various attempts were made to handle the issue of controlling the free boundary, via different types of regularizations, or relaxations.

The problem of optimal control of free boundaries in the case of an elliptic obstacle problem without any regularization was considered more recently in [13]. Here we propose another approach which, on the contrary to [13], extends to the parabolic case, as well.

This research was partially supported by Taft Memorial Foundation

2 Optimal Control of Free Boundaries in Elliptic Obstacle Problems

Let Ω be a bounded domain in R^n such that $\partial\Omega$ is locally in $C^{2,\alpha}$ for some $\alpha > 0$. Let A be elliptic operator defined by

$$Az = -\sum a_{ij}\frac{\partial^2 z}{\partial x_i \partial x_j} + \sum b_i \frac{\partial z}{\partial x_i} + cz. \tag{2.1}$$

We assume that

$$a_{ij} = a_{ji},\ a_{ij} \in C^{2,\alpha}(\Omega) \cap C^{1,\alpha}(\bar{\Omega}) \tag{2.2}$$

$$b_i, c \in C^{\alpha}(\bar{\Omega}) \tag{2.3}$$

$$\sum a_{ij}(x)\xi_i\xi_j \geq \lambda|\xi|^2 \quad \forall x \in \Omega,\ \xi \in R^n \quad (\lambda > 0) \tag{2.4}$$

$$c \geq 0. \tag{2.5}$$

Consider also the associated bilinear form

$$a(z,\zeta) = \int_\Omega \left\{\sum a_{ij}\frac{\partial z}{\partial x_i}\frac{\partial \zeta}{\partial x_j} + \sum\left(b_i + \sum \frac{\partial a_{ij}}{\partial x_j}\right)\frac{\partial z}{\partial x_i}\zeta + cz\zeta\right\} dx, \tag{2.6}$$

and we assume that a is coercive, that is,

$$a(z,z) \geq \beta\|z\|^2_{H^1_0(\Omega)} \quad \forall z \in H^1_0(\Omega) \quad (\beta > 0). \tag{2.7}$$

Let $u \in C^{1,\alpha}(\bar{\Omega})$ and $g \in C^{2,\alpha}(\bar{\Omega})$ be given. We also assume here that

$$u, g > 0 \quad \text{in } \bar{\Omega}. \tag{2.8}$$

Consider the obstacle problem: Find $z \in K$, such that

$$a(z, \zeta - z) \geq -\int_\Omega u(\zeta - z)dx \quad \forall \zeta \in K, \tag{2.9}$$

where K is the following closed convex set in $H^1(\Omega)$:

$$K = \left\{\zeta \in H^1(\Omega); \zeta - g \in H^1_0(\Omega), \zeta \geq 0 \text{ a.e.}\right\}. \tag{2.10}$$

Remark 2.1 *We have taken obstacle to be equal to* 0. *It is trivial to extend everything that follows to the case when obstacle is any* $C^{2,\alpha}(\bar{\Omega})$ *function* ψ, *such that* $\psi < g$ *on* $\partial\Omega$ *(see Remark (2.2)).*

It is very well known (see e.g. [7]) that the obstacle problem has a unique solution z, and

$$z \in W^{2,p}(\Omega) \cap W^{2,\infty}_{loc}(\Omega)\ \forall p < \infty. \tag{2.11}$$

Moreover, by the fundamental theorem of L. A. Caffarelli ([3, 4]), in some cases, it is possible to claim smoothness of the free boundary $\partial\{z > 0\} \cap \Omega$ (see also Chapter 2. of [7]). We shall say that if the assumptions of the Caffarelli's theorem are fulfilled that the free boundary is *regular*.

The stability of the free boundaries with respect to the perturbations in data has been studied in [12, 5]. The result there is, roughly speaking, that if the solution of the obstacle problem has a regular free boundary, then the obstacle problem with small perturbation of data has a regular free boundary which is *close* to the unperturbed one.

We shall state the perturbation formula in the case when the perturbations are in the right hand side u of the obstacle problem. Many other kinds of perturbations are possible to consider by the same method.

Let u be such that $z = z^u$ the corresponding solution of the obstacle problem has a regular free boundary $\Gamma^u = \partial\{z^u > 0\} \cap \Omega$. It has been observed (see [8] and reference given there) that then the obstacle map $u \mapsto z^u$, which is in general

(2.12) Lipschitz from $L^2(\Omega)$ to $H^1(\Omega)$,

is actually differentiable at u, in a sense that

$$\frac{z^{u+\lambda v} - z^u}{\lambda} \rightharpoonup w^{u;v} \text{ weakly in } H_0^1(\Omega), \tag{2.13}$$

as $\lambda \longrightarrow 0$, where

$$w^{u;v} = \begin{cases} \delta^{u;v} & \text{in } \{z^u > 0\} \\ 0 & \text{in } \{z^u = 0\} \end{cases}, \tag{2.14}$$

and where $\delta = \delta^{u;v}$ is the unique solution of the elliptic equation

$$\begin{aligned} A\delta &= -v \text{ in } \{z^u > 0\} \\ \delta &= 0 \text{ on } \partial\{z^u > 0\}. \end{aligned} \tag{2.15}$$

Let ν^u be the unit normal to Γ^u exterior to $\{z^u > 0\}$. Also, let ν_A^u be a conormal, i.e., $\nu_{Ai}^u = \sum a_{ij}\nu^u{}_j$.

We have

Theorem 2.1 (Perturbation Formula for Regular Free Boundaries; Elliptic Obstacle Problem Case) *The following perturbation formula holds*

$$\frac{I_{\{z^{u+\lambda v}>0\}} - I_{\{z^u>0\}}}{\lambda} \rightharpoonup -\frac{1}{u}\frac{\partial\delta^{u;v}}{\partial\nu_A^u}d\Gamma^u, \tag{2.16}$$

weakly in $H^{-1}(\Omega)$, *as* $\lambda \longrightarrow 0$. *Also,*

$$-\frac{1}{u}\frac{\partial\delta^{u;v}}{\partial\nu_A^u}d\Gamma^u = -\frac{\dfrac{\partial\delta^{u;v}}{\partial\nu^u}}{\dfrac{\partial^2 z^u}{(\partial\nu^u)^2}}d\Gamma^u \tag{2.17}$$

Proof: It is easy to see that z^u solves the nonlinear *equation*.

$$(2.18)\qquad \begin{aligned} Az^u + uI_{\{z^u>0\}} &= 0 \ \text{ a.e. in } \Omega \\ z^u &= g \ \text{ on } \partial\Omega, \end{aligned}$$

where I_D is the characteristic function of the set D, i.e., $I_D(x) = \begin{cases} 1 & \text{if } x \in D \\ 0 & \text{if } x \notin D \end{cases}$
Although it is not essential for what follows, it is worth remarking that because of (2.8, 2.7) the nonlinear operator in (2.18) is monotone, so that the uniqueness follows easily.

So, writing (2.18) in the weak form for $z^{u+\lambda v}$ and z^u, subtracting and dividing by λ, we get, for every $\varphi \in H_0^1(\Omega)$,

$$-a\left(\frac{z^{u+\lambda v} - z^u}{\lambda}, \varphi\right)$$

$$(2.19)\qquad = \int_\Omega vI_{\{z^{u+\lambda v}>0\}}\varphi dx + \int_\Omega u\frac{1}{\lambda}\left(I_{\{z^{u+\lambda v}>0\}} - I_{\{z^u>0\}}\right)\varphi dx.$$

Using (2.13) and the stability theorem of [5] we can pass the limit $\lambda \to 0$ in (2.19) to conclude

$$-a(w, \varphi)$$

$$(2.20)\qquad = \int_\Omega vI_{\{z^u>0\}}\varphi dx + \lim_{\lambda\to 0}\int_\Omega u\frac{1}{\lambda}\left(I_{\{z^{u+\lambda v}>0\}} - I_{\{z^u>0\}}\right)\varphi dx.$$

Now from (2.14, 2.20) we conclude that

$$(2.21)\qquad \lim_{\lambda\to 0}\int_\Omega u\frac{1}{\lambda}\left(I_{\{z^{u+\lambda v}>0\}} - I_{\{z^u>0\}}\right)\varphi dx = -\int_{\Gamma^u}\frac{\partial\delta^{u;v}}{\partial\nu_A^u}\varphi d\sigma,$$

and hence, since $\varphi \in H_0^1(\Omega)$ implies $\frac{\varphi}{u} \in H_0^1(\Omega)$,

$$(2.22)\qquad \lim_{\lambda\to 0}\int_\Omega \frac{1}{\lambda}\left(I_{\{z^{u+\lambda v}>0\}} - I_{\{z^u>0\}}\right)\varphi dx = -\int_{\Gamma^u}\frac{1}{u}\frac{\partial\delta^{u;v}}{\partial\nu_A^u}\varphi d\sigma,$$

and hence (2.16) is proved.

Notice that, since on Γ^u

$$(2.23)\qquad z^u = |\nabla z^u| = 0,$$

we have

$$(2.24)\qquad u(x) = -Az^u(x) = a^{ij}z_{x_ix_j}(x) \ \ \forall x \in \Gamma^u,$$

in the sense of trace from the $\{z^u > 0\}$-side. Here we have used regularity of Γ^u, as well. Furthermore, fix $x^0 \in \Gamma^u$ and notice that, without loss of generality, we can cuppose that $\nu^u(x^0) = e_n$. Then it is easy to see, using (2.23), that

$$(2.25)\qquad \sum_{i,j} a^{ij}z_{x_ix_j}(x^0) = a^{nn}z_{x_nx_n}(x^0),$$

and, since $\delta^{u;v} = 0$ on Γ^u, we deduce also

$$\sum_{i,j} a^{ij} \delta^{u;v}{}_{x_i} \nu^{uj} = a_{nn} \delta_{x_n}. \tag{2.26}$$

We conclude that, in general,

$$\frac{\sum_{i,j} a^{ij} \delta^{u;v}{}_{x_i} \nu^{uj}}{\sum_{i,j} a^{ij} z_{x_i x_j}} = \frac{\delta_{\nu^u}}{z^u{}_{\nu^u \nu^u}} \quad \text{on } \Gamma^u, \tag{2.27}$$

and hence (2.17) follows.

Remark 2.2 *It is easy to see that in the case of general obstacle $\psi \in C^{2,\alpha}(\bar{\Omega})$ Formula (2.16) becomes*

$$\frac{I_{\{z^{u+\lambda v}>0\}} - I_{\{z^u>0\}}}{\lambda} \rightharpoonup -\frac{1}{u + A\psi} \frac{\partial \delta^{u;v}}{\partial \nu_A^u} d\Gamma^u, \tag{2.28}$$

weakly in $H^{-1}(\Omega)$, as $\lambda \to 0$.

Remark 2.3 *Various optimal control problems can be now formulated and solved easily. We postpone that untill the next section.*

3 Optimal Control of Free Boundaries in Parabolic Obstacle Problems

3.1 Perturbation Formula

For simplicity we shall consider the heat operator. Results can be extended to the case of general parabolic operator.

As before, let Ω be a bounded domain in R^n such that $\partial\Omega$ is locally in $C^{2,\alpha}$ for some $\alpha > 0$. Let $Q_T = \Omega \times \{0 < t < T\}$. We assume

$$\begin{aligned} u, g, D_x g, D_x^2 g, D_t g &\in C^\alpha(\bar{Q_T}) \\ D_t u &\in L^\infty(Q_T). \end{aligned} \tag{3.1}$$

We also assume

$$u, g > 0 \quad \text{in } \bar{Q_T}. \tag{3.2}$$

Consider the obstacle problem: Find $z \in K$, such that

$$\begin{aligned} &\int_\Omega [z_t(\zeta - z) + \nabla z \cdot \nabla(\zeta - z)]\, dx \\ &\geq -\int_\Omega u(\zeta - z) dx \quad \text{for a.e. } t \ \forall \zeta \in K, \end{aligned} \tag{3.3}$$

where K is the following closed convex set in $H^1(Q_T)$:

(3.4) $$K = \{\zeta \in H^1(Q_T); \zeta = g \text{ on } \partial_p Q_T,\ \zeta \geq 0 \text{ a.e.}\}.$$

∂_p represents the parabolic boundary, in the above case $\partial_p Q_T = \Omega \times \{0\} \cup \partial\Omega \times \{0 < t < T\}$, and the boundary value is taken in the trace sense.

It is well known (see e.g. [7]) that the obstacle problem (3.3) has a unique solution z, and

(3.5) $$\begin{gathered} z, D_x z, D_x^2 z, D_t z \in L^p(\Omega) \text{ for a.e. } t \quad \forall p < \infty, \\ z, D_x z, D_x^2 z, D_t z \in L^\infty_{loc}(Q_T). \end{gathered}$$

Moreover, by the Caffarelli's theorem [3], in some cases, it is possible to claim smoothness of the free boundary $\partial\{z > 0\} \cap Q_T$.

The stability of the free boundaries in parabolic case is discussed in [7]. The result is similar as in the case of an elliptic obstacle problem.

As before we denote z^u the solution of the obstacle problem (3.3) corresponding to the right hand side u. It is known (see e.g. [1]) that the map $u \mapsto z^u$ is

(3.6) Lipschitz from $L^2(Q_T)$ to $C([0,T]; L^2(\Omega)) \cap L^2(0,T; H^1(\Omega))$.

Hence, it is not difficult to prove

Lemma 3.1 *Let u be such that the corresponding free boundary $\partial\{z > 0\} \cap Q_T$ is regular, and that the problem (3.9) admits an unique solution. Then*

(3.7) $$\frac{z^{u+\lambda v} - z^u}{\lambda} \rightharpoonup w^{u;v} \quad \textit{weakly in } L^2(0,T; H^1(\Omega)),$$

as $\lambda \to 0$, where

(3.8) $$w^{u;v} = \begin{cases} \delta^{u;v} & \textit{in } \{z^u > 0\} \\ 0 & \textit{in } \{z^u = 0\} \end{cases},$$

and where $\delta = \delta^{u;v}$ is the unique solution of the parabolic equation

(3.9) $$\begin{gathered} \delta_t - \Delta\delta = -v \quad \textit{in } \{z^u > 0\} \\ \delta = 0 \quad \textit{on } \partial_p\{z^u > 0\}. \end{gathered}$$

Let $\Gamma^u = \partial\{z^u > 0\} \cap Q_T$, and let for every $t \in (0,T)$, $\Gamma^u(t) = \partial\{z^u(\cdot,t) > 0\} \cap \Omega$. Also let $\{z^u > 0\}(t) = \{z^u(\cdot,t) > 0\} \cap \Omega$.

We have

Theorem 3.1 (Perturbation Formula for Regular Free Boundaries; Parabolic Obstacle Problem Case) *Under the previous assumptions the following perturbation formula holds*

(3.10) $$\frac{I_{\{z^{u+\lambda v} > 0\}} - I_{\{z^u > 0\}}}{\lambda} \rightharpoonup -\frac{1}{u}\frac{\partial \delta^{u;v}}{\partial \nu_x^u} d\Gamma^u(t)dt,$$

weakly in $H^{-1}(Q_T)$, as $\lambda \to 0$.

Proof: It is easy to see that z^u solves the nonlinear equation

$$(3.11)\qquad \begin{aligned} z_t^u - \Delta z^u + uI_{\{z^u>0\}} &= 0 \text{ a.e. in } Q_T \\ z^u &= g \text{ on } \partial_p Q_T, \end{aligned}$$

So, writing (3.11) in the weak form for $z^{u+\lambda v}$ and z^u, subtracting and dividing by λ, we get, for any $\varphi \in H^1(0,T;H_0^1(\Omega))$ such that $\varphi|_{t=T}=0$,

$$\begin{aligned} \int_{Q_T} \frac{z^{u+\lambda v} - z^u}{\lambda} \varphi_t dx dt - \int_{Q_T} \nabla \left(\frac{z^{u+\lambda v} - z^u}{\lambda} \right) \cdot \nabla \varphi dx dt \\ = \int_{Q_T} vI_{\{z^{u+\lambda v}>0\}} \varphi dx dt \\ (3.12)\qquad + \int_{Q_T} u\frac{1}{\lambda}\left(I_{\{z^{u+\lambda v}>0\}} - I_{\{z^u>0\}}\right) \varphi dx dt. \end{aligned}$$

Using (3.7) and the stability theorem of [7] we can pass the limit $\lambda \to 0$ in (2.19) to conclude

$$\begin{aligned} \int_{Q_T} w^{u;v} \varphi_t dx dt - \int_{Q_T} \nabla w^{u;v} \cdot \nabla \varphi dx dt \\ = \int_{Q_T} vI_{\{z^u>0\}} \varphi dx dt \\ (3.13)\qquad + \lim_{\lambda\to 0} \int_{Q_T} u\frac{1}{\lambda}\left(I_{\{z^{u+\lambda v}>0\}} - I_{\{z^u>0\}}\right) \varphi dx dt. \end{aligned}$$

Now from (3.8, 3.13) we conclude that

$$\begin{aligned} \lim_{\lambda\to 0} \int_0^T \int_\Omega u\frac{1}{\lambda}\left(I_{\{z^{u+\lambda v}>0\}} - I_{\{z^u>0\}}\right) \varphi dx dt \\ (3.14)\qquad = -\int_0^T \int_{\Gamma^u(t)} \frac{\partial \delta^{u;v}}{\partial \nu_x^u} \varphi d\sigma_x dt, \end{aligned}$$

and hence

$$\begin{aligned} \lim_{\lambda\to 0} \int_0^T \int_\Omega \frac{1}{\lambda}\left(I_{\{z^{u+\lambda v}>0\}} - I_{\{z^u>0\}}\right) \varphi dx dt \\ (3.15)\qquad = -\int_0^T \int_{\Gamma^u(t)} \frac{1}{u}\frac{\partial \delta^{u;v}}{\partial \nu_x^u} \varphi d\sigma_x dt. \end{aligned}$$

Remark 3.1 *If instead of the heat operator $\frac{\partial}{\partial t} - \Delta$ we are dealing with the general parabolic operator $\frac{\partial}{\partial t}+A$, then the differentiation formula (3.10) becomes*

$$(3.16)\qquad \frac{I_{\{z^{u+\lambda v}>0\}} - I_{\{z^u>0\}}}{\lambda} \rightharpoonup -\frac{1}{u}\frac{\partial \delta^{u;v}}{\partial \nu_A^u} d\Gamma^u(t) dt,$$

weakly in $H^{-1}(Q_T)$, as $\lambda \to 0$.

3.2 An Optimal Control Problem

We shall consider here only the problem of minimizing the noncoincidence set. So let

$$\Phi(u) = \int_{Q_T} \left[I_{\{z^u>0\}} + \frac{\epsilon}{2} u^2 \right] dx dt, \tag{3.17}$$

for $\epsilon > 0$. The set of controls is

$$\mathcal{C} = \{u \in C^{1,\alpha}(\bar{Q_T}); u > 0\}. \tag{3.18}$$

We compute the directional derivative

$$\begin{aligned} \Phi'(u;v) &= \lim_{\lambda \to 0} \int_{Q_T} \frac{1}{\lambda} \left[I_{\{z^{u+\lambda v}>0\}} - I_{\{z^u>0\}} \right] dx dt + \int_{Q_T} \epsilon u v dx dt \\ &= -\int_0^T \int_{\Gamma^u(t)} \frac{1}{u} \frac{\partial \delta^{u;v}}{\partial \nu_x^u} d\sigma_x dt + \int_{Q_T} \epsilon u v dx dt. \end{aligned} \tag{3.19}$$

Let $p = p^u$ be defined by

$$\begin{aligned} p_t + \Delta p &= 0 \ \text{ in } \{z^u > 0\} \\ p &= -\frac{1}{u} \ \text{ on } \Gamma^u \\ p &= 0 \ \text{ on } \partial_{bp}\{z^u > 0\} \setminus \Gamma^u \\ p &= 0 \ \text{ in } \{z^u = 0\}. \end{aligned} \tag{3.20}$$

Here ∂_{bp} represents the backward parabolic boundary. Then

$$\begin{aligned} \Phi'(u;v) &= \int_0^T \int_{\Gamma^u(t)} p^u \frac{\partial \delta^{u;v}}{\partial \nu_x^u} d\sigma_x dt + \int_{Q_T} \epsilon u v dx dt \\ &= \int_0^T \int_{\{z^u>0\}(t)} [(\Delta \delta^{u;v}) p^u - \delta^{u;v} \Delta p^u] dx dt + \int_{Q_T} \epsilon u v dx dt \\ &= \int_0^T \int_{\{z^u>0\}(t)} [(\delta_t^{u;v} + v) p^u + \delta^{u;v} p_t^u] dx dt + \int_{Q_T} \epsilon u v dx dt \\ &= \int_{Q_T} (p^u + \epsilon u) v dx dt, \end{aligned} \tag{3.21}$$

and we have proved

Proposition 3.1 *Let $u \in \mathcal{C}$ be such that corresponding free boundary Γ^u is regular, and that problems (3.9,3.20) admit unique solutions. Then Φ is differentiable at u, and*

$$\Phi'(u;v) = \int_\Omega (p^u + \epsilon u) v dx, \tag{3.22}$$

where p^u is defined in (3.20).

Remark 3.2

$$\nabla \Phi(u) = p^u + \epsilon u \ \left(\in L^2(Q_T)\right). \tag{3.23}$$

References

[1] V. Barbu, *Optimal Control of Variational Inequalities*, Research Notes in Mathematics 100, Pitman, London, 1984.

[2] V. Barbu and S. Stojanovic, Controlling the moving boundary of the parabolic obstacle problem, *Appl. Math. & Optimiz.* **27**, No. 3, 213-230, (1993).

[3] L. A. Caffarelli, Regularity of free boundaries in higher dimensions, *Acta Math.* **139**, 155-184 (1977).

[4] L. A. Caffarelli, Compactness methods in free boundary problems, *Comm. in P.D.E.'s* **5(4)**, 427-448 (1980).

[5] L. A. Caffarelli, A remark on the Hausdorff measure of a free boundary, and the convergence of coincidence sets, *Bollettino U. M. I.* (5) **18**-A, 109-113 (1981).

[6] A. Friedman, *Partial Differential Equations of Parabolic Type*, Prentice-Hall, Englewood Cliffs, NJ, 1964.

[7] A. Friedman, *Variational Principles and Free-Boundary Problems*, John Wiley & Sons, New York, 1982.

[8] A. Friedman, Optimal control for variational inequalities, *SIAM J. Control & Optim.* **24**, 439-451 (1986).

[9] F. Mignot, Contrôle dans les inéquationes variationnelles elliptiques, *J. Functional Analysis* **22**, 130-185 (1976).

[10] F. Mignot and J. P. Puel, Optimal control in some variational inequalities, *SIAM J. Control & Optim.* **22**, 466-476 (1984).

[11] J. P. Puel, Some results on optimal control for unilateral problems, *Control of Partial Differential Equations*, A. Bermudez, Ed., Springer-Verlag, New York, 1989.

[12] D. G. Schaeffer, A stability theorem for the obstacle problem, *Advances in Math.* **16**, 34-47 (1975).

[13] J. Sokolowski and J. P. Zolesio, *Introduction to Shape Optimization*, Springer-Verlag, Berlin, (1992)

[14] J. P. Yvon, *Contrôle optimal de systèmes gouvernés par des inéquationes variationnelles*, Thèse, Paris, 1973.

A Stability Theorem for Linear–Quadratic Parabolic Control Problems

Fredi Tröltzsch Faculty of Mathematics, Technical University of Chemnitz-Zwickau, Chemnitz, Germany

1 INTRODUCTION

In the past years, Sequential Quadratic Programming (SQP) methods have shown to be very useful for solving optimal control problems. In Tröltzsch (1994 a,b) their convergence was proven for the application to simplified nonlinear parabolic control problems. To show similar results for more general classes of parabolic problems, certain stability results for linear-quadratic approximations of the nonlinear problem are needed. This note presents a stability estimate, which seems to be very useful for the convergence analysis of SQP methods.

To have an application in mind, let us regard the following nonlinear control problem:

(NP) Minimize

$$\begin{aligned} f(\theta,u) &= \textstyle\int_\Omega \varphi(x,\theta(T,x))dx + \int\int_Q \Psi(t,x,\theta(t,x))dxdt \\ &\quad + \textstyle\int\int_{S_T} \chi(t,x,\theta(t,x),u(t,x))dSdt \end{aligned} \tag{1.1}$$

subject to the nonlinear initial-boundary value problem

$$\begin{aligned} \theta_t(t,x) &= \Delta\theta(t,x) && \text{on } \Omega \\ \theta(0,x) &= \theta_o(x) && \text{on } \Omega \\ \partial\theta/\partial n &= b(t,x,\theta(t,x),u(t,x)) && \text{on } \Gamma, \end{aligned} \tag{1.2}$$

$t \in [0,T]$, and subject to the constraints on the *control* u,

$$b_1(t,x) \le u(t,x) \le b_2(t,x) \quad \text{on } S_T. \tag{1.3}$$

Here, $\Omega \subset I\!R^n$ is a bounded domain with sufficiently regular boundary S, $S_T = [0,T] \times S$, $Q = [0,T] \times \Omega$, $\theta_o \in C(\bar{\Omega})$, $\varphi = \varphi(x,\theta)$, $\Psi = \Psi(t,x,\theta)$, $\chi = \chi(t,x,\theta,u)$, $b = b(t,x,\theta,u)$ are real-valued functions, twice continuously partially differentiable with respect to θ and u. Moreover, bounded and measurable real-valued functions b_1, b_2 are given on S_T, such that $b_1(t,x) \le b_2(t,x)$ a.e. on S_T.

2 OPTIMALITY CONDITIONS

In the paper, the following notations are used: $\|\cdot\|_{p,M} := \|\cdot\|_{L_p(M)}$, $\|(x,y)\|_{p,M} :=$ $:= \|x\|_{p,M} + \|y\|_{p,M}$, $\|(\theta,u)\|_{V,p} := \|\theta(T)\|_{p,\Omega} + \|\theta\|_{p,Q} + \|\theta\|_{p,S_T} + \|u\|_{p,S_T}$ for $(\theta,u) \in C(\bar{Q}) \times L_\infty(S_T)$. For certain perturbations π we define a norm $\|\pi\|_p$ in section 3. The natural inner product of $L_2(M)$ is written $(\cdot\,;\,\cdot)_M$. Moreover, we introduce the space $W(0,T) = \{\theta \in L_2(0,T;H^1(\Omega)) : \theta_t \in L_2(0,T;H^1(\Omega)^*)\}$ endowed with its natural norm (cf. Lions (1968)). In $W(0,T) \times L_2(S_T)$ we introduce $\|(\theta,u)\|_{W,2}^2 = \|\theta\|_{W(0,T)}^2 + \|u\|_{2,S_T}^2$.

Let a bilinear form a be defined on $H^1(\Omega)^2$ by

$$a(\theta,\eta) = (\nabla\theta\,;\,\nabla\eta)_\Omega.$$

The *Lagrange function* associated to (NP) is

$$\mathcal{L}(\theta,u,y) = f(\theta,u) - \int_0^T [(\theta_t\,;\,y)_\Omega + a(\theta,y)]\,dt + \int_0^T (b(\cdot,\theta,u)\,;\,y)_S\,dt.$$

This function is defined and twice differentiable with respect to (θ,u) in a certain sense. We shall not discuss this and refer to Tröltzsch (1994 a). The linear parts of $\mathcal{L}$ are differentiable as a linear function on $W(0,T)$, while its nonlinear part can be formally differentiated with respect to θ and u in $C(\bar{Q}) \times L_\infty(S_T)$ according to the rules of differentiating Nemytski operators.

Let $\bar{v} = (\bar{\theta},\bar{u})$ denote a fixed locally optimal reference pair for (NP). The standard first order necessary optimality conditions are

$$\mathcal{L}_\theta(\bar{\theta},\bar{u},\bar{y}) = 0 \tag{2.1}$$

$$\mathcal{L}_u(\bar{\theta},\bar{u},\bar{y})(u-\bar{u}) \ge 0 \qquad \forall u \in U^{ad}, \tag{2.2}$$

where $U^{ad} \subset L_\infty(S_T)$ is defined according to (1.3). We shall denote derivatives of φ, Ψ, etc. at the pair $\bar{v}$ by a bar and by according subscripts, for instance $\bar{\varphi}_{\theta\theta}(x) = \frac{\partial^2}{\partial\theta^2}\varphi(x,\bar{\theta}(T,x))$ etc. With this notation, formula (2.1), the *adjoint equation* for $y = \bar{y}$, reads

$$\begin{aligned} -y_t &= \Delta y + \bar{\Psi}_\theta \\ y(T) &= \bar{\varphi}_\theta \\ \partial y/\partial n &= \bar{b}_\theta\, y + \bar{\Psi}_\theta. \end{aligned} \tag{2.3}$$

The derivative $\mathcal{L}_u$ can be identified with a function of $L_\infty(S_T)$, which is denoted by $\mathcal{L}_u(t,x)$. We have

$$\mathcal{L}_u(t,x) = \bar{\chi}_u(t,x) + \bar{b}_u(t,x)\,\bar{y}(t,x).$$

Thus the first order condition (2.2) implies $\bar{u}(t,x) = b_1(t,x)$, if $\mathcal{L}_u(t,x) > 0$, $\bar{u}(t,x) = b_2(t,x)$, if $\mathcal{L}_u(t,x) < 0$. Here, we mention also the form of $f'(\bar{v})v$, namely

$$f'(\bar{v})v = (\bar{\phi}_\theta\,;\,\theta(T))_\Omega + (\bar{\psi}_\theta\,;\,\theta)_Q + (\bar{\chi}_\theta\,;\,\theta)_{S_T} + (\bar{\chi}_u\,;\,u)_{S_T} \tag{2.4}$$

for $v = (\theta, u) \in C(\bar{Q}) \times L_\infty(S_T)$. Moreover, we need the second order derivative $\mathcal{L}_{vv}(\bar{v}, \bar{y})[v_1, v_2] =: \bar{\mathcal{L}}_{vv}[v_1, v_2]$,

$$\bar{\mathcal{L}}_{vv}[v_1, v_2] := f_{vv}(\bar{v})[v_1, v_2] + (\bar{y}\,;\,\bar{b}_{vv}[v_1, v_2])_{S_T}, \tag{2.5}$$

where f_{vv} is defined through the expression

$$\begin{aligned} f_{vv}(\bar{v})[v_1, v_2] &= \textstyle\int_\Omega \bar{\varphi}_{\theta\theta}(x)\theta_1(T,x)\theta_2(T,x)dx + \int\int_Q (\bar{\psi}_{\theta\theta}\theta_1\theta_2)(t,x)dxdt \\ &\quad + \textstyle\int\int_{S_T} [\bar{\chi}_{\theta\theta}\theta_1\theta_2 + \bar{\chi}_{\theta u}(\theta_1 u_2 + \theta_2 u_1) + \bar{\chi}_{uu} u_1 u_2](t,x)dSdt \\ \bar{b}_{vv}[v_1, v_2] &= \bar{b}_{\theta\theta}\theta_1\theta_2 + \bar{b}_{\theta u}(\theta_1 u_2 + \theta_2 u_1) + \bar{b}_{uu} u_1 u_2. \end{aligned}$$

Let us take $\sigma > 0$ arbitrarily small but fixed and define $I_\sigma = \{(t,x) \in S_T : |\mathcal{L}_u(t,x)| > 0\}$. We assume the *sufficient second order condition* (SSC): There is a $\delta > 0$ such that

$$\mathcal{L}_{vv}(\bar{\theta}, \bar{u}, \bar{y})[v, v] \geq \delta \|u\|^2_{2,S_T} \tag{2.6}$$

for all $v = (\theta, u)$ with $u(t,x) = 0$ on I_σ and

$$\begin{aligned} \theta_t &= \Delta\theta \\ \theta(0) &= \theta_o \\ \partial\theta/\partial n &= \bar{b}_\theta \theta + \bar{b}_u u. \end{aligned} \tag{2.7}$$

The linear mapping $u \mapsto \theta$ is continuous from $L_2(S_T)$ to $W(0,T)$, hence (2.6) and (2.7) imply

$$\mathcal{L}_{vv}(\bar{\theta}, \bar{u}, \bar{y}) \geq \tilde{\delta}(\|\theta\|^2_{W(0,T)} + \|u\|^2_{2,S_T}) = \tilde{\delta}\|(\theta, u)\|^2_{W,2} \tag{2.8}$$

with some $\tilde{\delta} > 0$.

From now on we regard the following perturbed *linear-quadratic approximation* of (NP), which is related to the investigation of sequential quadratic programming methods: Let $\pi = (d, e) = (d_\Omega, d_Q, d_S, d_u, e) \in \Pi := C(\bar{\Omega}) \times L_\infty(Q) \times L_\infty(S_T)^3$ be a given perturbation.

(LQ_π) Minimize

$$F(v, d) = f'(\bar{v})(v - \bar{v}) + \frac{1}{2}\bar{\mathcal{L}}_{vv}[v - \bar{v}, v - \bar{v}] - d(v - \bar{v}) \tag{2.9}$$

subject to $v = (\theta, u)$ and

$$\begin{aligned} \theta_t &= \Delta\theta \\ \theta(0) &= \theta_o \\ \partial\theta/\partial n &= b(\bar{\theta}, \bar{u}) + \bar{b}_\theta\,(\theta - \bar{\theta}) + \bar{b}_u\,(u - \bar{u}) + e, \end{aligned} \tag{2.10}$$

$u \in U^{ad}$, and $u(t,x) = \bar{u}(t,x)$ for all $(t,x) \in I_\sigma$ (d, e are given fixed).

The linear functional d in (2.9) is defined by

$$d(v) = (d_\Omega\,;\,\theta(T))_\Omega + (d_Q\,;\,\theta)_Q + (d_S\,;\,\theta)_{S_T} + (d_u\,;\,u)_{S_T}.$$

The derivatives f' and $\mathcal{L}_{vv}$ are introduced formally through the expressions (2.4), (2.5). We do not have to discuss their sense. In view of our assumptions, the derivatives $\bar{b}_\theta$, $\bar{b}_{\theta\theta}$ etc. are bounded and measurable functions. The following result is a standard conclusion of strong convexity due to (SSC):

THEOREM 1 *For all $\pi \in \Pi$ the problem (LQ_π) admits a unique minimizer v_π.*

3 STABILITY THEOREM

First, we show a stability result with respect to the L_2-norm.

THEOREM 2 *Let perturbations $\pi_1,\ \pi_2 \in \Pi$ be given, v_1, v_2 be the associated unique solutions of (LQ_π). There is a positive constant L, which does not depend on $\pi_1,\ \pi_2$ such that*

$$\|v_1 - v_2\|_{W,2} \leq L\,\|\pi_1 - \pi_2\|_2. \tag{3.1}$$

Before proving this theorem, we introduce the Lagrange function associated to (LQ_π) by

$$\begin{aligned} \mathcal{L}(\theta, u, y, \pi) &= F(\theta, u, d) - \int_0^T [(\theta_t\,;\, y)_\Omega + a(\theta, y)]\, dt \\ &\quad + (\bar{b} + \bar{b}_\theta\,(\theta - \bar{\theta}) + \bar{b}_u\,(u - \bar{u}) + e\,;\, y)_{S_T}. \end{aligned} \tag{3.2}$$

From $\mathcal{L}_\theta = 0$ the associated adjoint states $y_1\,, y_2$ are obtained,

$$\begin{aligned} -y_{i,t} &= \Delta y_i + \bar{\Psi}_{\theta\theta} \cdot (\theta_i - \bar{\theta}) \\ y_i(T) &= \bar{\phi}_\theta - d_{i,\Omega} + \bar{\phi}_{\theta\theta} \cdot (\theta_i(T) - \bar{\theta}(T)) \\ \partial y_i / \partial n &= \bar{b}_\theta\, y_i + \bar{\chi}_\theta - d_{i,S} + (\bar{\chi}_{\theta\theta} + \bar{y}\bar{b}_{\theta\theta}) \cdot (\theta_i - \bar{\theta}) \\ &\quad + (\bar{\chi}_{\theta u} + \bar{y}\bar{b}_{\theta u}) \cdot (u_i - \bar{u}), \end{aligned} \tag{3.3}$$

$i = 1, 2$.

Proof of Theorem 2: From the first order necessary condition,

$$\begin{aligned} 0 &\leq \mathcal{L}_v(\theta_i, u_i, y_i, \pi_i)(v - v_i) \\ &= (f'(\bar{v}) - d_i)(v - v_i) + \bar{\mathcal{L}}_{vv}[v_i - \bar{v}, v - v_i] \\ &\quad - \int_0^T \{(\theta - \theta_{i,t}\,;\, y_i)_\Omega + a(\theta - \theta_i, y_i) - (\bar{b}_\theta(\theta - \theta_i) + \bar{b}_u(u - u_i)\,;\, y_i)_S\}dt, \end{aligned} \tag{3.4}$$

$i = 1, 2$. We insert $v = v_2$ in the first and $v = v_1$ in the second variational inequality. After adding them, we arrive at

$$\begin{aligned} 0 &\leq (d_2 - d_1)(v_2 - v_1) + \bar{\mathcal{L}}_{vv}[v_1 - v_2, v_2 - v_1] \\ &\quad - \int_0^T \{(\theta_{2,t} - \theta_{1,t}\,;\, y_1 - y_2)_\Omega + a(y_1 - y_2, \theta_2 - \theta_1) \\ &\quad - (y_1 - y_2\,;\, \bar{b}_\theta(\theta_2 - \theta_1) + \bar{b}_u(u_2 - u_1))_S\}dt \\ &= (d_2 - d_1)(v_2 - v_1) - \bar{\mathcal{L}}_{vv}[v_1 - v_2, v_1 - v_2] - (e_2 - e_1\,;\, y_1 - y_2)_{S_T}. \end{aligned} \tag{3.5}$$

Now $\bar{\mathcal{L}}_{vv}$ is added on both sides of (3.5), and (SSC) applies to obtain

$$\begin{aligned} \delta\|v_1 - v_2\|_{W,2}^2 &\leq c\|\pi_1 - \pi_2\|_2\,\|v_2 - v_1\|_{V,2} + \|e_2 - e_1\|_{2,S_T}\,\|y_1 - y_2\|_{2,S_T} \\ &\leq c\|\pi_1 - \pi_2\|_2\,\|v_1 - v_2\|_{W,2} + \|\pi_1 - \pi_2\|_{2,S_T}\,\|y_1 - y_2\|_{2,S_T}, \end{aligned} \tag{3.6}$$

as $|d(v)| \leq c\|d\|_2\,\|v\|_{V,2}$ and $\|v\|_{V,2} \leq c\,\|v\|_{W,2}$. From (3.3), parabolic L_2-regularity, and the uniform boundedness of $\bar{\phi}_\theta, \bar{\Psi}_{\theta\theta}$ etc. we derive

$$\begin{aligned} \|y_1 - y_2\|_{2,S_T} &\leq c_1\,\|v_1 - v_2\|_{V,2} + c_2\,\|\pi_1 - \pi_2\|_2 \\ &\leq c_1\,\|v_1 - v_2\|_{W,2} + c_2\,\|\pi_1 - \pi_2\|_2, \end{aligned} \tag{3.7}$$

thus (3.6) implies

$$\delta \|v_1 - v_2\|_{W,2}^2 \leq \tilde{c}_1 \|\pi_1 - \pi_2\|_2 \|v_1 - v_2\|_{W,2} + \tilde{c}_2 \|\pi_1 - \pi_2\|_2^2. \tag{3.8}$$

If $\|\pi_1 - \pi_2\|_2 \leq \|v_1 - v_2\|_{W,2}$, then (3.1) follows from (3.8). In the opposite case we have $\|v_1 - v_2\|_{W,2} \leq 1 \cdot \|\pi_1 - \pi_2\|_2$. Therefore, (3.1) is an easy conclusion. □

The main aim of this paper is to derive (3.1) in the L_∞-norm. This can be done combining the full regularity of parabolic operators with a detailed discussion of necessary optimality conditions. In the proof of the next theorem, we shall employ regularity properties in domains with sufficiently smooth boundary S. They are based on the well known *variation of constants formula*

$$\theta(t) = S(t)\theta_o + \int_0^t AS(t-s)N(\bar{b}_\theta\, \theta(s) + \bar{b}_u\, u(s) + b(\bar{\theta}, \bar{u}))ds \tag{3.9}$$

and the estimates

$$\|AS(t)N\|_{L_p(S)\to W_p^\sigma(\Omega)} \leq c\,t^{-(1-(\sigma'-\sigma)/2)} \tag{3.10}$$

$0 < \sigma < \sigma' < 1 + 1/p$,

$$\|S(t)\|_{L_p(\Omega)\to W_p^\sigma(\Omega)} \leq c\,t^{-\sigma/2}, \tag{3.11}$$

where $A = -\Delta + I$, defined on $\{\theta \in W_p^2(\Omega) : \partial\theta/\partial n = 0 \text{ on } S\}$, N is the so-called Neumann operator and $S(t) := e^t\, exp(-At)$ in $L_p(\Omega)$. We refer for the details to Amann (1986). (The equation (2.7) can be transformed by $\theta(t) = e^t w(t)$ to $w_t = \Delta w - w$). Further, we make use of properties of the weakly singular integral operator K,

$$(Kz)(t) = \int_0^t k(t,s)z(s)\, ds, \tag{3.12}$$

where k is a continuous real function on $0 \leq s < t \leq T$, $|k(t,s)| \leq c\,(t-s)^{-\alpha}$, $\alpha \in (0,1)$: K transforms continuously $L_p(0,T)$ into $L_{p'}(0,T)$, if

$$\frac{1}{p'} > \alpha - 1 + \frac{1}{p} \tag{3.13}$$

and $L_p(0,T)$ into $C[0,T]$, if

$$p > \frac{1}{1-\alpha}. \tag{3.14}$$

The following *generalized Legendre-Clebsch condition* is needed to derive the L_∞-estimate: There is a constant $\lambda > 0$ such that

$$\bar{\chi}_{uu}(t,x) + \bar{y}(t,x)\bar{b}_{uu}(t,x) \geq \lambda \quad \forall (t,x) \in S_T. \tag{3.15}$$

THEOREM 3 *Suppose that $\bar{v} = (\bar{\theta}, \bar{u})$ satisfies the first order conditions (2.1), (2.2) together with (SSC) and the generalized Legendre-Clebsch condition (3.15). Suppose further that the boundary S of Ω is sufficiently smooth and $\theta_o \in C(\bar{\Omega})$. Then there are positive constants ε_L, c_L such that*

$$\|v_1 - v_2\|_{C(\bar{Q})\times L_\infty(S_T)} \leq c_L \|\pi_1 - \pi_2\|_\infty \tag{3.16}$$

holds for arbitrary perturbations $\pi_i \in \Pi$ with $\|\pi_i\|_\infty < \varepsilon_L$, $i = 1, 2$.

Proof: a) Preparations

We first mention the variational inequality $\mathcal{L}_u(\theta_i, u_i, y_i, \pi_i)(u - u_i) \geq 0$ for all $u \in U^{ad}$:

$$(\bar{\chi}_u - d_u + y_i \bar{b}_u + (\bar{\chi}_{\theta u} + \bar{y}\bar{b}_{\theta u})(\theta_i - \bar{\theta}) + (\bar{\chi}_{uu} + \bar{y}\bar{b}_{uu})(u_i - \bar{u})\,;\, u - u_i)_{S_T} \geq 0 \qquad (3.17)$$

for all $u \in U^{ad}$. A standard pointwise discussion yields

$$u_i(t,x) = P(t,x)(\bar{u} - \frac{\bar{\chi}_u - d_u + (\bar{\chi}_{\theta u} + \bar{y}\bar{b}_{\theta u})(\theta_i - \bar{\theta}) + \bar{b}_u y_i}{\bar{\chi}_{uu} + \bar{y}\bar{b}_{uu}})(t,x) \qquad (3.18)$$

with the projection operator $P(t,x) : I\!R \to [b_1(t,x), b_2(t,x)]$. Note that P is Lipschitz with constant 1. Moreover, $P(t,x)u(t,x)$ is measurable, if u is measurable.

b) $L_2(S_T)$-estimate $\to$ $L_p(S_T)$-estimate with some $p > 2$:

From (3.1),

$$\|u_1 - u_2\|_{2,S_T} \leq l\, \|\pi_1 - \pi_2\|_2 \,. \qquad (3.19)$$

The difference of states $\theta_1 - \theta_2$ satisfies

$$\begin{aligned} (\theta_1 - \theta_2)_t &= \Delta(\theta_1 - \theta_2) \\ (\theta_1 - \theta_2)(0) &= 0 \\ \frac{\partial(\theta_1 - \theta_2)}{\partial n} - \bar{b}_\theta(\theta_1 - \theta_2) &= \bar{b}_u(u_1 - u_2) + (e_1 - e_2). \end{aligned} \qquad (3.20)$$

By parabolic regularity, right hand sides in the boundary condition contained in $L_2(S_T)$ are transformed continuously into $L_p(0,T; W_2^\sigma(\Omega))$, provided that

$$\frac{1}{p} > \frac{\sigma}{2} - \frac{1}{4} \qquad (3.21)$$

(cf. section 4). To achieve boundary data of $\theta_1 - \theta_2$ in $L_p(S_T)$ we need $W_2^\sigma(\Omega) \subset W_p^{\frac{1}{p}+\varepsilon}(\Omega)$ ($\varepsilon > 0$ arbitrarily small), hence we require according to the Sobolev embedding theorem

$$\sigma - \frac{n}{2} \geq \frac{1}{p} + \varepsilon - \frac{n}{p}\,. \qquad (3.22)$$

From (3.21), (3.22) the chain $2/p + 1/2 > \sigma > 1/p - n/p + n/2$ follows. It is satisfied, if

$$p < \frac{n+1}{n-1}\,. \qquad (3.23)$$

Summarizing up the first step,

$$\|\theta_1 - \theta_2\|_{p,S_T} \leq c\,(\|u_1 - u_2\|_{2,S_T} + \|e_1 - e_2\|_{2,S_T}) \qquad (3.24)$$

is obtained (in what follows, c denotes a generic constant). In the adjoint equation, the term $(\theta_1 - \theta_2)(T)$ must be defined in $L_p(\Omega)$, hence we need $\theta_1 - \theta_2 \in C([0,T], W_2^\sigma(\Omega))$. This holds true for $\sigma < 1/2$ (cf. section 4) and $W_2^\sigma(\Omega) \subset L_p(\Omega)$. The latter takes place, if $\sigma = 1/2 - \varepsilon \geq n/2 - n/p$, i.e.

$$p < 2\frac{n}{n-1}\,. \qquad (3.25)$$

We take $p = p_1 := 2 + 2/(n-1)$. Then (3.23), (3.25) are jointly satisfied, $p_1 > 2$, and

$$\|\theta_1 - \theta_2\|_{C([0,T],L_{p_1}(\Omega))} \le c\,(\|u_1 - u_2\|_{2,S_T} + \|e_1 - e_2\|_{2,S_T}) \tag{3.26}$$

holds. Note that our gain of smoothness is $\gamma = p_1 - 2 = 2/(n-1)$. Next we discuss $y_1 - y_2$. From (3.3),

$$\begin{array}{rcl} -(y_1 - y_2)_t &=& \Delta(y_1 - y_2) - (d_{1,Q} - d_{2,Q}) + \bar{\psi}_{\theta\theta} \cdot (\theta_1 - \theta_2) \\ (y_1 - y_2)(T) &=& -(d_{1,\Omega} - d_{2,\Omega}) + \bar{\phi}_{\theta\theta} \cdot (\theta_1(T) - \theta_2(T)) \\ \frac{\partial(y_1 - y_2)}{\partial n} - \bar{b}_\theta(y_1 - y_2) &=& -(d_{1,S} - d_{2,S}) + (\bar{\chi}_{\theta\theta} + \bar{y}\bar{b}_{\theta\theta})(\theta_1 - \theta_2) \\ && +(\bar{\chi}_{\theta u} + \bar{y}\bar{b}_{\theta u})(u_1 - u_2). \end{array} \tag{3.27}$$

Let us write $y_1 - y_2 = y_Q + y_\Omega + y_S$, where y_Q belongs to the right hand side of the heat equation (3.27) with $y_Q(T) = 0, \partial y_Q/\partial n - \bar{b}_\theta y_Q = 0$, y_Ω solves the homogeneous heat equation in (3.27) with homogeneous boundary data, and y_S is the remaining part (belonging to the inhomogeneous boundary data). Parabolic regularity applies to show

$$\|y_\Omega\|_{L_{p_1}(0,T;W_{p_1}^{\frac{1}{p_1}+\epsilon}(\Omega))} \le c\,(\|\bar{\phi}_{\theta\theta}\|_{\infty,\Omega}\|(\theta_1 - \theta_2)(T)\|_{p_1,\Omega} + \|d_{2,\Omega} - d_{2,\Omega}\|_{p_1,\Omega}). \tag{3.28}$$

In the same way,

$$\|y_Q\|_{L_{p_1}(0,T;W_{p_1}^{\frac{1}{p_1}+\epsilon}(\Omega))} \le c\,(\|\theta_1 - \theta_2\|_{p_1,Q} + \|d_{1,Q} - d_{2,Q}\|_{p_1,Q}) \tag{3.29}$$

$$\begin{array}{rcl} \|y_S\|_{L_{p_1}(0,T;W_{p_1}^{\frac{1}{p_1}+\epsilon}(\Omega))} &\le& c\,(\|\theta_1 - \theta_2\|_{p_1,S_T} + \|u_1 - u_2\|_{2,S_T} \\ && +\|d_{1,S} - d_{2,S}\|_{p_1,S_T}) \end{array} \tag{3.30}$$

holds. (3.29) can be derived from the usual variation of constants formula for parabolic equations with inhomogeneous right hand side. Thus

$$\begin{array}{rcl} \|y_1 - y_2\|_{p_1,S_T} &\le& c\,(\|(\theta_1 - \theta_2)(T)\|_{p_1,\Omega} + \|\theta_1 - \theta_2\|_{p_1,Q} \\ && +\|\theta_1 - \theta_2\|_{p_1,S_T} + \|u_1 - u_2\|_{2,S_T} \\ && +\|\pi_1 - \pi_2\|_{p_1}). \end{array} \tag{3.31}$$

In view of (3.18), (3.15),

$$|(u_1 - u_2)(t,x)| \le \frac{1}{\lambda}(|d_{1,u} - d_{2,u}| + c|\theta_1 - \theta_2| + c|y_1 - y_2|)(t,x) \tag{3.32}$$

holds on S_T, hence

$$\|u_1 - u_2\|_{p_1,S_T} \le c\,(\|\pi_1 - \pi_2\|_{p_1} + \|\theta_1 - \theta_2\|_{p_1,S_T} + \|y_1 - y_2\|_{p_1,S_T}). \tag{3.33}$$

Now

$$\|u_1 - u_2\|_{p_1,S_T} \le c_1\,\|\pi_1 - \pi_2\|_{p_1} \tag{3.34}$$

follows from (3.24), (3.31), (3.26), (3.19), and $\|\cdot\|_{2;M} \le c\|\cdot\|_{p_1,M}$ for any set M.

c) $L_{p_k}(S_T)$-estimate $\to$ $L_{p_{k+1}}(S_T)$-estimate for $p_{k+1} > p_k$

We have shown

$$\|u_1 - u_2\|_{p_k,S_T} \le c_k\,\|\pi_1 - \pi_2\|_{p_k} \tag{3.35}$$

for some $p_k \geq 2+\gamma$. This estimate yields together with the system (3.20)

$$\|\theta_1 - \theta_2\|_{C([0,T],W^{\sigma}_{p_k}(\Omega))} \leq c\,\|\pi_1 - \pi_2\|_{p_k}, \tag{3.36}$$

if

$$\sigma < 1 - \frac{1}{p_k} \tag{3.37}$$

(cf. section 4). To get traces in $L_{p_{k+1}}(S_T)$, where $p_{k+1} > p_k$, we need $W^{\sigma}_{p_k}(\Omega) \subset W^{\frac{1}{p_{k+1}}+\varepsilon}_{p_{k+1}}(\Omega)$, hence $\sigma - n/p_k \geq 1/p_{k+1} + \varepsilon - n/p_{k+1}$, i.e.

$$\sigma > \frac{1-n}{p_{k+1}} + \frac{n}{p_k}. \tag{3.38}$$

Put $p_{k+1} = l\,p_k$. (3.37), (3.38) have a non-void intersection, if

$$p_k > n+1-\frac{n-1}{l}. \tag{3.39}$$

If $p_k \geq n+1$, then this holds for all $l > 0$. In the case $p_k > n+1$ we are able to finish our bootstrapping procedure (cf. d)). Assume $p_k < n+1$. We know $p_k > 2$, hence (3.39) holds true for $l = 1$, and we may satisfy it for some $l > 1$. From (3.39),

$$l < \frac{n-1}{n+1-p_k}, \tag{3.40}$$

hence

$$\begin{aligned} p_{k+1} - p_k &= l\,p_k - p_k < \frac{(n-1)p_k}{n+1-p_k} - p_k \\ &= p_k \frac{p_k - 2}{n+1-p_k}. \end{aligned} \tag{3.41}$$

On the other hand, the right end of this chain is greater than $2\gamma/(n+1-2)$, thus the optimal gain of smoothness would be $p_{k+1} - p_k = 2\gamma/(n-1)$. We may ensure at least the gain

$$p_{k+1} - p_k \geq \frac{\gamma}{n-1}. \tag{3.42}$$

In the case $p_k = n+1$ we may take l and thus p_{k+1} arbitrarily large. Continuing our bootstrapping process, from (3.36), (3.38),

$$\|\theta_1 - \theta_2\|_{C([0,T],W^{\frac{1}{p_{k+1}}+\varepsilon}_{p_{k+1}}(\Omega))} \leq c\,\|\pi_1 - \pi_2\|_{p_k}, \tag{3.43}$$

is found. Next we may improve (3.31) to

$$\begin{aligned} \|y_1 - y_2\|_{p_{k+1},S_T} \leq\ & c\,(\|(\theta_1-\theta_2)(T)\|_{p_{k+1},\Omega} + \|\theta_1 - \theta_2\|_{p_{k+1},Q} \\ & + \|\theta_1 - \theta_2\|_{p_{k+1},S_T} + \|u_1 - u_2\|_{p_k,S_T} \\ & + \|\pi_1 - \pi_2\|_{p_{k+1},S_T}). \end{aligned} \tag{3.44}$$

Continuing as before, in view of (3.15), (3.18), (3.43), and (3.44)

$$\|u_1 - u_2\|_{p_{k+1},S_T} \leq c_{k+1}\,\|\pi_1 - \pi_2\|_{p_{k+1}}. \tag{3.45}$$

is derived.

d) $p_k > n+1$

This case is obtained after finitely many steps. In (3.37), all $\sigma < 1 - \frac{1}{n+1+\varepsilon}$ can be taken. The inequality $\sigma > n/p_k$ can be fulfilled in this case, ensuring $W^{\sigma}_{p_k}(\Omega)) \subset C(\bar{\Omega})$. In view of this, controls from $L_{p_k}(S_T)$ are transformed into the state-space $C([0,T], C(\bar{\Omega}))$. Now (3.36) admits the form

$$\|\theta_1 - \theta_2\|_{\infty,\bar{\Omega}} \leq c\,\|\pi_1 - \pi_2\|_{p_k} \leq c\,\|\pi_1 - \pi_2\|_\infty. \tag{3.46}$$

The estimation of $y_1 - y_2$ is along the lines of the preceding steps. Owing to the maximum principle (proved by means of the integral equation method),

$$\|y_\Omega\|_{C(\bar{Q})} \leq c\,(\|\bar{\phi}_{\theta\theta}\|_{\infty,\bar{\Omega}}\,\|(\theta_1 - \theta_2)(T)\|_{\infty,\bar{\Omega}} + \|d_{2,\Omega} - d_{2,\Omega}\|_{\infty,\bar{\Omega}}) \tag{3.47}$$

follows from (3.27), cf. (3.28). This leads to

$$\begin{aligned}\|y_1 - y_2\|_{\infty,S_T} \;\leq\; & c\,(\|(\theta_1 - \theta_2)(T)\|_{\infty,\bar{\Omega}} + \|\theta_1 - \theta_2\|_{\infty,Q} + \|\theta_1 - \theta_2\|_{\infty,S_T} \\ & +\|u_1 - u_2\|_{p_k,S_T} + \|\pi_1 - \pi_2\|_\infty).\end{aligned} \tag{3.48}$$

In view of (3.15) (3.32), (3.46), (3.48), (3.35)

$$\begin{aligned}\|u_1 - u_2\|_{\infty,S_T} &\leq c\,(\|\pi_1 - \pi_2\|_\infty + \|u_1 - u_2\|_{p_k,S_T}) \\ &\leq c\,(\|\pi_1 - \pi_2\|_\infty + \|\pi_1 - \pi_2\|_{p_k}) \\ &\leq c\,\|\pi_1 - \pi_2\|_\infty\end{aligned} \tag{3.49}$$

holds. (3.49), (3.46) yield the desired result. □

4 REMARKS

It can be shown that the second order condition (SSC) considering the active set of $\bar{u}$ is sufficient for local optimality of $\bar{u}$ for (NP) with respect to the topology of L_∞. This information was not needed to prove Theorem 2. The theorem expresses more or less the following simple fact: (SSC) implies local optimality of $\bar{u}$ for the problem (NP) with constraint $b_1 \leq u \leq b_2$, where b_1 and b_2 are redefined such that $b_1 = b_2 = \bar{u}$ on I_σ.

However, we are able to show more: The (global) solution v_π of (LQ_π) remains *locally* optimal for this problem (local in L_∞-sense), if the restriction $u(t,x) = \bar{u}(t,x)$ is deleted in (LQ_π), provided that $\|\pi\|_\infty$ is sufficiently small. Moreover, it is the only local minimizer in a sufficiently small L_∞-neighborhood of v_π. The radius of this neighborhood is uniform with respect to all sufficiently small perturbations π. Therefore, the main stability theorem 3 is a result concerning a set of local minimizers of (LQ_π) without the restricton $u(t,x) = \bar{u}(t,x)$ on I_σ. The discussion of these facts would go beyond the scope of this paper.

Let us comment some of the estimates used in the proof of Theorem 3. First, we derive (3.21): Apply (3.10) for $p = 2$ and $\sigma = 3/2 - 2\varepsilon$, where ε is arbitrarily small. Then

$$\begin{aligned}\|AS(t)N\|_{L_2(S)\to W_2^\sigma(\Omega)} &\leq c\,t^{-(1-(3/2-2\varepsilon-\sigma)/2)} \\ &= t^{-\alpha},\end{aligned} \tag{4.1}$$

where $\alpha = 1 + \sigma/2 - 3/4 + \varepsilon$. According to the transformation property (3.13) of K, the (Bochner) integral operator

$$u(t) \mapsto \int_0^t AS(t-s)Nu(s)\,ds \tag{4.2}$$

transforms continuously $L_2(0,T;L_2(S))$ into $L_p(0,T;W_2^\sigma(\Omega))$, if

$$\frac{1}{p} > \alpha - 1 + \frac{1}{2} = \frac{\sigma}{2} - \frac{1}{4} + \varepsilon.$$

This inequality is satisfied for sufficiently small ε, if $1/p > \sigma/2 - 1/4$, i.e. (3.21).

Moreover, we comment (3.37). Regard the operator (4.2), defined on $L_{p_k}(0,T;L_{p_k}(S))$. Then (3.10) applies,

$$\|AS(t)N\|_{L_{p_k}(S)\to W^\sigma_{p_k}(\Omega)} \le ct^{-\alpha},$$

where $\alpha = 1 - (1 + 1/p_k - 2\varepsilon - \sigma)/2 = 1/2 - 1/(2p_k) + \sigma/2 + \varepsilon$. The operator (4.2) is continuous from $L_{p_k}(0,T;L_{p_k}(S))$ to $C([0,T],W^\sigma_{p_k}(\Omega))$, if

$$p_k > \frac{1}{1-\alpha}$$

(cf. (3.14)). This amounts to $1 - 1/p_k > \sigma$.

Finally, it should be mentioned that the theory holds true for any uniformly elliptic differential operator A instead of $-\Delta$, which allows to derive the estimates (3.10), (3.11).

REFERENCES

[1] H. Amann (1986). "Parabolic evolution equations with nonlinear boundary conditions", In *Proc. Sympos. Pure Math. 45*, Part I, Nonlinear Functional Analysis, pp. 17–27, (F.E. Browder, ed.).

[2] H. Goldberg and F. Tröltzsch (1993). Second order sufficient optimality conditions for a class of nonlinear parabolic boundary control problems, *SIAM J. Control and Optimization 31*: 1007–1025.

[3] J.L. Lions (1968). Contrôle optimal de systèmes gouvernès par des équations aux dérivées partielles, Dunod, Paris.

[4] F. Tröltzsch (1994). An SQP method for the optimal control of a nonlinear heat equation, *Control and Cybernetics 23*: 267–288.

[5] F. Tröltzsch. "Convergence of an SQP-method for a class of nonlinear parabolic boundary control problems", In *Control and Estimation of Distributed Parameter Systems: Nonlinear Phenomena*, ISNM Vol. 118, pp. 343–358 (W. Desch, F. Kappel, K. Kunisch, eds.), Birkhäuser , Basel 1994.

ARTS AND CRAFTS IN LATE IMPERIAL RUSSIA

MODERN ARCHITECTURE AND CULTURAL IDENTITY

Modern Architecture and Cultural Identity will comprise monographic studies of important movements and buildings by European and American architects created roughly between 1850 and 1950. Unlike the first histories of modernism, which stressed the international aspects of modern architecture, recent scholarship has attempted to clarify the delicate balance achieved by architects working in a modernist idiom who maintained, nonetheless, a strong allegiance to their cultural roots. This series has been developed in response to this trend and will explore the complex interplay between modern identity and local, regional, national, and related cultural traditions.

ARTS AND CRAFTS IN LATE IMPERIAL RUSSIA

Reviving the Kustar Art Industries, 1870–1917

WENDY R. SALMOND
Chapman University

CAMBRIDGE UNIVERSITY PRESS

In memory of my parents,
George and Dorothy Salmond

Published by the Press Syndicate of the University of Cambridge
The Pitt Building, Trumpington Street, Cambridge CB2 1RP
40 West 20th Street, New York, NY 10011-4211, USA
10 Stamford Road, Oakleigh, Melbourne 3166, Australia

First published 1996

Printed in the United States of America

Library of Congress Cataloging-in-Publication Data

Salmond, Wendy R.

Arts and crafts in late Imperial Russia (Modern architecture and cultural identity) / Wendy R. Salmond.
p. cm.
Includes bibliographical references.
ISBN (invalid) 0-521-41576-4 (hc)
1. Arts and crafts movement – Russia (Federation) – History – 19th century. 2. Arts and crafts – Russia (Federation) – History – 20th century. I. Title.
NK975.S26 1996
745'.0947'09034 – dc20 95-46512
CIP

A catalog record for this book is available from the British Library

ISBN 0-521-41576-4 hardback

Contents

List of Illustrations viii
Acknowledgments xi
Map of Workshop Locations xv

Introduction 1

1 The Abramtsevo Workshops 15

2 The Solomenko Workshops 46

3 The Moscow Kustar Museum and the Sergiev Posad Workshops 80

4 Princess Maria Tenisheva and the Talashkino Workshops 115

5 Between Two Revolutions: The Kustar Art Industries, 1905–1917 144

Epilogue 187

Appendix **A**: Chronology of Events 191

Appendix **B**: Genealogical Chart 211

Notes 213
Selected Bibliography 249

List of Illustrations

Map of workshop locations xiv

1. Ornamental design, by Elena Polenova, ca. 1897–8 2
2. *Blütenbaum* (Blossoming Tree), by Hermann Obrist, ca. 1896 3
3. Birch dining room suite, by P. I. Subbotin, 1913 4
4. White salon, by Charles Rennie Mackintosh, 1902–3 5
5. Family of kustari weaving rush baskets, ca. 1913 18
6. Studio at Abramtsevo, by Viktor Gartman, 1873 23
7. Church at Abramtsevo, by Viktor Vasnetsov, 1881–2 24
8. Carved backboard of a peasant cart, late nineteenth century 25
9. Page from *Histoire de l'ornement russe,* compiled by V. I. Butovskii, 1870–3 29
10. Drawing of an izba in Kostroma province, by V. Dal', early 1870s 30
11. Design for a cupboard, by Elena Polenova, ca. 1885 32
12. Design for a column cupboard, by Elena Polenova, ca. 1885 33
13. "Cat and Owl" door, by Elena Polenova, early 1890s 40
14. Cupboard made at the Sergiev Posad art joinery workshop, ca. 1907 44
15. *Point de Moscou* lace, 1880s 50
16. Naboika patterns, two traditional and one designed by Natalia Ia. Davydova, ca. 1900 56
17. Interior of the Pavilion of the Far North, by Konstantin Korovin, 1896 58
18. *Ivan Tsarevich and the Fire Bird,* by Elena Polenova, 1896 60
19. *The Beast,* by Elena Polenova, 1896 63

20. Embroidered purses, by Elena Polenova, 1897–8 64
21. Embroidered wall hanging, by Natalia Ia. Davydova, late 1890s 65
22. Sketches for a dining room in the Russian style, by Elena Polenova, 1897–8 67
23. Sketch for a dining room in the Russian style, by Elena Polenova, 1897–8 68
24. Cover to *Mir iskusstva,* by Maria V. Iakunchikova, 1899 73
25. *The Little Girl and the Wood Sprite,* by Maria V. Iakunchikova, 1900 75
26. Carved kustar toys made at Sergiev Posad, early nineteenth century 85
27. Matreshka doll, by Sergei Maliutin, 1891 86
28. Carved goods made by Vasilii Vornoskov and family, ca. 1913 89
29. *L'izba des koustary* at the 1900 Paris Exposition Universelle, by Konstantin Korovin 91
30. Interior of *l'izba des koustary,* 1900 92
31. View of the First All-Russian Kustar Exhibition, St. Petersburg, 1902 94
32. Maple dining room suite, by Ivan Fomin, 1902 98
33. Carved shelf, by Natalia Ia. Davydova, 1902 99
34. "Merchant's Wife" money box, by Nikolai Bartram, ca. 1904–14 107
35. Carved sideboard, by Suvorov, 1913 109
36. Dining room in the Firsanova house, by A. G. Iakimchenko 110
37. Dining room in the Pertsov house, by Sergei Maliutin, 1905–7 111
38. Moscow Kustar Museum 112
39. Teremok at Talashkino, by Sergei Maliutin, ca. 1903 121
40. Armchair, by Sergei Maliutin, ca. 1903 123
41. Sofa with embroidered upholstery, by Princess Maria Tenisheva, ca. 1904 125
42. Furniture covered with art embroideries, by Princess Tenisheva and Aleksei Zinov'ev, ca. 1904–5 130
43. Two carved chairs, by Aleksei Zinov'ev, ca. 1904 132
44. Carved book cupboard, by Nikolai Roerich, ca. 1904 137
45. Embroideries made at the School of Folk Art, St. Petersburg, ca. 1912 151
46. Cushions made from naboika fabrics printed by Egor Maleev, 1913 153

ARTS AND CRAFTS IN LATE IMPERIAL RUSSIA

47. Oak furniture suite, by A. N. Durnovo, ca. 1913 155
48. Poltava Zemstvo House, by V. S. Krichevskii, 1907 156
49. Sideboard, by V. I. Cherchenko, ca. 1913 157
50. Embroideries from Verbovka and furniture from Sunki, ca. 1913 160
51. Women's cloaks, made at Skoptsy and Sunki, ca. 1913 161
52. Abramtsevo furniture and Solomenko carpets in the Empire style, ca. 1913 175
53. Winning entry for fashion designs in a Russian style, 1916 179
54. Three embroideries, by Ksenia Boguslavskaia, 1916 181
55. Two designs for handbags, by Nadezhda Udal'tsova and Olga Rozanova, 1916–17 182

Acknowledgments

I WOULD LIKE to thank the following colleagues, friends, and institutions for the contribution each has made to the research and writing of this book. For the formative roles they played in my early graduate career I owe a profound debt of gratitude to Peter Stupples of Otago University, New Zealand, who first introduced me to Abramtsevo and Talashkino in his pioneering seminar on modern Russian art in 1976 and who encouraged me to explore the applications of social and cultural history to Russian art; to John Clarke, Linda D. Henderson, and Leslie O'Bell of the University of Texas at Austin, who provided me with a model of scholarship coupled with humanity and integrity; and to Dr. John E. Bowlt, a generous and supportive mentor who has done so much to shape my understanding of Russian modernism.

In my efforts to reconstruct the forgotten lives and personalities of key members of the kustar art revival I was enormously helped and encouraged by their descendents. During the fall 1981 semester in Moscow, the many kindnesses I received from Elena ("Elka") Polenova and her family continued a tradition of hospitality deeply rooted in the Mamontov and Polenov family circles of the late nineteenth century. Four years later, during a summer research trip to France and Switzerland, I was welcomed with the same warmth and generosity by members of the Iakunchikova-Weber and Polenov families, including Marina Kelepovskaia, the late Irène Weber, Paco, Marielle, and Maribelle Herrera, and most especially Alexandre and Nicole Liapin. Finally, a visit to the late Irina Tamara at Princeton in 1986 provided invaluable insights into the life of her great-aunt, Maria F. Iakunchikova.

Among the colleagues and friends who have supported this project with advice and resources I want to thank William C. Brumfield, Sarah Burke, Linda Docherty, Nicola Gordon Bowe, Alison Hilton, Kirill Katkov, Karen Kettering, Ilya Levin, Nicoletta Misler, John Norman, Anne and Bill Odom, Sarah Sherrill, Galina Smorodina, Isabel Tirado, and Richard and Susan Wright. I am especially indebted to Dr. Mikhail Kiselev for his unstinting support and for his incomparable love and knowledge of the Russian fin-de-siècle. I want also to thank Adrienne Alexander for her skill, generosity, and patience in preparing the manuscript; Denise Weyhrich for bringing her graphic designer's expertise to bear on the map and the genealogical chart in Appendix B; and Gina Wilkinson of the Chapman University Library for locating a number of obscure items. Finally, I owe a special debt of gratitude to Beatrice Rehl at Cambridge University Press for her sustained commitment to this project over five years of revisions, and to Dr. Richard Etlin for agreeing to include my book in the Modern Architecture and Cultural Identity series and for his excellent editorial contributions. Finally, I want to thank Diana Gillooly for the many ways in which she improved the manuscript in its final stages.

The early stages of my dissertation research were supported by a University Fellowship, two Summer Grants, and an Art Historian Travel Grant from the University of Texas at Austin. Chapman University supported my revision of the manuscript with three Summer Research Grants, and the University of Illinois Russian Research Center granted me a Summer Fellowship. In the course of my research I have been able to work in a number of outstanding libraries, including the Library of Congress, the New York Public Library, the Library of the University of Illinois at Champaign-Urbana, the Lenin Library in Moscow, Hillwood Museum Library, and the Institute of Modern Russian Culture. I thank the staff of each of these institutions for the help they have given me.

Preliminary versions of Chapters 2 and 3 have appeared as articles: "The Solomenko Embroidery Workshops" in the *Journal of Decorative and Propaganda Arts* and "Reviving Folk Art in Russia: The Moscow Zemstvo and the Kustar Art Industries" in *Art and the National Dream* and in Italian translation in *Ricerche di Storia dell'Arte*. I would like to thank Dr. John E. Bowlt, Dr. Nicoletta Misler, and Dr. Nicola Gordon Bowe for providing me with the opportunity to publish these early results of my research.

Unless otherwise noted, all translations from French, German, and Russian sources are my own. The Library of Congress system has been followed in the transliteration of Russian words, with the exception of names with generally recognized spellings in English, such as Benois, Exter, and Roerich.

All dates in this book are rendered according to the Julian calendar, used in Russia between 1699 and 1918, which was twelve days behind the Western calendar in the nineteenth century and thirteen days behind in the twentieth.

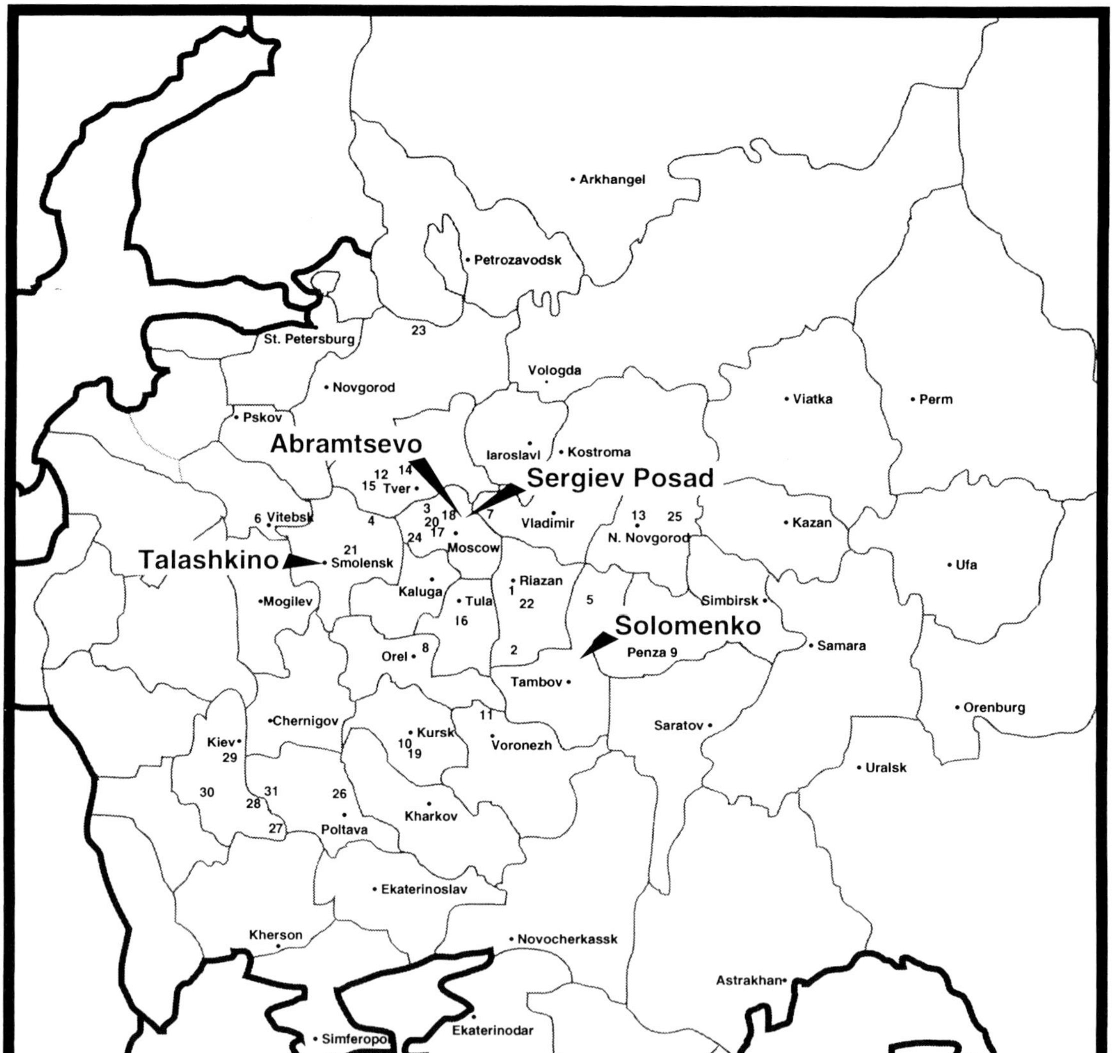

Arkhangel
Petrozavodsk
St. Petersburg
23
Novgorod
Vologda
Pskov
Viatka
Perm
Abramtsevo
Iaroslavl
Kostroma
12
14
15
Tver
Sergiev Posad
3
18
7
6
Vitebsk
4
20
17
Vladimir
13
25
N. Novgorod
Kazan
24
21
Moscow
Talashkino
Smolensk
Ufa
Kaluga
Riazan
1
22
5
Mogilev
Tula
16
Simbirsk
Solomenko
Orel
8
2
Penza 9
Samara
Tambov
Chernigov
11
Orenburg
Kiev
29
Kursk
10
19
Voronezh
Saratov
Uralsk
30
28
31
27
26
Poltava
Kharkov
Ekaterinoslav
Kherson
Novocherkassk
Astrakhan
Simferopol
Ekaterinodar
Stavropol

Map of Workshop Locations

The map opposite shows the approximate location of thirty-five of the most prominent kustar workshops active throughout European Russia, ca. 1876–1917. The workshops are listed in chronological order of their date of origin.

Abramtsevo carpentry workshop founded by E. G. Mamontova, Dmitrovskii uezd, Moscow province (f. 1876)

Solomenko embroidery workshop founded by M. F. Iakunchikova, Morshchanskii uezd, Tambov province (f. 1891)

Sergiev Posad toy and carpentry workshops operated by the Moscow provincial zemstvo (f. 1890)

Talashkino workshops founded by Princess M. K. Tenisheva, Smolensk province (f. 1900)

1. Vysokoe embroidery workshop founded by S. P. Kaznachaeva, Mikhailovskii uezd, Riazan province
2. Muraevnia embroidery workshop founded by Princesses M. N. Shakhovskaia and S. P. Dolgoruka, Dankovskii uezd, Riazan province
3. Spasskoe embroidery workshop founded by Princess A. V. L'vova, Klinskii uezd, Moscow province
4. Embroidery workshop founded by Princess M. A. Urusova, Sychevskii uezd, Smolensk province
5. Bykovo-Gora lace and embroidery workshop founded by Princess A. N. Naryshkina, Chernigovskii uezd, Tambov province
6. Weaving and carpet workshops founded by Countess A. A. Moll' near Vilnius, Vitebsk province

7. Bogorodskoe toy-carving workshop, Aleksandrovskii uezd, Vladimir province
8. Lace workshops organized by Princess A. D. Tenisheva, Mtsenskii uezd, Orel province
9. Vazerki embroidery workshop founded by Princess M. A. Shakhovskaia, Mokshanskii uezd, Penza province
10. Semenovka wood-carving workshop founded by N. D. Bartram, L'govskii uezd, Kursk province
11. Gorozhanka lace and embroidery workshop founded by E. Chokolova, Zadonskii uezd, Voronezh province
12. Novyi Torzhok gold-embroidery school-workshop founded by D. D. Romanov, Novotorzhskii uezd, Tver province
13. Semenov and Khokhloma painted woodware production, Semenovskii uezd, Nizhnii Novgorod province
14. Vyshnii Volochok School for Instructors in Spinning and Weaving founded by M.Z.i G.I., Tver province
15. Sosnovitsy School for Instructors in Spinning and Weaving founded by M.Z.i G.I., Novotorzhskii uezd, Tver province
16. Circle for Work Assistance founded by Countess S. A. Bobrinskaia at Bogoroditska, Tula province
17. Petrovskii Circle for Work Assistance, founded by Princess L. V. Golitsyna on the Golitsyn estate, Zvenigorodskii uezd, Moscow province
18. Kudrino woodworking workshops founded by the Vornoskov family, Moscow province
19. Carpet workshops in Kostora, Sudzha, and Miropol founded by the Sudzha district zemstvo, Kursk province
20. Ligachev School for Instructors in Art Joinery, Ligachevskii uezd, Moscow province
21. Aleksino embroidery production organized by Kh. S. Baryshnikova, Dorogobuzhskii uezd, Smolensk province
22. Borisovka, Ekaterinskii, and Andreevskii lace schools, Skopinskii uezd, Riazan province
23. Borovenets naboika school founded by Iu. P. Diagilev, Valdaiskii uezd, Novgorod province
24. Egor Maleev's naboika workshop, Kriukovo, Vereiskii uezd, Moscow province
25. Lyskovo furniture workshop founded by the Nizhnii Novgorod provincial zemstvo, Nizhnii Novgorod province
26. Degtiar School for Weaving Instructors founded by the Poltava provincial zemstvo, Poltava province

27. Verbovka embroidery workshop founded by N. M. Davydova, Chernigovskii uezd, Kiev province
28. Sunki embroidery workshop founded by Princess N. Iashvil, near Smela, Cherkasskii uezd, Kiev province
29. Elenovka weaving and carpet workshops founded by Varvara Khanenko, Vasil'kovskii uezd, Kiev province
30. Zozovo lace and embroidery workshop founded by Iu. N. Gudim-Levkovich, Kipovetskii uezd, Kiev province
31. Skoptsy embroidery and carpet workshops founded by E. Semigradova and E. Pribyl'skaia, Pereiaslavskii uezd, Poltava province

Introduction

ENSURING THE SURVIVAL of native traditions and national identity in the face of modernization is a challenge that every developed or developing country has faced in the past century and a half. In Imperial Russia, during the last decades of its existence (ca. 1861–1917), this collision of tradition and modernity was accompanied by an unprecedented revival and reassessment of native Russian traditions – most notably the music of the Mighty Five, the social realism of the Wanderers, and the exotic repertoire of the Ballets Russes. Over the years, however, one thread has gradually been dropped from the colorful fabric of cultural expression by which we in the West have come to recognize the face of Russia.[1] That thread is the kustar art revival, a movement that attempted to rescue, revive, and sometimes even reinvent the arts of the humble Russian *kustar,* or peasant handicraftsman, by adapting them to the needs of modern urban life and the laws of a new market economy.

In the space of fifty years the kustar revival achieved extraordinary results. Through the intervention of professional artists in the lives of thousands of peasant men and women, it radically altered the production of kustar goods for market. It brought about a change in taste and consumption that saw the demand for artistic peasant-made goods (laces, embroideries, toys, carved furniture, and bric-a-brac) pass from the peasantry themselves to an elite moneyed class. In a country where the profound gulf separating the intelligentsia and the common people was a well-established fact of the national psyche, it harnessed representatives of these two groups together in a collaborative experiment that mirrored the tensions of a country undergoing the pains of modernization. Finally, it helped to shape and market an image of Russia that the Western world holds dear to this day. This book is the first

Figure 1. Ornamental design, by Elena Polenova, 1897–8 (Reprinted from *Mir iskusstva,* 18–19 [1899])

account of how this forgotten cultural revolution took place, of the groups and individuals who engineered it, and of the consequences that their intervention in the lives of the kustar population were to have for the survival and perception of "folk art" in twentieth-century Russia.

Translated into the familiar terminology of European modernism, the kustar art revival could well be described as a Russian Arts and Crafts movement, Imperial Russia's answer to the vernacular revivals that emerged throughout Western Europe, England, and the United States as antidotes to the growing internationalization and mechanization of late-nineteenth-century life. In Moscow no less than in London, Dublin, Warsaw, Vienna, or Oslo, the ideas of John Ruskin and William Morris found enthusiastic supporters.[2] Like their contemporaries in nations farther west, educated Russians showed a strong faith in social reform through art, a commitment to healing the breach between the high art of painting and the minor decorative arts and crafts, and a concern for maintaining a distinct cultural identity in an era when national traditions were falling victim to the increasing homogeneity of the machine age. Stylistically, too, the kustar revival proclaimed its kinship with the *style moderne* – the Russian term for Art Nouveau in all of its national variants. Embroideries produced by peasant women in the kustar workshops of Solomenko bore distinct affinities to Hermann Obrist's contemporaneous experiments with floral abstractions (Figs. 1–2), just as furniture made in the Ligachev School for Instructors in Art Joinery in Moscow province echoed

Figure 2. *Blütenbaum* (Blossoming Tree), by Hermann Obrist, ca. 1896. Yellow, russet, and light green silk embroidery on dark brown silk, 321 cm × 2100 cm. (Courtesy of Die Neue Sammlung, Staatliches Museum für angewandte Kunst, Munich)

Charles Rennie Mackintosh's famous high-backed chair, seen in Russia for the first time in 1902 (Figs. 3–4). However much critics deplored the artifice of crossbreeding Paris with Novgorod, Viatka, or Kazan, the resulting kustar *style russe moderne* was immediately recognized abroad as Russia's "new decorative movement" and became for a brief but influential moment its most potent symbol of tradition reconciled to progress, of vernacular Russian forms integrated at last with the dominant Western style.

On the basis of these broad similarities alone, it will be readily apparent that a history of the kustar art revival fills a substantial gap in our expanding understanding of Western culture's confrontation with modernity, framing the dilemmas of educated Russia within a common Western experience. My purpose in writing this book was not primarily to establish common ground between Russia and the rest of the industrialized world, however. I wanted instead to understand how Russia's own particular brand of Arts and Crafts movement was shaped by a unique amalgam of artistic, social, political, and economic factors that constituted Russian life during "the twilight of the tsars." If in France Art Nouveau expressed the search for a lost golden age when French craftsmen led the world as arbiters of taste, craftsmanship, and originality, the *style russe moderne* that emerged from the kustar workshops of Abramtsevo, Solomenko, and Talashkino reminded modern Russians of a national past that had been buried for nearly two centuries beneath an alien Western culture and kept alive only in the arts and crafts of the peasantry.[3] Whereas in Germany the promotion of the applied arts took place in a spirit of harmony with industry and the new machine culture, the handmade crudeness of kustar goods ostentatiously proclaimed Russia's "separate path," the

Figure 3. Birch dining room suite designed by P. I. Subbotin and made at the Ligachev School for Instructors in Art Joinery, 1913. (Reprinted from *Russkoe nardnoe iskusstvo* [Petrograd, 1914])

populist notion that Russia alone of the European nations might bypass the stage of industrial capitalism entirely.[4] And unlike the Poles and the Irish, for whom the renaissance of handicrafts was directly associated with the struggle for political autonomy and home rule, Russians saw in the revival of peasant crafts both a way to maintain a precious status quo in the face of disruptive social change and a tool for promoting apparent concord within an empire of disparate and discontented nationalities.[5] Economically backward yet with aspirations to join the family of Western nations as a major power, torn in its cultural allegiances between East and West, juggling the nostalgic pull of tradition with the demands of modernity, Russia revealed its Janus-headed nature with poignant clarity in the history of the kustar art revival.

In the second half of the nineteenth century, when this narrative begins, Russia was widely perceived to be one of the last medieval countries in Europe. An agricultural country where peasants constituted the overwhelming majority of the population, Russia was also the last nation in Europe to abolish serfdom, with the Emancipation Act of

Figure 4. White salon, by Charles Rennie Mackintosh, shown at the Exhibition of Architecture and Industrial Art in the New Style in Moscow, 1902–3. (Reprinted from *Mir iskusstva,* 3 [1903])

February 1861. Since the reforms of Peter the Great in the early eighteenth century, Russia's tiny educated elite had thoroughly assimilated Western ways. In the words of the painter Viktor Vasnetsov, "We live in houses built on European models – at least we aspire to, we wear French fashions, we eat either like the French or the English. Our entire environment – dishes, plate, furniture – is foreign, with not even a corner left for independent national creativity."[6] An inevitable side effect of this wholesale Europeanization was a widespread contempt for all kinds of domestic production, but especially for the crafts and customs of the peasantry. Brought face to face with a display of peasant-made lace, complained critic Vladimir Stasov in 1870, the Russian public would be likely to "swear that there's nothing in it but clumsy and savage chinoiserie, until those whom they consider 'the masters' who set the tone announce and demonstrate by their example that these products are good and that one can admire them without damaging one's reputation."[7]

The industrial revolution came late to Russia, but with the disappearance of free serf labor in 1861 it began to make rapid progress, ushering in the decline of the landed gentry and the emergence of a merchant class that owed its wealth to mechanized textile production and the expansion of the railroad. As communications improved and consumption of factory goods increased, Russia experienced all of the changes that its more advanced neighbors had undergone decades before: increased mobility between city and country, a sharp rise in the number of peasants engaged in seasonal work and factory labor, and rapid changes in the cultural life of the rural population. With the coming of the Nikol'skaia Railway Line to the town of Torzhok in the 1870s, for instance, "not only has the famous leather industry declined, but the distinctive dialect, manner and customs of the Novo-Torjians . . . are things of the past. Its picturesque costumes are only found in books, and what is more serious . . . its rosy-cheeked, bright-eyed maidens have disappeared."[8] In other villages, concerned observers reported, "The old materials, patterns, colors, even the very working methods are giving way to cheap rubbish bought in city stalls. Bad German designs are showing up in the countryside through the intervention of middle-class women from the towns, aided by cheap illustrated publications."[9]

Moreover, it was becoming abundantly clear to all those familiar with rural life that the peasantry themselves were learning to despise their own traditional culture and, for reasons of upward mobility as much as economic necessity, were eagerly embracing the new factory products and city dress. Only the educated classes seemed to appreciate and mourn what was being irrevocably lost in the process. If the peasant, as the last tacit symbol and keeper of national identity, gave up the old ways without a fight, then how would modern Russia distinguish itself culturally from the rest of the Western world, especially since the international marketplace increasingly demanded that manufactured goods be distinguished by some mark of national distinctiveness? What was to become of national traditions in such a climate? Should they be left to their fate and allowed to "wither away" in a Darwinian survival of the fittest? Or could they be protected from the forces of change and with proper husbanding adapted to the new demands of modern life?

Educated Russians had begun to study and collect examples of what they considered pure or classic folk art (*narodnoe iskusstvo*) quite early in the nineteenth century, in a spirit of romantic nationalism fostered by the Napoleonic invasion and defeat of 1812. As in other countries struggling to construct a tangible national identity at this period, it was above all to ornament that Russian architects turned for the seminal ingredients of a revived "Old Russian style" in architecture and the applied arts. Motifs derived from the daily life of the peasant – from wood carving, embroideries, and laces – became especially important elements of this new ornamental language, because through them a direct and unbroken line could be traced that linked Russians on the brink of modernity back to their pre-Western roots.

With the coming of the industrial revolution to Russia around mid-century, however, complications began to appear in this homogeneous concept of folk art. While collectors, scholars, historians, and artists still clung to the romantic ideal of peasants ornamenting the objects of their everyday life for their own use and pleasure in accordance with centuries-old tradition, a very different aspect of peasant artistic production was beginning to supplant it, in the form of the kustar crafts (*kustarnye promysly*). These comprised all those "cottage" or "home" industries by which agricultural peasants supplemented their income during the long winter months. The majority of peasant kustari scraped out a living by producing rudimentary tools and merchandise (everything from barrels, nails, and knives to plows, window frames, and rope) for sale to a rural market.

But some kustar crafts – the production of lace, icons, toys, block-printed fabric (*naboika*), and carved wooden goods – retained an intrinsic decorative component that clearly revealed their origins in the classic folk art of the preindustrial era. By the 1870s connoisseurs of folk culture were coming to realize that the last practitioners of many folk crafts were in fact kustari eking out a bare subsistence from what was left of the skills of their forebears. While the majority of peasants might no longer cling voluntarily to the old ways (i.e., no longer produced "folk art" for their own use), around 1870 it began to occur to a number of landowners and philanthropists that the best artistic traditions of a rapidly disappearing past might be kept alive in the form of carefully controlled kustar production tailored to fit the needs and taste of a more affluent, urban clientele.

It is no coincidence that the first private workshops for retraining kustar men and women in these declining crafts were set up in the 1870s, when the ideals of populism (*narodnichestvo*) brought educated Russian youth in droves to the countryside to spread reform and revolution among the peasantry and when utopian communities were established to unite peasant and intellectual in emulation of the village commune. Unlike either the phalansteries of the Petrashevskii circle or the "To the People" movement of 1874, however, kustar training workshops had the intrinsic advantage of being founded on an economic base that augured well for their survival. The mechanism that grew out of these private initiatives thus hinged on the fact that kustar production contained the commercial incentives that alone could persuade kustari to resist their own desire for change in favor of a steady income at home in the village. Not unreasonably, better pay was the best incentive to convince a peasant woman to supplement her income by making lace rather than cutting peat, or to make a youth give up a waiter's job in the city for a carpenter's bench in the provinces.

As these first isolated instances evolved into a national movement of considerable size, a distinct shift in the evolution of the kustar industries occurred. While the more utilitarian crafts (the production of nails, barrels, sacking) fell victim to mechanization and competition from factory goods, the more "artistic" not only survived, but quickly developed into objects of "conspicuous consumption" by virtue of their aesthetic, handmade, and national qualities. This evolutionary jump resulted entirely from a reform movement engineered from above, as educated Russians attempted to shape the tide and rate of social change through the controlled cross-fertilization of folk art and modern taste.

The first chapter of this book opens at Abramtsevo, the country estate of the Mamontov family to the northeast of Moscow. It was here in 1885 that the artist Elena Polenova volunteered her services as artistic director of a small carpentry workshop for local peasant boys, established to keep rural youth from drifting to the city in search of jobs. During her eight years with the workshop, Polenova created a viable arts and crafts industry that was rooted in the folk art of wood carving but attuned to the aesthetic tastes of the urban middle class. The "Abramtsevo experiment" was the first successful organizational model for how "folk art" (made for personal use) might successfully metamorphose into "kustar art" (made for sale to an urban consumer). Whereas in England John Ruskin had believed that the language of Gothic art could be laboriously relearned by piecing together elements of ornament, Polenova ultimately came to the conclusion that the roots of a new Russian grammar of ornament lay within the individual artist and that it was possible to relearn, not simply the dead external forms and syntax of peasant art, but the creative process itself. Because folk traditions lived on in the Russian countryside, albeit in a form that many observers thought debased, the process of collaboration or intervention that was introduced at Abramtsevo was considered to be an entirely organic and natural process. What began as a form of practical populism – keeping peasants on the land and preserving the social status quo – was gradually transformed into an aesthetic concern, as Polenova's own desire for satisfying creative self-expression led her to abandon traditional folk motifs for new ones of her own invention. The Abramtsevo kustar workshop thus records a vital transitional moment in the development of a modern consciousness among Russian artists, as the social concerns and collective goals of the 1860s gave way to an increasingly individualistic and introspective approach to the artist's craft.

Chapter 2 charts this rapid transition from "good works" to "works of art" in the context of women's work (*zhenskie promysly*) – the embroidery, lace, and weaving workshops presided over by a large number of "lady patronesses" and women artists throughout the provinces of the Russian Empire. This division by gender was quite natural for kustar production, as it was for the patriarchal order of peasant life itself. But it was also no accident that women from the noble and merchant classes figured so prominently in the kustar cause, since the arenas of charitable good works and the decorative arts were both considered acceptable spheres of activity for ladies. The model kustar workshop thus became a place where women could play active artistic and social roles, particularly women artists whose cultivated taste played a key role in adapting traditional peasant crafts for luxury consumption.[10]

Of the dozens of women's kustar workshops in operation at the turn of the century, the Solomenko embroidery workshop, founded during the famine of 1891, reflects with particular clarity the unforeseen consequences that the intervention of professional artists from schools such as the Stroganov School of Technical Drawing in Moscow. This chapter provides a panoramic overview of the diversity that characterized the kustar schools and workshops of the prewar decade. It culminates in a detailed discussion of the Second All-Russian Kustar Exhibition, held in 1913 to coincide with the tercentenary celebrations of the Romanov dynasty, and thus a visible symbol of Imperial Russia on the brink of sweeping change.

This systematic promotion of kustar art by official agencies in the prewar era offers vivid insight into how Russia, like all Western nations, projected its image abroad both at the "international examinations" (Vladimir Stasov's term) that were the world's fairs and in the export of kustar crafts and souvenirs. Of crucial importance in this regard was the receptivity of the Western consumer to the aura of intangible values that the kustar-produced toy, embroidery, or carved whatnot was purported to possess. To those responsible for meeting the growing demand for kustar luxury goods abroad it was clear that the real value of a toy from Sergiev Posad, a hanging cupboard from Abramtsevo, or an embroidered runner from Solomenko lay not in any intrinsic quality of workmanship or materials but in the emotional resonances conjured up by the concept of the Russian peasant. While American customers thus came to the kustar objet d'art fresh from reading the anarchist writings of Prince Peter Kropotkin or seeing a stage production of Tolstoi's *Resurrection,* and English critics made envious comparisons between the Russian kustar and the English artisan, for their part the various kustar export organizations deliberately cultivated lavish ornament and technical crudeness as tokens of kustar art's authenticity as folk art. By 1913, however, luxury kustar goods had priced themselves beyond the reach of all but the well-to-do. As one observer commented of kustar textiles, "It is mostly the educated classes who can understand and appreciate their peculiar beauty, and the peasants themselves cannot have them at their disposal."[13] Kustar art had effectively taken the place of folk art for the majority of educated Russians and Westerners, leaving only a frustrated minority to mourn the substitution of ersatz for authentic.

What can we learn from this story of survival through accommodation, of vernacular traditions cut to fit the demands of modern urban society? First, it offers a new perspective on the complex relationship between educated and peasant Russia at a period of maximum stress and change. The kustar workshop was frequently a place where mistrust and miscommunication marked the relations between helper and helped, bridged only by guaranteed economic incentives. Cooperation had to be bought if peasants were to contemplate economically risky innovations or turn their backs on new job opportunities in order to return to the hand loom, the lace pillow, and the hatchet. Far from being an objective portrait of peasant life, however, the story I have reconstructed reflects how one specific aspect of it looked from the perspective of artists, landowners, and others concerned with exerting some form of benign control over the rate and kind of change they witnessed in the Russian countryside. It is inevitably one-sided, being essentially a study of how a certain cross section of educated Russian society reacted to the prospect of modernization and the steps they took to grapple with it. Consequently the beneficiaries of their concern – some two million kustari overall, a fraction of them involved in the luxury crafts – remain as shadowy to us, their desires as inexplicable, their fears as frustrating as they were to their benefactors.[14]

Second, this study suggests that the history of kustar revival, indeed the very concept of *kustarnichestvo,* merits a prominent place in the history of modern Russian art. It shows that the West's admiration for Russian culture began well before the arrival of the Ballets Russes in Paris or the triumph of Constructivism at the 1925 Exposition des Arts Décoratifs. As critic Georgii Lukomskii pointed out in 1928, "The works of the kustari constitute, along with the triumph of the Ballets Russes, the only export articles which have propagated the renown of Russian art abroad."[15] Not only does an acknowledgment of the kustar's existence oblige us to relinquish the romantic notion of "folk art" as monolithic, pure, and timeless, it also allows us to appreciate more fully the complexities and contradictions that accompanied the primitivist impulse that informed so much of the Russian avant-garde's worldview. The unusual prominence of women artists in the development of Russian modernism becomes less mysterious when seen in the light of this long tradition of women's work in the kustar arts, suggesting that the decorative and industrial arts deserve much more serious attention from Russian art historians than they have received. Last but not least, the Constructivist movement of the 1920s acquires an important prerevolutionary precedent in the kustar revival, which brought artists into contact with real life, real objects, and real people (*narod*) at the most fundamental level of production.

Finally, this book formulates general questions about the evolutionary powers of peasant culture in the modern era and implicitly draws an analogy to the way that the consumption of "primitive" culture still functions today as an antidote to the pressures and homogeneity of modern life. In late Imperial Russia the kustar revival was fueled by a belief that folk culture was still alive, that it had the inner dynamism to evolve and adapt itself to external change and modernity, and that it was wrong to imprison it in the past, in the form of dusty museum artifacts. This optimistic belief was frequently expressed in biological metaphors – sowing seed in fertile soil, grafting old wood onto new stock, reproduction by cellular division – that reflected the policies of benign intervention and controlled growth advocated by kustar activists. Working against this optimistic and positivistic belief, however, was the inescapable fact that the initiative must always come from outside, since the kustari themselves had no commitment to the preservation of folkways for their own sake. When all was said and done, the kustar revival was the creation of an educated elite whose attempts to engineer a truce between the forces of tradition and progress turned the kustar form of production into a highly controlled form of souvenir industry. It is not in the European Arts and Crafts movements of the turn of the century that we will find the most fitting parallels to Russia's kustar art revival, therefore, but rather in all those souvenir industries that have flourished throughout the Western world in the past century as a poignant reminder that modern man has not after all been able to survive without some tokens of tradition to brighten his brave new world.

1

The Abramtsevo Workshops

In the academic debate over who will emerge the victor in the battle between factory and kustar industry, artistic kustar production occupies one of the strongest positions, for the simple fact that artistic activity doesn't need heavy machinery, large engines, and the extensive appliances of the factory.

N. Elfimov

IN THE COMPLEX STRUCTURE of Russian peasant society at the close of the nineteenth century, the kustar, or peasant handicraftsman, occupied an uncertain and ambivalent position. Public opinion swung between two extremes. Was he the heir to centuries of folk culture or simply a primitive form of proto-industrialization? Was he Russia's only hope for the future or a source of national shame? Was he a precious symbol of country life or a symptom of agriculture's decline?[1] The officially sanctioned definition of kustar industry as "the small-scale family organization of production of goods for sale, common among the peasant population of Russia as a supplement to agriculture" did little to answer these questions.[2]

To the extent that the educated urban public thought about the kustar at all in the postemancipation period, it was with indifference or distaste:

In the mind of our public a kustar item will be one that can satisfy only the modest tastes of the rural consumer. According to the city dweller, the kustar makes gigantic boots for the Ukraine that would make any Frenchman or even a German shudder. He models pottery, simple milk pitchers, and earthenware dishes for rural use. He knocks together benches and tables with rickety legs for peasant huts. He weaves rough canvas and homespun skirts for fashion-conscious country belles. He makes felt boots, sledge runners, and carts for his peasant brother. He forges countless oven forks, nails, and horseshoes, and carves fairground wooden toys for five kopeks a pair.[3]

This had not always been the case, however. The word *kustar* had first come into use in the early eighteenth century and was thought to be a corruption of the German word *Künstler,* thus originally implying a skilled craftsman.[4] Many kustar crafts had their origins in the natural economy of Muscovite Russia, when households produced only what they needed for their own use. Others emerged in the wake of Peter the Great's attempts to foster factory-type manufacturing, as peasants took the expertise they had learned in the factory back to their villages where they set up their own cottage industries.[5] Under serfdom many kustar crafts were turned into high-quality commercial manufactories through a system of private estate workshops. According to one observer, in the late eighteenth and early nineteenth centuries, "There was scarcely a landowning family or farm that didn't make use of peasant labor in the form of kustar production. Making the peasants spin, weave, embroider, knit, and so on during the long winters, serf landowners in part met their own everyday needs, in part sold goods to the towns, thereby increasing the income from their holdings."[6] Women's crafts in particular were cultivated as a way of satisfying the landowners' preference for foreign-style goods, and serf women learned to produce skillful imitations of cashmere shawls and French laces that were sold in St. Petersburg and Moscow as the genuine article, packaged in London and Paris boxes.[7] Serf lacemakers and spinners were a favorite subject for early-nineteenth-century Romantic painters, and despite the brutal conditions under which they worked, the models for Vasilii Tropinin's *The Lacemaker* (1823) and Aleksei Venetsianov's *Girl Carding Flax* (1822), *Morning of a Lady Landowner* (1823), and *Woman Spinning* (1839) were nevertheless viewed as integral parts of a stable social and economic organism.

The kustar's difficulties began with the Emancipation Act of February 1861, which was as catastrophic in its consequences for kustar production as it was for agriculture, with which it was symbiotically linked. Russia's huge peasant population emerged from the Great Reforms with their freedom, but also with smaller and less fertile land allotments that made it harder to produce raw materials for home consumption, with crushing dues to pay their former owners, and with a disproportionate percentage of the nation's tax burden. In a growing consumer economy where money was a necessity, the production of goods for sale to an outside market became essential, while paradoxically it became cheaper to buy factory-made goods for personal use than to make them at home. For more and more peasants, especially those in the northern and central provinces where the land yielded a poor or unreliable living and the winters were long, kustar industry became an essential form of crisis management. But as crop failures and droughts became more commonplace throughout the second half of the century, the axiom that "The lower the income from farming, the stronger the tendency for the broad range of rural supplementary crafts and trades"[8] came to apply increasingly to the black soil regions of central Russia.

The kustar's greatest natural enemies in postreform Russia were the factory and a governmental policy for promoting rapid industrial growth that was to be funded primarily by the taxes and agricultural production of the peasantry. By the 1870s those kustar crafts where manual labor could be easily replaced by machines (nail making, hand weaving, textile printing, and dyeing) were already on the verge of extinction. Others managed to survive, either because of their extreme labor cheapness, because modern technology had yet to devise a way to replace human beings, or because the notion of hand production was an important part of the craft's luxury value. But even in industries like woodworking, icon painting, and lacemaking that could expect to withstand mechanization for some time to come, an exploitative middleman system kept entire villages trapped in a state of debt and dependency from which they could not escape unaided. Because they produced for an unspecified, often distant market and not just their immediate community, most of Russia's kustar population depended on middlemen (*skupshchiki*) to mediate between them and the customer, to tell them what would sell in far-off cities, and to provide patterns and raw materials. Kustar labor had almost no monetary value, since any work was better than none in the long winter months of enforced idleness; and with no labor laws to protect the kustar, an average work day of fourteen to eighteen hours was the norm, with children working alongside their parents (Fig. 5). Under these conditions the kustar's only advantage over the machine was the extraordinary cheapness of his labor.

All of these factors help to explain why, by the latter half of the nineteenth century, "kustarnichestvo" had entered the language as a common derogatory term for anything country-bumpkinish, primitive, or ill made. But in academic and economic circles there was a second, ideological level of meaning implicit in the word that began to emerge in the 1870s along with the spread of populism. The very backwardness and technological primitiveness of kustar industry came to acquire an absolute symbolic value in a nation undergoing the tion of kustar industries throughout the empire. The resulting sixteen-volume report, published in 1874, estimated that 7.5 million peasants depended on kustar production for survival and warned that many of these industries were in immediate need of assistance.[15] Under pressure from the minister of state domains, who brought the matter to the attention of Alexander III, the Ministry of Finance's Council on Trade and Industry appointed its own commission in 1872 to organize more fact-finding expeditions, and between 1874 and 1886 a second vast body of statistical information on the kustar industries was published, confirming both the importance of kustarnichestvo and the extent of its decline.[16]

To those who argued that events be allowed to run their course, or that the kustar could take care of himself, proponents of a government-sponsored revival retorted that

> leaving the kustar to his own devices will inevitably lead to the decline of our agricultural economy. The kustar is well aware of all the shortcomings of his trade. He knows that city goods of the same kind are more sophisticated and better suited to the customer's taste. He knows that he is dependent on the middleman . . . and he has a general idea of the changes in the conditions of production that would be beneficial to him. But by himself he can do nothing to bring those changes about.[17]

Despite the official signs of concern noted above, however, the Russian government was extraordinarily slow to embark on practical measures that would rescue the kustar from his predicament. Physically isolated in St. Petersburg from the problems of peasant life, the government agencies that might have come to the kustar's aid (the Ministries of Finance, State Domains, and Agriculture) spent almost twenty years bickering over whether the government should intervene in kustar affairs, which agency should take responsibility for them, and who should fund their revival, before finally embarking on a practical course of action in 1888.[18] Local government bodies in the provinces, the so-called zemstva, undertook their own fact-finding missions at the local level but were just as slow to invest in practical solutions. Consequently, for nearly a decade it was almost exclusively a handful of determined private landowners who, motivated by populist and philanthropic ideals, plunged into the problem of practical kustar reform with their own kind of very concrete localized aid. Although easily dismissed as mere Band-Aid solutions to an incurable disease, such instances of private estate reform actually created an important prototype for subsequent government measures, and in one notable case, the Abramtsevo kustar carpentry workshop, established a precedent for the intervention of professional artists in kustar production that was to dominate the future course of all successful efforts to save the kustar from extinction.

In many ways the kustar carpentry workshop established in 1876 at Abramtsevo, the estate of the Mamontov family in Moscow province, was conceived as a kind of hands-on populism very much in tune with the times. Situated some seventy kilometers northeast of Moscow, not far from the Troitse-Sergieva Lavra and the Khotkovo convent, Abramtsevo already had a rich history before it was purchased in 1870 by railway magnate Savva Mamontov and his wife, Elizaveta.[19] In the 1830s and 1840s it had been the home of the Slavophile writer Sergei Aksakov and the center of a literary circle that included Gogol and Turgenev.[20] These cultural and nationalist associations were not lost on the new owners, both of them typical representatives of the Moscow merchant class, with its strong tradition of social service and good works.[21] Both were scions of families that had made their considerable fortunes through industry – in his case through the construction of the Moscow-Troitse Railway and other ventures, in hers through the Sapozhnikov textile mills – and both belonged to a generation that considered that wealth and privilege entailed certain moral responsibilities toward the masses. Yet despite their similar backgrounds, the young couple's aspirations to good works were manifested in very different ways. Savva Mamontov's commitment to the masses, such as it was, was to enlighten them through art. His patronage of the major Russian painters of the 1880s and 1890s, his Private Opera Company (f. 1885), and his interest in art ceramics all partook vaguely of Dostoevskii's maxim "Beauty will save the world." His wife's philanthropy was of a more pragmatic kind and was inspired by her awareness of the local rural population's lack of health care, elementary education, and economic stability. It was thanks to Elizaveta Mamontova that in the early 1870s a hospital and an elementary school were opened on the estate to serve the peasant families of the area and that in 1876 a joinery workshop was added where pupils who had graduated from the school might learn a useful trade. The following passage, written by the first biographer of the Abramtsevo kustar workshop, outlines the ethical and philanthropic considerations underlying the workshop's initial foundation; it could, however, as easily describe any of a dozen similar efforts undertaken at this period, and its arguments and rationale reflect the populist sentiments of the 1870s and 1880s in favor of "peaceful cultural work among the peasantry, strengthening artifacts that could still be found in everyday use throughout the countryside. Increasingly dissatisfied with the patent failure of her workshop to serve her original philanthropic intentions, Mamontova decided to reorganize it along more artistic lines. To do so, she enlisted the help of a recent addition to the Mamontov circle, the artist Elena Dmitrievna Polenova.[28]

Born in Petrozavodsk in 1850, Elena Dmitrievna was the youngest in a family steeped in Russian history and culture. Her father, D. M. Polenov, was a respected archaeologist and secretary of the Imperial Archaeological Society, her mother, Maria Alekseevna, wrote and illustrated books for children, and her elder brother, Vasilii, had already embarked on a successful painting career. Polenova's childhood and adolescence were not especially happy, according to her brother-in-law, who remembered her as a shy, silent, and apathetic teenager.[29] Summers were spent at Imochentsy, her family's estate in the remote northern province of Olonets, or in Tambov province with her grandmother, from whom she learned a good deal about Russian history and folklore. But during the long winter months she led a solitary and rather joyless life with her family in St. Petersburg, enlivened however by lessons in drawing and watercolor from the academician Pavel

Figure 7. Church at Abramtsevo, by Viktor Vasnetsov, 1881–2. (Courtesy of William Craft Brumfield)

Chistiakov, as well as from Ivan Kramskoi, a leading member of the Association of Traveling Exhibitions, or Wanderers.

The single most important influence on Polenova in these formative years was her elder sister, Vera, a typical "woman of the sixties" consumed with her generation's passion for social service and self-sacrifice. During the Turko-Bulgarian war (1877–8) the two sisters served as volunteer nurses in Kiev,[30] and when they returned to St. Petersburg they attended medical courses for women with the intention of opening a clinic for local peasants at Imochentsy. By the age of twenty-eight, then, Polenova had tasted "the enormous pleasure a person receives from the realization that she has rendered direct aid to another in trouble" and realized that "to return to my former life, that is, to deprive myself of working for society in some form or another would be like depriving myself of healthy and nourishing food."[31] In St. Petersburg she immediately volunteered her services as a drawing teacher in a charity school for girls, where she introduced classes in sewing, tailoring, and drawing that transformed an elementary school into "a big professional school for women, with well-equipped workshops."[32]

The passion for social service did not rule out art, however, and in the winter of 1879 Polenova enrolled at the School of the Society for the Encouragement of the Arts in St. Petersburg, where she took classes in watercolor and ceramics, both considered eminently suitable media for a woman artist.[33] As the society's first pensioner to be sent abroad, she studied ceramic techniques in the Paris studio of the Rus-

Figure 8. Carved backboard from a peasant cart, late nineteenth century, Abramtsevo Museum. (Reprinted from N. V. Polenova, *Abramtsevo* [Moscow, 1922])

sian ceramist Evdokim Egorov and at the Deck factories,[34] and on her return to Russia opened classes in faience painting as well as continuing to study watercolor at the society's school. But the death of her sister, Vera, in 1881 proved so devastating that in October of the following year she left St. Petersburg and moved to Moscow, where her brother, Vasilii, now lived.

It was through Vasilii that Polenova met the Mamontovs, and she was immediately accepted into their circle as a talented and serious painter.[35] Lacking a formal education in oil painting and life drawing, Polenova worked primarily in watercolor, and her first exhibited works – flower studies and landscapes – were bought by Savva Mamontov and the collector Pavel Tret'iakov. Through the open houses that Vasilii held each Thursday for his friends and pupils (he was on the faculty of the Moscow School of Painting, Sculpture, and Architecture), she met many of the young artists who would become her colleagues in the Moscow Association of Artists a decade later, and she quickly became a participant in the amateur theatricals for which the Mamontovs were renowned, helping to design and make costumes.

But despite the active tenor of her new Moscow life, Polenova's rather uncompromising personality and lack of social graces, as well as her marked desire for some form of social service, made the company and interests of the devout and gentle Elizaveta Mamontova more congenial to her. Whereas Savva Ivanovich, "with his enterprises, his extravagance, his picnics, cavalcades and idleness, his entourage of artists and actors,"[36] revealed to Polenova a world of high art, idealism, and personal laurels that she timidly aspired to enter, his wife, Elizaveta, drew her into a more familiar and morally sustaining sphere of social responsibility and self-abnegation. Trained in the modest and womanly decorative arts, and yearning for a life spent in service to others, Polenova was the perfect solution to Mamontova's dilemma regarding her carpentry workshop. When Mamontova organized expeditions with her children into the surrounding countryside to collect examples of peasant carving and textiles, Polenova went along armed with a sketchbook in which she noted down ornamental motifs and fragmentary views of peasant life. It was probably Viktor Vasnetsov, a longtime member of the Mamontov circle, who first suggested to Polenova that she use her sketchbook and the growing Abramtsevo collection of peasant artifacts as raw material from which to create furniture designs that could be produced by the pupils in the carpentry workshop. Early in 1885 a prototype was made by buying an ordinary kitchen cupboard and painting it "in the style of Old Russian painting. Vasnetsov decorated the door on the main facade with stylized birds [on a ground] of stylized plants . . . so that a cupboard worth 1 ruble 30 kopeks became much more valuable."[37]

By this time Mamontova and Polenova were evidently seriously discussing the future of the kustar carpentry workshop on the estate, and the experiment with the cupboard suggested a way to reorganize the workshop along more artistic and, most important, more profitable lines. By the end of May 1885 they had agreed that Polenova should become the workshop's artistic director and create a range of new furniture designs closely modeled on authentic examples of folk art. At the same time, the two women would set up a sister workshop for peasant girls that would produce modified examples of peasant embroidery and weaving. Polenova presented the idea of this "furniture and costume enterprise" to Savva Mamontov and received his blessing, as she reported to her sister-in-law, Natalia:

> [He] encourages us to do both, and advises us to put them on a broader footing. I told him about my idea for producing this kind of furniture in the Abramtsevo workshop and then having artists supervise its decoration by some of the local painters at Troitse[-Sergieva Lavra], the ones who decorate pottery there – the more so because they say this production has totally stopped providing them with a living. Then we would put [the furniture] up for sale at a kustar exhibition. This pleased him very much.[38]

The girls' workshop operated for little more than a year before Mamontova was compelled to close it down for want of trained personnel. During that time it was typical of a dozen or so other private estate workshops already scattered throughout Russia in the 1880s:

> [Mamontova and Polenova] taught the girl pupils at the school to work the old Russian patterns, masses of which were collected in neighboring villages and in the adjoining province of Vladimir. In addition they distributed work to be done at home. [The girls] sewed cloths from homespun linen, and aprons from *pestriad* (striped linen) and naboika; they experimented with embroidering towels using old stitches and so on. Under Elizaveta Grigorievna's influence the local peasant women once again took up weaving linens and pestriad, plaiting sashes, and all manner of other peasant women's work that had been abandoned because of the factories.[39]

The boys' workshop offered much broader scope for intervention. Although wood carving had long been an integral part of peasant life in Dmitrievskii uezd, where Abramtsevo was located, furniture making had never been a traditional kustar industry there. Ever since the great Moscow fire of 1812 had created a "furniture famine" in the capital, the kustar furniture market had been concentrated in Zvenigorodskii and Moskovskii uezds to the south. If successful, the Abramtsevo carpentry workshop would thus introduce a new industry to the region, created from scratch out of what remained of folk art and folk traditions in the Russian countryside.

Polenova embarked on the task of designing art furniture for the Abramtsevo pupils with a clear sense of purpose and what might be called a fully formed design philosophy. In a letter written at the end of 1885, she explained the considerations that guided her in selecting motifs for the workshop:

> We have made it a condition to resort as little as possible to using publications and various kinds of printed material in general. . . . [O]ur aim is to seize hold of folk art that is still living and give it a chance to develop. The material that turns up in published sources is for the most part dead and forgotten. Consequently the thread has been broken and it's terribly difficult to mend it artificially. When a peasant is asked to copy from thirteenth- and fourteenth-century artifacts that are unknown to him and long forgotten, it may seem to him just as foreign as copying a Moroccan or ancient Greek artifact. This is why we are mostly looking for our inspiration and models by going around the huts and looking closely at the things that make up their environment, trying naturally to exclude the more recent foreign additions.[40]

The "thirteenth- and fourteenth-century artifacts" she mentions here were in fact the medieval manuscript illuminations whose interlace patterns were a major source for the so-called Old Russian style that was very much in vogue in polite Russian society at this time; their source – the influential albums and facsimiles published by Russia's two leading schools of industrial art, the Stroganov School of Technical Drawing in Moscow and the School of the Society for the Encouragement of the Arts in St. Petersburg (Fig. 9). Both schools were committed to the creation of a viable Russian style for the decorative arts that would make Russian manufactured goods more competitive on the European market, and for this reason they were generously subsidized by the Ministry of Finance and the Council on Trade and Industry. As the director of the Stroganov School, Viktor Butovskii, pointed out in the preface to *The History of Russian Ornament* (1870), it was not enough to find

> traces of ancient Russian art that bear witness to its independence, and to have rescued from neglect a body of materials showing a remarkable and contradictory ornamentation. These findings must be propagated, public

sympathy must be roused by these treasures of ancient Russian art which have been unknown for so long, and above all this art must be made accessible to industrialists and to artists.[41]

In addition to the bookish interlace style, Russian artists and architects also drew for ideas on the courtly life of the seventeenth-century Moscow boyars and on peasant art in all its manifestations, particularly embroideries, lace, and wooden architecture (Fig. 10). For the most part, the Old Russian style of the 1880s, as interpreted by architects like Ivan Ropet and Viktor Gartman, was a compilation of motifs from a wide range of periods and media, linked by little more than their common Russian origins. But in enthusiastically embracing the national cultural heritage, early practitioners of the Russian style came up against the problem of sources. The collecting and study of Russian art was a comparatively recent phenomenon, especially in the case of peasant art, and despite the activities of the Imperial Archaeological Society, the Academy of Arts, the Society for the Encouragement of the Arts in St. Petersburg, and the Stroganov School in Moscow, the

Figure 9. Page from *Histoire de l'ornement russe,* compiled by V. I. Butovskii (Paris, 1870–3). (Courtesy of the Getty Center Resource Collections)

number of publications and collections available by the mid-1880s was still very small. In the words of the architect Nikolai Sultanov,

> The need to work in the Russian style and the demand for it not only exist, but are growing all the time, and yet there is almost no material available. What alternative do our artists who have to compose Russian wooden decorations have? Naturally, only one – to make use of some kind of published materials, even if they don't have anything to do with wood carving but are taken from other branches of Russian ornamentation. And so in the "Russian" works of our artists we see transposed into wood, ornamental motifs embroidered on linen taken from [Stasov's] *Russian Folk Ornament* . . . or else motifs from manuscript illuminations, taken from [Butovskii's] *History of Russian Ornament.*[42]

Polenova's decision to avoid the hackneyed motifs that had been circulating for over a decade and to draw instead on the new, unpublished material from the Abramtsevo collection and her own sketchbooks provided the Abramtsevo kustar workshop with the invaluable commercial advantage of exclusivity and novelty – "the fact that these

Figure 10. Drawing of a wooden izba from Kostroma with details of exterior decorations, by Vladimir Dal', early 1870s. (Reprinted from *Zodchii,* 2 [1872])

things can be bought only from us," as she put it. An even more radical break with the past at Abramtsevo was the substitution of country kustari for the urban craftsmen (*remeslenniki*) to whom designs in the Russian style were usually entrusted. In St. Petersburg, where both the production and the clientele for such goods were concentrated, the kind of furniture illustrated in the journals *Zodchii* (The Architect) and *Motivy russkoi arkhitektury* (Motifs of Russian Architecture) was carried out in the professional workshops of Büchtger, Shtange, and Shutov, while glass and porcelain in the same intricate interlace pattern were made at the Imperial Glass and Porcelain Factories.[43] For the average educated Russian, accustomed to associate the kustar with the least admirable aspects of the Russian peasantry, entrusting peasants with the production of artistic furniture designed by a professional artist was a very dubious proposition.

In her first five years as artistic director of the Abramtsevo kustar workshop (1885–90), Polenova assumed a role that was essentially editorial, her own creative personality firmly subordinated, she believed, by the strict guidelines she had set herself. Rarely inventing motifs of her own, she was content to select from the store of ornamental fragments that she and Mamontova had accumulated during their collecting expeditions, refining them where necessary through the filter of her own taste to make them palatable to a more affluent, urban market. Many of her decisions were of a purely practical nature, involving the transposition of ornamental motifs from their original, exclusively rural source to objects with a broader application – for example, adapting the carving on the handles of laundry beetles to decorate brackets for a wall shelf, transferring a rosette from a roughly carved sideboard to a compact bedside table, or embellishing a simple hanging cupboard with floral chip-carving (Fig. 11). Some items (saltboxes, tea caddies, little boxes) could be copied in their entirety, while others needed only slight modifications (a child's chair enlarged into a full-size armchair). Polenova was also receptive to the pupils' own skills and ideas, as when they "made themselves little cupboards to hold their tea things. . . . On a visit to the workshop Elena Dmitrievna sketched a corner with two of these cupboards in it. It was decided to take one of them as a model and repeat it."[44]

Less frequently, Polenova allowed herself the challenge of imaginatively stepping inside the mental world of the anonymous peasant craftsman by trying to create as he might, with the ornamental fragments at her disposal. The outstanding example of this approach was the so-called column cupboard, which she designed in 1885 (Fig. 12). Using photographs, watercolor sketches from her album, and actual

Figure 11. Design for a cupboard, by Elena Polenova, ca. 1885. (Reprinted from *Mir iskusstva,* 1 [1899]. Courtesy of the Getty Center Resource Collections)

pieces from the Abramtsevo museum, Polenova selected a number of motifs from a wide array of sources:

Its basic shape was inspired by a cupboard that V. D. Polenov had made, the details Elena Dmitrievna took from the museum and her sketchbooks. The lower part with the sliding door came from a shelf in the village of Komiagino, the handle from a painted beetle found in the village of Valishchevo in Podolskii uezd. The top band came from the front of a cart and the column was found in the village of Bogoslovo in Iaroslavl province. The vase with the rose painted on the first cupboard was from V. D. Polenov's sketchbook, drawn from the swings on Maidens' Field.[45]

That such designs were a serious attempt on the part of an educated Russian painter to emulate the creative and emotional processes of some anonymous, illiterate Russian peasant is made clear in a letter that Polenova wrote to her sister-in-law Natalia, describing two new cupboard designs for the workshop: "The door of the column cupboard with stars, moon, wild strawberries, and flowers represents untamed nature. It's a meadow cupboard. My corner cupboard, though, is a two-story house – upstairs on the sill there's a flower in a pot,

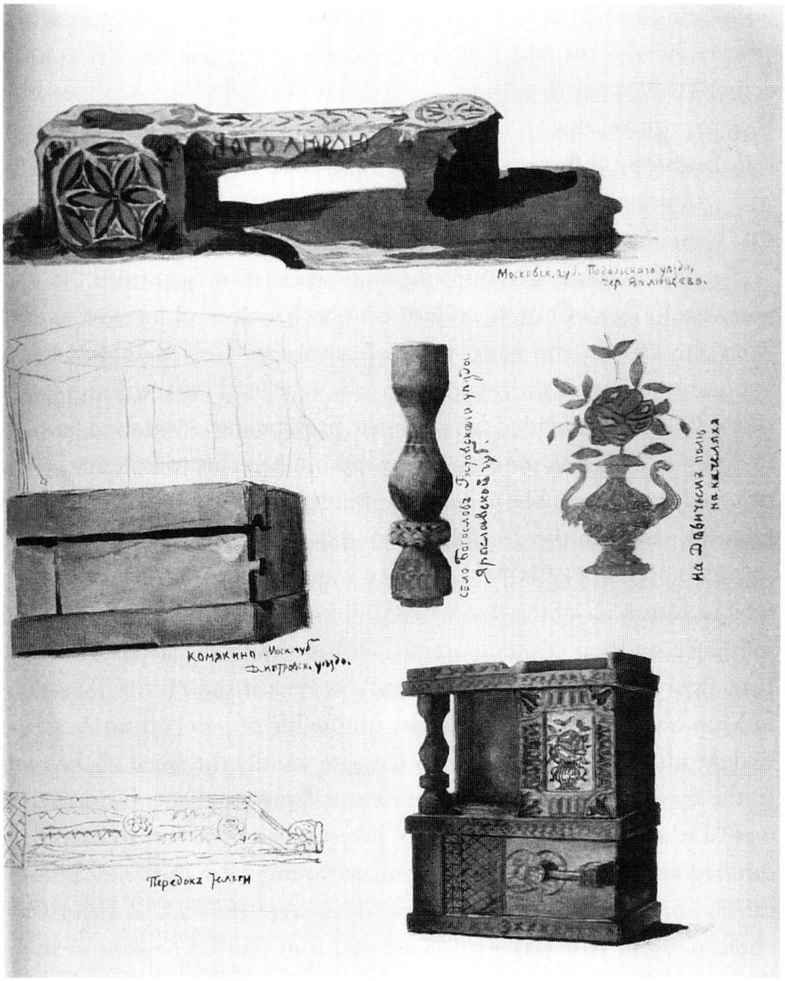

Figure 12. Page from Elena Polenova's sketchbook showing the evolution of the column cupboard, 1885. (Reprinted from N. V. Polenova, *Abramtsevo* [Moscow, 1922])

while through the lower window you can see the sunset."[46] The somewhat self-conscious naïveté of this statement, with its overtones of retreat to a child's imaginary world, suggests that Polenova was attempting a poetic reincarnation of folk art and folklore in the only way she could imagine, through her own childhood memories of nature. A direct link that would seem to confirm this idea is the project she was working on throughout her involvement with the workshop, to collect and illustrate Russian fairy tales. Her first, so-called Abramtsevo cycle was completed in 1889, although only one of the tales, *The War of the Mushrooms,* was published, and consisted not only of texts taken from the great folklorist Afanas'ev, but also of variants on traditional fairy tales that she had herself collected from peasants. As she later explained to Vladimir Stasov, her designs for the Abramtsevo workshop were inseparable in her own mind from her illustrations,

1885. The problem of markets (*sbyt*) was one that nobody committed to the improvement of kustar industry could ignore, for without guaranteed sales all technical and artistic improvements were pointless. The very nature of kustar production – the fact that the kustar worked not for a known customer but for a distant clientele whose needs and tastes he did not always understand – meant that those who wished to improve it had to assume the duties of the middleman themselves. As kustar supporters never tired of pointing out,

> Many crafts are in a particularly depressed state because they have an inadequate, capricious, and precarious market for their goods, or else there is quite a large market but the goods reach the consumer through the hands of a middleman who retains a considerable, sometimes excessive, part of the money earned. . . . To expand the market and to put the kustar in direct contact with the consumer – this is one of the most important goals whose attainment can not only speedily improve the position of the kustar, but also secure the success of all other measures.[53]

The owners of private workshops needed to be particularly ingenious in this regard, finding clients among their personal acquaintances and risking their own capital so as to convince kustari that the improved goods they were encouraged to produce would find guaranteed sales.[54] For the first five years the Abramtsevo workshop was in production Elizaveta Mamontova assumed the role of benevolent middleman between kustar and client herself, supplying raw materials on credit and often paying the most needy kustari immediately on receipt of their goods.

The workshop's first customers were mostly family friends like the painters Il'ia Repin and Viktor Vasnetsov, the collector Pavel Tret'iakov, and the singer Fedor Shaliapin. But as the production expanded and graduates of the workshop went home to their villages to work independently, the need to find a reliable retail outlet in Moscow became pressing. In this early period of meager government funding and support for kustar industry, such outlets were few and far between. In May 1885, the very month the Abramtsevo workshop was reformed, a major new outlet for kustar goods opened in Moscow near the Nikitskii Gates, the Moscow provincial zemstvo's Kustar Museum, equipped with warehouses and a commercial outlet for sales direct to the public. When Mamontova and Polenova brought samples of the new Abramtsevo repertoire to the Kustar Museum, however, they were greeted with suspicion and their goods accepted for sale only grudgingly.[55] There were questions raised as to whether the pupils in the workshop even qualified as kustari at all, given the professional training they received. Moreover, items such as the column cupboard accorded ill with public expectations about what authentic kustar goods should look like and showed the Kustar Museum's own "unimproved" stock (knitted scarves and stockings, locks, knives, and trays) in a most unflattering light. Equally unsatisfactory was a short-lived arrangement between Mamontova and her husband's sister-in-law, Maria Mamontova, who sold Abramtsevo furniture at a very high commission out of her Moscow toystore, Children's Education, on Leontievskii Lane.

Only in December 1886 did Mamontova finally move into her own premises on Povarskaia Street, which she shared with the toystore of a Madame Vereshchagina. Here the stock included "goods from the joinery workshop, peasant handicrafts, embroideries, linen, pestriad, naboika prints, belts, and birch-bark containers ordered from a sample that the . . . head of the workshop had happened to bring back from his native province of Vladimir."[56] Natalia Polenova reported that the Moscow public seemed eager to visit this "new kind of store, with its artistic kustar goods made exclusively by the Russian folk, long forgotten and supplanted by factory wares,"[57] and by the end of the second day sales had reached 360 rubles. At the same time the workshop exhibited at the charity bazaars that took place each December before Christmas and occasionally worked on individual commissions.[58] By the beginning of 1889 Elizaveta Mamontova was able to take over the entire shop on Povarskaia Street and hang outside a signboard that read "Sale of Carved Wooden Objects Worked by Pupils of the Joinery Workshop in the Village of Abramtsevo, Moscow Province, Dmitrievskii uezd."

After almost five years under Polenova's artistic leadership, with a stable Moscow outlet and a growing clientele among the well-to-do, the Abramtsevo kustar workshop was clearly a success, but an isolated one nonetheless. Whereas industrialists and the Ministry of Finance had long ago realized the commercial rewards that artistic refinement and a national ornament could bring to manufactured goods, those responsible for upgrading the kustar industries had still not learned this fundamental economic truth as late as 1889, when an exhibit of kustar products was sent to the Paris Exposition Universelle. As the official catalogue defensively conceded, "The collection, got together by subscription, is far from complete and gives only a vague understanding of an industry that occupies millions of hands."[59] The impression it made was uniformly depressing and only added fuel to the argument that kustar industry had no place in the life of a civilized, industrialized nation. One French journalist noted contemptuously the following kustar goods, displayed in cracked glass vitrines smeared

Figure 13. "Cat and Owl" carved door, designed by Elena Polenova and made at the Abramtsevo workshop, early 1890s. (Courtesy of Izobrazitel'noe Iskusstvo, Moscow)

door and wrote to Polenova with characteristic hyperbole: "Publish it in any 'ouvrage' you please and every intelligent person will swear that it's the work of some amazingly talented, anonymous master of our ancient Rus. This is something unimaginable for me, the stuff of legends. The ancient Russian fairy tales and poems have been resurrected!"[68] But although Stasov could not have suspected it at the time, this and a handful of other furniture designs done in the early 1890s pointed to the future rather than the past. Less and less interested in the ostensible ethnographic purity of the sources at her disposal, Polenova embarked on a search for a grammar of ornament that was above all emotionally convincing, based on her own internal relationships to the Russian countryside, her own perceptions of what it meant to be Russian.

It is not easy to account for this abrupt change in Polenova. A letter to a friend written in March 1889 suggests that she was even then interested in exploring more personal and intimate forms of expression. Of two ideas for future paintings she outlined, the first was to be a piece of historical genre painting, depicting "a bright sunny summer day, a courtyard in the estate of a prosperous grandee of the sixteenth or seventeenth century, a trained bear dancing in the middle with his trainer, a goat, and a drummer, to the left of a well a young lad is sitting while a pair of bear cubs are rolling on the grass beside him."[69] The second painting, however, of three children in a modern-day nursery, was to be a recollection of her own rather grim childhood. But while her sympathies were very much with this second theme, she also felt that "one shouldn't show such intimate things to the public." In fact, when *The Nursery* was exhibited in 1892 at the annual Wanderers exhibition in St. Petersburg, critics reacted with hostility to the disquieting lack of sentimentality in the scene, finding "neither the childish naïveté, the carefree joy, nor the enchanting smile that composes all the poetic charm of a child's little face."[70]

Certainly, the fact that the year 1890 marked the decisive break between her "old" and "new" styles was to some extent symptomatic of similar metamorphoses occurring in Russian culture as a whole. Within the Mamontov circle itself the shift could be felt in the unspoken tensions between Viktor Vasnetsov, then midway through painting the interior of the new Cathedral of St. Vladimir in Kiev, and Mikhail Vrubel, who had lost the commission to the older artist and had to be content with designing some of the ornamental bands on the aisles. Polenova was herself very much aware that "the epoch when it was easy and simple for people to band together in small groups and work together, advancing in a friendly crowd toward their goal," was at an end, and that a new era of artistic independence, individuality, and isolation was at hand.[71] While her psychological need to put herself at the service of others continued to be a strong motivating force in all her activities throughout the 1890s, it manifested itself in encouraging other artists to develop their potential, rather than in the volunteer social work among the masses that had sustained her during the previous decade.[72]

Other issues arose to complicate the apparently simple relationship that had bound Polenova to the workshop and to Elizaveta Mamontova. Although as late as 1890 Polenova was still enthusiastically devising new prototypes for the workshop, that March she complained to

Mamontova that although overall the boys were working fairly well, "they've really lost their independence and resourcefulness, and this is holding up production."[73] The restrictions on her own creativity that her self-imposed guidelines for designing kustar furniture necessitated were becoming irksome, and she felt an increasing need to assert her own personality, to get credit for her own work and not just for skillfully editing the work of others. "It's very hard to distinguish where what belongs to the people ends and what is my own begins," she later explained to Stasov. "And it's something I would like to make clear, not out of some kind of false modesty, but simply because it's much nicer to be praised for what I myself have created, rather than for something I've been able to make use of."[74] Polenova's interest in the workshop visibly cooled over the next few years, and her withdrawal came as a shock to Elizaveta Mamontova, who after the sudden death of her son Andrei in 1891 found herself more than ever in need of her friendship. According to art historian Dora Kogan, "At the time nobody could explain her sudden and apparently inexplicable cooling off toward Elizaveta Grigorievna, and she clearly felt out of her element at Abramtsevo."[75] Although for the next three years Polenova produced a handful of new designs – all of them using her new ornamental motifs – her energies were increasingly diverted in other directions, as she tried her hand at genre scenes in oil (*Guests, The Nursery*) and a second series of fairy-tale illustrations.

Whatever the reasons for her withdrawal, by 1893 Polenova was no longer visiting or designing for the Abramtsevo workshop, and from this time onward until her death in 1898 practically broke off all contact with it and with Elizaveta Mamontova. So ended the "Abramtsevo period" in Polenova's artistic career, although it was not to be the end of her crucial contribution to the transformation of the kustar industries, as we will see in the next chapter. Still less was it the end of her influence on the kustari of the Abramtsevo region, for Polenova left behind her not only a valuable repertoire of salable designs for furniture, but also the first successful working model for kustar reform specifically through artistic intervention and collaboration. With government funding still very tight by the mid-1890s, the efforts of private citizens like Mamontova played a far larger part in kustar reform than the government cared to admit. Perhaps because of its antecedents under serfdom, the estate workshop appeared to be the natural organism in which to "concentrate forces on assisting a few branches of kustarnichestvo, and within given branches to help perhaps only a specific sector of the population."[76] Given the monumental task of reforming the habits and overcoming the prejudices of millions of demoralized kustari, all on a shoestring budget, it was precisely small, carefully controlled training workshops such as that at Abramtsevo which promised immediate and gratifying results.[77]

In Polenova's absence the workshop continued to flourish under the direction of Egor Zelenkov, a former pupil.[78] The most popular of her designs – the column cupboard, the hanging shelves, mirrors, tables, and various carved caskets – were regularly reproduced both in the Abramtsevo workshop itself and in dozens of peasant huts and villages throughout the district. The workshop's visibility and fame increased dramatically throughout the 1890s, and it won gold medals at the Chicago World's Columbian Exposition in 1893, the Nizhnii Novgorod All-Russian Arts and Industries Exhibition in 1896, the Paris Exposition Universelle in 1900, and the First All-Russian Kustar Exhibition in 1902. Financially too, the workshop and its graduates thrived, while kustari in neighboring villages adopted the Polenova repertoire with alacrity and sold their work through the Moscow Kustar Museum.

Success was not without its side effects, however. For Mamontova, whose motives had been purely philanthropic from the start, "the more business grew and the number of present and past pupils increased, the more difficult it became to preserve a moral link with them, to direct them."[79] Isolated incidences of moral turpitude among peasants whom she had helped – two former pupils caught drunk at a fair, a woman stealing wood from the Abramtsevo forest – seemed not only acts of personal betrayal, but signs that the workshop and school had not after all had the ennobling effect she had hoped for. For others more concerned about the workshop's artistic integrity, expansion brought with it a decline in quality and an element of mechanical mass production.[80] Without constant artistic supervision and a steady flow of new designs, the kustari who made Abramtsevo furniture had nothing to gain economically by sticking to the letter of the law as regards Polenova's original designs. Now independent of the workshop and working in their own homes, they copied from copies of items like the column cupboard, blurring little by little the precision of Polenova's original vision (Fig. 14). What Natalia Polenova dismissed as "meaningless combinations of various types [of Polenova furniture] devoid of talent and with a veneer of debased marketplace taste" now appeared on the market, produced by kustari who had set up on their own or joined the Soiuz (Union) Association, a joint-stock company founded in Moscow in 1900 to trade in kustar goods.[81]

Figure 14. Cupboard made at the Sergiev Posad art joinery workshop, ca. 1907. (Reprinted from *Niva,* 39 [1907])

Despite the very tangible economic benefits that the Abramtsevo workshop had thus brought to the peasant population of Dmitrievskii uezd, for those who had undertaken this labor of love the larger goals of improving the peasant population's quality of life either ethically or aesthetically (rekindling traditions of popular art) proved more elusive. Without Polenova's continued presence, the mass production of her designs "deteriorated," at least in relation to the standard she had set, in a way that was endemic to the kustar industries. For, as one observer explained,

> If they do not receive new models, even the best kustar workers either repeat the same pieces year after year . . . or they copy those goods made by other workers that happen to come their way. Therefore they produce only the goods that have already been made in their own region for years. Not only are they all the same, they . . . also lose their original pleasing design and bright coloring, and in general become crude and tasteless.[82]

By the end of the century a noticeable shift could be detected in the expectations that the public had of "improved" kustar art, thanks to the success of the Abramtsevo experiment. Polenova furniture had become an economically viable industry in Moscow province and a desirable commodity among the bohemian and well-to-do circles of Russian society. Yet the more it achieved its primary goal of financial solvency for the kustari themselves, the more it was reproached (at least in artistic circles) for standing in one spot and becoming like the factories it had set out to challenge: mechanical, mass-produced, generic. As the kustar art industries moved ever closer to the industrial arts at the hands of government ministries, zemstva, and artists alike, the economic program with which the kustar revival had begun its activities and which at Abramtsevo had been the raison d'être for Polenova's activity there fell into the background while issues of stylistic innovation came to the fore. Still, throughout the next two decades Abramtsevo was to be the model and the standard against which other kustar workshops with artistic pretensions measured themselves and which they attempted to rival.

2

The Solomenko Workshops

There is no doubt that if competent people had not taken [the resurrection of peasant embroideries] in hand, soon there would not only be no examples of old work left, but there wouldn't even be the consciousness that Russian peasant women knew how to make refined handicrafts long before the "intelligentsia" began teaching them how.

A. N. Pogosskaia

IN THE COURSE OF THE 1890S the Abramtsevo experiment of intervention by professional artists in the production of kustar art began to take root in several different locations and branches of kustar industry. Locally, as we have seen, it spread out in ever-widening circles into the surrounding villages of Dmitrievskii uezd in the province of Moscow. But it was also transplanted to other provinces via a network of close family and artistic connections. One of the earliest direct offshoots of the Abramtsevo workshop was the kustar embroidery workshop that Maria Fedorovna Iakunchikova organized at her Solomenko estate in the province of Tambov.[1]

The establishment of training workshops where peasant women were encouraged to return to their abandoned handicrafts as an economic buffer against adversity was peculiarly the province of women from the upper classes. While the detailed statistical surveys of the 1870s, with their bleak picture of kustar industry's escalating decline, might have shocked the agricultural lobby in St. Petersburg, the landowning gentry, and particularly their wives, saw this decline and its effects at first hand. Thus, it was almost exclusively the wives of landowners with estates in the depressed black soil provinces who, for a variety of philanthropic reasons, established the first kustar workshops for women in the 1870s and 1880s. One of the first of these was S. P. Kaznachaeva's drawn-thread embroidery workshop in the village of Vysokoe in Mikhailovskii uezd, Riazan province, established in 1878 to counteract the results of devastating fires and periodic famines that drove women to seasonal jobs like peat cutting. Adapting the embroidery patterns of the Riazan peasant costume (red on white drawn-thread work) to decorate tablecloths, napkins, and so on, by 1881 Kaznachaeva was giving work to three hundred peasant women.[2] This initiative was repeated throughout the 1880s by a dozen or so noblewomen, following a similar pattern of adapting traditional forms of craft to modern, primarily urban needs. In 1880, for instance, Princess M. N. Shakhovskaia and Princess S. P. Dolgoruka established a school-workshop at Muraevnia in Dankovskii uezd, Riazan province, in response to a severe crop failure.[3] In Moscow province, Princess A. V. L'vova introduced needlework to the women of Spasskoe in Klinskii uezd, in an effort to lure them away from lucrative but "morally harmful" day jobs loading and unloading freight wagons on the nearby railway. Two more workshops were opened in 1885, one by Princess M. A. Urusova for spinning and weaving in Sychevskii uezd, Smolensk province, the other an embroidery workshop set up by Princess A. N. Naryshkina at Bykovo-Gora in Chernigovskii uezd, Tambov province.[4]

Such actions were undoubtedly motivated by that spirit of philanthropy and noblesse oblige so characteristic of educated Russian women in the postreform era.[5] But there were other factors involved in the proliferation of workshops for peasant women in the 1880s and 1890s. Nobody familiar with the economic and social structures of peasant life could fail to recognize how much their stability depended on the labor of women.[6] As agriculture increasingly failed to meet the needs of the peasant population and more and more women turned to better-paying seasonal labor and factory work, many observers predicted the breakup of the peasant family and the ultimate ruin of agriculture itself. Promoting kustar industry was seen by many as the ideal, indeed the only, solution to this problem, for as one economist noted:

> If the support and development of kustar industries among the male peasant population are important as a way of maintaining a link with the land and supporting agriculture, then it is even more important to support and develop among the female peasant population such occupations as do not distract them in other directions, do not separate them from the land and

The women's aid effort was spearheaded by the Grand Duchess Elizaveta Fedorovna, elder sister of the future tsarina Alexandra and wife of the governor general of Moscow. Among the measures taken by her relief committee in Moscow was the organization of charity bazaars and a commercial outlet at 7 Stoleshnikov Lane, selling "a thousand and one things dear to the feminine heart," or in other words, the handiwork produced by peasant women in the famine-stricken districts.[28] But despite the low prices asked for these goods, one American reporter wrote that "it was questionable whether they would meet with a ready sale in Moscow, where such things are common and not especially sought after."[29] Although Moscow was a city famous for its philanthropists, especially among the merchant class, the glut of women's handicrafts from workshops in the famine areas taxed Muscovite charity to the limit and created a fierce and futile competition:

> In one instance they buy lace because they know it was made by starving people. In another embroideries don't sell because there are the same goods on sale made by people who are even more starving. And so goods made by the starving and the not-so-starving get caught in a competition from which the most needy emerge victorious.[30]

The lack of cooperation among workshop organizers, as they struggled to find outlets for the backlog of kustar handicrafts they had caused to be produced, and the sameness of those goods, made two facts clear. Charity was an unreliable basis on which to build long-term stable employment for peasant women, and, rather than appeal to the customers' pity, improved kustar handicrafts must appeal to their desire for novelty and cultivate a preference for the hand over the machine and for the national over the foreign, by providing goods that were visibly related to a tradition of Russian folk art.

Nobody was more alive to these laws of the marketplace than Maria Fedorovna Iakunchikova, who in 1891 opened her own embroidery workshop at Solomenko, her estate in Morshanskii uezd, Tambov province. Maria Fedorovna was a member of the Mamontov clan by birth and well before the establishment of her workshop was a well-known figure in the Moscow art world.[31] The niece of Savva Mamontov, she was a frequent visitor at Abramtsevo during the 1880s and took part in many family theatricals, including a production of Ostrovskii's play *Snegurochka* (The Snow Maiden) in which she played the title role wearing a costume designed by Viktor Vasnetsov.[32] In the mid-1880s she married the eldest son of Nikolai Iakunchikov, whose textile mills at Nara-Fominsk in Moscow province were among the largest in the country, and in the process acquired a considerable fortune that would later subsidize her kustar workshop.

Like the Mamontovs, the young couple were enthusiastic patrons of the arts, financing trips by artists to Italy to copy Renaissance paintings for the Museum of Fine Arts in Moscow and compiling an excellent art library. According to Igor Grabar, they were "cultivated and had a taste for refinement. This applied particularly to Maria Fedorovna, a talented woman who knew what was what in art and how to distinguish the real from the false, the serious from the vulgar."[33] She was a childhood friend of the painter Valentin Serov and, in addition to buying several of his early paintings, was herself painted by him twice during the 1880s.[34] As well as consolidating dynastic ties, her marriage also strengthened her artistic connections, bringing her into closer contact with Elena Polenova (now her "half-sister-in-law") and with her husband's half-sister, the painter Maria Vasil'evna Iakunchikova, both of whom she would later recruit to design for her workshop at Solomenko.[35]

With such strong links to the Mamontovs and their circle, it must have seemed natural to Iakunchikova to choose a "fine artist" to supervise her new workshop, rather than send for a graduate of the Mariinskii Lace School as other landowners were doing. She was certainly familiar with the history of the carpentry workshop at Abramtsevo, and no doubt reasoned that what Elena Polenova had accomplished there – that is, turn an act of philanthropy into an artistic and commercial success – another artist could achieve at Solomenko. Probably she asked the Mamontovs and their artist friends to recommend a likely candidate, for she finally selected Natalia Iakovlevna Davydova, a young auditor in Konstantin Korovin's painting class at the Moscow School of Painting, Sculpture, and Architecture.

Davydova was just eighteen years old, with very little formal artistic training, when she took up her post as artistic director at Solomenko.[36] She began by following the example of workshops like Kaznachaeva's, showing peasant women in the surrounding villages how to revive traditional Tambov patterns and stitches by using their own clothing as models,

> encouraging them to copy the old designs from their chemises onto squares of linen which could be used for table covers or onto lengths which could be made into curtains. They wove their own linen, spun their own thread, coloring them with vegetable dyes, principally indigo and madder red (the color of their skirts); the only other color they knew was a green, extracted from saw-wort; white they left its natural tint.[37]

rapid industrialization and trade protectionism, the exhibition provided entrepreneurs like Savva Mamontov with a forum in which to advertise their contributions to the nation's future prosperity. In the Pavilion of the Far North, designed and decorated by Konstantin Korovin, Mamontov couched the opening-up of the Siberian tundra by his Moscow-Iaroslavl Railway Company in the seductive language of Korovin's cycle of eerily beautiful paintings depicting the peoples and landscape of the remote northern provinces (Fig. 17). Even Mamontov's decision to build a second makeshift outbuilding for two huge decorative panneaux by Vrubel, after their expulsion from the Fine Arts Pavilion, seemed to imply that Russia's industrialists had the vision to support daring experiments in artistic modernism in the manner of Courbet and Manet, in the teeth of conservative opinion.[44]

By contrast, Russia's biggest natural industry, agriculture, was virtually ignored by the exhibition's organizers, and only at the Congress of Trade and Industry, which convened in August at the conclusion of the exhibition, did the anti-industrial lobby express its outrage that

Figure 17. Interior of the Pavilion of the Far North showing Konstantin Korovin's cycle of decorative panneaux commissioned by Savva Mamontov, 1896. (Reprinted from *Niva,* 39 [1896])

"nobody had as yet paid any attention . . . to the difficult position of Russian agriculture."[45] As for the kustar industries, agriculture's traditional helpmate, although they were given a pavilion of their own, designed by the architect V. Suslov, they received none of the publicity accorded Mamontov's Pavilion of the Far North, still less the avant-garde notoriety accorded the "broken lines, capricious blotches, and unnatural colors" of Vrubel's two decorative panneaux. Instead, visitors went to the kustar pavilion to be amused by such oddities as a piano over which a Viatka peasant had labored for five years and which wouldn't play, a replica of Leonardo da Vinci's *Last Supper* carved in palmwood, and a sewing machine that embroidered souvenir handkerchiefs. These eccentric examples of the modern kustar's talents were supplemented by the handiwork of various schools and workshops run by philanthropic groups and "lady patronesses," where "hundreds of thousands of children are learning to make things that are downright useless and anti-artistic."[46] As examples of what could happen when the wrong people were placed in charge of such enterprises, one reviewer singled out Circassian children from Kazan province who produced "roosters, woolen flowers like no flower on earth, and thoroughly poor stitchery using loud, tasteless materials," and various schools run by the Society for the Improvement of Popular Labor, which exhibited satin and velvet cushions with roses embroidered on them.[47]

Against this backdrop of misguided energies and dubious taste, Maria Iakunchikova proposed to set a huge embroidered panneau (3.6 m by 2.7 m) designed by Elena Polenova, with the obvious intention of giving her kustar workshop the same cachet of artistic modernism that her uncle had so circumspectly bought for his railway company. Polenova had earlier turned down a similar commission from Iakunchikova, but when her mother's death at the end of 1895 left her in difficult financial circumstances she was obliged to reconsider. Since severing her connection with the Abramtsevo workshop two years before, she had been spending much of her time mobilizing members of the Moscow Association of Artists to organize an exhibition of paintings on historical and biblical themes geared toward a popular audience.[48] She had also begun a new cycle of fairy tales, using original material she had gathered during her trip to Kostroma province in the summer of 1889, and it was one of these new fairy tales, *Ivan Tsarevich and the Fire Bird,* that she hoped to adapt with a minimum of effort for the Nizhnii Novgorod exhibition. Early in 1896 she described how the design was progressing in a letter to Vladimir Stasov, with whom she had recently become acquainted:

I show a dark night with clouds above, and through them the moon and stars are visible. In the middle is a tree with golden fruits. On a branch a fiery bird trembles. Around the tree fantastic flowers and grasses coil and intertwine. Beneath, in the roots of the tree, hares hide, and lower still there are swamp grasses, pebbles, and algae – all highly stylized.[49]

Although no reproduction of the completed panneau seems to have survived, Polenova's description, with its emphasis on the fantastic and on a high degree of stylization, suggests that it was similar to another of her illustrations to *The Fire Bird* that she recycled later that year (Fig. 18). In December Stasov asked her to contribute a drawing for an address to be presented to the sculptor Mark Antokolskii on the occasion of his twenty-fifth jubilee.[50] Polenova chose the scene where Ivan holds aloft the feather he has plucked from the tail of the magical Fire Bird, and she depicted him dressed in a boldly patterned blue and white coat, wedged flat against the foreground between a

Figure 18. *Ivan Tsarevich and the Fire Bird,* by Elena Polenova, 1896. (Reprinted from *Peterburgskaia zhizn',* 219 [January 1897])

huge and rather sinister plant of campanula blossoms on writhing stems and a wraithlike cypress. In juxtaposing this spare, graphic design with Elizaveta Bem's sentimental and painterly *Allegory of Sculpture* (two children building a snowman), Stasov could not have illustrated more clearly, if inadvertently, the rift that was beginning to open up between the defenders of realism and the forces of modernism in Russia – a rift that extended even into the ranks of "lady artists."

As Polenova herself admitted, there was a simple technical explanation for the boldness of her new graphic style as it emerged in the fall of 1896. After her first attempt to publish her illustrated fairy tales in 1889 had proved a disappointing technical failure, she had finally found a publisher who would print her new Kostroma cycle in color, provided she use the lithographic technique made popular twenty years before by English illustrators like Walter Crane. When she solicited Stasov's opinion on her new drawings for the tales *The Cunning Peasant, How the Bear Lost His Tail,* and *Red and Ginger,* he objected vociferously to the reduced range of colors this necessitated ("What's the point of using five or six colors when these days you can and should use three or four times as many? Why the uncalled-for economy of means, why this imitation of folksy clumsiness and meagerness?"),[51] and tried to convince her that "the simplification of colors" required by the lithographic process was an affectation imported from abroad. But this was not his greatest concern as he watched Polenova the painterly realist disappear, to be replaced by an artist dangerously enamored of stylization:

> Her treatment of the landscape, hitherto so charming and truthful, was transformed into a kind of muddle or enigma. Even her very ornament came to seem at times precious, mannered, and almost tasteless. The unhappy conceit of constant stylization, i.e., of distorting ornament, removing it from its source, treating it capriciously, "turning everything on its head" (to use her own expression) dealt her art a terrible blow and did great harm.[52]

The vein of hostility that runs beneath the scrupulously polite surface of Polenova's correspondence with Stasov throughout 1897 and 1898 was symptomatic of a similar ideological conflict that had informed the debates on ornament in England during the 1840s and 1850s. Stasov, with his distaste for "the repetitions, centers, axes, or symmetry that are so boring in any art,"[53] fully shared John Ruskin's horror of nature conventionalized into ornament, while Polenova, like Owen Jones, A. W. Pugin, and Christopher Dresser before her, threw herself passionately into the invention of ornamental forms capable of conveying "internal sensations" and "abstract themes and ideas."[54]

The wealth of ornamental designs that Polenova produced between ca. 1894 and 1898 seems not to have been initially conceived as self-sufficient projects for execution at the Solomenko workshops, however, but rather was the by-product of her intense interest in allegorical or symbolic painting. In her search for new ways to express internal sensations or psychic states she was joined by her friend, the painter Maria Vasil'evna Iakunchikova, then living more or less permanently in Paris on account of her health. Throughout the 1890s the two women corresponded regularly, Iakunchikova reporting on new trends in the Paris art world (recent work by Redon, Carrière, and the Rose + Croix group), Polenova updating her on the internecine squabbles of her own Moscow circle. A deep mutual understanding of each other's aspirations and abilities developed between the two women, together with a shared fascination with subjective experiences that Russian realism had so long ignored.

Not surprisingly, there were striking similarities in the experimental paintings each artist attempted in the mid-1890s. Iakunchikova's color etching *Terror* (1893–5) was an attempt to express how "one winter night in the Medon Forest [she] experienced a sensation of acute terror at the sight of the large stars swaying between the bare branches,"[55] an episode of heightened sensory awareness not unlike Edvard Munch's experience of "nature's great scream" in his 1895 painting *The Scream*. Polenova, in an 1895 sketch inspired by a line in Edgar Allan Poe's poem *The Raven* ("the sobbing of late autumn"), tried to capture "the melancholy of loneliness" in a "nocturnal landscape, the view of a roof in the city, a smoking chimney in the foreground, and a person's face with an expression of oppressive melancholy blowing into it."[56] A similar mood of fin-de-siècle yearning and despair permeates both Iakunchikova's etching *The Irreparable* (1893–5) and Polenova's watercolor *The Beast,* in which grim reality in the form of a snakelike monster threatens youth's innocent dreams (Fig. 19). (Polenova actually worked on this painting in Iakunchikova's Paris studio during the spring of 1895.)

But as the poet Max Voloshin would later point out, it was not the "unnecessary literary signature" of these decorative pictures that stirred a responsive chord in the viewer, but the settings in which these narratives took place. In an article on Iakunchikova that appeared in *Vesy* (The Scales) in 1902, the year of her death, Voloshin wrote: "The plant is the highest symbol of life and joy in painting. Ornament created from the lines and stems of plants is an eternal

Figure 19. *The Beast,* by Elena Polenova, 1896. (Reprinted from *Mir iskusstva,* 18–19 [1899]. Courtesy of the Department of Special Collections, Stanford University Libraries)

reminder of the lost paradise of nature. Iakunchikova comes to plant ornament at the very summit of her career." The same must certainly be claimed for Polenova. Never quite secure in her mastery of the human body – that basic ingredient of the Western painting tradition that her education had denied her – she made plants her expertise and sought to express through them the wealth of emotional content usually reserved for the figure. Describing the haunting effect that Polenova's *The Beast* had on her, the painter Ekaterina Iunge was disturbed not so much by the figures of the reptilian monster and the girl, as by their claustrophobic setting, which was "not nature, as one would expect, but something incomprehensible and unimaginable, a backdrop of stylized plants, of strident patches and slabs, that completely block out the air."[57] Polenova's subjects were the wildflowers familiar to every Russian, the very foundation of a national ornamental language, but emotionally charged and attuned to internal psychic vibrations that were the antithesis of the robust physical and psychological

Iaroslavl-Vologda railroad. In 1895 Chokolova opened her own kustar workshop at Gorozhanka, her estate in Voronezh province south of Tambov, and hired a graduate of the Mariinskii Lace School to teach *point de Moscou* needle lace and carpet weaving. But she also commissioned Polenova to design art embroideries such as the bright green lambrequin with a blue eagle that was shown at the Moscow Association of Artists exhibition in 1902. Inspired by the example of the new Solomenko embroideries, Chokolova herself designed carpets on floral themes that were later exhibited in the World of Art exhibitions at the turn of the century, where they found customers among members of the imperial family.[58]

The most controversial of Polenova's followers, however, and the one who was to push the stylistic potential of kustarnichestvo to its limits, was Aleksandr Golovin.[59] Justly celebrated as one of Russia's great theater designers, particularly during his long collaboration with the director Vsevolod Meierkhold, Golovin spent the first decade of his professional career as Polenova's protégé, fully involved in all of her projects and closely associated with the decorative and kustar arts. The two artists met in January 1890, soon after Golovin's graduation from the architecture section of the Moscow School of Painting, Sculpture, and Architecture. Golovin was then living in abject poverty, eking out a living by decorating walls and ceilings for the Moscow interior decorator A. A. Tomashko and by painting flowers on satin panels for screens. Throughout the 1890s he formed a deep attachment to and dependence on Polenova, thirteen years his senior, sharing her Moscow studio on Chudovskii Lane, choosing Polenovaesque themes for his paintings (e.g., *In the Icon Workshop: An Old Russian Icon Painter,* 1894), and accompanying her to Spain and France during her convalescence from a mysterious illness in 1897–8.[60]

Golovin designed for three Mamontov-Iakunchikova enterprises throughout the 1890s – the embroidery workshop at Solomenko, the Abramtsevo carpentry workshop, and the Abramtsevo Ceramic Factory, established in 1890 by Savva Mamontov with the help of Mikhail Vrubel and Petr Vaulin.[61] But his professional association with Polenova culminated in the commission to design a "Russian dining room" for Maria F. Iakunchikova's country house at Nara outside Moscow.[62] Though it has gone unnoticed in histories of Russian modernism, the Iakunchikova dining room was in fact a project of some importance (Fig. 22). Not only was it the first fully resolved interior in a definitively neo-Russian style intended for actual living rather than the stage; its ornamental lushness, medieval details, leaded windows, and low rafters proclaimed its kinship with an international movement to-

Figure 22. (*facing page*) Two sketches for a dining room in the Russian style for Nara, the estate of Maria F. Iakunchikova, by Elena Polenova, 1897–8. (Reprinted from *Mir iskusstva* 21–22 [1899])

ward neovernacular styles (in particular the work of Baillie Scott comes to mind). Last but not least, it was the first example of an interior produced entirely by improved kustar labor, and as such could lay claim to represent the genuine resurrection of national decorative traditions.

The walls of this long narrow room (10.3 m by 5.2 m) were to be covered with alternating ornamental strips and panels, painted or embroidered on linen that was spun, woven, and dyed at Solomenko (Fig. 23). Golovin contributed a *Fairy Garden* panel painted on linen and a large appliqué panel illustrating the changeling Swan Maidens from Russian folklore in "an exquisite blending of delicate mauves, greens and blues"[63] (later duplicated in majolica for the facade of the Metropol Hotel in Moscow). To the left of this, hidden in an obscure nook, hung the *Fire Bird* panel that Polenova had designed for Nizhnii Novgorod.[64] All the remaining wall space was filled with narrow

Figure 23. Sketch for a dining room in the Russian style for Nara, the estate of Maria F. Iakunchikova, by Elena Polenova, 1897–8. (Reprinted from *Mir iskusstva,* 21–22 [1899])

bands of stylized dandelions, crocus, harebells, mimosa, and other flowers, densely packed and magnified to create the disquieting, rather suffocating impression of a space vibrating with its own organic life.[65] In its completed state it was to include tiled stoves and carved furniture from the Abramtsevo ceramic and furniture workshops.

In 1899 the Iakunchikova dining room was written up in the March issue of the London magazine *The Artist* by Netta Peacock, an English journalist who had become friendly with Polenova, Iakunchikova, and Golovin in Paris during preparations for the forthcoming Exposition Universelle.[66] Charmed by the room's "fairy tale element . . . the Russian's love of all that is unaccountable," Peacock praised "the naturally artistic temperament of the two authors of this interesting undertaking [which] stamps the whole with an undeniable distinction."[67] No doubt it was this exotic "Russianness" that blinded Peacock to a problematic aspect of the decor that, on more familiar territory, she had no qualms about criticizing. Six months later, when she reviewed the English Arts and Crafts Society's autumn exhibition in London for the new St. Petersburg journal *Art and Industrial Art,* she voiced a popular sentiment when she noted that "many of the decorative and industrial goods are not adapted to the normal demands of contemporary life. The medieval and renaissance styles are so ruined that, while they are very beautiful and natural for their own time, they are quite incapable of expressing our civilization or the real life of the present moment."[68] Just such an anachronism did the Iakunchikova dining room seem in the eyes of many Russians, for whom the peasant *izba* was no more appropriate a model for modern living in Russia than the Gothic cathedral was for England. Painter and critic Igor Grabar may well have had this project in mind when he wrote several years later: "Contemporary life has evolved in such a way that in one's day-to-day environment one wants a degree of coziness and calm, one wants to eat and lie down in comfort, not to be faced with walls with lines and colors that endlessly jump out at you like the glass of a kaleidoscope."[69]

During the last two years of her life Polenova continued to work on a variety of projects for Solomenko as well as for her own pleasure, despite severe headaches, dizzy spells, and eye strain caused, as her family thought, by a fall from a carriage in April 1896. Simultaneously, she found herself swept up in the virulent struggles that divided the Russian art world across generational and ideological lines at the turn of the century, struggles in which the Solomenko workshops, for which she was doing some of her most original decorative work, were inevitably implicated.

The immediate cause of contention was the fact that two new art journals were scheduled to begin publication in St. Petersburg at the end of 1898. *Iskusstvo i khudozhestvennaia promyshlennost'* (Art and Industrial Art) was to be funded by the conservative Society for the Encouragement of the Arts (Polenova's own alma mater) and edited by its secretary, Nikolai Sobko, a close friend of Stasov.[70] Its rival, *Mir iskusstva* (World of Art), was the brainchild of Sergei Diagilev, an ambitious young man of good family with a relatively recent passion for the visual arts.[71] While Sobko's journal vowed to "fight against routine, to familiarize the public with healthy ideas that would put it on guard against aberrations in the arts and the industrial arts, and to develop the idea of Nationality in Art,"[72] Diagilev called for an all-out attack on old established artists and a unification of the young and talented: "Twenty-five years ago the Wanderers appeared," Diagilev proclaimed. "They were unable to shoulder their burden, they grew old. And yet our art has not declined, there exists a group of young artists scattered in various cities who, if they banded together, could show that Russian art is fresh and original, and capable of bringing much that is new to the history of art."[73]

Yet for all their differences both journals were equally determined to champion the decorative and industrial arts in Russia as an artistic category at the forefront of international attention. As the most notable representative of a nationalist school of decorative art, Elena Polenova's support had a strategic importance that was second only to that of Viktor Vasnetsov, also the unwilling object of the two journals' bickering. Stasov, who had followed Polenova's career since 1893 and thus far maintained a quite cordial correspondence with her, was confident that she would side with him in his campaign to fight the "decadence" of Diagilev and his friends. When she firmly turned down his invitation to contribute to *Art and Industrial Art,* therefore, and also refused to be numbered among his chosen band of helpers at the forthcoming Paris Exposition Universelle, Stasov was furious: "Polenova has refused once and for all to work with us," he fumed to a family member in July 1898, "pleading illness (maybe that's true, I heard she's gone mad and is being treated in Paris!!) as well as work she'd promised to do for 'others.' Well, these others are none other than the decadents, beginning with her friend Iakunchikova, who moved to Paris a long time ago and is raging about there in a clique of Russian and French *decadents.*"[74]

Polenova had indeed promised Diagilev her full support when the two had first met in Paris in April 1897. She was impressed by the young man's ideas and enthusiasm and wrote as much to Stasov:

> I've seen little of Diagilev himself and I don't know him well, but I find his artistic tastes, the direction of his journal, and the nature of his exhibitions very much to my liking. Moreover, many of my friends will be contributing to his journal and taking part in his exhibitions. . . . I was won over by the fact that he is concerned above all to escape from that terrible soporific artistic rut in which a great many would like to remain.[75]

Diagilev's true feelings about that aspect of Polenova's work that dealt with kustar production per se can only be guessed at. Certainly his motives in promoting the decorative and kustar arts so heartily in the months leading up to the appearance of the first issue of *World of Art* were far from disinterested. He was well aware of the vested interest that the journal's two financiers, Savva Mamontov and Princess Maria Tenisheva, had in improving the industrial and kustar arts in Russia, but it seems highly unlikely that he ever intended "his" journal to serve as a vehicle for broad popular enlightenment or kustar revival as Mamontov wanted. Shortly before the first issue came out in November 1898, Mamontov was quoted in a newspaper interview as promising that "special attention will be given to ceramics, majolica, ironwork, cast iron, furniture (in particular), fabric design, and mosaic, all of which will undoubtedly improve the taste of the masses."[76] But as Lev Bakst wrote to Aleksandr Benois, Mamontov had even more specific plans to turn their journal into a practical aid for the kustar:

> Mamontov . . . sees the magazine as "a support" for the kustar, a practical way of making information available to him [the kustar] by way of artistic engravings, samples, and supplements. As he put it, "Just as the tailor hunts for twenty kopecks to buy *Niva* [a popular weekly journal] so that he can cut out the patterns, so I would like the worker to try to get hold of our journal and enrich kustar craft with more refined models." . . . What a crude lack of mutual understanding at the magazine's very inception![77]

The idea of making the *World of Art* an unofficial organ of the kustar revival came up again in an interview with an anonymous "young artist" who declared that "to direct kustari and workers in all spheres of the craft and kustar industries the best artistic talents will put together designs and provide models. . . . The journal will be national in character and will develop and use popular motifs for ornament, but in the fine arts it will give complete freedom to the individuality of the artist."[78]

Mamontov was soon disabused of his confident vision – how many kustari, after all, could afford to consult the luxurious *World of Art* in hopes of finding a new design to work from? Nevertheless, Diagilev's interest in the new Russian decorative art movement that Polenova represented and his desire to "align [*World of Art*] with the new branch of industrial art developing in Moscow and in Finland" appear to have been quite genuine. He may also have been influenced by two members of his own family who were directly involved in kustar affairs. His elder brother Iurii not only operated a private school for the revival of naboika printing in the village of Borovenets in Novgorod province but was also sufficiently involved in kustar affairs and politics to be subsequently appointed director of the Ministry of Agriculture's Kustar Museum in St. Petersburg.[79] Moreover, Iurii's future wife, Tatiana Lugovskaia, was associated with the Mariinskii Lace School in St. Petersburg and was a friend of Elena Polenova, whose influence can clearly be seen in Lugovskaia's design for a carpet reproduced in *World of Art* in 1899.[80]

Diagilev had barely won his tug-of-war with Stasov for Polenova's support when, after several years of failing health, the artist died of a brain tumor in November 1898, a few weeks after the first issue of *World of Art* appeared. But having secured Polenova's promise of support in her lifetime, Diagilev had little difficulty in winning the confidence of her family with his evidently sincere appreciation of her talent. "We are giving her best things to Diagilev for his publication," wrote Polenova's sister-in-law to Maria V. Iakunchikova. "She sympathized with it and with him, and he had a high regard for her. Besides, there is still no other fresh publication in Russia."[81] As a result, Diagilev was able to organize a posthumous section of Polenova's work at the first World of Art exhibition in St. Petersburg in January 1899 (the so-called International Exhibition of Paintings), selecting precisely those late works that had given Stasov such cause for concern – her "incomprehensible" allegorical pictures (*The Wind Moans, The Fire Bird, Candles*) and her Solomenko embroideries, which were shown alongside Tiffany vases and Lalique jewelry.[82]

Although *World of Art* never became the kustar aid of which Savva Mamontov had dreamed, it was generous in its reproductions of embroideries and furniture from the Abramtsevo, Solomenko, and Gorozhanka kustar workshops, and the work of Polenova, Natalia Davydova, Aleksandr Golovin, and Maria V. Iakunchikova was well represented in photographs, articles, and reviews. Their versions of a neo-Russian style, so intimately connected to the needs of kustar production and the stylistic peculiarities of kustarnichestvo, were easily adapted for use as borders and vignettes in a number of the earlier issues, and a cover design by Iakunchikova, used throughout 1899, featured floral motifs that she would later recycle for an important Solomenko commission (Fig. 24). The three workshops were also well represented at many of the World of Art exhibitions and at the exhibitions of the Moscow Association of Artists.

There is little doubt that the World of Art circle found in Polenova and her small circle of fellow kustar designers the closest thing to a national decorative movement – an Art Nouveau or *style russe moderne* – that Russia could produce. But there is perhaps another reason why Polenova's decorative work – her ornamental designs and illustrations – should have endeared her to the young and cosmopolitan members of the World of Art group. After her death two very "symbolist" qualities in her working method for creating ornament were recollected and often repeated. According to one source Polenova possessed the gift of synaesthesia and

Figure 24. Cover to *Mir iskusstva,* 13–24 (1899), by Maria V. Iakunchikova. Watercolor, gouache, and gold paint on paper, 34.4 cm × 27.8 cm. (Courtesy of the Department of Special Collections, Stanford University Libraries)

> could compose a great many ornamental motifs in a short space of time, especially after good music. As she lay listening to the sounds, patterns came to her with great clarity. . . . The combinations of colors and lines are so subtle that they can be admired as one would admire a good painting. They are not simply patterns, they are synaesthetic, just as if they contained mysterious thoughts.[83]

Another friend described how many of her designs came to her in dreams: "Elena Dmitrievna set herself the task of seeking in nature forms by which to convey internal sensations – the desire for symbolic depiction. Day and night her head worked. After the impressions of the day she would often dream of fantastic combinations of flowers imbued with life, and on waking would rush to put them down on paper."[84] Even the spiteful conclusion of her former well-wisher Stasov – that her later work was the result of mental illness – added a certain piquancy to the dozens of decorative motifs she produced prior to her death, and anticipated the mystique that would later attach to Mikhail Vrubel's controversial painting *Demon Downcast* when it was shown at the World of Art exhibition in 1902.

As preparations for the kustar section at the 1900 Paris Exposition Universelle geared up throughout 1899, it became evident that even here the forces of tradition and change were to fight for dominance. Since late 1897 a small group of Stasov's women protégées had been planning the construction and decoration of a *teremok,* or traditional women's chamber, in which to display the achievements of the women's kustar revival. Stasov had expected Polenova to join Natalia Shabel'skaia, Elizaveta Bem, Varvara Shneider, and Tatiana Semechkina in this project, and was deeply offended by reports that, after first refusing to design the teremok, Polenova later changed her mind on condition that she be paid 14,000 rubles for her services.[85] Henceforth it was Shabel'skaia, not Polenova, whom Stasov would consider "one of the principal models and *vehicles for* a fitting, legitimate, and correct way for Russian industrial art to approach modern national activity."[86]

Times had changed since the last Paris Exposition Universelle of 1889 when the kustar exhibits had proved so unflattering to Russia's image abroad. Now Maria F. Iakunchikova was appointed chief organizer of the kustar section, the Grand Duchess Elizaveta Fedorovna was its imperial patron, and the painters Golovin and Korovin were put in charge of designing the kustar pavilions. Throughout 1899 Maria Fedorovna and Natalia Davydova were busy assembling kustar exhibits from all over the empire; that May they were joined at Nara by Maria Vasil'evna Iakunchikova with her newborn son.

Still grieving over Polenova's death, Maria Vasil'evna threw herself into plans for the kustar pavilion with a sense of personal responsibility for finishing what her friend had begun. In addition to designing a variety of toys and furnishings for the Moscow zemstvo's kustar workshops at Sergiev Posad (see Chapter 3), she oversaw the production of two wall hangings at Solomenko. One was a translation into appliqué of Elena Polenova's illustration to *Ivanushka-Durachok and the Fire Bird,* the other a huge (3.6 m by 3.0 m) tapestry-size panneau of her own design, showing a little peasant girl gathering mushrooms in a forest, while a *leshchii,* or wood spirit, spied on her from behind a tree (Fig. 25). Iakunchikova traveled to Solomenko to "supervise the dyeing of the linens, the difficulty of obtaining the exact shades of green required being most successfully overcome. The cutting-out and placing of all the pieces necessary to the carrying-out of this large piece of work seems an almost impossible task, looking at it from the purely technical point of view: all of this the artist did herself, leaving simply the outline to be worked by Nastasiia Ivantchouka of Solomenko

Figure 25. *The Little Girl and the Wood Sprite,* embroidered and appliquéd panneau, 300 cm × 360 cm, designed by Maria V. Iakunchikova and made at the Solomenko workshops for the 1900 Exposition Universelle in Paris. (Reprinted from *Dekorative Kunst,* 6 [September 1900])

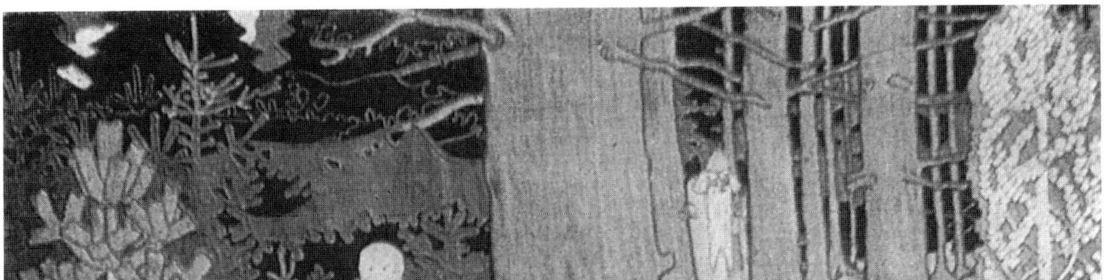

Guro wrote. "Here each ornament is a picture. It is marvelous music, an absolute symphony in colors. It is an entire world; the most profound creation of her soul. Precisely because nothing is defined, there are no boundaries to the imagination."[93] Such opinions were sensitive monitors of the shift that had occurred in Russian society since the great famine year of just a decade before. Why Polenova designed embroideries and furniture for kustar workshops was of far less interest to visitors than the realization that her most abstract and decorative designs could be more expressive, more intuitive, and more "symbolist" than oil painting.

Yet despite the interest in Polenova that her retrospective exhibition aroused, the year 1902 seemed to be less the beginning of a new era in the history of the decorative arts in Russia than the logical conclusion to a line of stylistic development that had begun at Abramtsevo seventeen years before. The sense that an era had ended was hinted at obliquely by Diagilev when he reviewed the Exhibition of Architecture and Industrial Art in the New Style that opened in Moscow that November:

> Till now all of Russia's industrial art seems to have been exhausted by the works of the Mamontov and Iakunchikova workshops, in which so many talented Russian people have taken part. These enterprises created an "epoch" of sorts in Russian industrial art, and it seemed as though nothing else new was being done and that the whole Western industrial art movement had, for the time being, left our artistic forces unmoved.[94]

The style that Polenova had established for kustar art through her pioneering work with the Abramtsevo and Solomenko kustari had become an acceptable, even preferred style for the further development and modernization of the kustar art industries – the more so as their potential as export items and souvenirs became realized. But in the sphere of the decorative and industrial arts those artists with an interest in the application of art to life recognized that this version, at least, of the neo-Russian style had serious shortcomings. When a selection of furniture and majolica designed by Golovin was exhibited at the 1902 Moscow Association of Artists exhibition, more than one critic felt that some fine line of taste and common sense had been transgressed. In reviewing the artist's designs the architect E. Baumgarten wrote that, while such works were "perhaps suitable for export abroad . . . it was a pity that the coarseness of the artist's taste might be quite erroneously attributed to all Russian art. Compositions by Vasnetsov, Vrubel, and Korovin that have appeared at other exhibitions suffer from a suggestion of the same coarseness and a complete disregard for constructive forms, so that the real place for such works is the theater stage and not everyday life."[95]

In 1902 Igor Grabar returned to Russia from an extended trip abroad to find disturbing developments in the new decorative movement since the Paris Exposition two years earlier. Among them was a penchant for self-conscious and exaggerated crudeness in design and execution. Certainly, Grabar conceded,

> there is a certain charm in clumsiness, in the naive awkwardness with which the kustar carves his simple patterns [and] beneath this clumsiness an artist's soul is often concealed. I admire folk toys, a beautiful shaft bow, marvelous birch-bark boxes. But then, this is the essence of sincerity, without an ounce of falsity. But when an artist with refined feelings, who has been brought up differently and above all whose lifestyle is different, imitates this talented kustar even down to his weak points, this I refuse to understand. We must ultimately admit that this imitation of the kustar contains a liberal dose of what is called "love for the charming little peasant," or *paysannerie.*[96]

3

The Moscow Kustar Museum and the Sergiev Posad Workshops

One almost envies the koustar peasant his simplicity of mind, since he so well understands childhood's desires.

Twentieth Century Russia

THE PRECEDING CHAPTERS examined the ways in which a growing number of private citizens became involved in small-scale kustar reform throughout the 1880s, injecting their money, their aesthetic tastes, and their charitable impulses into tiny pockets of the peasant population, in the hope that the seed of their good intentions would take root and multiply. By contrast, the Russian government successfully resisted practical and fiscal involvement in kustar affairs for nearly two decades. In 1888, however, under increasing pressure from proagricultural quarters, the State Council officially proclaimed the national importance of the kustar industries by assigning their welfare to the Ministry of State Domains.[1] The ministry was under no illusion that the task of rescuing this branch of the national economy from the brink of disaster was anything but a daunting one. With a miserly budget of 35,000 rubles at its disposal and a powerful enemy in the minister of finance, it did its best to draft long-term policy and initiate short-term improvements. It granted modest subsidies to many of the existing private women's workshops, began organizing exhibitions and publications, and pushed hard for the establishment of a central kustar museum in St. Petersburg that would acquaint the public with examples of kustar goods and disseminate improved designs, prototypes, and tools to the centers of production. Still, as of 1894 the ministry was obliged to admit that "little had as yet been done to develop and improve kustar industry."[2] Viewed from the distant perspective of St. Petersburg, the efforts of individuals like Elizaveta Mamontova were drops in the ocean, random instances of charity that were powerless to effect real change. If the kustar industries were to hold their own against rapid growth in the industrial sector, it was not to the private sector that central government looked for help, but to the governing bodies (zemstva) of the individual provinces where such crafts were still practiced.

The zemstvo system had been established in the years immediately following the Emancipation Act of 1861 to administer the day-to-day affairs of the rural population – education, health, judicial matters, and the local economy. Although the protection and promotion of a province's kustar industries were not actually spelled out in the statute governing zemstvo duties (and were certainly not budgeted for), as public interest in the kustar question increased in the 1870s a handful of zemstva realized that this would prove to be an important part of their brief.[3] These were for the most part zemstva in those provinces where agriculture was a primary but most unreliable source of income and industry virtually nonexistent. In the remote eastern province of Perm, for example, the zemstvo approved the establishment of a kustar bank in 1888 (opened 1894) as a way of providing local kustari with affordable credit. The Tver zemstvo tried to set up artels and a system of trading depots, and Nizhnii Novgorod organized a workshop to supply footwear to commissaries.[4]

But the most ambitious measures were those taken by the Moscow provincial zemstvo, which plunged into the question of kustar reform in 1875 with a statistical survey of all kustar industries still practiced throughout the province.[5] When it was completed in 1882, the survey revealed that some 140,000 peasants in Moscow province were involved in some form of kustar industry, making a clear profit of 7.5 million rubles each year, that is, about 18 percent of the annual budget for the entire peasant population of the province. In purely economic terms, the Moscow zemstvo's dependence on kustar industry was perhaps less acute than that of zemstva in the remote northern provinces or the depressed black soil regions. Moscow province was after all the empire's fastest-growing industrial center, thanks in large part to the merchant dynasties of Moscow (the Iakunchikovs, Mamontovs, Morozovs, Soldatenkovs, Tret'iakovs, and other families), which owned primarily textile factories throughout the Moscow industrial region and fostered the growth of the railway system. But it was also the country's biggest center of kustar furniture and toy production. As late kustar industry."[15] There was therefore a certain poetic justice in his determination to lavish a fortune made from the textile industry on the promotion of factory-threatened kustar handicrafts. Added to all this were Morozov's love of art – particularly folk art – and his contacts among the leading artists of his day, which made him the principal conduit through which the kind of artistic improvements practiced at Abramtsevo began to filter into Moscow's kustar art industries on a fairly large scale.

Morozov's proposals for expanding the zemstvo's kustar-related activities were approved by the Moscow zemstvo assembly and in 1891 resulted in the adoption of the training workshop as the basic unit for reforming selected kustar industries throughout the province. After much research into the viability of various crafts, the zemstvo opened six carefully chosen training workshops, all but two of them geared toward what might be called luxury crafts. Five of them targeted lacemaking, basket weaving, metalwork, netting, and papier-mâché goods, while the sixth was set up in the heart of Russia's traditional toy center, the trading settlement known as Sergiev Posad.

Sergiev Posad lay in the shadow of one of Russia's most important and venerated monasteries, the Troitse-Sergieva Lavra about seventy kilometers northeast of Moscow. Located near the birthplace of Saint Sergei Radonezhskii, the Troitse Lavra was an important pilgrimage site and attracted pilgrims from every corner of the empire. It had also been renowned for wood and toy carving since at least the eighteenth century and by mid-century was the acknowledged center of kustar toy production for the entire country.[16] Like all kustar products, "Posad" toys (Fig. 26), as they were popularly called, depended on a complex middleman system to reach their numerous and far-flung customers. Pilgrims bought the toys through special stores within the monastery complex itself; peddlers hawked them on city streets, at fairgrounds, and marketplaces; and carriers acted as go-betweens for the isolated kustar and the big Moscow department stores (Aksenov, Vasil'ev, Tikhonov, Muir and Merilees), which then sold the toys at a 200–300 percent markup. The kustar toy maker was thus caught in a web of dependency, aggravated by the fact that, of all the kustar industries, toy making was most susceptible to the swiftly changing wind of fashion.[17] By the end of the nineteenth century, the colorful, crude, and often satirical Posad toys were seriously threatened by imported toys from Germany, whose "refinement and good taste" appealed to a largely Europeanized city clientele.[18] All these factors, together with the fact that the Sergiev Posad toy industry was one of the largest in the province, made it an obvious target for the zemstvo's intervention.

Figure 26. Carved kustar toys from Sergiev Posad, early nineteenth century. (Reprinted from L. G. Orshanskii, *Khudozhestvennaia i kustarnaia promyshlennost' SSSR* [Moscow, 1927])

Few details are available about the toy workshop's activity during the first nine years of its existence, that is, until its debut at the 1900 Exposition Universelle in Paris. We know that technical experts were brought in to work with the kustari, better and cheaper raw materials were made available, and new designs introduced. At the 1896 Nizhnii Novgorod exhibition, for example, the workshop exhibited elegant dolls copied directly from Parisian models.[19] In fact, as of 1896 all six model workshops operated by the Moscow zemstvo were using exclusively foreign models to improve the salability of their products, thus capitulating to the average Russian consumer's disdain for conspicuously home-produced goods. Conversely, as one observer pointed out, this short-term solution to broadening the market for kustar toys in Russia automatically disqualified them for any sort of export abroad, since "there is nothing characteristically Russian about them."[20]

The first new toy to be produced at Sergiev Posad designed entirely by a professional Russian artist was the matreshka doll (Fig. 27).

Contrary to popular belief today, the matreshka was not an ancient folk symbol or even a traditional kustar toy,[21] but was designed in 1891 by the young artist Sergei Maliutin for Maria Mamontova's Children's Education toy shop on Leontievskii Lane.[22] In response to Mamontova's commission for a doll inspired by a Japanese nesting toy depicting the Sage Furkumu, Maliutin created a barrel-shaped peasant girl dressed in white kerchief and sarafan, red cheeked, broad smiling, and clutching a black rooster under her right arm. This "matreshka" (the diminutive of Matrena, a common peasant woman's name) opened to reveal a girl, within which was a smaller girl, and so on, until at the very core lay a swaddled infant.

Maliutin's design was produced and sold at Children's Education until 1898, when the store closed down and its stock, including the matreshka, was transferred to the Moscow zemstvo's workshop at Sergiev Posad. The new toy immediately went into production at the training workshop, and as volume increased into the millions the painting and finishing of the dolls were entrusted to a separate workshop operated by the Ivanov brothers. Here a strict division of labor

Figure 27. First matreshka doll, by Sergei Maliutin, 1891. (Reprinted from *Matreshka* [Moscow, 1969])

was maintained – not unlike that of traditional icon workshops – as "whole batches of matreshki passed from one craftsman to the next. One lays down the outlines of the face, a second paints the eyes, a third paints the rooster or, for variety, a goose. Others paint in the matreshka's kerchief, cover her with lacquer, polish her, and assemble all the matreshkas into a single set."[23]

The appearance of the matreshka was just the first of many sweeping innovations in design instituted by the Moscow Kustar Museum at Sergiev Posad. By 1903 an entirely new workshop of major importance had been created out of one of the humblest and worst paid of all kustar occupations, that of the box makers (*iashchichniki*) who prepared boxes and miniature wheels, rakes, spades, and doll furniture as accessories for toy sets. Nikolai Bartram, the leading designer of toys for the museum during the 1900s, explained how this metamorphosis came about:

> People needed to be given a chance to earn a better living, and they were offered an entirely new form of artistic joinery work with a distinctive poker-work decoration. It was proposed and put into production by a talented young artist and student of Levitan, V. I. Sokolov. . . . Either folk ornament or landscape and narrative themes were transferred onto turned and joined sandboxes, boxes, chests, panels, and other utilitarian objects. The decoration was done in bright, clear local colors that followed the silhouette of the poker-worked lines. And out of these "box maker" kustari a workshop was organized that prepared various objects with geometric and relief carving. There was such a demand for all these goods that a new building had to be made for the art joinery workshop.[24]

Vladimir Sokolov, the artist responsible for this new kustar craft, had become acquainted with Sergei Morozov around 1895 through Isaak Levitan, the landscape painter. By this time Morozov had already established himself as Levitan's patron, providing him with a studio in Moscow and pressing on both him and his friend Anton Chekhov generous, if not always welcome, financial assistance.[25] With an eye to the needs of the Kustar Museum, Morozov cultivated Sokolov's acquaintance, inviting him to spend the summer at Uspenskoe, his estate in Zvenigorodskii uezd, and by 1902 he had managed to persuade the painter to become head of the new art joinery workshop at Sergiev Posad.

Sokolov's introduction of poker-work (pyrogravure) designs on boxes and other decorative items was even more of a novelty than the matreshka. Although he supposedly drew on lubki (popular woodblock prints similar to the French *images d'Epinal*) and painted distaffs for inspiration, the actual technique of poker-work was neither a traditional folk craft nor a kustar industry. On the contrary, like appliqué embroidery it was a quintessential fin-de-siècle arts-and-crafts technique for producing simplified flat decorative motifs in the manner of the Pont-Aven school or the decorative arts illustrated in the English art magazine *Studio*. Closer to home, Sokolov was surely aware of the experiments in this medium that Maria Iakunchikova had made in the last years of her life (ca. 1895–1900), some of them not only inspired by the Troitse-Sergieva Lavra but actually intended for production as three-dimensional toys at the Sergiev Posad workshop nearby.[26] The themes Sokolov chose – winter landscapes, onion domes, kremlin walls, Russian villages – found parallels not only in Iakunchikova's painting of this period, but also in a national school of painters and illustrators, among them Sergei Ivanov, Ivan Riabushkin, Apollinarii Vasnetsov, Korovin, Ivan Bilibin, and Vasilii Kandinskii.

A third instance of modernized, improved kustar production promoted by the Moscow Kustar Museum was the unusual case of Vasilii Vornoskov (1876–1940), a peasant whose native initiative and early experiences at the Abramtsevo workshop made him one of the very few individuals to straddle the segregated worlds of intelligentsia and narod, and to become what was referred to as a "kustar-artist," creating his own style of carving and developing it into a local industry.[27] Vornoskov was born in the village of Kudrino, four kilometers from Abramtsevo, and as a young boy entered the Abramtsevo training workshop while Polenova was still active there. He completed his training in 1887, and in 1892 began sending his Polenova-designed wares to the new zemstvo workshop at Sergiev Posad for sale. But by 1902 Vornoskov no longer wanted to copy other artists' designs and proceeded to create his own. He developed a foliate carving style quite distinct from the geometric chip-carving of the Abramtsevo repertoire, a dense, mazelike field of tendrils, curlicues, acorns, and fruits carved in low relief, then stained and polished to a shiny finish (Fig. 28). On Sokolov's advice he adapted various shapes and objects from exhibits in the Moscow History Museum, among them boxes with lids in the shape of birds and beasts modeled on antique saltboxes. When his two elder brothers died, Vornoskov became the head of a large extended family of over thirty, many of whom he trained to carve in his own distinctive "Vornoskov style," and in 1906 a separate workshop was set up.[28] Among the goods that he and his family produced were desk sets, dishes, troughs, and decorative *kovshi* (bird-shaped dippers) "hollowed from a whole piece of wood, with hyperbolically developed decorative handles in the shape of horse heads and birds, stained dark brown, and carefully polished" that were exported in large numbers to Germany and England.[29]

Such innovations were not accepted without resistance and suspicion by the kustari of Sergiev Posad. It was reported, for instance, that they "didn't understand the benefit of introducing new lines and produced the new designs unwillingly, asking extremely high prices for the improved goods."[30] But when resistance did occur, it was motivated less by any innate fear of change – a stereotypical "peasant conservatism" – than by basic economic concerns, as a report from 1890 makes clear:

> The kustar toy makers understand quite well the importance of new models for their production. Ask any one of them and you're bound to hear that getting good new designs is absolutely essential in the toy industry. But that doesn't mean that every toy maker will definitely decide to be the first to produce goods from these new designs. The fact is that when a toy maker begins work on a new design he must incur new expenses either in materials or in acclimatizing himself; second, once he's used to making one kind of product, a kustar has to expect a decline in productivity if he moves to another kind; and finally, he is scared by the

Figure 28. Display of decorative kovshi and icon frames by Vasilii Vornoskov and family, ca. 1913. (Reprinted from *Russkoe narodnoe iskusstvo* [Petrograd, 1914])

crowing a little to a scoffing public, an unsympathetic bureaucracy, and a cosseted industrial elite. What had long been the kustar industries' shame – their technical backwardness, their resistance to mechanization and progress, their lack of modernity – now became their glory in a climate of backlash against accelerated industrial progress. A German correspondent rhapsodized:

> In the Tauride halls a genuine popular spirit blows, strong and healthy like the smell of the earth; an air that makes the chest expand and the muscles relax. Here the spontaneous creativity of the people reigns, here we catch the Russian people working, not in stuffy city factories that wash away the originality and the very personality of the worker, but at home on their own land, under conditions that do not hamper the free development of native originality and inventiveness.[48]

Minister of Agriculture A. S. Ermolov put his finger on both the kustar's trump card and the Achilles' heel of large-scale industry in a speech at the closing gala banquet: "This is no dead machine at work here; the kustar invests his work with his skill, his knowledge, his soul, and from this kustar goods derive their distinctive quality. They reflect the personality of the worker, his environment, his life."[49]

At least one member of the artistic community refused to be seduced by the exhibition's window dressing, romantic rhetoric, and businesslike organization. After visiting the Tauride Palace's halls and seeing their chaste neoclassical lines overlaid with trumpery cardboard *kokoshniki* (women's headdresses) and fire birds, Aleksandr Benois wrote a scathing review for the *World of Art,* subtitled "Testimonium paupertatis." As a hard-line Petersburgian with little sympathy for Moscow and its Slavophile associations, Benois recanted his initial sympathy with the neo-Russian style practiced at Abramtsevo and Solomenko and announced that its day had finally passed:

> Now artists have cooled toward this masquerade, and all these workshops are moving toward a rapid decline. Abramtsevo is unrecognizable, Chokolova carpets have deteriorated into total ugliness, and even the best things of all, the Moscow zemstvo's goods (from the workshop of Mme. Davydova) are when all's said and done just charming dilettantism and not the development of popular art. . . . Our people are in utter decline, and it is essential that we keep this constantly in mind, because this is a terrible business.[50]

But despite Benois's dark and undoubtedly partisan warnings, the sympathy of Russian artists for some sort of neo-Russian style formally linked to kustarnichestvo was not yet entirely spent. In December 1902 a group of Moscow artists and architects joined forces to organize an exhibition devoted to the problem of uniting a vernacular decorative tradition with the "new style" (as Art Nouveau and Jugendstil were called in Russia) to form a *style russe moderne*. This Exhibition of Architecture and Industrial Art in the New Style was the first group effort by the recently formed New Society of Architects, and its principal organizer was the young architect Ivan Fomin, "a recent negator of nationalism, opponent of everything Russian, and admirer of [Otto] Wagner and Vienna, [who] had made a sudden about-face in favor of Russian terems [chambers], icon lamps, and Sirin birds."[51] Also indicative of this change of heart was the contingent of kustar enthusiasts whom Fomin invited to be part of the planning committee – namely, Natalia Davydova, Aleksandr Golovin, Konstantin Korovin, and Sergei Morozov.

Among the participating architects, interest was clearly divided between the straightforward imitation of European styles (Lev Kekushev, William Walcot, Ivan Zholtovskii) and the hybridization of Russian vernacular elements with the formal vocabulary of international Art Nouveau (Ivan Bondarenko, Fomin, Fedor Shekhtel.) The pro-European orientation was undoubtedly strengthened by the participation of two prestigious practitioners of the new style from abroad. Evidently through the intervention of the omnipresent Grand Duchess Elizaveta Fedorovna, who as a princess of Hesse-Darmstadt retained a keen interest in the Darmstadt art colony patronized by her brother,[52] Joseph Maria Olbrich spent three weeks in Moscow preparing a dining room and living room for the exhibition, while Charles Rennie Mackintosh designed a white salon that was rudely likened by one reviewer to "an operating room in a clinic or a design for a Moscow barbershop" (see Fig. 4).[53]

Far from earning the approval of Olbrich, the Russian pastiches of Vienna Jugendstil that made up the majority of the exhibits merely caused him to comment, "Das ist zu wienerisch!" The three rooms that Fomin designed, however, aroused his interest. These were a sitting room with Karelian birch furnishings made by the Levisson Firm of Moscow, a *gornitsa* (the name given to the clean part of a peasant hut) of white polished beech with a black ceiling, and a dining room fitted out with a gray maple dining-room suite priced at 2,500 rubles (Olbrich's cost 3,400 rubles) and made in the Moscow zemstvo's kustar art carpentry workshop at Sergiev Posad (Fig. 32).[54] The Sergiev Posad kustari had evidently won this prestigious commission thanks to the presence of Sergei Morozov on the planning committee, and the subtext of their participation was clearly that the humble art of the Russian kustar could, under the right management, satisfy the demands of the most discerning customer. The furniture itself was constructed of right angles and massed horizontal slabs, with a sparse ornamentation based on the triangle and the square, enlivened with surprising motifs such as polar bears (perhaps a reference to Korovin's panneaux for the 1896 Pavilion of the Far North at Nizhnii Novgorod) and helmeted Varangian knights.

Fomin's ensembles provoked a good deal of controversy and placed a severe strain on many visitors' perceptions of what a Russian interior should be. Some critics, the architect Vladimir Karpovich for example, admired the way in which Fomin had modernized the Russian style without actually imitating specific details of antique objects and how he had managed in the process to express the pessimism of modern life through color and form. "You wouldn't say that this room was intended for merry feasting," he wrote. "The overall blue-gray tone of the furniture, the calm and monumental forms of the mantel, the quiet murmur of the fountain, the medieval chandeliers – the simplicity of it all astonishes you, but it doesn't create a bright and cheery mood. On the contrary, it has rather a depressing effect."[55] Others had more conservative and pragmatic views on what constituted Russianness, with comfort occupying a prominent place in their minds: "Whoever manages to sit [on the backless divan] for more than five minutes deserves a prize," one reviewer wrote huffily. "As for the arm-

Figure 32. Maple dining room suite designed by Ivan Fomin and made at the Sergiev Posad workshop for the Exhibition of Architecture and Industrial Art in the New Style, Moscow, 1902. (Reprinted from *Mir iskusstva,* 3 [1903])

chairs with the docked abbreviated backs with white leather embossed with polar bears stuck on them, the reclining backs are uncomfortable and the arms and seats are incised with sharp lines that dig into your body. It's impossible to fit your legs under the chair, and who's going to sit motionless with their legs stuck out in front for an entire dinner?"[56]

For all those for whom Fomin's veiled allusions to the familiar vocabulary of a Russian style were too subtle or too diluted to be easily read, and contravened a notion that comfort, practicality, and a certain uncomplicated heartiness were integral elements of any Russian style, the Sergiev Posad workshops also offered an entire room of improved kustar goods, many of them furniture and smaller objects designed by Natalia Davydova (Fig. 33), Lev Brailovskii, and K. V. Orlov. Inexpensive and decorative – perhaps not unlike the goods shown at the Tauride exhibition earlier that year – for many these shelves, cupboards, boxes, and frames "saved the entire exhibition from being censured for its complete estrangement from everything Russian."[57] The promotional value of taking part in such a notable exhibition was obvious. Here "the public discovered for the first time . . . that in the settlement around the Troitse-Sergieva Lavra a workshop directed by [V. I.] Borutskii produces elegant carved goods and entire ensembles, primarily in the Russian style."[58] Conversely, the absence of the Abramtsevo workshops was commented on by more than one reviewer, and their importance reevaluated. "The Abramtsevo workshop? Certainly it was a great step forward in the development of our applied

Figure 33. Carved shelf designed by Natalia Ia. Davydova and made at the Sergiev Posad art joinery workshop for the Exhibition of Architecture and Industrial Art in the New Style, Moscow, 1902. (Reprinted from *Mir iskusstva,* 3 [1903])

art," M. Mikhailov conceded. "But don't you notice that Mamontov's name is quite irrelevant here, and that Polenova, for all her talent and her works, has exhausted all she had to give?"[59] For architect E. Baumgarten the Abramtsevo group's affinities with kustar art were only skin-deep, confined to the decoration of surfaces, the "stylization of landscape and many animals, and the combination of various materials (wood, glass, majolica, metal, etc.)."[60] And A. Dmitriev noted that "goods from the Abramtsevo workshop quickly lost their vitality and, thanks to an upsurge of imitators, brought an element of falsity to an extremely interesting and promising undertaking."[61]

As late as 1902, then, the revival of kustar art as part of an interest in forging a national form of the new style still clearly enjoyed some support and prestige in the Moscow art world as an arena for experimentation in the applied arts and for the aesthetic transformation of modern life. Despite complaints from St. Petersburg, the bastion of a more European culture, a number of artists and architects continued to believe in the Russian style in general and specifically in the possibility of making improved kustar art *the* Russian decorative art par excellence. Meanwhile, a similar conclusion had been reached in very different quarters and for different reasons. The need for elementary art education in the nation's schools and the cultivation of national content for Russian industrial development had been acknowledged by the industrial sector ever since Russia had begun to take its place alongside other nations at the world fairs of the 1850s and 1860s, where Russian goods were ridiculed for slavishly imitating European styles. To the extent that the Ministry of Agriculture and State Domains envisioned the future of kustar art industries per se, it was still in the larger context of agriculture and the rural economy. For the Ministry of Finance, however, which throughout the 1880s and 1890s had adamantly resisted the kustar's claims to governmental support, the potential of kustar arts and crafts for development into a distinctively national form of industrial art now acquired a new interest. In 1902 the Ministry of Finance sponsored a new Statute on Industrial Art Training Institutions designed to restructure the teaching of basic artistic skills in the empire's schools. Whereas the existing statute, enacted in 1888, had allowed only the lowest tier of schools, the so-called primary craft schools, to "promote kustar production among the peasantry,"[62] under the new statute Russia's kustar population was to enjoy sweeping new educational privileges that would help to erode the barriers that still separated rural and urban Russia:

> For the most gifted kustari who have gone through a course of industrial art education, beginning with drawing classes and ending with an advanced industrial art school, broad prospects for the practical application of their talents and knowledge present themselves. The interaction of art and popular industrial art will prove mutually beneficial for both parties, deepening that national self-awareness without which there can be no truly original art.[63]

At the pinnacle of this new hierarchical structure were Russia's three oldest schools of design – in Moscow the Central Imperial Stroganov School of Technical Drawing (f. 1861) and in St. Petersburg both the School of the Society for the Encouragement of the Arts (f. 1862) and the Baron Shtiglits School (f. 1876) – where designers for the nation's factories and teachers for its burgeoning schools were to be trained. Gradually, with encouragement and subsidies from the Ministry of Finance, a second tier of industrial art schools and workshops began to develop in important centers of traditional kustar production: the Gogol School of Industrial Art in Mirgorod, Poltava province, which was the center of major ceramics production (1896); the Ekaterinburg School of Industrial Art in the center of the Urals lapidary and jewelry industry (1902); a training workshop in Tula to develop artistic metalwork (1903); and a training workshop for gold- and silversmiths in Bol'shoe Krasnoe, Kostroma province (1904). But since all of these institutions were located in large towns and drew most of their pupils from the urban population, the ministry's intention was clearly to facilitate and speed up the desired transformation of kustar crafts into industrial arts, while at the same time assimilating the stylistic hallmarks of kustarnichestvo into a new national style for all facets of manufacturing.

Lowest down in the new hierarchy of art education, an assortment of schools and workshops came into being, established and funded by the Ministry of Agriculture and State Domains or by various zemstva. In Moscow province the district zemstvo of Cherkizovskii uezd made its own modest efforts to bring a basic art education to the kustari of Ligachev, one of the busiest and oldest centers for furniture production in Moscow province.[64] In 1902 it opened a drawing school to improve the local furniture trade by teaching kustari to copy new furniture designs provided by the zemstvo. But since classes in simple copying were not followed up by instruction on how to translate drawings into actual goods, the initiative met with little support from the kustari, who could see "no direct results gained from learning drawing, either in an improved artistic standard in the school's graduof the kustar industries had become entirely commercial in their orientation. When Sergei Morozov presented a proposal for restructuring the zemstvo's handling of kustar reform to a special commission that year, two irreconcilable viewpoints emerged. A minority of the commission's members, evidently siding with Morozov, claimed that the Moscow Kustar Museum's emphasis on marketing luxury kustar art goods for an elite or foreign market was deflecting attention and precious resources from all those other kustar industries that required only technical improvements. The Kustar Museum's carrot-and-stick policy, they argued, was merely perpetuating the kustar's childlike dependency and stifling the kind of self-help and initiative possible only when kustari banded together in artels and cooperatives. But the majority of the commission were in favor of continuing the zemstvo's direct commercial activities as the only reliable way of ensuring fundamental changes in kustar production. "The zemstvo's independent trading activity is the best way of verifying the activity of the zemstvo specialists who introduce improvements into kustar production," they pointed out. "Retail sales in particular are essential for popularizing kustar work and handicrafts among the public and on the market, since the market is set up to accommodate the factory and takes no account of how the kustar should proceed to make his work competitive with that of the factory."[71] Of course, the time would come when "Moscow province would be covered by a network of cooperatives" that would relieve the zemstvo of its commercial functions, but until that day arrived the zemstvo resolved to continue on the path it had set itself.

Taken at face value, the Moscow Kustar Museum's intensive marketing of luxury kustar crafts in the early 1900s contravened many of the ideals that guided the rejuvenation of vernacular traditions throughout Europe and England – for example, that an object be both designed and made by the same person, that it be a joy to the maker and the user. The museum's market-driven policies ought therefore to have been anathema to any artist seriously concerned with the survival of genuine folk culture in Russia. But this was not the case, at least for that generation of the Russian artistic intelligentsia reared on the social service ideals of the late nineteenth century. Only a very naive observer ignorant of the conditions of kustar production in Russia could have failed to see how inextricably market forces and consumer demand were intertwined with questions of style. Nobody understood this more clearly than Nikolai Bartram (1873–1931), who joined the staff of the Moscow Kustar Museum as an artist in 1904 and served as its director from 1907 until 1917.[72] Bartram's hopes for a renaissance among the toy makers of Moscow province were rooted in his faith that the public could be weaned from its infatuation with factory-made toys and taught instead to appreciate the sincerity of kustar toys, and his entire career was devoted to reshaping consumer taste in this direction:

> Making society aware of the significance of the toy means simultaneously saving the child's creativity from being invaded and polluted by market speculators and allowing the kustar to revive his art. When people realize that the kustar toy is the result of artistic thought and is made not for empty amusement but for joy, when they see how superior it is to factory toys, then the demand for kustar toys will naturally increase. And this demand will allow the kustar not to try to imitate the factory, but to devote himself to what is closer to his art.[73]

Bartram had first become interested in folk art while still a student at the Moscow School of Painting, Sculpture, and Architecture, and in 1900 was among the artists invited to illustrate Pushkin's fairy tales for the Sytin publishing company's jubilee edition. Upon graduating he researched folk toys at the Moscow History Museum under Ivan Zabelin and in the private collections of Aleksandr Benois, Petr Shchukin, Lev Orshanskii, and Natalia Shabel'skaia. Whether inspired directly by the example of Abramtsevo, or more generally by the growing number of kustar training workshops, in 1893 Bartram set up his own workshop in an abandoned estate near the village of Semenovka in L'govskii uezd, Kursk province. Over the next decade he began to put into practice his theories on the aesthetic education of children. "The participant should receive satisfaction and joy from his work, see its beginning and its end in his own hands, experience it," Bartram wrote. "I feel sure that it is the making of wooden toys that best answers these demands."[74] Arguing that a peasant boy would more enjoyably and thoroughly learn the rudiments of the joinery trade by beginning with wooden toys, Bartram gave his pupils characters from Russian fairy tales as subjects for their toys, which were then sold in sets in a special box complete with a short text of the relevant fairy tale. The pupils worked their way from simple silhouettes cut from board to toys turned on the lathe and decorated with patterns taken from old brocades, naboika, and embroideries in Bartram's own extensive collection.[75] The program culminated in the preparation of wooden furniture inspired by old Russian forms and also, no doubt, by Polenova's Abramtsevo repertoire: "broad settles with collapsible backs, shelves with ingenious little cupboards, mirrors in carved

Bartram was a tireless propagandist of his beloved kustar toys. Through a series of exhibitions – Art in the Life of the Child (St. Petersburg, 1908), Russian Toys of Past and Present, and How Toys Are Made (both held at the Moscow Kustar Museum, 1910) – he did a great deal to salvage Posad and Bogorodskoe toys from oblivion and to make them economically viable by giving them an aura of primitive chic that caught educated Russia's new interest in more authentic forms of emotional and aesthetic experience. The long-standing association of children and peasant cultures in the public imagination was given scholarly credibility in the miscellanies *Toys: The Joy of Children* (1911) and *The Toy: Its History and Significance* (1912), both under Bartram's editorship. In 1912 Bartram was also given the opportunity to explain his position on the renaissance of kustar toy production in the pages of the St. Petersburg journal *Apolon* (Apollo). "The kustar's creativity has not died," he claimed, "it is just sleeping until it is called to life, [but] it cannot emerge without outside help. Once the craftsman's income is guaranteed, he must have the chance to make what he himself wants to make, being carefully helped to select material and to develop his own independence. It is impossible to teach him to create, but he can be helped."[81] Prefacing Bartram's article was a brief, whimsical piece by Aleksandr Benois, who was himself a passionate collector of kustar toys. Though sympathetic to Bartram's crusade, Benois nevertheless felt torn between admiration and misgiving:

> In Moscow now they have undertaken to save the production of popular toys, for it truly is in decline, squeezed out by cheap factory stuff. . . . But I just can't find within myself the answer as to whether we should welcome this artificial salvation or not. Bartram, who heads this effort, is such a charming fanatic, such a toiler, expert, and artist, he has succeeded in doing so much that I cannot but wish him every future success. But then the doubt creeps in that perhaps this is "aestheticism," that it's being done for gentlemen enthusiasts and grown-up cultural philistines, that they will turn out to be the only consumers of this "popular children's art." And if that's so, then the whole thing becomes boring, somehow.[82]

Working for the ideal consumer – that is, the child – was a luxury that neither the Kustar Museum nor the Sergiev Posad toy makers could afford to indulge, however, and when all was said and done there was nothing boring about a good and steady income. Whatever the misgivings of Benois and other defenders of kustar art's purity, the Kustar Museum was proud of its achievements on behalf of Posad kustari, and listed them as follows:

> (1) It has improved the quality of toys by developing typical Russian prototypes, which increased sales both in Russia and abroad and thereby

raised the income of the kustari; (2) it has taught as many as three hundred kustar households how to produce high-quality goods; (3) it has moved many kustar carpenters over to the more lucrative carpentry and carving craft, thereby laying the foundation for a new art furniture and carving craft; (4) it has introduced poker-work on a broad scale, which has made wooden goods more interesting and thus more appealing to the consumer; and (5) it has made the kustari aware that in order to improve their material situation they must learn more sophisticated methods of production, follow innovations on the market, and unite with each other by organizing various cooperatives.[83]

Because of the growing interest in toys and the cult of the child throughout Europe in this period, we know a good deal about the toy industry at Sergiev Posad and the artists who worked there. Fewer details are available about the art joinery and furniture workshop, and the artists who designed for it. Until 1904 it was part of the toy workshop, and a number of kustari involved in producing carved furniture were in fact former toy makers promoted to this more lucrative and prestigious trade. As an independent enterprise serving some fifty kustari in Sergiev Posad and surrounding villages, the art furniture workshop kept a permanent personnel of about fifteen craftsmen and about ten apprentices, for whom special drawing classes were organized. Much of its output took the form of commissions for large items of birch or oak furniture, as well as icon cases, cupboards, shelves, and decorative bric-a-brac, all made from designs provided by the Kustar Museum in Moscow (Fig. 35).

Figure 35. Sideboard designed by the artist Suvorov and made at the Sergiev Posad art joinery workshop, 1913. (Reprinted from *Russkoe narodnoe iskusstvo* [Petrograd, 1914])

It seems likely that the art carpentry workshop at Sergiev Posad was the source of a large percentage of the furniture and interiors in the so-called Old Russian style that were produced in the early years of the century. Despite a rapid rise in consumer demand for neoclassical and Empire furnishings in the prewar decade, the Russian style was still considered appropriate for public spaces such as officers' clubs, banks, railway stations, and, most notably, churches. Judging from the lavishly illustrated architectural journals of the period, however, the demand for ostentatious nationalism among private citizens was more limited. The most memorable decorative commissions of the era – the cavernous dining room that A. G. Iakimchenko designed for the Firsanova home on Neglinnyi Proezd (Fig. 36)[84] and the apartments of the famous Pertsov house designed by Sergei Maliutin (Fig. 37)[85] – were popularly perceived as expressions of whimsical self-indulgence and escapism, rather than as viable designs for modern living. But if few educated Russians wanted total immersion in the Russian style in the form of complete interiors, there was obviously a

Figure 36. Dining room in the Firsanova house, Neglinnyi proezd, Moscow, by A. G. Iakimchenko. (Reprinted from *Ezhegodnik Obshchestva Arkhitektorov-khudozhnikov,* 1909)

brisk market for single pieces of furniture to add local color and piquancy to a room, and for smaller knickknacks that could serve as "poor man's Fabergé" – carved wooden frames, caskets, cigarette cases, and kovshi that made acceptable presentation gifts as well as lucrative export items. The prototype furniture designed by Viktor Vasnetsov's brother Apollinarii, a painter who specialized in poetic evocations of seventeenth-century Moscow, established a norm for the Old Russian style. His buffets, stools, cupboards, and chests were massive and generously decorated with plant and fruit motifs carved in high relief. While obviously intended for everyday use – Vasnetsov's own Moscow apartment was furnished with a complete suite in heavy oak – their purpose as showpieces to attract business and prestige for the Moscow Kustar Museum is evident in the publication of several Vasnetsov pieces in *International Studio* in 1912.[86]

By 1913 the success of the improvements introduced by the Moscow Kustar Museum at Sergiev Posad had manifested itself in several ways, most tangibly in monetary terms. The highest income – 900,000 ru-

Figure 37. Dining room in the Pertsov house, Moscow, by Sergei Maliutin, 1905–7. (Reprinted from *Ezhegodnik moskovskogo arkhitekturnogo obshchestva,* 1909.)

bles in 1913[87] – was earned by the toy workshop and the five hundred or so households connected with it.[88] But the art furniture and poker-work industries that grew out of the parent workshop and provided a livelihood for one hundred households were also thriving. "In the nineteen years of its existence 626,946 rubles 16 kopeks worth of goods have passed through this workshop, of which 62,062 rubles 3 kopeks worth was sent abroad."[89] Those kustari lucky enough to have adopted the new models and the markets that went with them benefited from a substantial increase in wages. A 1907 article on the Sergiev Posad workshops cites the monthly wage there as 40–50 rubles, whereas in 1896 an annual income from these same industries of 70–80 rubles was considered extremely high.[90] By 1913 the Moscow zemstvo counted no less than eight workshops and organizations practicing the new reforms.[91]

The rapid commercial boom in the kustar toy and furniture industries also encouraged a greater willingness to invest money in improvements. In 1906 the Moscow zemstvo was given a 10,000 ruble subsidy from the Chief Administration of Agriculture and Land Tenure (G.U.Z.i Z.) to build new premises for the art joinery workshop,

Figure 38. Moscow Kustar Museum, 7 Leontievskii Lane. (Reprinted from *Kustarnoe delo moskovskogo zemstva* [Moscow, 1913]. Courtesy of the Institute of Modern Russian Culture)

and in 1908 Sergei Morozov funded a special building for ten pupils and fifteen adult toy makers, which he donated along with the land to the Moscow zemstvo. Equally tangible was the expansion of the Kustar Museum's facilities in Moscow. Thanks once again to Sergei Morozov's generosity, in 1903 the museum moved to new premises at 7 Leontievskii Lane, an eighteenth-century building renovated in the Russian style by the architect S. Iu. Solov'ev and decorated inside by Konstantin Korovin with friezes on Old Russian themes (Fig. 38).[92] The main storehouse and collection point for incoming and outgoing kustar goods was housed in the basement, the museum shop was on the first floor, and on the second floor were the artists' headquarters.

In 1910 an important new department was added, called the Museum of Models. Created by Bartram with Morozov's blessing and financial support, the Museum of Models was intended "to help kustari improve the artistic side of their wares by fostering their own creativity and preparing special models and designs for kustar goods." It consisted of a collection of antique folk art, assembled with funds provided by the Moscow zemstvo, the G.U.Z.i Z., and Morozov, from which the museum's staff artists were expected to create a range of new designs for all of the workshops and cooperatives connected with the Moscow Kustar Museum. By "combining designs, copying antique artifacts with a number of variations, and generally persistently introducing into kustar goods the characteristic features of Russian antiquities,"[93] the museum hoped to inject back into kustar art the basic formal ingredients of an Old Russian style. The perfect foil to this new department was the existing Museum of Crafts, which comprised "the wares, tools, instruments, and work environment of kustar production." Since most of these exhibits had been acquired in 1882 at the All-Russian Arts and Industries Exhibition, and thus "showed what kustar industry was like in Moscow province thirty years ago," they provided a graphic reminder of the miracles wrought by the Moscow zemstvo since then, an inspirational "before and after."[94]

Statistics for 1913 showed that the Museum of Models' staff produced a total of 1,531 prototypes in both contemporary and antique Russian styles, using both in-house designs and samples bought from outside artists and other individuals. For example,

> it introduced models of antique Russian folk art to the wood carving, carpentry, and toy crafts, it replaced old types of Christmas tree decorations with new ones; it taught kustari working with papier-mâché to reproduce copies of old Russian miniatures on their goods, instead of the previous pseudo-Russian designs; it helped naboika production with models taken from genuine naboika fragments and board; and it gave

some kustari working on their own designs Bronze Age clay vessels to copy.[95]

No zemstvo understood the connection between artistic intervention in kustar production and increased profits more clearly than the Moscow provincial zemstvo, and it is easy to criticize the steady commercialization of kustar art that the Moscow Kustar Museum promoted. As a reviewer of Moscow province's kustar contribution to the 1913 All-Russian Arts and Industries Exhibition in Kiev put it, "This is a regular production, a factory that produces goods for all of Europe. . . . In Moscow they've already worked out a routine, even a sort of style of their own, a sort of Bartromitis, and it's all second-rate. Because what the Stroganov School can accomplish in a rich and ornate fashion, with the best and most expensive materials and woods, gets done here at an inferior level. But it's the same 'kustar *style russe moderne.*' "[96]

Still, there is no doubt that the Moscow zemstvo regarded its intermediary role in the buying and selling of kustar art as a necessary but temporary stage in the development of kustar self-reliance and cooperation.[97] When World War I broke out it was already tentatively letting go of the leading strings. In fact, under pressure from Sergei Morozov the Moscow zemstvo had been slowly moving toward a system of cooperatives and greater kustar independence since 1901, when an artel of twenty toy sculptors broke away from the main toy workshop with the Moscow Kustar Museum's blessing.[98] By 1913 there were thirty artels in operation throughout the province, all but two of which received low-interest loans from the zemstvo. These included the "Kustar-Artist" artel of poker-workers and carvers, formed in 1910; a women's artel that specialized in dressing dolls; an artel of carpenters and joiners; and in 1913 the toy workshop itself, which was handed over to a kustar cooperative with all its goods, tools, and materials, valued at over 17,000 rubles.[99]

Nevertheless, the specter of filthy lucre has cast a pall over the whole "zemstvo episode" in the kustar reform movement, particularly during the Soviet era when it was de rigueur to condemn the Moscow Kustar Museum and its artists for "polluting the pure spring of folk art" with the taint of capitalism and the decadent neo-Russian style.[100] As the following chapter will show, similar objections motivated the activities of one of the most famous kustar reform centers of the prerevolutionary period – the Talashkino workshops of Princess Maria Tenisheva.

4

Princess Maria Tenisheva and the Talashkino Workshops

THOUGH SIMILAR in many ways to the three kustar centers discussed so far, the complex of workshops that Princess Maria Tenisheva operated between about 1898 and 1905 at Talashkino, her estate in the province of Smolensk, stands conspicuously apart from them and from the kustar revival movement as a whole. Talashkino's isolation from Abramtsevo, Solomenko, and the Sergiev Posad workshops was in part self-imposed, in part involuntary. Whereas the three older enterprises were bound by strong family, social, or regional ties, the lavish kustar complex at Talashkino was created by a woman of quite another social stratum and background.[1] Though an aspiring artist herself, Tenisheva neither belonged to nor generated a close-knit artistic circle such as the Mamontovs gathered around them.[2] Her complex personality – a mixture of painful shyness and ambition, morbid sensitivity and tactlessness, gullibility and high-handedness – did little to endear her to her contemporaries. In the opinion of artist Anna Ostroumova-Lebedeva, she was simply "a very wealthy woman who has decided to patronize the arts, and though apparently she doesn't understand much about them, has the good sense to consult people who do."[3] Tenisheva was a generous and well-intentioned patron, but she also had an unfortunate gift for alienating many of the artists and celebrities whose association with Talashkino made it one of the most famous artistic centers of its day. Aleksandr Benois, Sergei Diagilev, Sergei Maliutin, Sofi Menter, Il'ia Repin, and Dmitrii Stelletskii were among those who enjoyed the princess's largesse and hospitality, only to fall out with her on closer acquaintance.

In its basic structure the Talashkino complex was closely modeled on Abramtsevo. It too had its school for local peasant children; its church; its collection of antiquities and folk art; its carving, embroidery and ceramic workshops; and its shop in Moscow. Summer guests found excellent facilities, stimulating company, portrait and decorative commissions, and handsome honoraria that were particularly welcome after 1899, when Savva Mamontov's bankruptcy brought his career as a Maecenas to an abrupt end. Yet Tenisheva clearly intended her estate to be like no other model kustar center, and all of her enterprises were operated with a disregard for cost that was quite impracticable for Abramtsevo or, indeed, any kustar enterprise that hoped to remain solvent by its own efforts. A comparative latecomer both to the cause of kustar reform and to the vogue for things Russian, Princess Tenisheva had no intention of confining herself to the dictates, policies, and financial scrimping of governmental or zemstvo bodies. Nor was she prepared to follow meekly in the footsteps of the Mamontov-Iakunchikova dynasty with its monopoly on modernized kustar art.

During an unhappy first marriage while still in her teens, Tenisheva had made modest forays into the art world as a singer, painter, and art collector, but these early efforts had been hampered by her lack of social position and funds. It was not until her second marriage in 1892 to Prince V. N. Tenishev, a wealthy industrialist from an old noble family, that her numerous plans and enthusiasms could be realized.[4] Her first two years of married life were spent in Briansk, where her husband's locomotive factories were located and where she expended some of her formidable energies for philanthropic service by organizing schools, clubs, cafeterias, and new accommodations for the workers and their families. She also donated a village school to the peasants in the area of Khotylevo, her summer estate near Briansk.

In 1896, however, Prince Tenishev bought his wife the estate of Talashkino outside the ancient city of Smolensk from her childhood friend Princess Sviatopolk-Chetvertinskaia,[5] and from this time on Tenisheva devoted her summers exclusively to her new estate's development, determined to make it a model in all respects. First, strongly influenced by Princess Chetvertinskaia's passion for agronomy and popular education, she established an agricultural school for local peasant children and orphans.[6] Strictly speaking, both the school and the workshops that later grew out of it were located not at Talashkino itself, but at Flenovo, a small neighboring estate that Tenisheva also acquired. No expense was spared in making the Flenovo school exemplary. With an annual budget of 15,000 rubles it boasted courses in animal husbandry and agriculture, a beekeeping museum, a meteorological station, a school uniform, and obligatory physical education. The pupils were educated and housed free of charge, and, as further incentive, those who received their diploma were exempt from military service and corporal punishment for life.[7] The Flenovo school soon spawned a number of workshops in which the pupils could learn a craft, the first two being a musical instrument workshop, which made balalaikas for the school orchestra, and an embroidery workshop.

In establishing kustar workshops at Talashkino in earnest around 1900, Tenisheva was of course keenly aware of Abramtsevo's pioneering reputation in this arena. As co-financer of the *World of Art* with Savva Mamontov, she witnessed the publicity and respect given to the Abramtsevo and Solomenko kustar workshops in its pages and at the World of Art exhibitions. At the 1900 Exposition Universelle in Paris Tenisheva was in an excellent position to witness the laurels won by the Mamontov circle, for her husband was appointed general commissar of the entire Russian section and she herself gained considerable public exposure as a society hostess. Tenisheva's own artistic contribution to the Russian section at the exhibition was a complete orchestra of twelve balalaikas decorated by Davydova, Golovin, Maliutin, Vrubel, and herself. These Talashkino balalaikas received added publicity from the presence in Paris of Andreev's Great-Russian Balalaika Orchestra, which Princess Tenisheva patronized and promoted.[8]

Since Polenova's death two years before, however, there was a growing sense that the "new Russian school of decorative art" had lost its momentum and that the field for her successor was wide open. In retrospect Princess Tenisheva preferred to remember her bid to supersede Abramtsevo as something quite unprecedented, claiming that "the *style russe,* as it had been treated till then, was completely forgotten. Everyone regarded it as something obsolete and dead, incapable of being resurrected and taking its place in contemporary art."[9] While this was scarcely an accurate view of the situation, Tenisheva's desire to "provide a wider range of designs, to deluge the market with new forms, to infuse a fresh new spirit" into kustar production no doubt also reflected the Russian public's desire for novelty in the range of kustar objets d'art available for sale:

> I didn't want to imitate other workshops, not even that at Abramtsevo, which turned out one and the same motif and repeated it a thousand times in all possible ways. All these little boxes, children's blocks, and shelves that we had seen at all the exhibitions and in the storeroom of the Moscow zemstvo had long ceased to be interesting in their uniformity and lack of fantasy.[10]

From a purely practical point of view, this insistence on originality and the unique art object was inherently antithetical to any kustar workshop's economic survival. For kustar revival to succeed economically, as it had done at Abramtsevo, Solomenko, and Sergiev Posad, large numbers of fairly uniform items were required, not just one-of-a-kind art pieces. This inherent contradiction between purely aesthetic ideals and practical considerations, between romantic idealism and mundane reality, provided a rather shaky foundation for the workshops and was ultimately to play its part in Talashkino's early demise.

While there is as yet no reliable chronology available for establishing exactly how and when each of the kustar workshops at Talashkino was set up, a pivotal date is 1900, the year Sergei Maliutin entered Princess Tenisheva's employ as artist-in-residence at Talashkino. Since her energies and resources to date had been focused on the problem of popular education, the potential for developing specifically kustar art workshops awaited a second Elena Polenova or, in Tenisheva's words, "a person with whom I might create the artistic atmosphere I lacked." Despite summer visits by artists like Aleksandr Benois, Il'ia Repin, Ian Tsionglinskii, and Mikhail Vrubel, prior to 1900 Tenisheva's immediate circle for the most part consisted of "people who were near and dear, but entirely uninitiated in the arts."[11]

The artist whom Tenisheva eventually found to enter her employ and live permanently at Talashkino as artistic director of her workshops was Sergei Maliutin. Born in 1859 into a Moscow merchant family and orphaned by the age of three, Maliutin spent his teenage years working as a clerk in the city of Voronezh before he was able to enroll as an auditor at the Moscow School of Painting, Sculpture, and Architecture in 1881. Like Nikolai Bartram, he found a mentor in Ivan Zabelin at the Moscow History Museum, and it was there that he made his first serious study of Russian antiquities and folk art. In addition to painting, teaching, and occasional commissions, Maliutin found time to indulge a new interest: designing furniture and ornamental motifs that might be used for embroideries or wood carvings. These sketches were carried out by independent kustar craftsmen, first for the use of his own family and then as commissions for friends and acquaintances.

With his growing reputation as an exponent of the Russian style, Maliutin quickly came to the notice of Savva Mamontov, and throughout the 1890s he was frequently involved in Mamontov projects without actually becoming part of the prestigious "Mamontov circle." In addition to his 1891 matreshka doll commission for Maria Mamontova, in 1896 Maliutin executed a large proportion of Korovin's famous panneaux for the Pavilion of the Far North at the Nizhnii Novgorod exhibition. The following year Mamontov invited him to work as a set designer for his Private Opera Company.[12] To mark the Pushkin jubilee of 1900, the publishing house of A. I. Mamontov issued several of the poet's fairy tales illustrated by Maliutin.[13] Finally, as an indication of his standing with the newly formed World of Art group, several of Maliutin's drawings and furniture designs were included in the first strongly nationalist numbers of the *World of Art*.

Maliutin was particularly suited to Princess Tenisheva's needs for yet another reason. At the age of forty he was already the head of a large and growing family and perpetually struggling to make ends meet. When Princess Chetvertinskaia visited Maliutin in Moscow to lay before him her friend's proposal, she found him and his family living in appalling poverty. The handsome monthly salary of three hundred rubles that Tenisheva was prepared to pay him could not be refused, therefore, and Maliutin arrived at Talashkino, not as a guest like his friends and fellow painters, but as an employee.

Maliutin's immediate duties at Talashkino were not to organize kustar workshops on a commercial basis, however, but to design and construct a church near the school at Flenovo.[14] Vrubel, who himself had good reason to appreciate the princess's generous patronage, acted as her intermediary in the negotiations to bring Maliutin to Talashkino and wrote to the artist explaining to him what his first project there would consist of.[15]

She asks you to make her a colored sketch of a small church that meets the following specifications: (1) a sheltered porch; (2) a crypt; (3) an ambulatory or vestibule for processions when the weather prevents them being held out of doors. In the Iaroslavl style. But you're to have complete freedom in decorating it: the church is intended to be lavishly faced with tiles and every kind of glazed details such as the roof tiles and even whole structural elements.[16]

That all of Maliutin's subsequent activities at Talashkino – the carving workshop, the ceramic workshop, the Moscow store The Source – grew out of this commission is confirmed by the legal documents that Maliutin and Princess Tenisheva exchanged in 1904, shortly after the artist left her employ:

Mr. Maliutin was invited by me at the beginning of 1900 to draw up designs for a church in the village of Flenovo in Smolensk district and province, in a style of his choosing, to supervise its construction according to his tastes and plans, and to organize an art workshop for such joinery and carving as was necessary for building the church. Mr. Maliutin, having established the workshop, whose primary task was to serve the church, subsequently assumed responsibility for expanding the workshop and making it a permanent establishment for the art joinery craft.[17]

This terse statement thus provides a secure date of 1900 for the foundation of the woodworking workshop at Talashkino. Furthermore, a letter written by Maliutin to the princess at the very beginning of their relationship suggests that the idea of setting up permanent kustar joinery and carving workshops at Talashkino was also his. Echoing Elizaveta Mamontova's reasons for opening the Abramtsevo carpentry workshop over two decades before, Maliutin advised Tenisheva "to open joinery and carving workshops, so as to train master craftsmen who, once the church was finished, would become kustari and could earn a living while still living at home in the countryside, rather than having to cart firewood five or ten miles into Smolensk."[18]

Maliutin came to Talashkino already equipped with a repertoire of decorative ideas and motifs developed over the past decade, with an interest in designing furniture, embroideries, theater sets, and costumes, and with practical experience in working with kustar carvers and joiners. In other words, the Maliutin style was almost fully formed and awaited only the excellent facilities and financial security of Talashkino to burst into bloom. To staff the new workshop "a notice was posted in the school announcing the recruitment of boys for [the workshop]. Immediately peasant boys with dreams of studying this 'art' trade that had previously been closed to them were dying to get into the workshop."[19] As it turned out, construction on the Flenovo church did not get under way until 1902, and Maliutin and the joinery workshop turned instead to the production of furniture of all kinds and to the construction of several new buildings on the estate: a teremok to house the school library (Fig. 39), a theater where the pupils put on plays under the princess's direction,[20] and an elaborate carved fence and entrance gates. A ceramic workshop was also established, initially to meet the needs of the Flenovo church, since Tenisheva wanted the bricks and tiles for its construction made locally of local materials, but later devoted to the production of a variety of art ceramics.

Tenisheva's firm belief that her kustar enterprise would find its competitive edge in stylistic variety – that "the more co-workers, initiative, and originality I had, the better the results would be" – brought an array of visiting artists into contact with the pupils in the

Figure 39. Teremok at Talashkino, by Sergei Maliutin, ca. 1903. (Reprinted from S. Makovskii, *Talashkino* [St. Petersburg, 1905])

various workshops. According to historian Igor Belogortsev, Valentin Serov and Viktor Vasnetsov "helped the embroidresses and lacemakers with advice and instructions"; Il'ia Repin "sometimes looked into the workshops and took part in improving folk motifs";[21] and Maria V. Iakunchikova and Natalia Davydova reputedly provided embroidery designs. Other artists known to have visited are S. N. Andrianov, Iakov Ovchinnikov, Dmitrii Stelletskii, and A. F. Zakharevich.[22] For many of these guests, their contribution appears to have been limited to a sketch presented to their hostess in return for her hospitality, or perhaps an afternoon spent decorating one of the whistles, shaft bows, or balalaikas that were bought directly from peasants in the surrounding villages.

Yet it was undoubtedly Maliutin's presence at Talashkino that won for Tenisheva's enterprise the attention she craved. Whatever criticisms were later leveled at Maliutin as a designer, he unquestionably

Prior to this arrangement, the peasant women who worked for the princess were using purchased materials for their embroideries, "English yarns that made the embroideries look banal."[41] We may therefore assume that the beautiful muted tones for which Talashkino embroideries were justly famous were the younger Pogosskaia's achievement.[42] It is even possible that the very idea of reviving the art of natural dyes was suggested to Tenisheva by her newfound friend, who had extensive experience in the women's kustar industries and was a long-standing admirer of another advocate of natural dyes, William Morris.[43]

Traditional methods of dyeing had been among the first kustar industries to fall victim to the mechanization of textile production in Russia. The availability of aniline dyes had virtually crippled kustar dye shops, as natural indigo was replaced with a cheaper chemical substitute, and madder virtually disappeared from use. Although chemical dyes faded quickly and their harsh colors offended the discriminating eye of connoisseurs, peasants welcomed them since they were a fraction of the cost of traditional dye stuffs with their labor-intensive procedures. Whereas a kustar dyer charged three kopeks per arshin for a simple indigo dye and four to five kopeks for separate colors, aniline dyes were just a quarter of the price.[44]

Assuming that Smolensk women resorted to chemical dyes because they could not afford the genuine article, Tenisheva plunged into the difficult and expensive task of setting up a natural dye shop. As in her other workshops, she spared no personal expense in defying the laws of the marketplace in the name of social reform through the elevation of popular taste. A French admirer of Talashkino embroideries, Denis Roche, reported that "Princess Tenisheva's task lies in weaning the village women off the crude and barbarous colors they are accustomed to using out of economic necessity and supplying them instead with colors that are simple and harmonious but still affordable and tied to ancient tradition. It was found that in making refined shades available to the peasant women, they too became more refined."[45]

In point of fact, the colors used in traditional Smolensk embroideries had always been exclusively bright orange, red, and yellow. Tenisheva herself admitted that "at first [the peasant women] didn't like our colors, they said they were 'drab,'" and only by overpaying them for their work could she prevail on them to use them.[46] Although her conviction that more subtly colored materials both represented and engendered a corresponding spiritual refinement in the producer undoubtedly derived from theories widespread among social reformers of the period, it may also have been confirmed by comparison with antique embroideries, whose gently faded patina of age could easily be misread as the mark of a lost golden age of folk culture. But whether or not exposure to the more refined tones yielded by vegetable dyes had the desired ennobling effects on Talashkino women, the consumers of these embroideries – artists, foreigners, and the urban elite – responded with aesthetic delight to their painterly complexity and variety. Again, to quote Denis Roche:

> The shades are a subtle transposition of genuine Russian shades. The most tender nuances of pink and "strawberry" tones vary the range of reds and red-browns. The blues are the deepest, the most concentrated, the densest, the most celestial, and the most languishing. The greens are "gay" and singing, or ambiguous and subdued, mignonette, lavender. The violets and mauves . . . are, like the blues, very rich in their full and muffled notes. The yellows turn quickly to "chamois," "beige," "gold," "gamboge," and are used for ornament rather than grounds.[47]

Most of the Talashkino embroideries were done at home as piecework, the women bringing their own homespun fabrics and threads to be dyed at the Talashkino dye shop, then taking them home again to embroider with traditional stitches and patterns. Apart from insisting that only approved colors be used, and that items of contemporary urban use be made, Tenisheva and her staff intervened only to the extent of praising excellent work and chastising sloppiness. At their peak the embroidery and dyeing workshops at Talashkino employed about two thousand women in two hundred villages, and a skilled worker could expect to earn ten to twelve rubles for a winter's work.[48]

In addition to these modified traditional embroideries, the workshop also produced a certain number of "art embroideries" designed by a variety of visiting artists and by the princess herself. It was here that Tenisheva's own taste and sense of whimsy were most clearly felt, in the "broken lines, seemingly unending stems, the heads of many-petaled flowers, vague intimations of beast and bird faintly reminiscent of the East"[49] (Fig. 42). Her embroidery designs also show a noticeable preference for Russian beauties in kokoshniki that call to mind Vasilii Kandinskii's early paintings on Russian themes. Although it has been suggested that Elena Polenova, Maria V. Iakunchikova, and Natalia Davydova all provided designs for the Talashkino embroidery workshops, no firm evidence of this connection has been produced, nor do any of the numerous articles and monographs on Talashkino reproduce designs for which they are identified as the artist.[50]

The first jarring note in the apparent harmony of Talashkino life occurred in September 1903, when without warning Maliutin moved back to Moscow with his family. The official reason was an invitation from the Moscow School of Painting, Sculpture, and Architecture to join its faculty. The invitation was formally delivered by two of Maliutin's friends, the painters Vasilii Baksheev and Aleksandr Korin, who according to Tenisheva "sowed doubts in his mind" as to whether he should be wasting his talents as a painter on the applied arts.[51] It has also been suggested that the princess's interference in artistic matters had long since begun to irritate the artist and that Maliutin resented his inferior social status at Talashkino. For her part Tenisheva blamed his disaffection on Anna Pogosskaia's socialist propaganda and her "terrorizing" influence and also hinted darkly at a psychological cause for Maliutin's erratic behavior: "Everything pointed to the fact that he was depressed by something. His paranoia grew more pronounced; he complained that his wife was trying to poison him. In short, he left terrified and confused."[52]

Whatever the reasons for this precipitous departure, Princess Tenisheva and her workshops suddenly found themselves without full-

Figure 42. Talashkino furniture upholstered with art embroideries designed by Aleksei Zinov'ev and Princess Maria Tenisheva, ca. 1904–5. (Reprinted from S. Makovskii, *Talashkino* [St. Petersburg, 1905])

time supervision and artistic guidance. In the three years and three months of his employment, Maliutin had almost single-handedly put Talashkino on the map as a viable center for kustar revival and the neo-Russian style. The buildings on the estate, the store in Moscow, and the goods produced in the carpentry and ceramic workshops all bore the unmistakable imprint of his unruly decorative imagination. It was imperative to the continued success of the Talashkino production that a new full-time artistic director be found.

"I thought it would be very difficult without Maliutin," the princess wrote, "but I quickly found a replacement." There was, as she later remembered, no shortage of candidates for the position since, thanks to The Source, the entire Stroganov School of Technical Drawing in Moscow was interested in Talashkino and she had only to choose from among the school's pupils.[53] The Stroganov School was an ideal recruiting ground in all respects. Not only was it one of the three premier industrial art schools in the country, but under the direction of Nikolai Globa it specifically promoted national traditions wedded to the *style moderne*.[54] Moreover, since the Stroganov prepared artists exclusively for a useful career in the industrial and decorative arts, prospective employers would presumably be free of the clash of wills that more temperamental "fine artists" might cause. To succeed Maliutin, Tenisheva finally selected Aleksei Zinov'ev, a Stroganov graduate and one-time employee of the Fabergé jewelry firm's Moscow branch, who was shortly joined at Talashkino in late 1903 by his schoolmate and inseparable companion V. V. Beketov.[55]

Perhaps inevitably, in view of their lack of experience and the numerous reminders of Maliutin's presence at Talashkino, Zinov'ev and Beketov at first succumbed to the force of his style. As critic Sergei Makovskii disapprovingly remarked, their first efforts showed "the same enthusiasms and the same weaknesses. The decorative fairy-tale element completely absorbs the industrial logic of the piece. Fantastical ornament weighs down the direction of the lines. They're too florid, crude, and discordant. All these wooden cupboards, shelves, frames, dishes, and so on are at best successful experiments in terem luxury à la Maliutin, but it would be difficult to find them a place in the modern home."[56] Much more promising, in Makovskii's eyes at least, were Zinov'ev's later works of 1904–5, which showed "a new understanding of industrial art" (Fig. 43):

> The forms do not blindly imitate peasant kustarnichestvo. To a significant degree they are already modern, *our* forms in harmony with the twentieth century. Yet at the same time the ornament, the upholstery, various random lines, trivial details that at first sight are scarcely noticeable, in short

everything that gives furnishing its physiognomy, is native and close to a popular sense of beauty.[57]

These pronouncements by Makovskii appeared in 1905 in a monograph that Princess Tenisheva had commissioned to publicize the Talashkino workshops – a sort of deluxe sales catalogue.[58] Son of the realist painter Konstantin Makovskii and future editor of the neoclassicizing journal *Apollo,* the cosmopolitan Makovskii was an odd choice for the task. Possibly he was recommended to Tenisheva by the painter and amateur archaeologist Nikolai Roerich, whose lyrical reminiscences of his visits to Talashkino prefaced the more scholarly main text and whose paintings of Russian architectural monuments had already aroused the young critic's mystical sympathies in late 1903.[59] By 1904, however, Makovskii was already absorbed in the vexed question of Russia's relationship to European culture, particularly as it concerned the environment of everyday life, and had little patience with utopian ideas about Russia's separate path. The tone of his comments in the Talashkino monograph was one of only qualified

Figure 43. Two carved chairs, by Aleksei Zinov'ev, ca. 1904. (Reprinted from S. Makovskii, *Talashkino* [St. Petersburg, 1905])

approval, therefore: "We have fallen in love with [folk art], forgetting that the resurrection of national motifs is not an end in itself, but that these motifs are just material to be used in creating a style that corresponds to the conditions of a European-wide culture."[60] In the end, he felt, Talashkino would be known, not as the source of a new style for modern living, but for "educating the tastes of the people" and developing "a sensitivity to ornament . . . that still lives among our peasantry today, despite the horrible ugliness of their lives, close to factory and railway centers, despite the fact that they have forgotten the artistic tenets of the past."[61] "There is a clear goal confronting the industrial art of our day – to be beautiful and at the same time comfortable and functional," Makovskii wrote, and a synthesis of Russian ornament and European culture was needed if Russia was to evolve its own national version of the new style.[62] Thus, while he could praise Talashkino embroideries for their "impression of harmony and peace – a sign of intimacy so essential to the contemporary style,"[63] he ultimately remained unconvinced that such vernacular forms of expression, however tempered by civilizing influences, could ever meet the needs of Russian citizens as they joined the modern world:

Naturally, I am far from thinking that the embroideries of Smolensk peasant women will ever replace expensive, refined fabrics . . . in our apartments. In general I don't at all wish to exaggerate the importance of peasant kustarnichestvo even in that new aspect which it is acquiring, resurrected under the influence of cultured people with a deep commitment to art. We shouldn't forget that this is only the "root," while we should go to "the tops." It is the "base"; we need the "crown."[64]

Makovskii's patronizing warnings against isolating Russian art from the mainstream of European culture fell on deaf ears at Talashkino. Already in the late 1890s, under the tutelage of noted archaeologists Adrian Prakhov and Vladimir Sizov, Princess Tenisheva had embarked on an ambitious program to build an outstanding collection of Russian antiquities. It was on Sizov's advice, too, that in 1897 Princess Tenisheva brought to Talashkino the amateur photographer Ivan Barshchevskii, whom she subsequently appointed curator and later director of her museum.[65] But the most decisive influence on shaping the princess's cultural allegiances was Nikolai Roerich, in whom she found both a mentor and a soulmate. "Of all the Russian artists I met in my life, with the exception of Vrubel, he was the only one with whom one could talk," she later reminisced. "He was cultivated and very well educated, a true European, neither narrow-minded nor one-sided. Well bred and pleasant to deal with, his company was indispensable, and he had a broad understanding of and interest in art. Our relations were those of . . . kindred spirits."[66] When Roerich and his wife made their first visit to Talashkino in late 1903, he was already quite well known for his paintings of pre-Christian Russian life and history. In addition, he occupied an official post as assistant to Nikolai Sobko, the secretary of the Society for the Encouragement of the Arts, contributed to the Symbolist journal *The Scales,* and enjoyed a reputation as a serious archaeologist.

While Sizov, Prakhov, and Barshchevskii were valuable advisors and sources of information on Russian archaeology and architecture, it was Roerich's more mystical and philosophical bent that fostered Tenisheva's growing disaffection with European culture and her search for a more authentic source of Russianness. As a protégé of Vladimir Stasov in the 1890s, Roerich had absorbed the older man's theories on the Eastern origins of Russian culture,[67] and his archaeological activities excavating burial mounds in Petersburg province had brought him face to face with some of the most ancient remains of pre-Christian Slavic culture. As a theosophist,[68] moreover, Roerich was keenly interested in strengthening the spiritual link between India and Russia at the expense of Russia's more recent Western orientation. In this he was joined by his friend the archaeologist Professor V. V. Golubev:[69]

> We have long dreamed about the foundations of Indian art. Involuntarily the question arose of the continuity of our ancient way of life and our art from India. In intimate conversations we often focused on the cradle of popular culture, our Slavic culture in particular. . . . The customs, the funeral mounds with their fences of everyday objects, building methods, details of headdress and clothing, all the monuments of wall painting, and finally the roots of speech – all these things were so close to our roots. In everything one felt that the initial path was the same. It's clear that if we are to delve deep into our roots, a real study of India will provide unique material. And we should be quick to study these popular treasures; otherwise it won't be long before English culture will extinguish much that was near to us.[70]

Princess Tenisheva came to share these views, exclaiming, "What are foreign thirteenth-century madonnas to me? What do I care for marble capitals? What do I care for the ingenious works of Benvenuto Cellini?"[71] As her passion for collecting Russian antiquities grew, she began to liquidate her collection of Western art to pay for new acquisitions, many of them collected by Barshchevskii on expeditions to remote provinces. One result of this passion for ancient Rus was the princess's Museum of Russian Antiquities, which officially opened in Smolensk in 1905 and was considered second only to the Moscow History Museum for the breadth and quality of its holdings.[72]

Tenisheva was also drawn to Roerich by a mutual interest in archaeology, and the two undertook excavations in the pre-Christian burial mounds for which the Smolensk area was noted. In a sense archaeology was the ideal metaphor for kustar reform as practiced at Talashkino after Maliutin's departure. In Roerich's eyes, the skills of the modern kustar made a poor showing beside the art of ancient Rus and were all of a piece with the general decline of Russian culture as its Eastern origins became more and more deeply buried beneath the alien deposits of an imported Western civilization. Rather than being a symbol of everything good in Russian culture, kustar art was just the pale reflection of a lost past:

> Out of a whole series of refined and wonderful art crafts there have come down to us only the crude allusions of kustar goods. Among the deposits that a lack of culture has left, it is difficult to discern exactly what time has preserved of these goods' original appearance. Unable to make sense of these layers and ignorant of the ways our art has been created, the public at large forms an impression of some kind of crude legacy, of something unworthy of the modern enlightened eye. To top it all off, these coarsened but still popular forms are often replaced by the most vulgar kind of modernism, and then all horizons are closed off.[73]

Just as in his archaeological digs, Roerich's response to kustar revival was to scour away the accretions of later civilizations, to radically reunite the peasant with his prehistoric ancestors in one fell swoop. In very similar terms Tenisheva professed her determination to unearth the hidden treasures within the modern peasant's soul, to "clean away what had accrued from a lack of culture, and in this good but neglected soil, any seed you please can grow." And again, "You can find everything in the Russian peasant, you just have to dig for it. . . . I loved to divine these natures, to work on them, to direct them. Yes, I love my people and believe that they are Russia's whole future; one need only honorably direct their strengths and abilities."[74]

Intermittent though it was, Roerich's presence at Talashkino in the wake of Maliutin's departure had a decisive impact on the stylistic shift that occurred in the workshops over the next two years. His influence on Zinov'ev can be seen in the increasingly "primitive" or archaizing quality of the younger artist's designs, and this was rooted in the sources on which they drew for inspiration. In his paintings, as Denis Roche pointed out, Roerich "attempted nothing less than to shift the center of what had been accepted as 'Russian.'" Whereas Moscow had been this geographical and symbolic center for painters like Viktor and Apollinarii Vasnetsov, "Roerich was among those who shifted it violently to the north."[75] If Maliutin's neovernacular ornamentalism was the logical conclusion of the style Elena Polenova had created fifteen years before, drawing on the iconography of fairy tales and *byliny* (legends) and the formal qualities of kustarnichestvo, Roerich and Zinov'ev rejected that relatively modern source of inspiration, the peasant izba, in favor of the archaeological artifacts of the Stone Age.[76] An important indicator of this shift in both time and place was Roerich's marked lack of interest in the vegetable world that had provided Polenova, Davydova, Golovin, and to some extent Maliutin with such rich material for both personal and national disclosure. As Max Voloshin later put it, "The vegetable kingdom is mysteriously alien to Roerich's work. . . . There are so many stones on his earth and so little soil that there is nowhere for a tree to grow."[77] Allusions to fossils and shards, amulets corroded from long burial in the earth, totemic birds and beasts, and stones themselves were part of what Soviet historian Nikolai Sobolev disapprovingly dubbed the "deliberate archaism" of later Talashkino design. So too were the materials used: "a light, dirty-green mordant, insets of colored stones from the Crimea, old metal coins that had turned green, some kind of tiny chains and other accessories of the like, finally a large and entirely superfluous amount of wood in all the carved pieces, small and large."[78] Among the few objects that Roerich is known to have designed for the Talashkino workshops is a book cupboard whose doors are entirely covered with the fossilized forms of prehistoric animals (Fig. 44).[79]

Despite the financial and artistic resources that Princess Tenisheva lavished on her workshops, and the publicity that she generated for them, the Talashkino kustar complex paradoxically proved to be one of the most short-lived of all attempts to regenerate kustar art industries in Russia between 1870 and 1913. To a large extent the failure of the workshops was a matter of sheer bad luck, since nobody could have predicted the destructive power that the 1905 Revolution was to have in Smolensk province. Peasant unrest was widespread, and incidences of arson on gentry estates, including Talashkino, were common.[80] Yet even without the violence that hit the countryside around Talashkino that year, the ideal world that the princess had so lovingly created there was unusually vulnerable to fluctuations in the outside world, thanks to its utopian resistance to the basic economic and social facts of peasant existence. A less idealistic or headstrong woman than Teni-

Figure 44. Carved book cupboard, by Nikolai Roerich, ca. 1904. (Reprinted from S. Makovskii, *Talashkino* [St. Petersburg, 1905])

sheva might have heeded the growing signs of discontent in her school and workshops, which by 1903 were already so obvious that the district police superintendent conducted a search of the Flenovo school. It is no doubt true, as Tenisheva records in her memoirs, that the revolutionary sentiment at Talashkino was fostered by the teachers at the school, who had contacts with Social Revolutionaries based in the Smolensk area. But at the same time, these "subversive elements" must have found a fertile soil awaiting them in the school and the workshops, for as generous as the princess undoubtedly was toward her charges, she distributed her philanthropy in a spirit of such magnificent condescension that a skilled propagandist could play upon it to good effect. Covert study groups would gather at the house of the

Pankovs, teachers at the school, on the pretext of studying the balalaika; propaganda leaflets were found in the school dormitories; and in March 1904 the pupils went on strike.

Sedition and freethinking were traced to yet another source, Anna Pogosskaia, whom Princess Tenisheva had installed in Maliutin's teremok as head of the experimental dye shop.[81] As Tenisheva bitterly recalled,

> Young and very flighty, with all the tricks of an American education, yet unkempt and slovenly, and most of the time dressed in a sarafan, she acted the part of a *narodnitsa* [populist], stirring up the teachers and condemning everything produced at Talashkino. Very soon her presence at Flenovo told upon the teachers' attitude toward me. . . . All her conversations had to do with equality, in the sense that if one person has something another doesn't, then that person must be hated, regarded as an enemy, and if possible destroyed.[82]

Tenisheva's first inkling of Pogosskaia's perfidy came in the summer of 1904, when she accidentally intercepted a letter in which Pogosskaia not only made fun of the princess's activities, the state of affairs in the school, and what she called the terrible, unbearable life of the teachers, but also described how she was siphoning off a percentage of the dyed threads and silks to send to her mother in Moscow. Her suspicions now fully aroused, Tenisheva conducted a complete audit of The Source, which revealed that "in the less than three months of its existence it already had a 3,000 ruble deficit."[83] Both mother and daughter were dismissed, but not before a quantity of costly Persian madder from the dye shop was found to be missing. In the wake of this scandal the dye shop was moved to the building that housed the ceramic workshop in Smolensk and placed under the management of Barshchevskii, curator of the princess's collections. The needlework section was transferred to the artists' studio, under the supervision of the princess's friend E. Sosnovskaia.

In Tenisheva's version of the incident the Pogosskaias figure as unprincipled confidence tricksters with a long history of similar swindles behind them.[84] Whether this was true or not, one can easily imagine how they might have justified their fleecing of the gullible and wealthy princess as simply robbing the rich to help the poor. Given the lavishness and inefficiency of the Talashkino operation – silks were dyed in uneconomically small batches to prevent mass production, for example – the Pogosskaias might well have reasoned that a proportion of this largesse could be better used elsewhere. The freethinking socialism that the younger Pogosskaia preached at Talashkino points to a widening gulf within the kustar revival between the paternalistic landowning gentry to which Tenisheva belonged and a new breed of political activists committed to radical social reform as a necessary prerequisite for true kustar revival. William Morris's insistence that "you cannot have happy villagers living in pretty houses among the trees doing pretty-looking work in their own houses or in the pretty village workshop between seed-time and harvest time unless you remove the *causes* that have made the brutal slum-dweller and the starving field-laborer"[85] was nowhere more applicable than at Talashkino. As followers of Morris, the Pogosskaias could not but condemn the inherent split between the paradise of Talashkino and the world outside its fairy-tale gates. Anna Pogosskaia senior surely had Talashkino in mind when she later wrote – as it turned out, prophetically: "[Industrial colonies] have always failed when there was no ground principle of service and mutual interest at their root, but gathered under the sole influence of some powerful magnetic personality. As soon as this personal influence was removed . . . the whole thing collapsed one way or another."[86]

For almost two years Princess Tenisheva turned a blind eye to the discontent brewing in her school and workshops, and as late as 1905 visitors to the estate reported with delight on what the princess showed them. The journalist N. Breshko-Breshkovskii had this to say of the "fairy-tale kingdom" at Talashkino:

> The pupils and craftsmen in the Talashkino school are all the children of peasants from neighboring villages. The interesting, sensible work ennobles their souls, and from Talashkino they take back home pure, bright impulses toward something beautiful and poetic, like an early morning daydream. The young lad who sees with amazement the artistic cupboard or table decorated with Sirin birds that his skillful hands produce will no longer be drawn to the filthy, stinking tavern with disgusting cursing filling the air. Whether he will be happy is another matter, but he won't be the same as the rest.[87]

In reality, by 1904 outbreaks of arson were beginning to occur on the estate, and the rapid deterioration in relations between Tenisheva and her "little folk," as she called the school pupils, was having a disastrous effect on the workshops:

> The slightest thing would cause the workers to rebel and express their dissatisfaction, they would come up with all sorts of excuses, walk out en masse in support of one of their fellows, then come back again but go about their work reluctantly. The work didn't go well. As soon as you turned your back they'd stop work and huddle together discussing something. I saw that things were getting out of hand. Matters were no better in the needlework workshop. The peasant women stopped coming to get

work, they were slow to bring in what they'd finished, and their embroidery was slipshod. Others would come only in secret and seemed terrified, but we could get nothing out of them. . . . We were obliged to stop work here too. We put everything in order, stored the finished work in boxes, and closed up the workshops completely.[88]

By early 1905, therefore, the Talashkino kustar workshops had ceased production: unfinished work was completed in Smolensk by a handful of Talashkino kustari supervised by Zinov'ev and Beketov, and by February all the work was completed and the remaining joiners and embroiderers were paid off. The Source remained in operation until 1906, when it too closed its doors.[89] Literally fearing for their lives Princess Tenisheva and Princess Sviatopolk-Chetvertinskaia fled to Paris, leaving the just-completed Museum of Russian Antiquities in Smolensk, full of priceless objects, to the mercy of a rioting mob.

This was by no means the end of Talashkino, however, for during the next three years of Princess Tenisheva's self-imposed exile in Paris – from 1905 to 1908 – she conducted a publicity campaign that was to establish the Talashkino kustar workshops as the most widely known example of their type in Europe. Soon after arriving in Paris she summoned Barshchevskii to join her with the rarest treasures from the Museum of Russian Antiquities. These she put on display at her Paris home to acquaint the French artistic and social elite with the finest that Russian culture could offer. When rumors of her collection reached the French minister of fine arts in 1907, Tenisheva was offered four halls in the Musée des Arts Décoratifs at the Louvre and all the display cases in the Pavillon Marsan in which to organize an exhibition. It was, as she later claimed, "the most successful exhibition of the whole season and it was widely discussed and written about. There were 78,000 visitors."[90] At the end of the year Tenisheva followed up this display of Russian antiquities with an Exposition d'Art Russe Moderne at the Galerie des Artistes Modernes, rue Caumartin. This second exhibition was much more than a companion piece to the first, however, for Princess Tenisheva planned it as nothing less than a direct challenge to Sergei Diagilev's growing prestige and authority as a champion of Russian art abroad. To her great chagrin, in 1906 Diagilev had made his first assault on the French capital with a retrospective exhibition of Russian art at the Paris Salon d'Automne.[91] The most notable aspect of the exhibition was the unprecedented emphasis that Diagilev and his World of Art friends laid on the period of westernization in Russian art that followed Peter the Great's reforms. By contrast, those aspects of Russian culture most familiar to the French public through various international expositions were either downplayed or entirely absent. As a reviewer sympathetic to Princess Tenisheva's endeavors later remarked,

> Misleading foreign public opinion with regrettable levity, after they had shown icons of Byzantine origin, [the exhibition entrepreneurs] made an abrupt leap to the eighteenth century. . . . As for modern artists, it was an official presentation of the cosmopolitan clan, fervent admirers of the outrageous formulae preached by the little chapels of dubious aesthetics in Munich and Paris.[92]

Adding insult to injury, Diagilev and his hanging committee consigned the paintings of Nikolai Roerich to the periphery of the exhibition. At the end of 1906 Princess Tenisheva wrote bitterly to Roerich: "Diagilev is moving in the best circles here, he's received with honor at the embassy, and he and his satellites Bakst and Benois are pushing themselves everywhere. . . . Yes, Diagilev and Co. have acquired a terrible power, and I don't know what weapons to fight them with."[93] The Exposition d'Art Russe Moderne was to act as an antidote to Diagilev's insinuating presence, while at the same time restoring her slighted friend to a central position in the public eye and reaffirming the true national characteristics of Russian art.

From the very beginning, however, the exhibition was fated to enjoy only a moderate success. Neither Golovin, Korovin, nor Mikhail Nesterov, all well known for working in a Russian style, agreed to participate, and Tenisheva was obliged to supplement the 135 works by Roerich ("the king of the show") with contributions by less prominent artists.[94] In addition to what remained of the stock from the Talashkino kustar workshops, the exhibition showed illustrations by Ivan Bilibin, sculpture by Konstantin Rausch von Traubenburg, church projects by Aleksei Shchusev and Vladimir Pokrovskii, and the princess's own champlevé enamels. Reviews focused on Roerich and the Talashkino embroideries, sales of which were so brisk – 3,000 francs in sales the first day – that stocks were soon sold out.[95] When the exhibition moved on to Prague it enjoyed a similar success, and the following summer a modified version was included in the London salon at the Albert Hall. This series of exhibitions was the most complete representation of ancient and modern Russian folk art to be seen in Paris since the Exposition Universelle of 1900, and Tenisheva claimed that it had a visible impact on the world of Parisian couturiers, jewelers, and textile manufacturers, thus predating Lev Bakst's orientalist sets and costumes for the Ballets Russes by two years.[96] The following year, however, the princess was obliged yet again to play second fiddle to Diagilev and the first of his hugely successful Russian Seasons.[97]

At stake here were not merely Princess Tenisheva's pride and her aspirations to be a seminal figure in the world of Russian art. The bitter competition she waged with Diagilev was to determine the true nature of Russian art in the European imagination. Where Diagilev and his circle touted a catholic balance of all spheres and periods of Russian art (with the exception of the Wanderers), Tenisheva, under the influence of Adrian Prakhov and especially Roerich, was moving toward an increasingly anti-European interpretation of Russian culture. She would later voice clearly this anti-Western attitude in her memoirs of 1916, but at the time of the 1906–7 exhibitions her stance found its most virulent public expression in the reviews of C. de Danilovich. Danilovich stressed the racial purity of the Talashkino workshops, presenting the Russian people as barbaric, childish, and ultimately unknowable:

> The achievement of the Talashkino ateliers cannot be compared with similar works of the Occident. Their originality is so strong, so surprising that, accustomed as we are to certain aesthetic postulates, to certain formulae of criticism, we feel that in order to judge them we require a special sense; we feel that it is impossible to measure this art by the same standard as that which we apply to the art of the Occident.[98]

It is likely that Princess Tenisheva's decision to return to Russia was prompted not only by homesickness, but also by the inescapable presence in Europe of her bête noire Diagilev and the frustration of her hopes to establish herself as a central figure there. In 1908 she arrived back in Smolensk, only to find the workshops abandoned and the estate "a graveyard." Although during her absence Zinov'ev and Beketov had continued to design embroidered and carved goods on a small scale, the distribution point for embroidery commissions was now in the nearby village of Grinevo, while most of the carving produced was in the form of prototypes that were then sold to the Moscow Kustar Museum for production at the flourishing Sergiev Posad workshops. Although Tenisheva was able to persuade the two artists to return to Talashkino – Zinov'ev to compose designs, Beketov to dye fabrics – even this semblance of normalcy ended tragically with Beketov's suicide in 1910.

In the long run, the cultural oasis on which Tenisheva had lavished so much love and money appears to have made little lasting impression on the surrounding area. The two thousand peasant women who had depended on Talashkino for their markets were apparently left to fend for themselves after 1906, and not until 1930 did a group of Talashkino embroiderers emerge as the collective Thirty Years of October.[99] Nor does there seem to be any evidence that either wood carving or ceramics caught on among the Talashkino peasants as an ongoing industry, although one historian has noted the infiltration of Talashkino motifs into the decoration of local peasant houses. For observers of the kustar industries' continuing struggle for survival against big-scale industry, Talashkino epitomized the ultimate powerlessness of such small philanthropic and utopian enterprises to halt the inexorable advance of Western capitalism and culture in Russia:

> Our industrial art is suffering from an influx of Western machine-made goods as a result of Russian society's total ignorance of the beauties of our handmade popular art. . . . It is time to admit that charitable enterprises such as Princess Tenisheva's will not improve our kustar production, which is being supplanted by the iron laws of industrial and factory development.[100]

But despite this contemporary pessimism, and the fact that no amount of money could have saved Talashkino from its fate, Princess Tenisheva was superbly successful in insuring that the memory of Talashkino, at least, would outlive the entire kustar revival movement and the dozens of workshops it engendered. By documenting the existence of her beloved workshops, by leaving tangible proof of their existence, not just in her own memoirs, but more crucially in the national and international art journals of her era, she was able to shape an image of kustar art revival in Russia that is now quite erroneously accepted as the norm. Just what "the norm" became in the last decade of the Russian Empire is the subject of the final chapter.

World War I, its carpets actually won a gold medal at a local agricultural exhibition, a fact that brought home to her "how undiscriminating the provincial public is, how different its tastes, and how cautious one must be in distributing awards which can lead to great confusion."[26]

In other schools and workshops instructors favored the neo-Russian style created by Elena Polenova and cultivated by the Stroganov School in Moscow. In the town of Torzhok in Tver province, long famous for its gold embroidery and decorative leather work,

> the teacher at the zemstvo kustar school turned out to be a "modernist" who replaced precise, disciplined design with a new kind, slipshod and decadent, the bright fresh range of colors with impressionistic vagueness, and the outline with stylization. In a word, total innovation. At first, thanks to the then-prevailing fashion for decadence, all this enjoyed great success, and the stylized dandelions à la Polenova (Varvara Schneider's expression) sold like hotcakes. . . . [But] the same thing happened to the kustari of Torzhok as happened to Maliavin's paintings of peasant women. The first canvases went to the Alexander III Museum, while no one wanted to buy the rest.[27]

In fact, the short-lived benefits to be gained by moving to such new patterns were frequently remarked on. For the peasant women of Katunskaia volost', Balakhninskii uezd, in Nizhnii Novgorod province, who produced drawn-thread work for 10 to 20 kopeks a day, "the more fashionable new designs pay better as long as a small number of people are producing them, but as this number increases the amount they earn decreases, until a new kind of stitch appears and higher prices are observed once again."[28]

Clashes of taste could occur between the helper and the helped even where purely technical improvements were proposed. When Professor A. E. Porai-Koshits, the G.U.Z.i Z. expert on dyeing technology, inspected the naboika workshop of Egor Maleev in Vereiskii uezd, Moscow province, he was dismayed by the archaic technology Maleev still clung to. A protégé of Maria F. Iakunchikova and Sergei Morozov, who provided him with boards carved from antique models at Sergiev Posad and who guaranteed him sales through the Kustar Museum in Moscow, Maleev used clay to thicken the dye during printing (Fig. 46). The result, Porai-Koshits wrote disdainfully, was "not a naboika but a sort of sculpture" that was quite unusable for anything but upholstery fabric, "and then only if there's a particular fondness for kustar goods. Only the fashion for these goods, and the very real artistic merit of the pattern itself, can explain the fact that such naboiki always find a market."[29] It was with difficulty that Porai-Koshits persuaded Maleev to try the artificial aniline indigo manufactured by the Baden Aniline and Soda Factory in Moscow.

Schools and training workshops were also responsible for the rapid spread of new urban fashions in the centers of kustar production. This was as true for the *style moderne* popular at the turn of the century as it was for the neoclassical revival of the prewar years. With the advent of zemstvo-operated training workshops for carpet weaving in the villages of Sudzhanskii uezd, Kursk province, production quickly became polarized between local customs and city demands. The weaving of carpets for the dowries of local peasant girls continued to be done on small looms at home, using traditional patterns and colors of the weavers' own choosing. Entirely different criteria governed the design of larger carpets intended for the city, under the supervision of a Mariinskii Lace School graduate and the artist V. P. Trofimov: "A peculiarity of Sudzha carpets is the fact that the patterns are not local, taken from old carpets or using local ornament, but are designed by artists in a contemporary style. The dye shades in carpets intended for the city are exclusively pale and faded; conversely, the local population likes very bright shades."[30] Of the thirty-six carpets woven in the

Figure 46. Cushions made from naboika fabrics printed by Egor Maleev, 1913. (Reprinted from *Russkoe narodnoe iskusstvo* [Petrograd, 1914])

novich, one of the two panneaux that had so scandalized the city's visitors just fifteen years before.[37]

But Moscow's strongest competition came from Little Russia, as the Ukraine was then called, where a network of zemstvo and private workshops brought a new note of ethnic and artistic diversity to the production and marketing of kustar art. The official kustar movement in the Ukraine was centered in Poltava, where the provincial zemstvo had invested quite generously in kustar reform since the 1890s.[38] In 1907 it opened a kustar outlet and followed this the next year with a Zemskii Dom (Zemstvo House) – a complex of museum, exhibition, and commercial facilities on a grand scale (Fig. 48).[39] The building was designed in a "Southern Russian style" by the architect V. G. Krichevskii, and the well-known Ukrainian painters N. S. Samokish and S. I. Vasilkovskii were enlisted to decorate it.[40] The Poltava zemstvo's new center was to be a model of kustar reform motives and tactics, including the exploitation of local ornament:

> The actual idea of establishing the Zemstvo House arose out of concern for the local kustari. The zemstvo very reasonably considered that one of the most important conditions for good sales of kustar goods is their originality and the uniqueness of local artistic traditions. To this end it sought out artistic models and put them into production, . . . founded a museum

Figure 48. Poltava Zemstvo House, by V. S. Krichevskii, 1907. (Reprinted from *Niva,* 18 [1908])

of ethnography and archaeology, and began to collect antiquities. Everything that came into the museum . . . was . . . sketched and photographed [and] made into albums. The zemstvo published them and distributed them among the kustari. . . . Finally, the zemstvo decided to popularize the Ukrainian past *en grand* [with] the establishment of a huge building for zemstvo organizations in the Ukrainian style.[41]

By 1911 the Poltava zemstvo was spending over 140,000 rubles per annum (including government subsidies) on efforts to improve kustar industry. Among the workshops it funded were the Gogol School of Industrial Art in Mirgorod, the Poltava Craft School, the Degtiar School for Weaving Instructors, and the Bol'shaia Budinskaia and Poltava carpentry workshops.[42] Under the artistic direction of V. I. Cherchenko, a graduate of the Stroganov School, the Poltava workshop produced goods in a "neo-Ukrainian" style that offered a distinctive alternative to the Great Russian traditions promoted by the workshops of the Moscow zemstvo (Fig. 49).

As for the women's kustar industries in the Ukraine, the major organizational force was the Kiev Kustar Society, established under imperial patronage in August 1906 as "the only center that unites, regulates, and directs the activity of individuals and institutions in Kiev, Podolsk, Chernigov, and Bessarab provinces."[43] The core of the

Figure 49. Maple sideboard designed by V. I. Cherchenko and made at the Poltava zemstvo woodworking workshop. (Reprinted from *Russkoe narodnoe iskusstvo* [Petrograd, 1914])

society's membership was a small group of women who, "possessing great artistic taste and a knowledge of local ornament, and also keeping constantly abreast of public demands and the dictates of fashion," were able to "direct the kustari to produce items, each one of which is in its own way a complete work of art in a Little Russian style."[44] Among them were Natalia M. Davydova, Princess Natalia Iashvil, Varvara Khanenko, and Aleksandra Exter, all of whom, like Elena Polenova, Maria Iakunchikova, and Maria Tenisheva before them, were connected to the world of high art rather than the traditions of gentry philanthropy, a fact that had a decisive impact on the orientation of the society's work.

The president of the society was the painter Natalia Mikhailovna Davydova, who operated an embroidery workshop at Verbovka in Chernigovskii uezd. Unfortunately, we know virtually nothing about the life or artistic career of this influential woman.[45] In all probability her husband was Vladimir ("Bob") Davydov, grandson of the exiled Decembrist Vasilii Davydov and nephew and literary executor of the composer Petr Tchaikovsky.[46] The Verbovka estate, located near the Davydov family's principal residence, Kamenka, near Kiev, had been bought in 1870 by Davydova's father-in-law and was a favorite retreat for Tchaikovsky throughout the 1870s and 1880s. It seems likely that she opened a kustar embroidery workshop at Verbovka soon after her husband's death in 1906 and in response to the success of the First Exhibition of Applied Art and Kustar Crafts in Kiev.[47] Subsequently an active participant in the avant-garde exhibitions and Constructivist experiments of the early Soviet period, Davydova was to bring Verbovka during the First World War into closer contact with the world of "high art" than any other workshop since Talashkino, as we will see.

Equally prominent in the society's activities was Princess Natalia Iashvil, who with her sister operated a kustar embroidery workshop at her Sunki estate near the town of Smela in Cherkasskii uezd, Kiev province, a property that had formerly belonged to the Raevskii family and had been a haunt of the poet Aleksandr Pushkin.[48] Like Maria Tenisheva, Iashvil had been an aspiring artist before her marriage into the nobility. (As Natalia Filipson she had studied watercolor with Polenova's drawing master Pavel Chistiakov.) But whereas Tenisheva married into a fortune, by the mid-1890s Iashvil was left an impoverished widow with two small children and a run-down estate. We can only imagine how by the early 1900s she had managed to build an excellent school for peasant children at Sunki and to open an embroidery workshop where "museum-quality pieces from the seventeenth and eighteenth centuries were used as models."[49] At least two artists are known to have designed for the Sunki workshops: Nikolai Prakhov, son of the irascible Adrian Prakhov,[50] and Mikhail Nesterov, who had been a frequent visitor to Abramtsevo in the 1880s and 1890s. Nesterov had first met Princess Iashvil while working with Viktor Vasnetsov on the decorations for the new Cathedral of St. Vladimir in Kiev, and for almost nine years he and his family lived on her farm at Kniagino, about three miles from Sunki. It was there that he completed almost all of his sketches for the Grand Duchess Elizaveta Fedorovna's Martha and Mary Convent in Moscow, as well as a compelling portrait of his patron and friend. It was Princess Iashvil, no doubt, who arranged for embroideries from the Sunki and other Ukrainian workshops to be shown at Nesterov's one-man exhibition held in Moscow in 1907.[51]

Other notable workshops affiliated with the society were those of Iu. N. Gudim-Levkovich at Zozovo, and the well-equipped weaving and carpet workshops (f. 1907 and 1911, respectively) operated by Varvara N. Khanenko in the village of Elenovka near Motovilovka. On a tour of inspection of Elenovka made in 1913, A. G. Dolivo-Dobrovol'skaia noted that "the population understands very well the significance and usefulness of the workshop, willingly sending their daughters to study weaving, which gives them a good income in the time they have free from field work."[52] As at Sunki, Khanenko's carpets were based on patterns taken from her own collection of antique Ukrainian carpets, as well as that of the Kiev Museum of History and Antiquities. All of these centers, in fact, cultivated a strong regional tradition of folk ornament characterized by bold and bright floral motifs, an artistic policy that also had ideological overtones, since it represented a rejection of "the overdone tones of eighteenth-century Polish [*panskie*] embroideries" favored by the Poltava zemstvo workshops (Fig. 50).[53] Until 1915, when Verbovka came within the orbit of the Petrograd avant-garde associated with Kazimir Malevich, the work of the Kiev Kustar Society proceeded very much "under the sway of museum influences," imitating Ukrainian folk embroideries and carpets in the Kiev Museum of History and Antiquities.

The one weak spot in this otherwise well-run production of traditional Ukrainian textiles was the lack of facilities for dyeing fabrics and yarn that would make the use of aniline dyes redundant and perfect the reproduction of traditional designs. As a result, peasant women working for these workshops were placed in a difficult position, caught between the aesthetic ideals of the Kustar Society and the realities of their own situation. Thus, it was reported by one observer that

Figure 50. Embroideries from the Verbovka kustar workshop and furniture made at Sunki, Kiev province, ca. 1912. (Reprinted from *Russkoe narodnoe iskusstvo* [Petrograd, 1914])

"in order to satisfy the Kustar Society's desire to have carpets woven to patterns it had provided, using the color schemes found in old carpets and of equal fastness, the kustari in one district could find no better solution than to unravel several old worn-out carpets and use the yarn to make new ones."[54]

One final private workshop among the cluster operating in the Ukraine deserves mention. Though not officially affiliated with the Kiev Kustar Society, the embroidery workshop run by Anastasia Semigradova in the village of Skoptsy in Pereiaslavskii uezd, Poltava province, appears to have been linked to it through its artistic director, Evgenia Pribyl'skaia (see Fig. 51).[55] Pribyl'skaia had studied landscape painting at the Kiev School of Art before becoming interested in folk art through exposure to the work of Elena Polenova and her circle and signing on at Skoptsy in 1910. It is tempting to speculate that the catalyst for bringing these women together in pursuit of a common aesthetic and social goal was the Kiev painter Aleksandra Exter, herself a 1906 graduate of the Kiev School of Art as well as a member of the

Figure 51. Embroidered cloaks made at the Skoptsy workshops, Poltava province, and a woman's jacket of dyed canvas made by the Sunki kustari, Poltava province, ca. 1913. (Reprinted from *Russkoe narodnoe iskusstvo* [Petrograd, 1914])

Kiev Kustar Society. Much work remains to be done in clarifying these complex and far-reaching personal and artistic relationships.

The taste and demand for all things Russian ran high in England, Europe, and America on the eve of World War I. The best-known manifestation of this was of course the phenomenal success of Sergei Diagilev's Ballets Russes, whose "primitive" ballets of 1910–14, beginning with *L'Oiseau de feu* and ending with *Le Coq d'or,* had an electrifying effect on Western audiences. But this was just the climax of a growing interest in *russerie,* to which export kustar goods contributed and from which they also benefited. Among the theories put forward for this rather sudden interest in Russian goods abroad, that of the economist Nikolai Kablukov is particularly interesting, in that it demonstrates a pragmatic awareness of the rise of both individualism and living standards that might have contributed to the European demand for objects perceived to be exotic and primitive. In terms reminiscent of Thorstein Veblen's theory of conspicuous consumption, Kablukov wrote:

> This demand among the well-to-do [for kustar goods] is brought about by a desire to be distinct from others, not to be confused with them, to stand out from the crowd, and so not to use mass-produced goods, but to show that I can allow myself what is beyond the reach of others. . . . With the rise in living standards among the masses, the increased productivity of labor, and the development of personality, the demands that arise

crude types of faces, their small Slavonic eyes, heavy with dreams and melancholy. They are the originals of the Russian peasants and workmen painted by the masters of romance, Dostoievsky, Tolstoi and Gorki, the author of *Quicksands*.[65]

By 1912 foreign demand for kustar goods was promising enough that the Russian Export House added a kustar division to its activities, while the Society for the Improvement of Popular Labor in Memory of the Tsar-Liberator Alexander II opened a depot in St. Petersburg (62 Sergievskaia Street) designed to increase sales both at home and abroad. Still, nobody involved in kustar exports could afford to forget that "national flavor" was the tenuous thread on which the whole success of their enterprise depended, "since these goods often fall out of favor with the consumer, and thus with the foreign market, as soon as they lose their distinctive national character. In this respect particular caution must be taken with kustar art goods, which are expensive and in demand only because they are distinctively Russian; otherwise they become unwanted rubbish, with which our markets and those abroad are already overflowing as it is."[66]

Important though sales of kustar goods abroad became in the prewar decade, the foreigner's stamp of approval also proved vital in winning them greater acceptance at home. As an authority on kustar exports pointed out, "We would perhaps have continued to be skeptical about the kustari and their goods to this day if, thanks to the exhibitions in Western Europe, foreigners had not given them the attention they deserved and reported favorably on them."[67] Nevertheless, it remained an immutable fact of Russian cultural and economic life that "the most profitable and expedient kind of production is to use foreign models for Russia and Russian models for export."[68] For example, while exports of Russian lace increased almost twofold between 1895 and 1913 (from 62,000 to 103,300 rubles), they were still minuscule when compared with the 3,899,800 rubles worth of foreign lace imported.[69] The deeply engrained contempt for domestic products that these figures reflect also caused many kustar goods to be sold in Russia under phony foreign or factory labels. A case in point was the experience of Countess A. A. Moll', a landowner in Vitebsk province who had operated and financed a network of kustar weaving workshops in and around Rezhitsa since 1889. "One of the best trading houses in Warsaw proposed to Madame Moll' that she deliver her fabrics with the proviso that her labels be taken off. But she refused, arguing that her goods would then be sold as factory-made and so would lose their individuality, at a time when kustar weaving in this region had attained a high standard and acquired a much-deserved reputation."[70] Of interest here is not so much the deviousness of the merchant as Countess Moll''s recognition of the role played by labels (in the broadest sense) in influencing the buying habits and aesthetic attitudes of consumers. While the design and quality of many kustar-produced goods might make them physically indistinguishable from their factory equivalents, the means of their production still had a powerful symbolic significance that was as undesirable to the average Russian consumer as it was seductive for a foreigner.

As the development of an export market for kustar goods proceeded apace, built on the twin precepts of national flavor and artistic refinement, a fundamental dilemma began to emerge regarding the domestic market. The question was essentially this: should those engineering the reform of kustar industries gear production to the needs of the countryside or the city? Country goods were defined as "for the most part crude, requiring no training, and mass [produced] for a wide circle of consumers," while products destined for the city were those that "demanded a comparatively high level of technical skill, sometimes of an artistic nature, providing expensive goods for comparatively well-to-do, and sometimes wealthy people."[71] If the latter crafts continued to receive the lion's share of government assistance, then what would become of the many humble utilitarian activities which, because they occupied the vast majority of Russia's kustar population, were still far more crucial to peasant subsistence than the elite luxury crafts?

A clear differentiation between artistic and utilitarian goods (i.e., those destined for urban and rural markets) became increasingly evident wherever outside intervention occurred in an area of kustar production. When Tambov held its first provincial exhibition of kustar women's work in 1912, for instance, the contrast between the haves and have-nots was clearly marked. The nine workshops where controlled assistance from above was practiced (including Solomenko and Bykova-Gora) were seen to "stand out even more sharply against the background of remaining kustar goods, where the peasants work independently, without any efforts being taken on their behalf."[72] Peasant women from a district where there was no training workshop to guide them exhibited "a great many brightly colored striped woolen girdles and skirt lengths, for which the best quality wool is normally used and which are beautifully made, but which are absolutely unusable because of their color."[73] Others were still using looms on which only the narrow fabric widths traditionally used in peasant households could be woven, a defect that severely restricted marketability beyond the confines of village life. At the same time, peasant women were fully aware of the appeal that their traditional clothing had for collectors of folk art and sold items from their dowries willingly. At the Berezovskaia Agriculture and Kustar Exhibition held in Ramona, Voronezh province, in 1912, women sold off whatever they had in their trunks, in particular the red-striped wool skirt ticking which, "although they had completely stopped wearing, they still wove for their dowries and willingly sold for high prices to lovers of finery."[74]

As the second decade of the century began and the gulf between these two sectors of kustar production (improved and unimproved, decorative and functional, city- and country-bound) steadily widened, it became increasingly clear which would survive and which would be left to wither away. Count P. S. Sheremetev presented a simple prescription for the future of the kustar industries in Moscow province that was to hold true for the entire empire: "Among the crafts it is desirable to pay special attention to those that bear an artistic imprint and are therefore capable, once they are improved, of competing with machine labor. Nails and kustar bobbins, however, cannot be supported."[75]

While G.U.Z.i Z. and other organizations were forging ahead in the selective popularization of kustar art, a vocal minority among the Russian intelligentsia was looking back with considerable misgivings on the path taken since the 1870s. Even those who accepted the idea that help for the kustari must come from outside – specifically from the educated classes – took exception to the form in which this help was being rendered, including the deliberate falsification of such traditional "kustar" qualities as formal crudeness and technical primitivism:

> Russian craftsmen are capable of producing splendid carved objects, but unfortunately one can point to instances where the influence of supervisors from the educated classes on the kustar production of carved goods in the Old Russian style has produced negative results. Under the influence of recent artistic trends, the carvers have taken a turn in the wrong direction, since those who have initiated the enterprise, in their enthusiasm for the inaccuracies of design found in the old models, have forced the craftsmen to work crudely, with deliberate asymmetry of proportions, and so they have trained them not to reproduce goods in the spirit of the old pieces, but to fake them.[76]

Concerned observers began to question the extent to which improved kustar goods could still be considered popular or folk art, as the export propaganda implied. What relationship did the new designs have to the experience of the people who made them? And what were the immediate and long-term effects on the kustari of producing objects designed by artists for a market the needs of which they scarcely comprehended? The increasingly purist approach to folk art that emerged in the prewar decade held all attempts to modernize and adapt folk art in suspicion, but especially those that involved the formal education of the kustar. As the critic and architect Georgii Lukomskii put it, "Folk art should not be isolated from the peasant hut and the forest, and transported to a musty, bleak, uncomfortable, state-run schoolroom, with monitors and blackboards. [The people] should not be forced to copy clay tuppenny roosters on expensive Whatman paper."[77] Sharing a large part of the blame for this falsification of folk art for economic gain were the workshops where the idea of artistic intervention had originated. Lukomskii voiced the sentiments of a new generation sympathetic to the emerging primitivist aesthetic when he complained in 1913: "Haven't we had enough of all these Russian forms in the new style? This whole kustar *style russe moderne* originating in the Abramtsevo workshop, in the whole atmosphere of Talashkino?"[78]

Some members of this younger, more skeptical generation had personal grounds for disliking the commercialization of kustar art. While an impoverished student at the Moscow School of Painting, Sculpture, and Architecture in 1907, the poet and painter Vladimir Maiakovskii supported himself doing job work for the Dazziaro store in Moscow. Maiakovskii described this experience in his autobiography as follows:

> Work. No money in the family. Had to do poker-work and drawing. I especially remember the Easter eggs. Round, they turn and scrape like doors. I sold the eggs to a kustar store on Neglinaia Street. Fifteen kopeks for ten. Ever since I've had an undying hatred for Bems, the *style russe,* and *kustarshchina.*[79]

That these were the sentiments not just of a few individuals but of an entire generation was borne out by the tone of the papers given at the All-Russian Congress of Russian Artists which convened in St. Petersburg in the winter of 1911–12. It was symptomatic of the widespread interest in the "kustar question" that an entire session was devoted to "Industrial Art and Kustar Industry," while the list of delegates included the names of many prominent artists and administrators associated with the movement.[80] Although hopes for a new era in industrial and kustar art in Russia still ran high, complaints were beginning to surface about the direction that the management of kustar affairs had taken, complaints in which both self-blame and finger-pointing were commingled, directed first and foremost at the kustar schools, "lady patronesses," and dangerously modern instructors. The delegate for the Russian Industrial Art Society, the sculptor Maria Ditrikh, warned against "educated women artists giving the people models foreign to them," while E. N. Polovtsova, a member of the Russkoe Zerno (Russian Grain) Society and herself in charge of several lacemaking workshops in Skopinskii uezd, Riazan province, delivered an attack on the running of lace schools in Russia that included a strong element of mea culpa. "In the schools and workshops we give them the models. They work 'for the boss'! Exclusively to make a living. They make not what they want to, but what we want them to, we the supervisors who keep up with demand in the market, we who are trying to improve industrial art."[81] With the benefit of several years' hindsight others reached the painful conclusion that those most concerned for the future survival of folk/kustar art in Russia had in fact proved to be the agents of its further deterioration. "All of these in many instances heroic efforts did not ultimately have the desired effect," an observer wrote in 1923, "but rather gave rise to a great many false tendencies in the Russian popular style, because they were based on a completely inadequate knowledge of sources. The initiators and instructors gave the kustari weak and artistically false models that unforgivably broke down and dispersed the organic link between construction and decoration in folk art and condemned the kustar to the dull and punctilious reproduction of anemic, corrupted motifs and themes from popular ornament and iconography."[82]

This new wave of interest in preserving the purity and authenticity of Russian folk art – rather than in refining its rough edges – is now best known to us through the neo-Primitivist paintings and manifestos of Natalia Goncharova, Mikhail Larionov, Kazimir Malevich, and Aleksandr Shevchenko.[83] But it was also reflected in an exhibition of Russian folk art organized by the Société des Artistes Russes as part of the Eleventh Paris Salon d'Automne of 1913, an event that the young critic Iakov Tugendkhol'd described as "the perfectly logical conclusion to that interest in our art that began abroad with the triumphs of Russian music, ballet, and theater design."[84] Comprising icons, toys, lubki, prianiki, and naboika prints from the private collections of Ivan Bilibin, Sergei Chekhonin, Mstislav Dobuzhinskii, Aleksandra Exter, Nikolai Kulbin, Larionov, Georgii Narbut, Roerich, and Sergei Sudeikin, the exhibition brought members of Russia's "left" and "moderate" artistic camps together as perhaps no other occasion could have done at this period.[85]

In his introduction to the catalogue Tugendkhol'd invoked the primitive ideal as exemplified by Paul Gauguin and stated that the exhibition's goal was to suggest "the true image of popular Russian art," and to show that "a love of folklore has become a factor in artistic progress."[86] Also included were embroideries and carpets from the workshops of Princess Iashvil, Evgenia Pribyl'skaia, Varvara Khanenko, and Natalia M. Davydova, and toys produced by the Sergiev Posad and Bogorodskoe kustari to Nikolai Bartram's designs. Although the inclusion of these improved exhibits seems incongruous, given the aims of the exhibition, it is partially explained – in the case of the embroideries – by the shared Ukrainian background of many of the participants, and in the case of the toys, by the new interest in the life of the child as the ultimate primitive experience, a field in which Bartram was an acknowledged expert. This may, incidentally, have been the exhibition that the choreographer Mikhail Fokine visited when he was "inspired by kustar toys, poses in old icons, the remains of wall paintings, and old Russian embroideries" to choreograph the 1913 ballet *Le Coq d'or.*[87]

None of these new ideas succeeded in affecting or altering the ponderous course of the official kustar movement, however. Once the connection between artistic improvements and increased profits had been fully realized, the mechanism seemed to have a life of its own, impervious to either aesthetic or ethical second thoughts. Taken out of the hands of individuals and now firmly centralized under the bureaucratic control of G.U.Z.i Z. with its network of workshops, schools, museums, and exhibitions, by 1913 the kustar art industries had almost completely metamorphosed into a highly regulated form of industrial art.

The public unveiling of this carefully constructed transformation took place at the Second All-Russian Kustar Exhibition, which opened in the halls of the St. Petersburg Botanical Gardens in March 1913. Not only was this to be the greatest celebration of the policy of kustar revival to take place in the imperial period, but as one of many events planned to mark the tercentenary of the Romanov dynasty throughout 1913, the exhibition had an extra role to play in upholding and illustrating the complex ideology of imperial power.[88] The organization of both the exhibition and the Congress of Workers in the Kustar Industries that accompanied it fell to G.U.Z.i Z. and an organizing committee chaired by the tsar's brother-in-law, Duke Peter of Oldenburg.[89] Much was expected from this exhibition, in both symbolic and practical terms, and for official kustar policy over the past four years to prove itself justified a full turnout of Russia's kustar art industries was essential. As the committee concluded, "Only a complete collection of kustar goods from all regions can give a picture

> Given the influence that the educated classes exert on peasant kustarnichestvo, it is naturally important that a turning point should occur in them themselves, the present and future mentors of kustar artists – toward native tastes and traditions. To be stable, such a turning point should begin with the impressions of childhood. . . . To envelop the life of a private family in the beauty of old Russian surroundings is within the power of only the more well-to-do. But the influence of this upper cultural stratum on the lower classes, and in particular on the development of folk art and kustar industry, is enormous. . . . Today we have Russian educated women, Russian artists, and a whole list of training workshops directing the work of the kustari, supplying them with designs, and so on. This is why it is so important . . . to surround the educated Russian family in its everyday milieu with impressions of Russian folk art. . . . In such a setting those talented people who will lead Russian folk art forward will be formed and strengthened.[95]

In calling for this trickle-down russification – what one critic jeeringly dubbed the current fashion for "cabbage and black bread" among the aristocracy – Prakhov tacitly admitted the government's difficulties in effecting a similar psychological transformation in the kustari themselves. In fact, the catalogue admitted, "the kustari themselves often find factory goods more beautiful [than] the usual local products. It must be said that kustari are often supported in this by local consumers, since in the provincial backwoods local ornament is not seen as valuable, is considered too commonplace and 'commonfolk,' and so factory goods have a great success."[96] Whether G.U.Z.i Z. liked it or not, its attempts to direct kustar industry and stabilize rural life were taking place at a time "when the Russian village, the hearth of our kustar industry, is undergoing a major transition from certain forms of agriculture and even lifestyle to other forms that harbor the possibility of a profound change in the very psychological foundations of our rural population."[97]

The government's progress in expanding its reform activities to the more remote kustar centers of the empire was reflected in the ratio of catalogue illustrations to the various geographical regions.[98] Eighty-four plates were assigned to Great Russia with Moscow at its center; thirty-nine to Little Russia with Poltava as its capital; twenty-two to the Caucasus and Asian region; seven to the southern provinces of European Russia; and six each to Poland and Finland. Times had changed since 1890, however, when it was still possible to view Russian colonialism in Central Asia as the bringing of light to benighted Asiatic peoples. At the Turkestan Exhibition held in Tashkent that year, the kustar crafts of the russified Turkmen were favorably compared with the unimproved art of Bukhara and Khivin, which, "although it is also distinctive and striking to the eye of someone unaccustomed to Asiatic life, is at the same time stuck in a centuries-old mold." The natives of Turkmenia, by contrast, were "already liberated from the limits of routine and habit" by Russian civilization.[99] Two decades later a similar comparison between the "improved" Great Russian kustar and his counterparts in other parts of the empire yielded a very different conclusion. Comparing Sudzha carpets with those made in Central Asia, the reviewer for the newspaper *Rech'* (Speech) found them beautiful but "ultramodernist and completely alien to any sort of popular motifs."[100]

> One's first impression is that the marches and non-Russian territories [of the empire] display a more clearly defined national character in their art than does native Russia. Lithuanian fabrics, belts, and embroideries, Polish patterns and colors, the Caucasus and Transcaucasus with their carpets, chased metal wares, and ceramics, the Kirgizhians, Little Russians, and Finns have exhibited scarcely anything that is market-oriented, lacking in style or a clearly marked physiognomy of their own.[101]

Whereas the non-Russian exhibits were still for the most part goods made by peasants for their own use (i.e., folk art in the classic sense), the majority of Russian kustari were now working "primarily for others, for the market, for sale, and for the most part they produce objects that you will never see on a peasant's back or in his izba." By way of example the reviewer cited a 6,000 ruble dining-room suite in a modernized "Russian" style and carpets intended exclusively for "the gentry."[102]

The complexity of the Great Russian kustar's position vis-à-vis his changing market was brought home by the revival of early-nineteenth-century Empire and neoclassical styles that had such a special resonance for Russians on the eve of World War I. The previous year, 1912, had marked the centenary of Russia's victory over Napoleon and the great Moscow fire of 1812, and since this latter event had directly stimulated the growth of kustar furniture making in Moscow province, a resurgence of the simple mahogany styles of that era was both timely and natural. With its restrained ornamentation and relatively straightforward construction, the Empire style was considered ideally suited to the skills of kustar carpenters, while it admirably answered the reigning taste for the Alexandrine period among the upper classes and in artistic circles. A similar nostalgia for the "good old days" surfaced in 1911, in response to the fiftieth anniversary of the abolition of serfdom. The cult of the *usad'ba* (country house), reflected in a general way in periodicals like *Starye gody* (Bygone Years) and

Stolitsa i usad'ba (Town and Country), had a special bittersweet association for all those aristocratic landowners now engaged in kustar reform on their own family estates. In a talk given to the Moscow provincial zemstvo assembly in 1911, Count P. S. Sheremetev complained bitterly about the one-sided attention currently being focused on the darker aspects of serfdom and the refusal to acknowledge "what was undoubtedly good from an artistic perspective" in the estate culture of his ancestors. In fact, as he quite rightly pointed out, "one could write an entire book describing all that many landowners, and especially lady landowners, are busy with in our own day in their family nests across the entire face of our fatherland. . . . Our estates thus continue the industrial arts enlightenment of former times and God grant them success, if they are only spared the senseless movement to imperil and burn them, supposedly in the interests of social justice."[103] As public tastes changed, the reintroduction of serf traditions and patterns into kustar production was seen as both a natural and a welcome alternative to the usual "folky" stock-in-trade. In Poltava, Kiev, and Ekaterinoslav provinces, zemstvo museums began collecting large quantities of serf-made carpets that were then copied or adapted (e.g., by the Degtiar School), while at Solomenko, where a carpet workshop had been opened in 1909 under the supervision of a Mariinskii-trained instructor, sixty peasant girls were now producing carpets in the Empire style. The ability of the new improved kustar to handle such an eclectic variety of styles and consumer tastes was demonstrated with particular clarity by the joint displays of the Abramtsevo and Solomenko workshops, still operating under the direction of Iakunchikova and Davydova. Of the two Abramtsevo displays illustrated in the catalogue, one featured mahogany furniture in the Russian Empire style of the early nineteenth century (Fig. 52), while the other contained furniture in the Russian style, much of it intended for religious purposes (royal gates, a candle cupboard, icon frames), as well as some items dating back to Elena Polenova's time. In both cases the furniture was set against a backdrop of Solomenko carpets in the Empire style.[104]

As far as the general public was concerned, the Second All-Russian Kustar Exhibition was an outstanding success. Almost two million visitors attended the exhibition between 10 March and 28 April, and "for two months the Russian kustar became the hero of the day."[105] One of its most remarkable achievements was the completion of a subtle, barely acknowledged shift in the public perception and definition of "folk art." In 1910 ceramics expert Nikolai Root had predicted that "kustar production in most of its branches, given essential cultural and aesthetic education, must in the future inevitably take the form of industrial art."[106] That transformation was made almost complete, not only by the exhibits themselves, which were in many cases identical to the work of Russia's leading schools of design, but especially by the rhetoric that accompanied them. The fact that the catalogue billed the event as an "exhibition of folk art" broke down a vital semantic barrier by implying that improved kustar art had effectively become synonymous with folk art.

But neither the pomp and rhetoric nor the "chic and brilliance" of the exhibition insulated it from criticism, nor was the packaging of aesthetically improved kustar art as folk art accepted without comment. In fact, the very artistry on which G.U.Z.i Z. congratulated itself as a sign of progress was widely criticized for its dubious authenticity. "If at the first exhibition [in 1902] you could still find genuine works of folk art, at the second you'll search for them in vain. The nature of this exhibition is such that it should rather be called an exhi-

Figure 52. Solomenko carpets and mahogany and Karelian birch furniture made at Abramtsevo, both in the Empire style, ca. 1913, exhibited at the Second All-Russian Kustar Exhibition in St. Petersburg, 1913. (Reprinted from *Russkoe narodnoe iskusstvo* [Petrograd, 1914])

bition of popular industry, because all the goods in it are geared to the tastes of the consumers and to a market among the population at large."[107] There were glaring examples of misused and misguided talent: a kustar called M. F. Sidorov who had devoted five years to building a model chapel without the aid of nails or glue, pupils at the Vyshnii-Volochok Training School for Instructors in Dyeing and Weaving in Tver province who spent one and a half years weaving Viktor Vasnetsov's famous painting *Three Bogatyrs,* and similar efforts expended on Ivan Bilibin's popular illustration to the Kievan bylina of the Volga (priced at 250 rubles) at the G.U.Z.i Z. school in the village of Sosnovitsy, Tver province. Such excesses lent weight to the view that "folk art has reached some sort of dead end, that the worker's imagination is sluggish, and that he mostly works from drawings of artistically impoverished patterns taken from the supplement to some illustrated magazine for family reading."[108]

The rosy picture of progress displayed in the exhibition's halls was especially galling to those who were familiar with the real condition of the kustar industries as a whole. Recent statistical evidence showed that the economic situation of the vast majority of kustari *excluded* from the 1913 catalogue (e.g., producers of nails, barrels, shoes, felt boots) had actually continued to deteriorate since the 1870s, thanks to a lack of markets, credit, and technical improvements, and to the increased costs of raw materials. Moreover, the overall number of kustari, which populist economists had long insisted was anywhere between 7.5 and 15 million, was recalculated on the basis of new zemstvo statistics at a mere 1.8 million in forty of the empire's provinces, while the 200–250 million rubles actually generated annually by kustar industry fell far short of earlier estimates of 2–3 billion.[109] Puncturing populist optimism still further was a 1913 report that only one-fiftieth of the kustar goods produced in Russia were sold through museums or artels and that at best some 20,000 of the 2 million or so kustari active by 1913 were receiving any protection from exploitation by middlemen.[110] Small wonder that economist Mikhail Tugan-Baranovskii labeled the whole exhibition a Potemkin village designed to impress and delude foreigners. In pure economic terms, he pointed out, those few kustar industries that actually received government and zemstvo aid (in 1913 alone 1,647,000 rubles from G.U.Z.i Z. and 25,000 from various zemstva) were operating in the red. "Under today's conditions any system can produce well-designed fabric," he wrote; "the question is how much it costs to produce it. In zemstvo workshops it's being done at a loss."[111] For both those who cared about the future of a pure, "classical" folk art and those concerned about the fate of all those kustari beyond the pale of the government's benevolence, the Second All-Russian Kustar Exhibition was a sham:

In all good conscience this can only theoretically be called a "kustar" exhibition. Of course, it did present work by kustari, but for the most part it was filled with exhibits from training workshops, various philanthropic institutions (under the patronage of high-placed individuals), and so on. But most important the exhibition did not include the one thing that could have been especially interesting . . . the kustar himself, his lifestyle, his social and kustar concerns, and his sufferings. There was a kustar present, but it was a "Potemkin" kustar, just as throughout the entire exhibition one sensed the hand of cunning decorators from various [government] departments out to make their careers.[112]

All of these objections notwithstanding, in several respects the chain of events that had begun at Abramtsevo in 1885 had borne positive fruit by 1913. A number of crafts, not to mention countless artifacts, had been rescued from oblivion and destruction thanks to the efforts of concerned individuals and organizations; small numbers of kustari throughout the empire now had reliable incomes and skills to combat the vagaries of the land and the economy; and Russia's political, economic, and social structure had to some extent bolstered itself against social change by strengthening the ties binding the kustar population to the land. For all of these reasons the kustar art industries had finally won the support of the government, and in 1913 there was no sign that the kustar boom would not continue. Yet to a growing number of Russian intellectuals none of these positive achievements was adequate compensation for what was irrevocably lost: the spontaneity, naïveté, and purity of genuine folk art. For this sector of the population, all that this benevolent and pragmatic intervention had produced was a pastiche of folk art cut to fit the stereotypes and preconceptions of foreigners and middle-class city dwellers. Summing up the effect of the kustar art he had seen at the Kiev All-Russian Arts and Industries Exhibition in 1913, Georgii Lukomskii wrote:

Finally you come to the conclusion that folk art – under the influence of the schools for instructors, the care of our lady patronesses (more and more of German origins or French education), under pressure of greater demand that creates a desire to make the material they work popular and above all salable – is increasingly being ruined in Russia. They are destroying the pure waters of the wellspring of its ideas, destroying the bright splendor of its many colors, killing all the half-childlike, naive emotion of artistic sentiment.[113]

When Russia mobilized for war against Austria in August 1914, both the ambitions of kustar reformers and the fears of their detractors were

As for the former owners of those estates, we can only assume that the majority either escaped the Revolution through emigration, like Princess Maria Tenisheva, who spent the last years of her life at "le Petit Talashkino," her St. Cloud home near Paris, or met a more violent end like Grand Duchess Elizaveta Fedorovna, who was murdered in 1918 along with other members of the imperial family.

Yet despite these radical reorganizations and the hardships caused by revolution, civil war, and famine, many kustar art centers managed to resume production almost immediately after 1917, thanks in large part to the Soviet government's support for kustar art exports as a ready source of foreign currency and a palatable symbol of national cultural values that could be exhibited at the world's fairs of the 1920s. At Abramtsevo, for example, the carpentry workshop was placed under the control of Narkompros in 1918 (the year of Savva Mamontov's death) and was hailed by its new director, K. V. Orlov, as "the oldest of all kustar training workshops," where "the distribution of technical and artistic aid to the kustari and a collaborative relationship between artist and kustar had their origins."[4] By the early 1920s items of furniture that Elena Polenova had designed in the 1880s were once again on sale at the shop of the former Moscow Kustar Museum, and small column cupboards could be bought for seven rubles at G.U.M., Moscow's main department store.

Toy production was also soon reestablished at Sergiev Posad (now renamed Zagorsk), where the former zemstvo workshop building became a museum of models, the Troitse-Sergieva Lava was turned into a museum displaying traditional folk carving, and a House of the Kustar (Dom Kustaria) was set up in the former Moderne theater to exhibit kustar goods.[5] As for the Moscow Kustar Museum, now the administrative center of Glavkustprom, kustari once again applied to it for artistic and technical advice, while the museum renewed its artistic influence on the province's kustar population, "improving their artistic literacy, developing their taste and feel for works of art created by the kustari themselves." After an interim period (1918–20) as the museum's director, Natalia Ia. Davydova was replaced by Nikolai Bartram, who continued to devote himself to the development of kustar toys with all of his prerevolutionary zeal.[6] When the museum's shop reopened in January 1922 it featured much of the same stock it had sold before the Revolution, and when foreigners like Walter Benjamin visited they found there Abramtsevo furniture, lush carvings from the workshops of the Vornoskov family, laces, weaving and embroideries, and the matreshka and other toys that continued to be among the most popular souvenir items.[7]

Although Solomenko was reportedly razed to the ground by local peasants during the Revolution, the two women who had brought it to national and international attention soon turned their energies in new directions. In 1924 Natalia Ia. Davydova joined with Maria F. Iakunchikova to organize a very successful embroidery artel at Tarusa, Kaluga province.[8] When Davydova died two years later, Iakunchikova emigrated to Paris, where she lived until her death in 1952(?). She is buried in the Russian cemetery at St. Geneviève-des-Bois, together with many other members of Russia's Silver Age.

After Princess Tenisheva's emigration Talashkino never again became a thriving center of kustar art production, although an embroidery artel was established nearby in the 1930s. Sergei Maliutin's fairytale teremok and church were left to molder throughout most of the Soviet era, while for the remainder of his life the artist turned almost entirely to painting brooding portraits of his contemporaries and his children.

Perhaps most successful in weathering the painful transitions of the 1920s were the women's kustar art industries as they had been revived in scores of zemstvo, government, and private workshops throughout the Russian Empire. In the early 1920s the kustar export organization Mosekust commissioned Natalia M. and Natalia Ia. Davydova, Aleksandra Exter, Nadezhda Lamanova, sculptor Vera Mukhina, Evgenia Pribyl'skaia, and Varvara Stepanova to design fashions for export and domestic use that would incorporate kustar fabrics, embroideries, and laces, as well as the geometric cut of traditional peasant dress. Displays of this experimental "kustar constructivist" clothing at the 1925 Exposition des Arts Décoratifs in Paris represented a rare, perhaps a final, moment of reconciliation between kustar tradition and avant-garde artistic experiment.

- The All-Russian Manufacturing Exhibition is held to coincide with the First Congress of Trade and Industry and features "a corner of a seventeenth-century dwelling."
- The kustar question is discussed at both the Congress of Agronomists in Moscow and the Congress of Manufacturers, Factory Owners, and Individuals Interested in National Industry in St. Petersburg.
- Savva and Elizaveta Mamontov buy Abramtsevo.
- Ivan Zabelin presents lectures to the Moscow Society of Lovers of the Arts entitled "The Significance of Russian Archaeology for Russian Historical Painting" and "Russian Architecture, Clothing, and the Domestic Environment in Russian Life of the Past."

1871
- The Imperial Russian Geographical Society forms a commission to compile a list of areas where kustar industry is widespread.
- The Ministry of Finance establishes a commission to study the kustar industries.
- Viktor Gartman designs an artist's studio at Abramtsevo.

1872
- The Moscow Polytechnic Exhibition includes a kustar section, whose contents later form the basis of the History and Polytechnic Museums.
- The architect Vladimir Dal' makes the first of his annual expeditions throughout the empire (1872–8) to study vernacular architecture.
- Vladimir Stasov's *Russian Folk Ornament* is published by the Society for the Encouragement of the Arts.
- The State Treasury gives a 30,000 ruble loan to the Pavlovo kustar artel.
- Ivan Ropet designs a bathhouse at Abramtsevo.
- Vladimir Stasov becomes head of the Art Department at the Public Library in St. Petersburg.
- The court architect I. Monigetti becomes head of the "Russian style" section of the journal *The Architect.*

1873
- A school and hospital are opened at Abramtsevo.
- The Russian pavilion at the Vienna Exposition Universelle includes a Russian house by Monigetti, Valberg, and Bogomalov, an imperial pavilion, and an izba.
- The duchess of Edinburgh, daughter of Alexander II, pres-

ents a collection of Russian laces to the South Kensington Museum, London.

1874 • The Imperial Russian Geographical Society sponsors publication of *Collection of Materials on the Kustar Industries in Russia.*

1875 • The Moscow Provincial Zemstvo Assembly allocates funds for an investigation of Moscow province's kustar industries.

• A commission is appointed under the aegis of the Council on Trade and Industry to study kustar industry.

1876 • Elizaveta Mamontova establishes a joinery workshop at Abramtsevo.

• A stand selling kustar laces and other textiles is included at the 1876 Philadelphia exhibition.

1877 • Eugène Viollet-le-Duc's *L'Art russe* is published in Paris.

1878 • At the Exposition Universelle in Paris the Russian Section includes wooden pavilions by Ivan Ropet.

• S. P. Kaznachaeva opens a drawn-thread workshop for peasant women in the village of Vysokoe in Mikhailovskii uezd, Riazan province; similar workshops are subsequently opened in Podlesnoe, Podviaz'e, Glinskoe, etc.

1879 • P. Martynov's *L'Art russe* is published in Paris.

1880 • Princesses M. N. Shakhovskaia and S. P. Dolgoruka set up women's kustar workshops in the village of Muraevnia, Dankovskii uezd, Riazan province.

• Sofia Davydova begins a two-year tour of ten provinces to study the kustar lace industry.

• Elena Polenova is sent to Paris for training in ceramics by the Society for the Encouragement of the Arts.

1881 • Construction begins on the Abramtsevo church.

• Vasilii Polenov and Il'ia Repin find the first piece of peasant carving for the Abramtsevo museum.

1882 • The Moscow provincial zemstvo publishes *Collection of Statistical Data on Moscow Province* and *Trades of Moscow Province.*

• The Mamontov circle performs *Snegurochka* at Abramtsevo.

• The All-Russian Ethnographical Exhibition is held in Moscow.

• N. Simakov's *Russian Ornament in Antique Examples of Indus-*

Art
Lat
Rus

- Nikolai Roerich is elected secretary of the Society for the Encouragement of the Arts.
- In November the Society in Aid of Manual Labor is founded in St. Petersburg, directed by the Administration for Labor Aid, with a shop at 30 Morskaia Street.
- In November Grand Duchess Elizaveta Fedorovna appoints Prince S. M. Golitsyn her representative in investigating ways of developing the export of kustar goods to France and Germany.
- In December the First International Exhibition of Ceramics is held at the Passage Arcade in St. Petersburg, including Copenhagen, Abramtsevo, Daum, and Tiffany wares, and kustar ceramics.

1902

- Z. S. Sokolova establishes a handwork outlet in the village of Ustano, Voronezh province, producing embroidery yardage from old liturgical texts and bead work that are sold at outlets in St. Petersburg and Moscow and later through Anna Pogosskaia's London shop.
- In March the First All-Russian Congress of Workers in the Kustar Industries is convened by M.Z.i G.I. in St. Petersburg, and coincides with the First All-Russian Kustar Exhibition held in the Tauride Palace.
- In June the Ministry of Finance passes a Statute on Industrial Art Training Institutions.
- In August the Talashkino theater, designed by Sergei Maliutin, is completed.
- The fourth World of Art exhibition includes a display of Talashkino wares, Abramtsevo majolica, and Frolov glass.
- In December M.Z.i G.I. opens its Permanent Kustar Depot in St. Petersburg (28 Liteinyi Prospekt).
- Princess A. N. Naryshkina and Countess M. F. Sheremeteva open a Central Kustar Outlet in St. Petersburg.
- A posthumous exhibition of Elena Polenova's work is held in the Moscow History Museum; Princess Tenisheva buys examples for the Alexander III Museum.
- V. Suslov becomes a professor at the Academy of Arts, where he reorganizes the ornament class and teaches a course on the history of Russian art and architecture.
- M. V. Iakunchikova dies 14 December.

- In December–January the Exhibition of Architecture and Industrial Art in the New Style is held in Moscow.
- Princess Tenisheva stages an exhibition of the Talashkino kustar workshops' work in Smolensk.

1903
- The First Floating Demonstration Exhibition for Artisans and Kustari is launched on the Volga River, subsidized by the Ministry of Finance.
- Princess Tenisheva opens The Source in Moscow to sell Talashkino goods.
- In January the tsar and tsarina host a ball at which all the guests wear antique or replica boyar costumes.
- Nikolai Roerich travels throughout Russia to study ancient art; his seventy-five oil sketches are exhibited at the Society for the Encouragement of the Arts and then in America.
- In September Sergei Maliutin leaves Talashkino.
- Prince Tenishev dies in Paris, and Nikolai Roerich arrives at Talashkino.
- The art joinery workshop at Sergiev Posad becomes independent of the toy workshop.
- Kostroma kustari take part in the Exhibition of Stone and Metal Goods held in St. Petersburg.
- In September the new Moscow Kustar Museum at 7 Leontievskii Lane is completed.
- The Moscow zemstvo establishes a drawing school for kustar carpenters in the village of Ligachev, Moscow province.
- The Contemporary Art exhibition, a show of interior design, takes place in St. Petersburg, organized by members of the World of Art group.

1904
- In January a Historical Exhibition of Works of Art in Moscow includes Bartram toys and Solomenko carpets.
- In the winter the School of the Society for the Encouragement of the Arts holds an Exhibition of Artifacts of Russian Antiquity, followed in February by an exhibition of goods from the Talashkino workshops organized in the same location by N. Roerich.
- In February the New Artists' Union holds an Art and Architecture Exhibition at the Academy of Arts, with exhibits

- The Skoptsy embroidery workshops are founded in Poltava province.
- M. G. Raevskaia opens a carpet workshop in the village of Krasnen'koe, Novokhonerskii uezd, Voronezh province.
- Count A. A. Bobrinskii's important six-volume *Russian Popular Wooden Goods* is published, its aim "to serve as a collection of samples for directors of training and other workshops producing wooden kustar goods."
- Princess Tenisheva officially hands her Museum of Russian Antiquities over to the Smolensk Division of the Imperial Archaeological Institute.

1911
- In February a major kustar exhibition is held in Berlin.
- An outlet for kustar goods is opened by the Society for the Improvement of Popular Labor in Memory of the Liberator-Tsar Alexander II in St. Petersburg (62 Sergievskaia).
- In May the First Demonstration Exhibition of Kustar Work is held in St. Petersburg, organized on private initiative.
- In January the School of Folk Art opens in St. Petersburg.
- In February G.U.Z.i Z. establishes a School for Instructors in Art Joinery in Ligachev, Moscow province.
- Russia's kustar section at the Turin International Exposition sells 35,000 lire worth of goods.
- The Grand Duchess Olga Aleksandrovna founds a carpet and weaving workshop in Staro-Zhivotinnoe, Voronezh province, working from old patterns provided by the Society for the Encouragement of the Arts.
- The Russian Grain Society is founded in St. Petersburg with its own kustar committee.
- The Moscow Kustar Museum holds an exhibition-bazaar of women's handwork; a new wing is added to the building on Leontievskii Lane; the museum takes part in the Tsarskoe Selo Jubilee Exhibition, where it wins the tsar's prize for organizing and marketing kustar industries.
- In December the All-Russian Congress of Russian Artists takes place in St. Petersburg, with sessions on industrial art and kustar industry, and exhibitions of icons and antiquities, industrial art and kustar art, and toys.

1912 • The Kustar Section of the Russian Export House opens in St. Petersburg.

• In May the first specialized periodical devoted to kustar industry begins publication – the *Annual of Kustar Industry* edited by E. D. Maksimov.

• In 1912 and 1913 traveling exhibitions of kustar goods take place throughout England organized by R. Martens.

1913 • Throughout the year events celebrating the tercentenary of the Romanov dynasty are held.

• In March–April the Second All-Russian Kustar Exhibition takes place in St. Petersburg.

• In October the Ideal Home Exhibition in London features a Russian Village stocked with kustar goods.

• The Bogorodskoe zemstvo art school opens, teaching drawing, wood technology, modeling, and carving under the direction of Academy-trained V. I. Bystrenin and the painter K. I. Lintblat.

• A Zemstvo Congress is held in Moscow to discuss the marketing of kustar goods.

• In May an All-Russian Arts and Industries Exhibition is held in Kiev, including rooms and a chapel decorated by the Stroganov School and goods from the Kiev Kustar Society.

• In Pskov the van der Fleet School of Applied Art is founded with I. F. Root as director.

• Sergei Chekhonin enters the employ of G.U.Z.i Z. as a specialist on kustar art and becomes director of the Rostov enamel workshop, where he revives the art of enameling.

• The Société des Artistes Russes organizes an exhibition of Russian folk art at the Salon d'Automne in Paris.

• The Sergiev Posad toy workshop is reorganized as an artel.

1914 • The Talashkino workshops close down entirely.

• Princess Tenisheva exhibits her champlevé enamels in Rome at the Exposition d'Art Précieux, is awarded the Diplôme d'Honneur, and is elected an honorary member of the Rome Archaeological Society.

• In Berlin the German Lyceum Club holds an Exhibition of Russian Folk Art.

- An Exhibition of Ancient Russian Art is held in the Delovoi Palace, organized by the Moscow Archaeological Society.

1915
- The exhibition The Russian Countryside and Kustar Goods is held in London.
- The Society for the Renaissance of Artistic Rus is founded in Moscow, its members including prominent kustar activists.
- The First Exhibition of the Contemporary Decorative Art of Southern Russia is held at the Lemercier Gallery, Moscow.

1916
- Nikolai Bartram instigates the formation of the Union of Workers in Decorative Art and Industrial Art.

1917
- The Mariinskii Lace School is evacuated to Riazan.
- An Exhibition of Contemporary Art is held at the Mikhailova Salon in Moscow.

1918
- A School for Art Woodwork opens in Semenov, Nizhnii Novgorod province, later called the Semenov Professional Technical School.
- On 24 March Savva Mamontov dies in Moscow.
- In October K. V. Orlov becomes head of the Abramtsevo workshop, which is placed under the direction of IZO Narkompros.
- The Mutkovo carving artel is formed.
- The Shtiglits School becomes the State Labor Training Workshops for the Decorative Arts; the Stroganov School is merged with the Moscow School of Painting, Sculpture, and Architecture to form the Free State Workshops (SVOMAS).
- IZO Narkompros reopens the Talashkino woodwork workshop under Aleksei Zinov'ev's direction, but closes it again in 1920.
- Olga Rozanova is appointed head of the Industrial Arts Subsection of IZO Narkompros until her death at the end of the year.
- Natalia Ia. Davydova is made director of the Moscow Kustar Museum.

1919
- Narkompros establishes industrial art workshops in Viatka.
- In April Lenin and Kalinin sign the decree On Measures to

Aid Kustar Industry, which initiates the collectivization of small workshops and individual kustari into cooperatives and artels.

- In May the State Workshops of Industrial Models and Designs are set up attached to the Industrial Art Subsection of IZO Narkompros. Their work includes "branches of industry and kustarnichestvo where the intervention of an artist is required."
- In August the First All-Russian Conference on Industrial Art convenes, organized by the Industrial Art Subsection of IZO Narkompros. Those present include I. Averintsev, O. Brik, N. Erikson, N. Lamanova, A. Lunacharskii, K. V. Orlov, N. Punin, and D. Shternberg.
- The Talashkino Museum of Russian Antiquities is merged with the Smolensk Museum of History and Ethnography.

1920
- The exhibition The Kustar for the Famine Areas opens.
- A conference of artists working in production and taking part in the Higher State Workshops (VKhUTEMAS) is convened in Moscow.
- In November the Toy Museum opens on Smolenskii Boulevard in Moscow, and is later moved to Prechistenka.

1921
- The Kudrino artel is founded.
- An exhibition of kustar industries is held in Berlin.
- N. Globa organizes a kustar and industrial art *tekhnikum* in Moscow.
- The Moscow Kustar Museum is placed under the authority of Glavkustprom (Chief Administration of Kustar Industries).
- In December an exhibition of Russian peasant art is held at the Moscow History Museum, curated by N. Voronov.

1922
- At the end of May the Moscow Kustar Museum opens its doors to the public again, with A. A. Vol'ter as director.
- The House of the Kustar opens in Sergiev Posad.

1923
- In March the First All-Russian Exhibition of Agriculture and Kustar Industry opens in the Central Park of Rest and Culture in Moscow, sponsored by the Academy of Artistic Sciences (GAKhN).
- An All-Russian Exhibition of Industrial Art opens in St. Petersburg, organized by GAKhN.

Notes

INTRODUCTION

1. The almost complete neglect of the kustar art revival throughout most of the Soviet period is in itself a fascinating phenomenon. After barely a decade of enthusiastic interest in and promotion of all aspects of artistic kustarnichesvo, by the mid-1920s a definite ideological reaction set in against this quintessential symbol of Russia's prerevolutionary, agrarian society. As a movement dominated by leading figures of the Old Regime and by artists identified with Symbolism and early modernism, it was soon routinely condemned by Soviet art historians for its destructive impact on peasant art in Russia, an assumption that has survived largely unexamined and unchallenged to this day. Not least effective in ensuring the invisibility of this once highly visible movement has been the term *kustar* itself. Initially a very specific term borrowed from everyday Russian life and used to describe a particular means of rural production, it has gradually become an archaism that conveys only the vaguest pejorative connotations. Although Soviet art historians soon replaced the anachronistic concept of the kustar art industries with the generic "popular arts and crafts" (*narodnye khudozhestvennye promysly*), and Western scholars have translated it as the familiar "handicrafts," I believe it is essential to call the kustar art revival by its proper name if we are fully to appreciate this peculiarly Russian variant on an international movement.

2. As evidence of this interest Grigorii Sternin has noted that "in 1900 alone [Ruskin's] works and lectures were published four times in Moscow. The same year Ruskin's long essay on the Preraphaelites was published in the magazine *World of Art*." G. Iu. Sternin,"Zarubezhnoe iskusstvo v Rossii," *Russkaia khudozhestvennaia kul'tura vtoroi poloviny XIX–nachala XX vekov* (Moscow, 1984), 97. For the translation of Ruskin's article, see Dzhon Reskin, "Prerafaelitizm," *Mir iskusstva,* 2 (1900). For a paraphrase of William Morris's theories on the decorative arts, see Z. Vengerova, "Vozrozhdenie dekorativnogo iskusstva Villiam Morris," *Arkhitekturnyi muzei,* 1 (1903), 1–9. On Morris in Russia, see Kirill Makarov, "Morris i sud'by dekorativnogo iskusstva," *Dekorativnoe iskusstvo SSSR,* 8 (1984), 27–32.

3. On the decorative arts in France, see Debora L. Silverman, *Art Nouveau in Fin-de-siècle France: Politics, Psychology, and Style* (Berkeley, 1989); Nancy J. Troy, *Modernism and the Decorative Arts in France: Art Nouveau to Le Corbusier* (New Haven and London, 1991).

4. On Jugendstil's role in the development of industrial design in Germany, see Gert Selle, *Jugendstil und Kunst-industrie. Zur Ökonomie und Ästhetik des Kunstgewerbes um 1900* (Ravensburg, 1974).

5. On the revival of peasant crafts in Poland, see David Crowley, *National Style and Nation-State: Design in Poland from the Vernacular Revival to the Inter-*

khod – the exodus of peasants from their villages to towns and other provinces in search of work – which in turn was linked to the sharp decline in kustar industry in many provinces. A major incentive and attraction was the high wages offered for factory work, but as one zemstvo correspondent phrased it, "the prospect of becoming a waiter, a shop assistant, a lackey, looks positively enticing compared with the hard lot of the agriculturist burdened by intolerable payments." See Tugan-Baranovsky, *Russian Factory in the Nineteenth Century,* 407–12. In the case of Mamontova and many other private landowners, the desire to staunch the flow of rural youth to the city was caused less by anti-industrial sentiment per se than by a belief that the factory was the source of moral delinquency.

24. The phrase is M. Kurchinskii's ("Russkaia kustarnaia promyshlennost'," *Zhizn',* 1 [1901], 88–9), but echoes earlier findings: "In Moscow province we find the furniture trade employs almost 2,000 hands in 87 villages, alongside the production of guitars and walking sticks. The latter two areas provide income to just a few dozen peasants. We think that the former deserves far greater attention on the researcher's part" (Isaev, "K voprosu o kustarnoi promyshlennosti," 104–5).

25. For a fascinating discussion of Slavophile and populist theories about wood and wood carving, see A. I. Skvortsov, *Russkaia narodnaia propil'naia rez' ba* (Leningrad, 1984). An excellent discussion of the ideological connotations of wood carving and peasant architecture for professional architects may be found in E. I. Kirichenko, *Russkaia arkhitektura 1830–1910-kh godov* (Moscow, 1978). On the quest for a national style in Russian architecture, see also William Craft Brumfield, *The Origins of Modernism in Russian Architecture* (Berkeley, 1991), 32–9.

26. The many famous artists who formed the "Mamontov circle" include Valentin Serov, Il'ia Repin, Viktor Vasnetsov, Vasilii Polenov, Konstantin Korovin, and Il'ia Ostroukhov. Like their patron and friend Savva Mamontov they appeared to have played almost no active part in the activities of the Abramtsevo kustar workshop, and became interested in the applied arts belatedly, when Mamontov founded his Abramtsevo Ceramic Factory in 1890.

27. Polenov's design, with its intricate interlace pattern inspired by medieval manuscript illuminations, was typical of the fashionable furniture "in the Russian style" produced in the St. Petersburg workshops of Büchtger, Shtange, and Shutov to designs by architects A. L. Gun, I. S. Kitner, I. A. Monigetti, and others. Their works may be seen in the architectural journal *Zodchii* (The Architect), which in 1872 started a special "Russian Style" section devoted to architecture and the applied arts. On this period in the development of a Russian style for the decorative arts, see Evgenia Kirichenko, *Russian Design and the Fine Arts, 1750–1917* (New York, 1991), 91–133.

28. Detailed biographies and primary sources on Polenova may be found in V. V. Stasov, "Elena Dmitrievna Polenova. Biograficheskii ocherk," *Stat'i i zametki ne voshedshie v sobraniia sochinenii,* vol. 2 (Moscow, 1954), 221–68; Sakharova, *V. D. Polenov. E. D. Polenova.* A doctoral dissertation by N. M. Beloglazova, "Tvorchestvo E. D. Polenovoi." Avtoreferat (Moscow, 1973), has yet to be published.

29. Stasov, "E. D. Polenova," 224.

30. For Vera Polenova's detailed memoirs of this period, see "Kruzhok 'Krugloi bashni.' Iz vospominanii V. D. Khrushchevoi, 1877–78," *Vestnik Evropy,* 1 (February 1901), 240–84.

31. Sakharova, *V. D. Polenov. E. D. Polenova,* 536.

32. The school was run by the "Liteiny-Tauride Circle" of the Society for the Relief of Needy Women. Among its committee members was the prominent feminist and advocate of women's education N. V. Stasova, who supervised Polenova's teaching activities at the school. On Stasova and "the woman question" in Russia, see Richard Stites, *The Women's Liberation Movement in Russia: Feminism, Nihilism, and Bolshevism, 1860–1930* (Princeton, N.J., 1978).

33. Many Russian women in search of artistic training in the nineteenth century found it not in the Imperial Academy of Arts, but in the three major schools of design: in Moscow the Stroganov School of Technical Drawing, in St. Petersburg the Shtiglits School and the School of the Society for the Encouragement of the Arts. In 1896 only 34 of the Academy's 388 pupils were women, compared to 84 women of the 189 enrolled at the Shtiglits School. *Brokgauz-Efron entsiklopedicheskii*

slovar Rossii (St. Petersburg, 1898; reprint, Leningrad, 1991), 396.

34. Evdokim Alekseevich Egorov (1832–91) and his wife, Maria Kirillovna (?), kept a ceramic workshop in Paris in the 1870s and 1880s, where they gave lessons in painting on porcelain to visiting Russian artists. Il'ia Repin, who in 1874 had dabbled in ceramics in this workshop along with other pensioners from the Academy of Arts, wrote that Egorov was "making a great fuss about getting hold of two or three peasants and teaching them his craft. Then, when they returned to their village . . . they would pass it on to their fellow workers as well as fostering examples of good taste." K. N. Pruslina, *Russkaia keramika* (Moscow, 1974), 48–9.
35. At this time Vasilii Polenov had cemented an already close friendship with the family by marrying Elizaveta Mamontova's cousin, Natalia Iakunchikova (1858–1931).
36. M. V. Nesterov, *Vospominaniia* (Moscow, 1985), 93–4.
37. Sakharova, *V. D. Polenov. E. D. Polenova,* 352. Since Vasnetsov left for Kiev in 1885 to begin work on the interior of St. Vladimir's Cathedral, he designed only a few pieces for the workshop, none of which seem to have become part of the commercial repertoire. Despite his reputation as the force behind the Abramtsevo workshop, only sixteen projects for furniture are listed in N. Morgunov and N. Morgunova-Rudnitskaia, *V. M. Vasnetsov. Zhizn' i tvorchestvo* (Moscow, 1962), 450–1.
38. Sakharova, *V. D. Polenov. E. D. Polenova,* 354–5. There seems to be no evidence that Mamontov gave his wife's kustar enterprise anything more tangible than moral support. At the time the workshop was reorganized Mamontov was already investing substantial sums in his own Private Opera Company. It was not until 1890, when a ceramics workshop got under way at Abramtsevo, that Mamontov showed any personal interest in the decorative arts.
39. Polenova, *Abramtsevo,* 60. The decline of the kustar weaving and printing industries is vividly described in Tugan-Baranovsky, *Russian Factory in the Nineteenth Century.*
40. Sakharova, *V. D. Polenov. E. D. Polenova,* 362–63. Polenova's words invalidate the claim, sometimes made by Soviet critics, that she and her circle were ignorant of the realities of peasant life and art, and that their work came "not so much from direct contact with folk art, as from a museum study of mainly ancient carved artifacts dating back to the seventeenth or even the sixteenth century." N. N. Sobolev, *Russkaia narodnaia rez'ba po derevu* (Moscow-Leningrad, 1934), 456–7.
41. V. I. Butovskii (1815–81), *Histoire de l'ornement russe du Xe au XVIe siècle* (Paris, 1870–3), 6. The first volume contained one hundred facsimiles from Greek and Russian manuscripts in the Moscow Syndic Library and the Imperial Library in St. Petersburg, while the second contained one hundred "didactic" plates showing the ornamental designs that might be derived from these sources. When asked by a skeptical English visitor "upon what ground designs evidently Byzantine in style had been appropriated by Russia, [Butovskii's] reply was that a MS. when in Russian characters was presumed to be of Russian origin." J. Beavington Atkinson, *An Art Tour to Russia* (London and New York, 1986), 257.
42. N. Sultanov, "Russkie dereviannye porezki," *Zodchii,* 2 (1881), 125.
43. In the 1870s members of the imperial family had been the leading patrons of the Russian style. I. A. Monigetti (1819–78), editor of the "Russian Style" section of the journal *Zodchii,* was architect to the imperial household and designed among other things glass and porcelain in the interlace style for the Imperial Glass and Porcelain Factories. *Zodchii* published such important commissions as a boudoir designed in the Russian style by A. Shreter for the Grand Duchess Maria Pavlovna, a tabernacle designed by D. Grimm for the duchess of Edinburgh, a box at the Manège designed for the emperor by N. Sultanov, a dairy for the empress on the imperial farm at Il'inskoe designed by A. Rezinov, and a dining room for the St. Petersburg palace of the Grand Duke Vladimir Aleksandrovich.
44. Polenova, *Abramtsevo,* 49.
45. Ibid., 57–8. Vasilii and Natalia Polenov also created their own collection of folk art, but according to their daughter E. V. Sakharova it consisted "primarily of kustar goods" that were "more decorative than utilitarian. They were made for sale at fairs – the 'Mushroom Market'

held on the Moscow embankment in the first week after Shrovetide, the Palm Sunday Fair, as well as other bazaars in the provinces." E. V. Sakharova, *Gosudarstvennyi muzei V. D. Polenova. Putevoditel'* (Tula, 1963), 49–50.

46. Polenova, *Abramtsevo,* 58.
47. N. Borok [N. V. Polenova], "E. D. Polenova," *Mir iskusstva,* 18–19 (1899), 105.
48. *Niva,* 4 (1909), 79. Zabelin's magnum opus was the monograph *Domashnii byt russkogo naroda v XVI i XVII vekakh,* of which the first volume (*Domashnii byt russkikh tsarei v XVI i XVII stoletiiakh*) appeared in 1862 and the second (*Domashnii byt russkikh tsarits v XVI i XVII stoletiiakh*) in 1872. For biographical information, see the introduction to I. E. Zabelin, *Domashnii byt russkikh tsarei v XVI i XVII stoletiiakh. Kniga pervaia. Gosudarev dvor, ili dvorets* (Moscow, 1991), 5–35.
49. Quoted in Borok, "E. D. Polenova," 106. V. O. Kliuchevskii (1841–1911) occupied the chair of history at Moscow University from 1879 until his death.
50. Polenova, *Abramtsevo,* 44.
51. Ibid., 50.
52. Ibid., 77. Now in the Tret'iakov Gallery in Moscow, *The Icon Workshop* won second prize at the Society for the Encouragement of the Arts' annual exhibition in St. Petersburg.
53. *Kratkii obzor kustarnykh promyslov* (St. Petersburg, 1896), xxx.
54. On the awareness of markets as a crucial issue in private workshops, see N. Kablukov, "Obshcheekonomicheskoe znachenie zhenskikh kustarnykh promyslov i sposoby sodeistviia im," *Novoe slovo,* 5 (February 1896), 73.
55. A year later, in a letter to Mamontova, Polenova expressed her own reservations about the activities of official bodies on the kustar's behalf: "Yesterday I read in the newspaper about the ministry [of State Domains?] project for the kustar industries. Not only are we not their allies, we will have to fight against them and intensify our efforts to save the precious remains of folk art from their destructive intrusions, with their banal, clichéd Petersburg and European models. Our enterprise could be really interesting." Quoted in G. Iu. Sternin, ed., *Abramtsevo* (Leningrad, 1988), 149.
56. Polenova, *Abramtsevo,* 67.
57. Ibid.
58. In the winter of 1888, e.g., Princess Natalia Naryshkina, a lady-in-waiting to the empress, commissioned Polenova to design a set of thirty carved and painted mirrors, fashioned after Vologda flails and distaffs in the Abramtsevo museum, to be presented to the tsarina and the grand duchesses at her annual ball in St. Petersburg.
59. *Exposition universelle de Paris 1889. Catalogue officiel de la Section Russe* (Paris, 1889), 273.
60. V. Dedlov, *Franko-russkie vpechatleniia* (St. Petersburg, 1890), 164.
61. Polenova wrote to Mamontova after her interview with Morozov in 1890: "[The kustari] can get materials from us and pay us for them once their work is sold. Moreover [the museum] can pay the most needy in cash as we designate. That way we will have a lot less trouble and expense, and it won't be as hard on the workers as if we straight away stopped assisting them *and left them to fend for themselves as they want*" (my emphasis). Polenova, *Abramtsevo,* 96.
62. In 1894 Polenova counted nineteen students and twenty-eight graduates of the Abramtsevo workshop (Stasov, "E. D. Polenova," 241). However, the official 1898 census of the region in which the workshop was located recorded only twelve joiners, seven of whom worked at the workshop and five on large furniture commissions for Mamontova with materials supplied by her. *Ukazatel' Moskovskogo otdela II vserossiiskoi kustarnoi vystavki v S.-Peterburge 1913 g. Kustarnoe delo moskovskogo zemstva* (Moscow, 1913), 25.
63. Graduates of the Abramtsevo workshop organized kustar artels or cooperatives in the following outlying villages: Khotkovo (1910), Mutovka (1918), Kudrino (1921), Levkovo (1931), and Akhtyrka (1932). In 1936 the Kudrino, Mutovka, and Akhtyrka artels merged under the name of the latter, which in turn merged with the Khotkovo artel in 1940. In 1960 the artel became the Khotkovo Factory of Carved Art Goods. *Promysly RSFSR. Spravochnik* (Moscow, 1973), 198.
64. Polenova, *Abramtsevo,* 94.
65. The sketch, entitled *Terem* and now in the collection of the Abramtsevo Museum, has been identified as a set design for N. A. Rimskii-Korsakov's

opera *Tsarskaia Nevesta* (The Tsar's Bride), which was staged by the Mamontov Private Opera Company in 1899. *Abramtsevo. Khudozhestvennyi kruzhok. Zhivopis'. Grafika. Skul'ptura. Teatr. Masterskie* (Leningrad, 1988), 129, 311.

66. P. K. Suzdalev, *Vrubel'. Lichnost'. Mirovozzrenie. Metod* (Moscow, 1984), 403. The sketch is now in the Abramtsevo Museum.
67. Vladimir Stasov (1824–1906) is best known as a music critic who championed the "Mighty Five" composers. As an art critic he supported the realism and social commentary of the Wanderers and was the most vocal opponent of the new "decadent" trends that he associated with Sergei Diagilev and the World of Art group. At Stasov's request, in 1894 Polenova compiled an album of the twenty-one designs she considered her most important work for the Abramtsevo workshop, of which the "Cat and Owl" door was one.
68. Quoted in Sakharova, *V. D. Polenov. E. D. Polenova,* 507.
69. Ibid., 419.
70. Quoted in N. P. Sobko, *Slovar' russkikh khudozhnikov,* vol. 3, no. 1 (St. Petersburg, 1899), 330.
71. See her letter to S. V. Ivanov, 15 December 1891, in Sakharova, *V. D. Polenov. E. D. Polenova,* 476–7.
72. "Most of all I would like not to lose two abilities – the ability to help, inspire, and provide support and impetus to the work of other artists. This trait in me is a positive one: it's become as necessary for me as water to a fish, without it I would find it hard to live, it's not a hardship for me but a pleasure." Quoted in A. S., *Elena Dmitrievna Polenova, 1850–1898* (Moscow, 1902), 51.
73. Polenova to Mamontova, 13/25 March 1889, in Sakharova, *V. D. Polenov. E. D. Polenova,* 419.
74. Polenova to Stasov, quoted in A. S., *E. D. Polenova,* 32–3.
75. Dora Kogan, *Mamontovskii kruzhok* (Moscow, 1970), 171. The late Dora Kogan is to my knowledge the only historian of the Mamontov circle to have tackled this difficult question of Polenova's abandonment of Mamontova and the psychological issues connected with the two women's involvement with folk and kustar art.
76. A. A. Isaev, *O kustarnom komitete i o sodeistvii kustarnoi promyshlennosti* (St. Petersburg, 1896), 8.
77. An exact though incongruous analogy can be made to the model kustar cobbling workshops established by the local zemstvo in Shadrinksii uezd, Perm province. "On completing the course each of the pupils sets up shop in his own home and works directly from there. . . . There occurs a process that in physiology is called 'reproduction by division': out of a single cell other cells separate off." "Kustarnoe proizvodstvo obuvi v Shadrinskom u'ezde Permskoi gubernii," *Niva,* 38 (1907), 629.
78. Egor Abramovich Zelenkov (1872–1938) lived at Abramtsevo from the age of three and at sixteen was put in charge of the workshop. "He bought materials, supplied the tools, trusted no one else with packing the finished goods onto the five- or seven-horse-drawn cart to be transported to the station at Khotkovo." Galina Dain, "Istoriia i problemy promysla: Abramtsevo-Kudrinskaia rez' ba," *Dekorativnoe iskusstvo SSSR,* 6/355 (1987), 42.
79. Polenova, *Abramtsevo,* 97.
80. When Natalia Ia. Davydova, later to take charge of design matters for Abramtsevo after Polenova's death, saw the modern versions of Polenova's furniture designs at her retrospective exhibition in Moscow in 1902, she was so shocked by the degree of deviation from the originals that she had every item remade exactly to Polenova's original drawings, and she herself painted and varnished them. Reported in Polenova, *Abramtsevo,* 103.
81. By 1904 the Soiuz Association had established trading contacts with forty zemstva, thirty-eight consumer societies, and four artels. On its wide-ranging activities, see A. Kolkotin, *Kustarnyi vopros v Rossii* (St. Petersburg, 1905), 107–10.
82. M.Z.i G.I., *Obzor deiatel'nosti pravitel'stva na pol'zu kustarnoi promyshlennosti (1888–1902)* (St. Petersburg, 1902), 52–3.

1. For a brief introduction to the Solomenko workshops, see Wendy Salmond, "The Solomenko Embroidery Workshops," *Journal of Decorative and Propaganda Arts*, 5 (Summer 1987), 126–43.
2. A total of 7,000 rubles worth of goods was sold at the 1882 All-Russian Arts and Industries Exhibition in Moscow. In 1888 Kaznachaeva founded a school in the village of Podlesnoe to control the quality of work produced in the area, but for personal reasons transferred control of it to the Ministry of State Domains. K. G. Berens, "Vyshival'nyi promysel v Rossii," *Kustarnaia promyshlennost' v Rossii. Zhenskie promysly* (St. Petersburg, 1913), 410–13.
3. The number of women working for the workshop grew from 7 in 1880 to 140 in 1902, while sales for the same period totaled 53,100 rubles, 32,000 of which went directly to the kustari (Berens, "Vyshival'nyi promysel," 413). By an extraordinary coincidence Muraevnia also attracted the attention of Olga Tian-Shanskaia, daughter of the famous geographer, whose detailed ethnographic study of the daily life and beliefs of the population of Muraevnia and surrounding villages was published in 1903. See Olga Semenova Tian-Shanskaia, *Village Life in Late Tsarist Russia*, ed. David L. Ransel (Bloomington and Indianapolis, 1993).
4. For further information on the phenomenon and significance of women's kustar workshops, see Berens, "Vyshival'nyi promysel"; N. A. Kablukov, "Obshcheekonomicheskoe znachenie zhenskikh kustarnykh promyslov i sposoby sodeistviia," *Novoe slovo*, 5 (February 1896), 55–86.
5. See Richard Stites, *The Women's Liberation Movement in Russia: Feminism, Nihilism, and Bolshevism, 1860–1930* (Princeton, N.J., 1978). Also Adele Lindenmeyr, "Public Life, Private Virtues: Women in Russian Charity, 1762–1914," *Signs*, 3 (Spring 1993), 562–91.
6. On the nature of women's work in the Russian countryside, see Rose L. Glickman, "Peasant Women and Their Work," in Beatrice Farnsworth and Lynn Viola, eds., *Russian Peasant Women* (New York and Oxford, 1992), 54–72. On women involved in kustar industries in Moscow province, see Judith Pallot, "Women's Domestic Industries in Moscow Province, 1880–1900," in Barbara Evans Clements, Barbara Alpern Engel, and Christine D. Worobec, eds., *Russia's Women: Accommodation, Resistance, Transformation* (Berkeley, Los Angeles, and London, 1991), 163–84. Also see Barbara Alpern Engel, *Between the Fields and the City: Women, Work, and Family in Russia, 1861–1914* (Cambridge, 1994).
7. Kablukov, "Obshcheekonomicheskoe znachenie," 65–6.
8. See, e.g., A. P. Bakhrushin, *Kto chto sobiraet (Iz zapisnoi knizhki 1892 g.)* (Moscow, 1916).
9. The degree to which Russian art was indebted to the arts of Western Europe, Byzantium, and particularly India and Asia was a critical question in the 1870s, as architects and scholars attempted to reconstruct the nation's complex art history. In 1872 architect Lev Dal' requested funds from the Academy of Arts to travel to India, arguing that "defining the similarity between Indian and Russian architecture is for us Russians a vital . . . question" ("O russkom stile," *Zodchii*, 10 [1895], 73). The same year Vladimir Stasov published his seminal *Russkii narodnyi ornament*, in which he attempted to prove that Russian folk ornament had its origins in Indian and Persian sources. This hypothesis was picked up and popularized by Eugène-Emmanuel Viollet-le-Duc in his *L'Art russe: ses origines, ses éléments constitutifs, son apogée, son avenir* (Paris, 1877), in which the French architectural historian concluded that "the East gave [Russian art] at least nine-tenths of its elements" (150). Fedor Buslaev responded to the extremes of this position in his review "Russkoe iskusstvo v otsenke frantsuzskogo uchenogo," *Kriticheskoe obozrenie*, 2 (1879), 2–20; 5 (1879), 1–24. A counterreview was written by V. I. Butovskii, director of the Stroganov School of Technical Drawing in Moscow and the man who commissioned Viollet-le-Duc to write his history. See V. I. Butovskii, *Russkoe iskusstvo i mnenie o nem E. Violle-le-Diuka, frantsuzskogo uchenogo arkhitektora i F. I. Buslaeva, russkogo uchenogo arkheologa. Kriticheskii obzor* (Moscow, 1879).
10. V. V. Stasov, *Russkii narodnyi ornament. Vypusk per-*

vyi, Shit'e, tkani, kruzheva (St. Petersburg, 1872), 1. In addition to the scholarly text exploring the Finnish and Oriental origins of Russian ornament, the album contained color reproductions of embroideries that were frequently reproduced or copied by convent and kustar embroidery workshops. For instance, at the 1893 World's Columbian Exposition, the Convent of the Holy Virgin at Serpukhov exhibited an armchair embroidered with a design taken from a sixteenth-century priest's vestment reproduced in Stasov's album. *Catalogue of the Russian Section, World's Columbian Exposition, 1893, Chicago* (St. Petersburg, 1893), 290.

11. V. V. Stasov, *Sobranie sochinenii,* vol. 2 (St. Petersburg, 1894), 952.
12. V. V. Stasov, "Nataliia Leonidovna Shabel'skaia. Biograficheskii ocherk," *Stat'i i zametki ne voshedshie v sobranie sochinenii,* vol. 1 (Moscow, 1952), 197.
13. Sofia Aleksandrovna Davydova (née von Hoyer, 1841–1915) studied drawing at the School of the Society for the Encouragement of the Arts in St. Petersburg. See V. V. Stasov, "Razbor sochineniia gospozhi S. A. Davydovoi: 'Russkoe kruzhevo. Issledovanie istoricheskoe, tekhnicheskoe i staticheskoe' (rukopis)," *Sobranie sochinenii,* vol. 2 (St. Petersburg, 1894), 1007–26.
14. In 1885 Stasov wrote to A. N. Shul'govskaia, "The production of lace in Russia is an *extremely important* [Stasov's emphasis] issue. These past few years I've taken great pains on its behalf, seeing how this popular and so original branch of industrial art is in decline in Russia as a result of general ignorance, indifference, and ridiculous French fashions." V. V. Stasov, *Pis'ma k deiatel'iam russkoi kul'tury,* vol. 2 (Moscow, 1962), 293.
15. A. N. Pogosskaia, "Kustari na Vserossiiskoi vystavke," *Novoe slovo,* 1 (October 1896), 10.
16. On the history of the Mariinskii Lace School, see Sophie Davydoff, *La Dentelle russe. Histoire technique statistique* (Leipzig, 1895), 19–22.
17. The prototype for the Mariinskii School was a similar school near Venice, under the patronage of Queen Margarita. Davydova notes that she got little support for her idea from the Kustar Commission in charge of kustar affairs in the early 1880s, and that it was Baroness Edita Fedorovna Rahden, a friend of the Grand Duchess Elena Pavlovna, who was instrumental in the school's foundation (Davydoff, *La Dentelle russe,* iv).
18. *Point de Moscou* had been created in the early 1870s by E. E. Novosiltsova (1834–1901), who in 1874 formed a Lace Society in Moscow composed of society women who were committed to spreading *point de Moscou* among the poorer working women of the city. Though some patterns were taken straight from Butovskii's *Histoire de l'ornement russe,* others were composed by artists such as one Dobrokhotov, a graduate of the Stroganov School and a pupil of the architect Viktor Gartman (Davydoff, *La Dentelle russe,* 23).
19. As late as 1914 the recent origins of the kustar embroidery industry were still remarked on: "The lace industries have existed from time immemorial, and the contemporary educated woman has approached them as an existing craft, helping to improve them technically. . . . [The art of embroidery] as an *industry* is still in its infancy; with few exceptions embroidery centers have arisen specifically as a result of educated intervention." *Russkoe narodnoe iskusstvo na vtoroi vserossiiskoi kustarnoi vystavke v Petrograde v 1913 g.* (Petrograd, 1914), 43.
20. Berens, "Vyshival'nyi promysel," 384.
21. M. Kurchinskii, "Russkaia kustarnaia promyshlennost'," *Zhizn',* 1 (1901), 91–2.
22. "By and large advertisements are of great artistic and educational importance for the countryside," Gorky wrote, "which means that a good advertisement for Siu and Broccard, printed on thick cardboard with good colors, gets a much warmer reception than the pictures printed by Posrednik [a publishing company that specialized in didactic literature for the people]." M. G., "Beglye zametki," *Nizhegorodskoi listok,* 30 June 1896, reprinted in *Iskusstvo,* 5 (1936), 143–4.
23. In his study of Russian literacy and popular literature during the same period, Jeffrey Brooks has shown very similar attitudes on the part of "critics, educators, political activists and others interested in the common people. . . . Their complaints about the commercial media illuminate the areas in which the emergent popular culture of the common people diverged most sharply from the values which the upper classes preferred them to hold." *When Russia Learned to Read: Literacy and*

Popular Literature, 1861–1917 (Princeton, N.J., 1988), xix.

24. For information on Natalia Leonidovna Shabel'skaia (née Kroneberg, 1845–1904), see V. V. Stasov, "Nataliia Leonidovna Shabel'skaia. Biograficheskii ocherk," *Stat'i i zametki ne voshedshie v sobranie sochinenii,* vol. 1 (Moscow, 1952), 195–7.
25. Items from Shabel'skaia's vast textile collection were first shown at an exhibition held in Moscow to coincide with the Eighth Archaeological Congress in 1890. They were exhibited again at a Red Cross exhibition in aid of famine victims in St. Petersburg in 1892, at the World's Columbian Exposition in 1893, and in conjuntion with the coronation of Nicolas II at the Moscow History Museum in 1896 (2,500 pieces). In 1910 Shabel'skaia's daughters published the first volume of a projected catalogue raisonné of the collection (*Sobranie russkoi stariny kniagini V. P. Sidamon-Eristovoi i N. P. Shabel'skoi,* vol. 1 [Moscow, 1910]), and selected textiles were also used to illustrate a special issue on Russian folk art published by the English magazine *Studio* (Charles Holme, ed., *Peasant Art in Russia* [London, 1912]). Part of the collection is now in the Brooklyn Museum, on which see Elizabeth Ann Coleman, *Pearls among the Gold: Russian Women's Festive Dress* (Brooklyn, N.Y., 1987); Mary B. Kelly, "The Shabelsky Collection," in *Goddesses and Their Offspring: Nineteenth- and Twentieth-Century Eastern European Embroideries* (Binghamton, N.Y., 1986), 45–52.
26. On the kinds of relief organized, see Richard G. Robbins, Jr., *Famine in Russia, 1891–1892: The Imperial Government Responds to a Crisis* (New York and London, 1975).
27. A. N. Pogosskaia, "Kustari na vserossiiskoi vystavke," *Novoe slovo,* 2 (November 1896), 9.
28. This depot did in fact become a viable commercial outlet for women's crafts throughout the empire: its profits rose from 10,708 rubles in 1892 to 76,500 rubles in 1909. E. D. Maksimov, *Ezhegodnik kustarnoi promyshlennosti* (St. Petersburg, 1912), 176.
29. W. C. Edgar, "Russia's Conflict with Hunger," *Review of Reviews,* 30 (July 1892), 696.
30. Kablukov, "Obshcheekonomicheskoe znachenie," 76.
31. Maria Fedorovna Iakunchikova (1864–1952) has been consistently confused with her husband's half sister, the painter Maria Vasil'evna Iakunchikova-Weber (1870–1902) – an understandable mistake since both women were closely linked to the Abramtsevo workshops. Contemporaries and acquaintances regularly confused the two in their memoirs.
32. "Masha was just terrible [in the role of Snegurochka]," wrote Elena Polenova to her sister-in-law of the Mamontovs' family production of *Snegurochka* in 1884 (Sakharova, *V. D. Polenov. E. D. Polenova,* 330). The cast wore authentic folk costumes from Tula province, adapted for the purpose by Elena Polenova. Maria Fedorovna's sister Sofia (1866–1920) was also a visitor to Abramtsevo and a participant in the family's domestic theatricals. Married to P. A. Tuchkov, a zemstvo official and landowner in Vereiskii uezd, Moscow province, Sofia shared her elder sister's commitment to good works among the peasantry and for a number of years "directed performances of plays from the Russian classic repertoire, mostly Ostrovskii, for the people." I. S. Zilbershtein and V. A. Samkov, eds., *Konstantin Korovin vspominaet* (Moscow, 1971), 796.
33. I. E. Grabar, *Moia zhizn'. Avtomonografiia* (Moscow-Leningrad, 1937), 148–9.
34. For a detailed account of Serov's 1888 portrait of Iakunchikova (now in the Malmo City Museum, Sweden), see I. E. Zilbershtein and V. A. Samkov, eds., *Valentin Serov v vospominaniiakh, dnevnikakh i perepiske sovremennikov,* vol. 1 (Leningrad, 1971), 232–3. On Serov's death in 1910 Iakunchikova, as one of his closest friends, contributed 5,000 rubles to a fund in aid of Serov's widow and children. Ibid., 271.
35. The web of relationships that this marriage created between the Mamontov, Iakunchikov, and Polenov families is extraordinarily difficult to unravel. Elena Polenova's brother Vasilii married Maria Fedorovna's sister-in-law, Natalia Vasil'evna Iakunchikova, who became the first biographer of the Abramtsevo workshops, while Maria Vasil'evna was the half sister of Maria Fedorovna's husband, V. V. Iakunchikov. Maria Vasil'evna's mother, Zinaida Nikolaevna, was also a Mamontov by birth. See Appendix B for a schematic genealogical chart showing these relationships.
36. Although a member of both the Mamontov circle and the World of Art group, Natalia Iakovlevna

Davydova (1873–1926) remains a mysterious and greatly underrated figure. She has often been confused with Natalia Mikhailovna Davydova, who operated embroidery workshops at Verbovka in the Ukraine (see chap. 5). For brief accounts of her life and work, see [N. D. Bartram], "Pamiati N. Ia. Davydovoi," *Izvestiia,* 20 April 1926; "Nataliia Iakovlevna Davydova," *Vestnik promyslovoi kooperatsii,* 5/34 (1926), 75; *Khudozhniki narodov SSSR. Biograficheskii slovar',* vol. 1 (Moscow, 1976), 272.

37. Netta Peacock, "The New Movement in Russian Decorative Art," *International Studio,* 13 (May 1901), 274.
38. [Bartram], "Pamiati N. Ia. Davydovoi."
39. In addition to Solomenko and S. A. Il'inskaia's workshop in the village of Vorontskova in Tula province, the Moscow and Tver zemstva developed the beading craft, which enjoyed a great success at the Leipzig Bazaar and was imitated by Munich craft schools. See V. Dudoreva, *Biser v starinnom rukodelii* (Moscow, 1923). Naboika block printing enjoyed a similar revival, thanks largely to the efforts of M. F. Iakunchikova and to S. T. Morozov through the Moscow Kustar Museum. See N. I. Sobolev, *Naboika v Rossii* (Moscow, 1912).
40. Pogosskaia, "Kustari na vserossiiskoi vystavke," 10.
41. The Russian Women's Section in Chicago included the work of convents, charitable societies, and workshops operated by such aristocratic ladies as the Princesses Shakhovskaia, Naryshkina, L'vova, and Urusova. Also featured were Abramtsevo furniture and toys from the Moscow toy store operated by Savva Mamontov's sister-in-law. For a detailed list of the exhibits in the Columbian Exposition, see *Catalogue of the Russian Section, World's Columbian Exposition, 1893, Chicago.*
42. The Women's Section was under the patronage of the empress, and the Ladies' Committee included V. Vyshnegradskaia, Princesses A. and E. Naryshkina, V. Kotchubei, and M. Shakhovskaia. The fostering of women's kustar crafts was cast entirely in the light of charitable good works. As the catalogue put it, "It is the charitable hand extended by the woman of noble lineage to the humble worker. It is the intellectual and moral raising up of the popular classes by those women who have felt the imperative need to place their hearts, their energy, and their fortune at the service of their fellowman." *Exposition d'Anvers 1894. Section russe. Travaux des femmes* (Brussels, 1894), 20.
43. The exhibition was organized by Maria Vasil'evna Iakunchikova and the ceramist M. K. Egorova with the aim of "giving artists the chance to sell their works, giving a percentage [of the profits] to the society, and making ourselves known." It included Egorova's ceramics, French and Czech embroideries, Abramtsevo furniture and decorative poker-work on wood, and a shipment of Solomenko embroideries imported from Russia at a cost of 358 francs in import duties (Sakharova, *V. D. Polenov. E. D. Polenova,* 512–13). One member recalled: "In four days over 600 francs worth have already been sold, mostly to the French and Americans. They've received an invitation to take part in an exhibition of the magazine *Art et Décoration.* A. S. Shatskikh, "Russkaia akademiia v Parizhe," *Sovetskoe iskusstvoznanie '21* (Moscow, 1987), 31.
44. On the controversy surrounding Vrubel's panneaux, see V. P. Lapshin, "M. A. Vrubel' na Vserossiiskoi vystavke 1896 goda. (Materialy k biografii khudozhnika)," *Iz istorii russkogo iskusstva vtoroi poloviny XIX–nachala XX vekov* (Moscow, 1978), 78–91.
45. For a summary of the congress and its main points, see Theodore von Laue, *Sergei Witte and the Industrialization of Russia* (New York and London, 1963), 133–7.
46. Pogosskaia, "Kustari na vserossiiskoi vystavke," 17, 19.
47. Ibid., 17.
48. According to E. M. Tatevosian, "The idea of organizing exhibitions for the people grew out of the Moscow Association of Artists. A program was worked out of subjects taken from the Bible and Russian history. . . . Elena Dmitrievna Polenova used the meetings to keep interest in the project alive. . . . The idea of an exhibition for the people was original, interesting, and quite new. It was proposed that it travel around the towns and villages of Russia. The principle of spreading art among the people is a good aim, but difficult to realize. . . . I even think that if Elena Dmitrievna had remained alive she might perhaps have been able to overcome all the obstacles and make the

exhibition happen." Quoted in Zilbershtein and Samkov, *Valentin Serov*, vol. 1, 135–6, n. 42. For Polenova's correspondence on the exhibition, see Sakharova, *V. D. Polenov. E. D. Polenova*, 518ff.

49. Quoted in A. S., *Elena Dmitrievna Polenova* (Moscow, 1902), 44.
50. Earlier in the year Polenova had contributed designs to two other jubilee testimonials, at Stasov's invitation: one for Elizaveta Bem (a motif from her fairy tale *The Little White Duck*), and the other for the architect Ivan Ropet (a perspective view of an old Russian town glimpsed in the distance from beneath an archway crowned with horses' heads).
51. Sakharova, *V. D. Polenov. E. D. Polenova*, 567.
52. Stasov, "E. D. Polenova," *Stat'i i zametki*, vol. 2, 267.
53. "Po povodu konkursa na ornamental'nye sochineniia," *Iskusstvo i khudozhestvennaia promyshlennost'*, 7 (1899), v.
54. For a brief history of these debates in England, see the chapter "Nature and Ornament" in Stuart Durant, *Ornament from the Industrial Revolution to Today* (Woodstock, N.Y., 1986), 25–62. See also the chapter "The Power of Ornament to Express Feelings" in C. Dresser, *The Art of Decorative Design* (London, 1862).
55. Maksimilian Voloshin, "Tvorchestvo M. Iakunchikovoi," *Vesy*, 1 (1905), 33.
56. Sakharova, *V. D. Polenov. E. D. Polenova*, 774 n. 1. This painting was shown in Il'ia Repin's Exhibition of Sketches and Experiments in Creative Art at the end of 1896. On the exhibition, see Alison Hilton, "The Exhibition of Experiments in St. Petersburg and the Independent Sketch," *Art Bulletin*, 70, no. 4 (December 1988), 676–98.
57. E. Iunge, "Vospominaniia ob E. D. Polenove," *Russkaia starina*, 150 (June 1912), 545.
58. Chokolova carpets and wall hangings were a regular feature at both the World of Art and the Moscow Association of Artists exhibitions ca. 1898–1902. At the 1900 World of Art exhibition Chokolova carpets were purchased by the Grand Duchess Maria Pavlovna and the Grand Duke Petr Nikolaevich (*Mir iskusstva*, 10 [1900], 65). Three of her carpets are illustrated in *Mir iskusstva*, 10 (1899), 188–90. At the 1902 First All-Russian Kustar Exhibition in St. Petersburg, Chokolova won a gold medal for "founding and cultivating lacemaking, for her display of silk embroideries that retain antique patterns, and for her excellently executed laces." G.U.Z.i Z., *Otchety i issledovaniia po kustarnoi promyshlennosti v Rossii*, vol. 8 (St. Petersburg, 1907), 327.
59. For general information on Aleksandr Iakovlevich Golovin (1863–1930), see Sergei Makovskii, "A. Ia. Golovin," *Apolon*, 4 (1913), 19–21; E. F. Gollerbakh, *A. Ia. Golovin. 1863–1930. Vstrechi i vpechatleniia. Vospominaniia khudozhnika* (Leningrad-Moscow, 1940); S. A. Onufrieva, *A. Ia. Golovin* (Moscow, 1977). On his paintings from the 1890s, see S. A. Onufrieva, "O nekotorykh rannikh rabotakh A. Ia. Golovina," *Muzei 7* (Moscow, 1987), 88–100.
60. In a letter to Golovin shortly after Polenova's death, Maria V. Iakunchikova wrote: "I want to extend a hand of sympathy to you in our shared grief. No one can mourn her as you and I. It's as if we'd been orphaned, only this word can express . . . the full force of our loss." Quoted in M. F. Kiselev, *Mariia Vasil'ievna Iakunchikova 1870–1902* (Moscow, 1979), 241.
61. On the Abramtsevo Ceramic Factory, see K. Pruslina, *Russkaia keramika* (Moscow, 1974), 46–74.
62. Iakunchikova was also the first important patron of Golovin's paintings. At the sixth Moscow Association of Artists exhibition, held in March 1899, she bought his painting *The Lament of Iaroslavna* for 700 rubles (see Onufrieva, "O nekotorykh rannikh rabotakh A. Ia. Golovina," 92). An unflattering portrait of the Iakunchikovs was later drawn by Anton Chekhov, who spent the summer of 1903 at Nara working on *The Cherry Orchard:* "A more disgracefully idle, absurd, and tasteless life than that at the white house would be hard to find. These people live exclusively for pleasure. . . . There are only two decent people here worthy of respect: N. Ia. Davydova and Maksim [the servant]." A. P. Chekhov, *Pol'noe sobranie sochinenii i pisem*, 2d series, vol. 20 (Moscow, 1951), no. 3936. To celebrate the first performance of *The Cherry Orchard* at the Moscow Art Theater, Iakunchikova presented Chekhov with a model of an old Russian town and an embroidered hanging, both designed by Davydova. See Mariia Chekhova, *Dom-muzei A. P. Chekhova v Ialte* (Moscow, 1963), 21, 53.

63. Netta Peacock, "A Log House Dining Room in Russia," *The Artist,* 24 (January–April, 1899), 6.

64. Ibid., 5–6. Polenova designed at least three versions of the Fire Bird theme for embroidery, of which only the version shown at the Paris Exposition of 1900 appears to have been published. Whether the first embroidery commissioned for the Nizhnii Novgorod exhibition was subsequently installed in Iakunchikova's Russian dining room in 1898 is unclear. A sketch on the same theme in the collection of Mr. and Mrs. Nikita Lobanov-Rostovsky, previously attributed to Polenova, is almost certainly an imitation of her style by the graphic artist A. G. Iakimchenko. It was installed in the dining room of a Madame Firsanova's Moscow home ca. 1909. See *Ezhegodnik Obshchestva arkhitektorov-khudozhnikov* (Moscow, 1909), 149.

65. Many previously unidentified ornamental designs by Polenova were intended for this room. For instance, the ornamental design for one of the embroidered panels above the doors, which Netta Peacock entitles "Flowers Saluting the Rising Sun," is published in O. I. Arzumanova et al., *Muzei-zapovednik Abramtsevo* (Moscow, 1984).

66. Peacock, "A Log House Dining Room in Russia," 1–7. Netta Peacock wrote a total of four articles on what she called "the new decorative movement in Russian art" that Polenova had helped to create, and these contain much of the available information on Solomenko. Sergei Diagilev had intended to feature the Iakunchikova dining room in an 1899 issue of *World of Art,* but this plan did not materialize. He also had rather halfhearted plans to include Netta Peacock's reminiscences of Polenova ("Dan' pamiati Eleny Polenovoi") in the magazine, but these eventually were published in the rival journal *Art and Industrial Art,* 18 (1900), 411–13. See I. S. Zilbershtein and V. A. Samkov, eds., *Sergei Diagilev i russkoe iskusstvo. Stat'i, otkrytye pis'ma, interviu. Perepiska. Sovremenniki o Diagileve v dvukh tomakh,* vol. 2 (Moscow, 1982), 38–39.

67. Peacock, "A Log House Dining Room in Russia," 4.

68. N. Peacock, "Londonskie vystavki rannei oseni 1899 g.," *Iskusstvo i khudozhestvennaia promyshlennost',* 15 (December 1899), 160.

69. I. Grabar', "Neskol'ko myslei o sovremennom prikladnom iskusstve v Rossii," *Mir iskusstva,* 3 (1902), 54.

70. Nikolai Petrovich Sobko (1851–1906) was a close friend of Stasov, whose nickname for him was "Popka." It was generally known that Stasov was the real power behind the journal and that he intended to use it as a weapon against Diagilev and the "decadents." The journal was not a success and ceased publication in 1902, when it was replaced by *Khudozhestvennye sokrovishcha Rossii* under Aleksander Benois's editorship.

71. Sergei Pavlovich Diagilev (1872–1929), publisher of the *World of Art,* creator of the World of Art exhibitions and the Ballets Russes. For a comprehensive collection of writings by and about Diagilev, see Zilbershtein and Samkov, *Sergei Diagilev.*

72. "Avant-propos," *Iskusstvo i khudozhestvennaia promyshlennost',* 1–2 (October–November 1898), vii.

73. Quoted in Zilbershtein and Samkov, *Valentin Serov,* vol. 1, 452.

74. V. V. Stasov, *Pis'ma k rodnym,* vol. 3, pt. 1 (Moscow, 1962), 252. It is not clear whether Stasov refers here to Maria Vasil'evna or Maria Fedorovna. While the former lived permanently in Paris throughout the 1890s, it was the latter who, as main organizer of the Kustar Section at the 1900 Paris Exposition, was more likely to have enlisted Polenova's participation.

75. Polenova to Stasov, 22 May 1898, in Sakharova, *V. D. Polenov. E. D. Polenova,* 579. Polenova and M. V. Iakunchikova were among the artists invited by Diagilev to take part in the first World of Art exhibition in 1898 and become members of the society. His formal letter of invitation is published in Zilbershtein and Samkov, *Sergei Diagilev,* vol. 2, 24–5.

76. Zilbershtein and Samkov, *Sergei Diagilev,* vol. 1, 301.

77. Quoted in G. Iu. Sternin, "O rannikh godakh 'Mir iskusstva'," *Russkaia khudozhestvennaia kul'tura vtoroi poloviny XIX–nachala XX vekov* (Moscow, 1984), 153–4. Such specific publications as Mamontov had in mind were not new: from 1889 to 1897 *Sel'skii vestnik* (Village Messenger) dispersed information about kustar production to literate peasants, including plans and drawings, and in 1896 the Ministry of Agriculture and State Domains published five volumes of *Kustar Crafts by N. K. Rom,* dealing with everything from lathe

new Museum of Models in 1910; a 100,000-ruble fund to promote cooperatives among Moscow kustari; building and equipping a new workshop at Sergiev Posad; and financing a foreign training trip for a basket-weaving teacher. "In a word," a 1913 report concluded, "Mr. Morozov's role in the development of kustar industry in Moscow province is so considerable that one can safely say that without [his patronage] the Moscow provincial zemstvo could not have attained that position which has given it its leading reputation and made it first among all the empire's zemstva working in aid of kustar industry." "Moskovskaia guberniia," *Otchety i issledovaniia po kustarnoi promyshlennosti v Rossii,* vol. 11 (Petrograd, 1915), 50.

15. Quoted in Jo Ann Ruckman, *The Moscow Business Elite: A Social and Cultural Portrait of Two Generations,* 1840–1905 (De Kalb, Ill., 1984), 160.
16. On the history of toy production at Sergiev Posad, see D. I. Vvedenskii, *U sergievskogo igrushechnika* (Moscow, 1926).
17. On the traditional conditions of toy production at Sergiev Posad, see N. V. Ponomarev, *Vystavka igrushek v S.-Peterburge. Kustarnoe proizvodstvo igrushek v Rossii* (St. Petersburg, 1890).
18. Aleksandr Benois, a great enthusiast of Posad toys with their memories of childhood and the nursery, was particularly critical of "Mamas [who] prefer overdressed, cloying dolls from various 'foreign stores,' boring Froebel games and suchlike nonsense to simple-hearted, undemanding, thoroughly genuine kustar toys." A. N. Benua, "Russkie narodnye igrushki v otkrytkakh Aleksandra Benua," *Otkrytoe pis'mo,* 1 (1904–5), 199–210. Benois's *Russkie narodnye igrushki* was published in 1904 by the firm of Golike and Vil'borg.
19. One reviewer wrote of these toys: "It is hard to believe that both these elegant, almost Parisian dolls and primitive crude dolls could be worked by one and the same craftsman. The fact is that at the French exhibition in Moscow the zemstvo acquired a collection of Parisian dolls which it then distributed to the kustar workshops." *Niva,* 39 (1896), 977–8.
20. A. Pogosskaia, "Kustari na vserossiiskoi vystavke," *Novoe slovo,* 1 (1896), 2.
21. There have been several attempts to find the matreshka's origins in ancient folk culture, but the myth of the matreshka as an authentic kustar toy had taken hold as early as 1907: "Matreshki are rather reminiscent of the well-known nesting eggs [and] were actually developed from them. Some shrewd kustar took it into his head to replace the egg with a fat conical *baba* in a kerchief with a rooster under her arm." "Kustarnoe tsarstvo v Sergievskom posade," *Niva,* 39 (1907), 642.
22. Children's Education was founded in 1883 by Maria Alekseevna Mamontova (?–1904), wife of the publisher Anatolii Mamontov and Savva Mamontov's sister-in-law. This close family relationship was evidently the reason why this store was one of the first outlets for Abramtsevo goods in Moscow. The Mamontovs frequently received both Russian and foreign artists, writers, and performers in their home, which was perhaps where Mamontova first met Sergei Maliutin. See M. K. Morozova, "Moi vospominaniia," *Nashe nasledie,* 6 (1991), 95. Children's Education was represented in the Women's Section of the Russian Section at the World's Columbian Exposition held in Chicago in 1893, along with the Moscow Kustar Museum, Abramtsevo, and a number of other kustar workshops operated by noblewomen. See no. 620 in the section on women's work in *Catalogue of the Russian Section, World's Columbian Exposition, 1893, Chicago* (St. Petersburg, 1893).
23. "Kustarnoe tsarstvo v Sergievskom posade," 642.
24. E. S. Ovchinnikova and A. N. Izergina, eds., *N. D. Bartram. Izbrannye stat'i. Vospominaniia o khudozhnike* (Moscow, 1979), 104–5.
25. In July 1897 Chekhov described a visit to Levitan at Uspenskoe: "The other day I was at the estate of the millionaire Morozov; a house like the Vatican, footmen in white piqué waistcoats with gold chains across their stomachs, tasteless furniture, [expensive] wines, the host with no expression on his face – I fled." *Perepiska A. P. Chekhova v dvukh tomakh,* vol. 1 (Moscow, 1984), 176.
26. In 1899 Iakunchikova produced *The Village* in this technique, a collection of whimsical wooden izby and onion-domed churches within a white stone kremlin wall. The same year this panneau was translated into a toy village of painted wood, made at the Sergiev Posad workshop and exhibited the next year at the 1900 Paris Exposition Universelle.
27. For biographical information on Vornoskov, see

E. N. Shulgina, "Vasilii Petrovich Vornoskov," *Sbornik trudov NIIKhP,* 5 (1972) 106–20; V. S. Voronov, "Mastera narodnogo iskusstva," in *O krest'ianskom iskusstve. Izbrannye stat'i* (Moscow, 1972), 286–7.

28. By 1913 there were at least four workshops operated by members of the Vornoskov family, with an annual turnover that ranged from 750 rubles to 7,000 rubles. *Ukazatel' moskovskogo otdela II vserossiiskoi kustarnoi vystavki v S.-Peterburge 1913 g. Kustarnoe delo moskovskogo zemstva* (Moscow, 1913), 106–7.
29. Voronov, "Mastera narodnogo iskusstva," 286.
30. N. D. Bartram, ed., *Igrushka. Ee istoriia i znachenie* (Moscow, 1912), 212.
31. Ponomarev, *Vystavka igrushek v S.-Peterburge,* 17.
32. V. A. Pereleshin, "Kustarnye promysly moskovskoi gubernii," in *Trudy s'ezda deiatelei po kustarnoi promyshlennosti v S.-Peterburge 1910 g.,* vol. 1, pts. 1 and 2 (St. Petersburg, 1910), 240.
33. By this time the Moscow Provincial Zemstvo Assembly was prepared to invest quite liberally in the Kustar Museum's future: in 1899 it gave the museum 50,000 rubles in working capital, a 25,000-ruble loan to develop its intermediary functions in carrying out commissions, and a further 22,504 rubles for the upkeep of the museum and its workshops. I. M. Ivanov, *Russkie kustari* (St. Petersburg, 1902), 83. By way of comparison, the average total expenditure on kustar improvements in thirty-four provinces for 1900 was put at 500,000 rubles, or 19 percent of the 2,301,000 rubles spent on the rural economy (2.6 percent of the total zemstvo budget).
34. On Elizaveta Fedorovna, see E. M. Almedingen, *An Unbroken Unity: A Memoir of Grand Duchess Serge of Russia, 1864–1918* (London, 1964); Mary Grace Swift, "Grand Duchess Elizabeth: No Nobler Woman in Russian History," *New Oxford Review* (September 1987), 6–11. Korovin may well have come to the grand duchess's attention through her secretary, Nikolai Aleksandrovich Zhedrinskii (1852–1930), who was a hunting crony of the artist's.
35. R. I. Vlasova, *K. K. Korovin. Tvorchestvo* (Leningrad, 1969), 95. In the summer of 1894 Korovin and Valentin Serov were sent by Savva Mamontov on an expedition to the Russian north to sketch this virgin territory just as it was being opened up by the Moscow-Iaroslavl railway line. Mamontov's intention was that these impressions serve as the basis for the Pavilion of the Far North at the 1896 All-Russian Arts and Industries Exhibition held in Nizhnii Novgorod. Korovin's memoirs of the journey are reprinted in I. E. Zilbershtein and V. A. Samkov, eds., *Valentin Serov v vospominaniiakh,* vol. 1 (Leningrad, 1971), 320–38.
36. V. S. Serova, *Kak vyros moi syn* (Leningrad, 1968), 256. Korovin went on to design the decor for the 1901 World of Art exhibition in St. Petersburg (together with A. A. Baryshnikov and P. M. Zybin), and the 1902 First All-Russian Kustar Exhibition in St. Petersburg (see below).
37. Zilbershtein and Samkov, *Valentin Serov,* vol. 1, 368.
38. Maurice Normand, "La Russie à l'Exposition," *L'Illustration,* 5 May 1900, 282.
39. Quoted in Zilbershtein and Samkov, *Valentin Serov,* vol. 1, 368 n. 39.
40. S. A. Onufrieva, *A. Ia. Golovin* (Moscow, 1977), 19.
41. Iakunchikova was closely involved with the Posad workshop immediately before and after the 1900 Paris Exposition Universelle. In the collection of the late Stefan Weber at Sallanches, Switzerland, there are several sheets covered with drawings of kovshi, chairs, and buffets in the neo-Russian style of Polenova and Davydova. In 1900 Iakunchikova also began to design an alphabet based on the kustar toys exhibited in Paris, but this was never completed because of a sudden deterioration in her health. Iakunchikova had been closely involved in collecting folk art through the Abramtsevo circle, but this sudden burst of activity in designing kustar items seems to have been prompted by her sense of responsibility to finish what Elena Polenova had been unable to complete at her untimely death in 1898.
42. Normand, "La Russie à l'Exposition," 283.
43. Bartram, *Igrushka,* 222.
44. The reviewer for *Niva* wrote: "One needn't be a soothsayer to predict the success of the kustar exhibition. It fully deserves it, and what's more the kustar is the height of fashion now, and is being written and talked about a great deal. What has

caused this turnaround in public opinion? Probably the disappointment in big industry that we are now experiencing, in this period of industrial crisis, when in the wake of crashes and failures a depression has set in in all aspects of our economic life." "Vserossiiskaia kustarno-promyshlennaia vystavka," *Niva,* 11 (1902), 215.

45. "The number of kustari must be gauged subject to how one defines the term," the congress's organizers reasoned. If one included every peasant making goods for his own needs and only in part for sale, then practically the whole population qualified. "But if we consider kustari to be only those working for the market, then the total is about ten to twelve million." "Otchet o Vserossiiskoi Kustarno-Promyshlennoi vystavke v 1902 godu," in G.U.Z.i Z., *Otchety i issledovaniia po kustarnoi promyshlennosti v Rossii,* vol. 8 (St. Petersburg, 1907), 288. Similar confusion arose in screening exhibitors for the 1902 exhibition. The organizing committee thus considered such questions as: Is the exhibitor's class important, and can nonpeasants living in the countryside and making crafts qualify? Can urban artisans of the peasant class be kustari? If they make a big profit and hire workers? Ibid., 299.
46. *Obzor pravitel'stvennykh sodeistvii kustarnoi promyshlennosti* (St. Petersburg, 1913), 59.
47. "Otchet o vserossiiskoi kustarno-promyshlennoi vystavke v 1902 godu," 309.
48. This extract from an article in the newspaper *St. Petersburger Zeitung* was quoted in ibid., 352–3.
49. Ibid., 350.
50. A. N. Benua, "Kustarnye vystavki. 'Testimonium paupertatis,'" *Mir iskusstva,* 3 (1902), 48–50.
51. M. Mikhailov, "Ist zu wienerisch," *Iskusstvo stroitel'noe i dekorativnoe,* 3 (1902), 13. In 1897, after being expelled from the St. Petersburg Academy of Arts for his involvement in student disorders, Fomin had moved to Paris for a year to continue his studies. There he was strongly influenced by Art Nouveau, and especially by the Viennese architects Joseph Maria Olbrich and Josef Hoffmann. Returning to Moscow, he entered Lev Kekushchev's firm as an assistant, and the following year transferred to Fedor Shekhtel's office. Igor Grabar, *O russkoi arkhitekture* (Moscow, 1969), 337. The society's initial aim was to propagate Vienna *moderne* in Russia, as symbolized by its logo, an "emaciated hyena" designed by Mikhail Vrubel, who together with Korovin was an honorary member. When the society, under Fomin's influence, began to veer toward working in the Russian style, Vrubel's logo was replaced with the Sirin bird, while the sunken-chested beauties with flowing hair and hollow eyes that first appeared on the posters advertising the exhibition were replaced with the towers of the Kremlin.
52. With this exhibition the grand duchess emerges not merely as an aristocratic patron, but as someone with a specific interest in promoting a national version of the "new style." One critic noted that the term "neo-Russian style" really came of age when she sponsored a design competition at the Stroganov School of Technical Drawing for "a necklace in the neo-Russian style." Mikhailov, "Ist zu wienerisch," 22. The Darmstadt colony attracted lively interest among Russian artists, particularly those in the World of Art group. See, e.g., Sergei Diagilev, "Vystavki v Germanii," *Mir iskusstva,* 8–9 (1901), 142–52.
53. Mikhailov, "Ist zu wienerisch," 18. The same reviewer described Margaret Mackintosh's designs at the exhibition as "a page of sexual psychopathology, narrated in the language of wood, metal, paint, and contour lines." Ibid.
54. By way of comparison, the cost of a bedroom designed in the Russian style of the 1870s by A. P. Gun was placed at 5,000 rubles. See *Zodchii,* 7 (1875), ill. 24.
55. V. Karpovich, "Arkhitekturnaia i khudozhestvenno-promyshlennaia vystavka v Moskve," *Arkhitekturnyi muzei,* 2 (1903), 31.
56. Mikhailov, "Ist zu wienerisch," 21.
57. Ibid., 18.
58. Ibid., 19. Vladimir Ivanovich Borutskii was head of the toy workshop at Sergiev Posad.
59. Ibid., 17.
60. E. Baumgarten, "Iskusstvo u kustarei," *Zodchii,* 24 (16 July 1902), 276.
61. A. Dmitriev, "Vpechatleniia ot vystavki 'Arkhitektury i khudozhestvennoi promyshlennosti novogo stilia v Moskve dekabria 1902–ianvariia 1903," *Zodchii,* 9 (1903), 116. One such imitator was the architect N. A. Gal'nbek, president of the Society of Architects and Artists, who founded a

Society of Industrial Art in St. Petersburg in March 1904 (although it was already active in 1903 when it took part in a Floating Kustar Exhibition on the Volga River). The society's premises on Bol'shaia Ruzheinaia Street were also an exhibition space showing a small collection of antique kustar objects, as well as modern examples of applied art, of which the *World of Art* critic (probably Benois) wrote: "Reminiscences of banal 'decadence' try unsuccessfully to fuse with the imitation of old Russian art – à la Bilibin and Vasnetsov, naturally." See "Zametki," *Mir iskusstva,* 2 (1904), 52. Also "Khronika," *Zodchii,* 15 (1904), 181.

62. The higher-level schools were reserved for training workers in factory-based industry and urban trades. Of the forty-eight primary craft schools in operation in 1900, only three were located in rural areas. "Zhurnal soveshchaniia o nuzhdakh drevoobrabatyvaiushchei kustarnoi promyshennosti," in G.U.Z.i Z., *Otchety i issledovaniia po kustarnoi promyshlennosti v Rossii,* vol. 8 (St. Petersburg, 1907), 93.
63. The basic points of the statute are outlined in E. Baumgarten, "Khudozhestvenno-promyshlennoe obrazovanie," *Zodchii,* 28 (1902), 321–2.
64. The production of kustar furniture in Ligachev was begun in 1823 by Ivan Zenin, a serf whose master had arranged for him to learn carpentry, but the real stimulus for furniture making in Moscow province was the great Moscow fire of 1812, which created a huge demand for furniture.
65. N. A. Filippov, comp., *Kustarnaia promyshlennost' Rossii. Promysly po obrabotke dereva* (St. Petersburg, 1913), 26.
66. "Moskovskaia guberniia," in G.U.Z.i Z., *Otchety i issledovaniia po kustarnoi promyshlennosti v Rossii,* vol. 11 (Petrograd, 1915), 13.
67. A sample of the goods that Parisians could find at this store is illustrated in P. A., *Les Koustari Russes* (n.p., n.d.).
68. Ibid., 18.
69. "Kustarnoe tsarstvo v Sergievskom posade," 642.
70. E. D. Maksimov, ed., *Ezhegodnik kustarnoi promyshlennosti 1912 god,* vol. 1, no. 1 (St. Petersburg, 1912), 303.
71. "Moskovskaia guberniia," 21.
72. Biographical information on Bartram is taken from E. S. Ovchinnikova, "Nikolai Dmitrievich Bartram," *Sbornik trudov NIIKhP,* 5 (1972), 65–72; and Ovchinnikova and Izergina, *N. D. Bartram.*
73. N. D. Bartram, "O vozmozhnosti vozrozhdeniia v igrushke narodnogo tvorchestva," *Apolon,* 2 (1912), 57.
74. Ovchinnikova and Izergina, *N. D. Bartram,* 73.
75. Like many of his contemporaries Bartram was a committed collector of folk art in the field. Throughout the 1890s he traveled his native province of Kursk collecting women's headdresses, naboika, embroideries, clothing, and household objects, which he later donated to the Kursk Archaeological Society.
76. Ovchinnikova and Izergina, *N. D. Bartram,* 75.
77. Agnessa Eduardovna Lindeman (?) specialized in graphic art and embroidery. She studied under Il'ia Repin at the Tenisheva Art Studio in St. Petersburg. From 1908 until 1916 she taught art embroidery at the School of the Society for the Encouragement of the Arts. Igor Grabar described her work as a "combination of Larson and Elena Polenova" (Igor Grabar', *Pis'ma. 1891–1917,* vol. 1 [Moscow, 1974], 174).
78. In her memoirs Bartram's daughter recalls how three young "Stroganovtsy" – Musatov, Bel'skii, and Il'in – designed egg-shaped toys in the form of pheasants, peacocks, sparrows, Chinese ducks, and drakes that were actually produced at Sergiev Posad. Ovchinnikova and Izergina, *N. D. Bartram,* 106.
79. Between 1907 and 1911 over a hundred of Bartram's new designs were put into production, including variations on the matreshka, toys with moving parts, and toy sets such as a Ukrainian farmhouse (*khata*) with people and livestock.
80. Ovchinnikova and Izergina, *N. D. Bartram,* 114.
81. Bartram, "O vozmozhnosti vozrozhdeniia v igrushke narodnogo tvorchestva," 60–1.
82. Aleksandr Benua, "Igrushki," *Apolon,* 2 (1912), 54.
83. "Moskovskaia guberniia," 32.
84. See *Ezhegodnik Obshchestva arkhitektorov-khudozhnikov* (1909), 148–9.
85. Although the exterior of the Pertsov house remains one of Moscow's architectural landmarks, the interior was already gutted in the 1930s. The

catalogue for an exhibition of Maliutin's work in 1934 noted that "part of the carved oak walls and furniture of the apartment are currently in the State Tret'iakov Gallery." *Vystavka proizvedenii Maliutina. Katalog* (Moscow, 1934), 10.

86. See "Studio-Talk. Moscow," *International Studio*, 45 (February 1912), 333–5. Apollinarii Vasnetsov's furniture was also reproduced in the Russian architectural journal *Ezhegodnik Obshchestva arkhitektorov-khudozhnikov* (St. Petersburg, 1910), 32–5. These items and their present location are listed in L. A. Bespalova, *A. M. Vasnetsov* (Moscow, 1983), 210.

87. *Ukazatel' moskovskogo otdela II vserossiiskoi vystavki v S.-Peterburge 1913 g. Kustarnoe delo moskovskogo zemstva,* 68. As an example of foreign sales on the eve of the First World War, the matreshka doll, produced at 30–35 rubles per hundred, retailed in America for $1.00 each (Vvedenskii, *U Sergievskogo igrushechnika,* 57).

88. Compare this 1913 figure with V. I. Borutskii's 1910 estimate that of the 1,200 households in Sergiev Posad, only 250 families were directly affected by the zemstvo's toy workshop. V. I. Borutskii, "Ocherk deiatel'nosti uchebnoi masterskoi igrushek v Sergievskom posade," *S'ezd deiatelei po kustarnoi promyshlennosti 1910 goda* (St. Petersburg, 1910), 498.

89. Ibid., 68.

90. See "Kustarnoe tsarstvo v Sergievskom posade," 642; *Kratkii obzor kustarnykh promyslov. Illiustrirovannoe izdanie* (St. Petersburg, 1896), xiv.

91. Details of these and all other industries influenced by the Moscow zemstvo may be found in *Ukazatel' moskovskogo otdela,* 26. The statistics given here underscore the tiny number of kustari actually involved in these workshops – ten in "Kustar-Khudozhnik," seven in the Sergiev Posad carpentry artel, etc. Of the 153,000 kustari in Moscow province cited in a 1915 guidebook, only 4,000 to 5,000, or 1 in 30, were receiving zemstvo aid. A. Orlov, "Kustarnyi muzei moskovskogo gubernskogo zemstva," in E. A. Zviagintseva et al., eds., *Moskva. Putevoditel'* (Moscow, 1915), 522.

92. In her memoirs M. K. Morozova implied that the museum's new premises were actually the former home of publisher Anatolii Mamontov and his wife, Maria. M. K. Morozova, "Moi vospominaniia," *Nashe nasledie,* 6 (1991), 95. Sergei Morozov purchased the building in 1900 and had it rebuilt by S. Iu. Solov'ev in 1902–3 with a new street facade in the Russian style. The original eighteenth-century interior, with its central grand staircase, was left intact. In 1911 a shop was built onto the museum, designed by B. I. Bashkirov. See *Pamiatniki arkhitektury Moskvy. Belyi gorod* (Moscow, 1989), 123–4.

93. "Moskovskaia guberniia," 11.

94. Orlov, "Kustarnyi muzei moskovskogo gubernskogo zemstva," 523.

95. "Moskovskaia guberniia," 25.

96. G. Lukomskii, "Arkhitektura, khudozhestvennaia promyshlennost' i pechatnoe delo na Vserossiiskoi vystavke v Kieve," *Iskusstvo v iuzhnoi Rossii,* 7–8 (1913), 356.

97. Thus, Vasilii Borutskii maintained that the Moscow zemstvo's chief goal was to develop "the kustari's ability to organize themselves into cooperatives and various kinds of artels or consumer warehouse societies." V. I. Borutskii, "Kustarnye igrushechnye promysli Moskovskoi gubernii," in Bartram, *Igrushka. Ee istoriia i znachenie,* 226.

98. The financial support came from Morozov's 100,000 ruble fund opened in 1910, another 70,000 from the zemstvo in 1912 to mark the Emancipation Act of 1861, and a further 25,000 loan from G.U.Z.i Z. at 15 percent interest.

99. On the Moscow artels, see "Moskovskaia guberniia," 44ff.

100. See, e.g., G. N. Bocharov, "Moskovskii kustarnyi muzei i deiatel'nost' N. D. Bartrama," *Russkaia khudozhestvennaia kul'tura kontsa XIX–nachala XX vekov (1908–1917),* vol. 4: *Izobrazitel'noe iskusstvo. Arkhitektura. Dekorativno-prikladnoe iskusstvo* (Moscow, 1980), 462–72.

1. Maria Klavdievna Tenisheva (née Piatkovskaia, 1867–1928), the illegitimate daughter of a nobleman, grew up in a well-to-do St. Petersburg family and at sixteen married R. N. Nikolaev, by whom she had a daughter. In the mid-1880s she studied singing in Paris with the well-known voice teacher Marquesi and took drawing lessons at both the Académie Julien in Paris and the School of the Society for the Encouragement of the Arts in St. Petersburg. The major source of information on the princess are her memoirs, *Vpechatleniia moei zhizni* (Paris, 1933; reprint, Leningrad, 1991), written in 1916 before her final emigration to Paris the following year. They are, however, written in a mood of bitterness that makes her perspective on many events and individuals distorted and unreliable. For the only extensive information on the princess in English, see John E. Bowlt, "Two Russian Maecenases: Savva Mamontov and Princess Tenisheva," *Apollo* (December 1973), 444–53, and *The Silver Age: Russian Art of the Early Twentieth Century and the "World of Art" Group* (Newtonville, Mass., 1979), 39–46.
2. It is true that Tenisheva's St. Petersburg apartment, dubbed the "Hermitage," was the gathering place for members of the World of Art group in the months preceding the first issue of the *World of Art* in 1898. However, the general attitude of the group toward their benefactor was one of contempt rather than gratitude, and Sergei Diagilev spurned all Tenisheva's efforts to direct the policy of the journal. As for the Mamontov circle which congregated at Abramtsevo, there is no evidence that Princess Tenisheva was on friendly terms with the family, despite Belogortsev's claim that she visited the estate on several occasions. I. Belogortsev, "Talashkino," *Smolenskii almanakh 4* (Smolensk, 1948), 43.
3. A. P. Ostroumova to K. P. Trunevaia, 11 December 1897. Quoted in I. S. Zilbershtein and V. A. Samkov, eds., *Sergei Diagilev i russkoe iskusstvo,* vol. 2 (Moscow, 1982), 154.
4. Prince Viacheslav Nikolaevich Tenishev (1843–1903) was co-owner of the Briansk, Putilov, and Ekaterinoslav locomotive works, as well as of an electromagnetic factory in St. Petersburg. He was also a passionate music lover and an amateur ethnographer. In 1896 he established a program for collecting ethnographical data on the peasants of Central Russia and a special Ethnographical Bureau, whose records are now housed in the Leningrad Museum of Ethnography. Of Maria Klavdievna's marriage to the prince, Larisa Zhuravleva writes: "To a certain extent [it] was a marriage of convenience. In exchange for her youth she 'acquired' as she herself wrote, 'position, important connections, one of the oldest names, and enormous wealth.'" "K stoletiiu so dnia rozhdeniia M. K. Tenishevoi," *Materialy po izucheniiu smolenskoi oblasti,* 7 (1970), 363.
5. Princess Ekaterina Konstantinovna ("Kitu") Sviatopolk-Chetvertinskaia (née Shupinskaia) was Princess Tenisheva's lifelong companion and supporter. Both women suffered unhappy marriages and shared a belief that life "has no interest without striving toward lofty goals (serving one's country, consecrating oneself to art, etc.)." Konstantin Somov dismissed the two princesses as "followers of artistic fashion" (*modnitsy iskusstva*), but Aleksandr Benois, who knew both women well in the 1890s and assembled Tenisheva's collection of watercolors, considered Chetvertinskaia "the inspiration behind the great schemes of her more striking friend." Among these schemes he cited the collection of Russian and European watercolors and the agricultural school at Talashkino. See Aleksandr Benua, *Moi vospominaniia,* vol. 2 (Moscow, 1980), 52–4.
6. Larisa Zhuravleva gives a date of 1894 for the founding of the school (*Teremok* [Moscow, 1974], 31). However, a later date of 1896 is implied in Princess Tenisheva's memoirs, where she notes that one of the school's first pupils was an orphan who had lost his mother at the disaster on Khodynka Field, which took place during the coronation of Nicholas II in 1896. See Tenisheva, *Vpechatleniia,* 233.
7. For insight into the role of corporal punishment in peasant life, see James Mandel, "Paternalistic Authority in the Russian Countryside, 1856–1906," Ph.D. diss., Columbia University, 1978. On the rigors of military service, see John Bush-

64. Ibid.

65. Even before his arrival at Talashkino in 1897, Ivan Fedorovich Barshchevskii (1851–1946?) was well known for his photographs of ancient Russian architecture. His first photographs possibly date back to 1880, when the architect A. M. Pavlinov was commissioned by the Academy of Arts to tour Russia to study architectural monuments. In 1883 he became the academy's official photographer, as well as corresponding member of and photographer to the Moscow Archaeological Society, and he collaborated extensively with the architect V. V. Suslov and the restorer N. V. Sultanov. As with Maliutin, Princess Tenisheva's interest in him was timely, since Barshchevskii and his family were then living in near poverty in Iaroslavl. See M. G. Rogozina, "Kollektsiia fotografii I. F. Barshchevskogo," *Pamiatniki otechestva,* 2 (1983), 153–6.

66. Tenisheva, *Vpechatleniia,* 387.

67. See, e.g., Stasov's study "Proizkhozhdenie russkoi byliny," *Vestnik Evropy* (1868), in which he argued that Russia's legendary heroes such as Sadko and Dobrynia Nikitich had their direct prototypes in the heroic pantheons of India, Tibet, and Persia. "Our Novgorodian merchant Sadko is none other than the Indian King Iada in Russian form. . . . The music of Sadko in the underwater kingdom is the Buddhist hero's assertion of true faith in the kingdom of serpents and evil spirits." Quoted in E. Poliakova, *Nikolai Rerikh* (Moscow, 1985), 37–8.

68. On the theosophical movement in Russia, see M. M. Carlson, *"No Religion Higher Than Truth": A History of the Theosophical Movement in Russia, 1875–1922* (Princeton, N.J., 1993).

69. V. V. Golubev (ca. 1880–1945), archaeologist, expert on Oriental art, and collector of Persian, Indian, and Turkish miniatures that were sold to the Boston Museum of Fine Arts in 1911–12. According to Prince Shcherbatov, Golubev was at one time director of the Musée Guimet in Paris. See Sergei Shcherbatov, *Khudozhnik v ushedshei Rossii* (New York, 1955), 174.

70. N. Rerikh, "Indiiskii put'," *Sobranie sochinenii,* vol. 1 (Moscow, 1914), 258. Writing of the Talashkino church in 1905, Roerich noted how its ornamentation was to include references to the monuments of old Rus, "even distant paths. Even the temples of Ajanta and Lkhassa." Makowsky, *Talachkino,* 21.

71. Tenisheva, *Vpechatleniia,* 301.

72. On the museum's collection, see I. F. Barshevsky, *Objets d'art russe anciens faisant partie des collections de la Princesse Marie Tenichev exposés au Musée des Arts Décoratifs du 10 mai au 10 octobre, 1907* (Paris, [1907]); L. Zhuravleva, *Russkoe narodnoe reznoe i rospisnoe derevo v sobranii smolenskogo gosudarstvennogo istoricheskogo i arkhitekturno-khudozhestvennogo muzeia-zapovednika* (Moscow, 1985).

73. N. K. Rerikh, "Zakliatoe zver'e. (Emali kn. M. K. Tenishevoi)," *Niva,* 18 (1909), 311.

74. Tenisheva, *Vpechatleniia,* 251.

75. Denis Roche, "Nicolas Roerich," *Gazette des Beaux-Arts,* 39 (1908), 144.

76. Rerikh, "Zakliatoe zver'e," 310.

77. Maksimilian Voloshin, "Arkhaizm v russkom zhivopisi. (Rerikh, Bogaevskii i Bakst)," *Apolon,* 1 (1909), 51.

78. N. N. Sobolev, *Russkaia narodnaia rez'ba po derevu* (Moscow-Leningrad, 1934), 459.

79. This cupboard was to have been exhibited in a room of Talashkino goods at a projected International Exhibition of Decorative Art in Paris in 1905, in a Russian section organized by the Russian Art Circle in Paris and headed by Denis Roche. In addition to Roerich's room, Talashkino wares designed by Maliutin, Zinov'ev, Beketov, and Barshchevskii were also to be exhibited. Whether the exhibition took place is unknown. See *Zodchii,* 40 (1904), 450.

80. For a detailed study of how the 1905 disturbances affected southwest Russia (the right-bank Ukraine), see Robert Edelman, *Proletarian Peasants: The Revolution of 1905 in Russia's Southwest* (Ithaca, N.Y., 1987).

81. Tenisheva also held Pogosskaia responsible for Maliutin's defection: "She would often have interminable conversations with Maliutin, arguing and trying to convince him about something. And after each of these conversations he would walk about looking strange somehow, preoccupied, deeply agitated by something, not himself" (Tenisheva, *Vpechatleniia,* 342). Maliutin was still on good terms with the Pogosskaias in 1913, when he designed furniture for their London shop. See Abramova, *Zhizn' khudozhnika,* 162.

82. Tenisheva, *Vpechatleniia,* 342. Virtually no infor-

mation is available on the younger Pogosskaia: her mother described her as "an artist and pupil of our great Repin," and in a letter to me the painter and restorer Mr. Kirill Katkov wrote that she was a good friend of Ivan Bilibin, serving as the model for the heroine in his illustrations to the fairy tale *Vasilisa Prekrasnaia.*

83. Tenisheva, *Vpechatleniia,* 340.
84. According to Tenisheva, "Only later did I find out that Pogosskaia senior had pulled this same trick in many zemstva, sending the best stock to her eldest daughter in England where she worked as the agent of some store. She also swindled some other charitable society that had entrusted her with selling kustar goods in London" (ibid., 341). This account is difficult to reconcile with the respectable position the family continued to hold in London where, as one observer commented in 1916, for many "the knowledge of Russian art begins and ends with a visit to Mme Pogosky's depot for Peasant Industries in Bond St." Rosa Newmarch, *The Russian Arts* (New York, 1916), v. For further information on the family, see chap. 5.
85. William Morris, "Art and Its Products," *Art and the Beauty of the Earth,* 353.
86. A. L. Pogosky, *Fellowship in Work* (London, n.d.), 92. Based in London, where she and her children had a kustar depot in Bond Street, Pogosskaia had a direct connection to the English Arts and Crafts movement as a member of the International Union of Manual Labor, an organization founded in 1910 with Walter Crane as its president, Charlotte Despard its vice-president, and D. N. Dunlop its council representative. In a tract called *Idealy truda kak osnova schastlivoi zhizni* (The Ideals of Labor as the Basis for a Happy Life), published in 1913, Pogosskaia describes the effect of William Morris's speech at the First Arts and Crafts Exhibition held in London in 1889: "Perhaps without the venerable artist's fiery speeches ringing in my soul I would not have found the courage and long-suffering to keep on continuously searching" (105).
87. N. Breshko-Breshkovskii, "V skazochnom tsarstve," *Novyi mir,* 23 (1905), 266.
88. Tenisheva, *Vpechatleniia,* 402.
89. The Source was replaced by a similar store run by E. I. Silin, "a well-known authority on art antiquities. Attached to the store he has established a workshop producing art goods, and he intends to attract new talent to participate in this venture." *Zolotoe runo,* 6 (1906), 94.
90. Tenisheva, *Vpechatleniia,* 423. The exhibition was open from 10 May until 10 October 1907 and was accompanied by a catalogue written by Barshchevskii (see n. 72). For reviews of the exhibition, see C. de Danilowicz, "Art décoratif russe ancien. Collection de la Princesse M. Tenichef," *L'Art décoratif,* 17 (April 1907), 136–45; Denis Roche, "L'Exposition d'art russe ancien au Musée des Arts Décoratifs," *Gazette des beaux arts,* 38 (1 September 1907), 250–64.
91. For a detailed account of the planning of this exhibition, which took place from 15 October to 11 November 1906, see A. V. Tolstoi, "O russkoi vystavke 1906 goda v Parizhe," *Sovetskoe iskusstvoznanie '81,* 1 (Moscow, 1982), 279–90. Tolstoi remarks: "Judging Russian art as a whole on the basis of the art sections at the International Expositions and a few personal exhibitions abroad, [French artists and art experts] had assumed that it was provincial and exotic, without any connection to European processes of artistic development. The only value such art had in the eyes of foreigners lay in the fact that Russian paintings and statues conveyed some idea of Russian life and the 'mysterious Slavic soul'" (285).
92. C. de Danilowicz, "Une Exposition d'art russe moderne à Paris," *L'Art et les Artistes,* 34 (January 1908), 481.
93. Quoted in Zilbershtein and Samkov, *Sergei Diagilev,* vol. 1, 408.
94. *Exposition d'Art Russe Moderne.* Catalogue. Introduction by C. de Danilowicz (Paris, 1907).
95. For reviews of the exhibition, see Denis Roche, "Une Exposition d'artistes russes à Paris," *Art et décoration,* 22 (1907), 173–90; C. de Danilowicz, "Une Exposition d'art russe moderne à Paris," *L'Art et les artistes,* 34 (January 1908), 480–4; N. Gumil'ev, "Vystavka novogo russkogo iskusstva v Parizhe," *Vesy,* 11 (1907), 88.
96. "Both my Paris exhibitions had a strong influence on fashions and accessories. The following year I noticed the clear influence of our embroideries, our Russian clothing, sarafans, chemises, headdresses, and homespun coats. Even the term 'blouse russe' became popular. . . . Manufacturers of textiles and embroideries also turned up at the

rue Caumartin in search of whatever new ideas and patterns they could lay their hands on." Tenisheva, *Vpechatleniia,* 426.

97. On Diagilev's Russian Seasons, see M. Pozharskaia and T. Volodina, *The Art of the Ballets Russes: The Russian Seasons in Paris, 1908–1929* (New York, 1990); R. I. Vlasova, "'Russkie Sezony' vo Frantsii. 'Zolotoi Vek'," *Russkoe teatral'no-dekoratsionnoe iskusstvo nachala XX veka* (Leningrad, 1984), 102–69. That Tenisheva regarded Diagilev as a competitor for the honor of opening European eyes to the cultural wealth of Russia is further suggested by her activities in opera and music. In 1908, learning that Diagilev had reneged on his promise to assist the head of the Opéra Comique in staging *Snegurochka,* she made her collection of Russian costumes available to the directorate of the Opéra Comique and supervised its research. And just as Diagilev organized a concert of Russian music in conjunction with the exhibition in the Palais d'Elysées, so too Princess Tenisheva organized musical evenings at which, in addition to the music of Grieg and Chopin, works by Russian composers were performed. A balalaika orchestra was also formed to play Russian folk songs. Despite the success of *Snegurochka,* however, it was eclipsed by the spectacular triumph enjoyed by Diagilev's staging of *Boris Godunov* at the Grande Opéra.
98. C. de Danilowicz, "Talashkino: Princess Tenisheva's School of Russian Applied Art," *International Studio,* 32, no. 126 (August 1907), 139.
99. On the renaissance of the Smolensk embroideries in the Soviet period, see Belogortsev, *Talashkino,* 124–32.
100. "Khronika," *Iskusstvo i pechatnoe delo,* 4–6 (1909), 28.

Chapter 5 BETWEEN TWO REVOLUTIONS

1. N. Tseretelli, *Russkaia krest'ianskaia igrushka* (Leningrad, 1933), 226. By mid-decade Elizaveta Mamontova had virtually lost all interest in a workshop that had itself "lost interest in creative work in which each carpenter might feel himself a participant and turned into a routine business for teaching a craft." N. V. Polenova, *Abramtsevo. Vospominaniia* (Moscow, 1922), 103–4.
2. The majority of these workshops were for weaving, lacemaking, and basket weaving, with a smaller number devoted to carpets, joinery, cooperage, toys, ceramics, dyeing, etc. See G.U.Z.i Z., *Obzor pravitel'stvennogo sodeistviia kustarnoi promyshlennosti* (St. Petersburg, 1913), 37. Over the same five-year period subsidies for kustar exhibitions also doubled, from 23,875 to 57,005 rubles.
3. These figures, given by S. A. Davydova in a speech at the 1910 Congress of Workers in the Kustar Industries, are quoted in I. A. Kriukova, "Narodnoe iskusstvo i zemskie organizatsii," *Russkaia khudozhestvennaia kul'tura kontsa XIX–nachala XX vekov (1908–1917),* bk. 4 (Moscow, 1980), 450. Another source cites over 260 training workshops in Russia by 1908. See "Peasant Industry and Labour Conditions," *The Russian Year-Book 1911* (London, 1911), 270.
4. See G.U.Z.i Z., *Obzor pravitel'stvennogo sodeistviia,* 36.
5. "Zhurnal soveshchaniia o nuzhdakh goncharnoi kustarnoi promyshlennosti," in G.U.Z.i Z., *Otchety i issledovaniia po kustarnoi promyshlennosti v Rossii,* vol. 8 (St. Petersburg, 1907), 2.
6. Nikolai Root, "Khudozhestvenno-promyshlennoe obrazovanie v Rossii i za granitsei," *Zodchii,* 1 (3 January 1910), 7.
7. N. A. Filippov, *Kustarnaia promyshlennost' Rossii. Promysly po obrabotke dereva* (St. Petersburg, 1913), 32.
8. "Zhurnal soveshchaniia o nuzhdakh goncharnoi kustarnoi promyshlennosti," 13.
9. A. V. Prakhov, "Osnovy dlia organizatsii russkogo narodnogo khudozhestvennogo truda," *Trudy s'ezda deiatelei po kustarnoi promyshlennosti v Sankt Peterburge (1 ianvariia–15 fevral'ia),* vol. 1, pts. 1 and 2 (St. Petersburg, 1910), 622.
10. From P. P. Marseru's speech, reported in N. Elfimov, "Kustarnyi s'ezd," *Zodchii,* 10 (7 March 1910), 98.

11. Ibid.
12. Ibid.
13. Prakhov, "Osnovy dlia organizatsii," 654.
14. For details on the Stroganov curriculum, see *Otchet po imperatorskomu stroganovskomu Tsentral'nomu khudozhestvenno-promyshlennomu uchilishchu za 1909 g.* (Moscow, 1910).
15. Prakhov, "Osnovy dlia organizatsii," 650.
16. Ibid.
17. Elfimov, "Kustarnyi s'ezd," 99.
18. F. S. Golitsyn, *Kustarnoe delo v Rossii,* vol. 1, pt. 1 (St. Petersburg, 1904), 396–7.
19. The school's initial funding came from the tsarina (2,000 rubles) and G.U.Z.i Z. (11,000 rubles). Premises were found on Demidov Lane, where the School of the Society for the Encouragement of the Arts had its workshops. It was later moved to a site near the Alexander III Museum.
20. "Shkola narodnogo iskusstva," *Niva,* 6 (1914), 114–15.
21. Both Varvara Petrovna Shneider (1860–1941) and her sister Aleksandra were trained as watercolorists, and their St. Petersburg apartment (3 Malaia Masterskaia) was a meeting place for artists (Nikolai Roerich, Elizaveta Bem), scholars (geographer P. P. Semenov-Tianshanskii, Indianologist F. I. Shcherbatskii), and literary figures (literary historian F. D. Batiushkov, *Art and Industrial Art* editor N. P. Sobko). Their guest book is now housed in the manuscript section of Pushkinskii Dom, St. Petersburg. See L. V. Korotkina, *Rerikh v Peterburge-Petrograde* (Leningrad, 1985), 79–82. Varvara Shneider was also a close friend of Olga Semenova Tian-Shanskaia, with whom she compiled a large collection of folk costumes from Riazan province. See Tian-Shanskaia, *Village Life in Late Tsarist Russia,* ed. David L. Ransel (Bloomington and Indianapolis, 1993).
22. "Shkola narodnogo iskusstva," 114.
23. "Shkola narodnogo tvorchestva," *Solntse Rossii,* 205 (January 1914), 13.
24. Ibid.
25. A. G. Dolivo-Dobrovol'skaia, "Kustarnoe tkachestvo v guberniiakh: Kievskoi, Tambovskoi, Voronezhskoi, Vitebskoi, Kurliandskoi, Grodnenskoi i Novgorodskoi," in G.U.Z.i Z., *Otchety i issledovaniia po kustarnoi promyshlennosti v Rossii,* vol. 11 (Petrograd, 1915), 8.
26. Ibid., 9.
27. B. V. Kirichenko-Astromov, *Kustarnyi otdel na vsemirnoi turinskoi vystavke 1911 goda* (St. Petersburg, 1911), 20. Filip Maliavin specialized in richly colored paintings of peasant women in traditional dress.
28. K. G. Berens, "Vyshival'nyi promysel v Rossii," *Kustarnaia promyshlennost' v Rossii. Zhenskie promysly* (St. Petersburg, 1913), 396.
29. A. E. Porai-Koshits, "Kustarnye krasil'noe i nabivnoe proizvodstva," in G.U.Z.i Z., *Otchety i issledovaniia po kustarnoi promyshlennosti v Rossii,* vol. 9 (St. Petersburg, 1911), 104.
30. A. E. Porai-Koshits, "Krashenie priazhi v Sudzhanskom uezde, Kurskoi gubernii," in G.U.Z.i Z., *Otchety i issledovaniia po kustarnoi promyshlennosti v Rossii,* vol. 11, 331.
31. Alexander Pronin and Barbara Pronin, *Russian Folk Art* (South Brunswick, N.Y., and London, 1975), 181.
32. On Tatishchev, see V. V. Stasov, "Khudozhestvennye zametki o vystavke 1870 g. (v Solianom Gorodke)," *Sobranie sochinenii,* vol. 1 (St. Petersburg, 1894), 284ff.
33. M.Z.i G.I., *Obzor deiatel'nosti pravitel'stva na pol'zu kustarnoi promyshlennosti (1888–1903)* (St. Petersburg, 1902), 105.
34. "A. N. Durnovo," *Vestnik promyslovoi kooperatsii,* 1–2 (1924), 54.
35. V. M. Vishnevskaia, *Russkaia narodnaia rez'ba i rospis' po derevu* (Moscow, 1956), 72.
36. L. Suprin, "Grani talanta. O dekorativno-prikladnykh rabotakh S. V. Maliutina," *Iskusstvo,* 4 (1971), 42.
37. The museum was one of the many buildings slated for construction in 1913 to commemorate the tercentenary of the Romanov family. Shekhtel's projected designs are reproduced in *Ezhegodnik Moskovskogo Arkhitekturnogo Obshchestva* (1914–16), 86–7.
38. On the activities of the Poltava zemstvo, see Basilius Trepke, *Kustargewerbe im Gouvernement Poltawa* (Berlin, 1910), and "Poltavskaia guberniia," *Obzor deiatel'nosti zemstv po kustarnoi promyshlennosti,* vol. 3 (Petrograd, 1916), 69–74.
39. On the Poltava Zemstvo House, see V. V. Eisner, "Zemskii dom v Poltave," *Iskusstvo i pechatnoe delo,* 1 (January 1910), 11, and "Novyi zemskii dom v g. Poltave," *Niva,* 18 (1908), 836–7, 840.
40. "Together with Vasilkovskii [Samokish] painted a

Selected Bibliography

A. GENERAL WORKS CONSULTED

Belokonskii, I. P. *Zemskoe dvizhenie,* 2d ed. Moscow, 1914.

Benua, A. N. *Moi vospominaniia,* 2 vols. Moscow, 1980.

Bill, Valentine T. *The Forgotten Class: The Russian Bourgeoisie from the Earliest Beginnings to 1900.* New York, 1959.

Bowe, Nicola Gordon, ed. *Art and the National Dream: The Search for Vernacular Expression in Turn-of-the-Century Design.* Dublin, 1993.

Bowlt, John E. *The Silver Age: Russian Art of the Early Twentieth Century and the "World of Art" Group.* Newtonville, Mass., 1979.

"Two Russian Maecenases: Savva Mamontov and Princess Tenisheva." *Apollo,* December 1973, 444–53.

Brooks, Jeffrey. *When Russia Learned to Read: Literacy and Popular Culture, 1861–1917.* Princeton, N.J., 1986.

Brumfield, William Craft. *The Origins of Modernism in Russian Architecture.* Berkeley, 1991.

Burishkin, T. A. *Moskva kupecheskaia.* New York, 1954.

Butovskii, V. I. *Histoire de l'ornement russe du Xe au XVIe siècle d'après les manuscrits,* 2 vols. Paris, 1870–3.

O prilozhenii esteticheskogo obrazovaniia k promyshlennosti v Evrope i v osobennosti v Rossii. St. Petersburg, 1870.

Callen, A. *Angel in the Studio: Women in the Arts and Crafts Movement, 1870–1914.* London, 1979.

Clements, Barbara Evans; Barbara Alpern Engel; and Christine D. Worobec, eds. *Russia's Women: Accommodation, Resistance, Transformation.* Berkeley, 1991.

Emmons, T., and W. Vucinich, eds. *The Zemstvo in Russia: An Experiment in Local Self-Government.* Cambridge and New York, 1982.

Farnsworth, Beatrice, and Lynne Viola, eds. *Russian Peasant Women.* New York and Oxford, 1992.

Fedorov-Davydov, A. A. *Russkoe iskusstvo promyshlennogo kapitalizma.* Moscow, 1929.

G.U.Z.i Z. [Glavnoe upravlenie zemledeliia i zemleustroistva]. *Kustarnaia promyshlennost' Rossii,* 4 vols. St. Petersburg, 1913.

Obzor pravitel'stvennogo sodeistviia kustarnoi promyshlennosti. St. Petersburg, 1913.

Otchety i issledovaniia po kustarnoi promyshlennosti v Rossii, 11 vols. St. Petersburg, 1905–15.

Gartvich, A. *Shkola risovaniia v otnoshenii k iskusstvam i remeslam uchrezhdennaia v 1823 gody S. G. Stroganovym: Ee vozniknovenie i razvitie do 1860 goda.* Moscow, 1901.

Golitsyn, F. S. *Kustarnoe delo v Rossii v sviazi s umstvenno-dukhovnym razvitiem russkogo naroda,* vol. 1: *Dela v Rossii. Deiatel'nost' pravitel'stva, zemstv i chastnykh lits.* St. Petersburg, 1904.

Goloushev, S. S. "O vozrozhdenii russkoi khudozhestvennoi promyshlennosti." *Trudy vserossiiskogo s'ezda khudozhnikov dekiabria 1911–ianvaria 1912,* vol. 2, 211–12. Petrograd, 1912.

Gorbunova, M. K. *Sbornik statisticheskikh svedenii po moskovskoi gubernii. Zhenskie promysly,* vol. 4. Moscow, 1882.

Grabar, I. E. "Neskol'ko myslei o sovremennom prikladnom iskusstve v Rossii." *Mir iskusstva,* 3 (1902), 51–6.

Isaev, Andrei. "K voprosu o kustarnoi promyshlennosti v Rossii." *Russkaia zhizn',* 11 (1880), 73–125.

Ivanov, I. M. *Russkie kustari. Nauchno-populiarnyi ocherk.* St. Petersburg, 1902.

Kablukov, N. *Kursy po kooperatsii. Kustarnaia promyshlennost' i ee sviaz' s kooperativami.* Moscow, 1915.

"Obshcheekonomicheskoe znachenie zhenskikh kustarnykh promyslov i sposoby sodeistviia im." *Novoe slovo,* 5 (February 1896), 55–86.

Katalog vtoroi vystavki sovremennogo dekorativnogo iskusstva. Vyshivki po eskizam khudozhnikov, ispol'nennye krest'iankami sela "Verbovki." Moscow, [1917].

Kirichenko-Astromov, B. V. *Kustarnyi otdel' na vsemirnoi turinskoi vystavke 1911 goda.* St. Petersburg, 1911.

Kolkotin, A. *Kustarnyi vopros v Rossii. Opyt ob'ektivnogo issledovaniia.* St. Petersburg, 1905.

Kondakov, N. P. *Sovremennoe polozhenie russkoi narodnoi ikonopisi.* St. Petersburg, 1900.

Kriukova, I. A. "Narodnoe iskusstvo i zemskie organizatsii." in *Russkaia khudozhestvennaia kul'tura kontsa XIX–nachala XX vekov (1908–1917),* bk. 4, 447–61. Moscow, 1980.

Kurchinskii, M. "Russkaia kustarnaia promyshlennost'. (Po dannym poslednykh let: 1894–1899)." *Zhizn',* 1 (1901), 76–102; 3 (1901), 97–122.

"Kustari: The Peasant Industries of Russia." *Russian Review,* October 1916, 186–90.

"Kustarnaia promyshlennost'." In V. I. Kovalevskii, ed., *Rossiia v kontse XIX veka,* 511–17. St. Petersburg, 1900.

Kustarnaia promyshlennost', ee nuzhdy i mery dlia ikh udovletvoreniia. St. Petersburg, 1909.

Kustarnaia promyshlennost' Rossii. Zhenskie promysly v ocherkakh S. A. Davydovoi, E. N. Polovtsovoi, K. I. Berens, i E. O. Sviderskoi. St. Petersburg, 1913.

Liakhovskaia, M. "Vystavka sovremennogo dekorativnogo iskusstva." *Mir zhenshchin,* 15–16 (1916), 30.

Lukomskii, G. "Neskol'ko slov o narodnom iskusstve na kustarnoi vystavke." *Apolon,* 4 (1913), 39–41.

M. G. [Maxim Gorky]. "Khudozhestvennost' i promyshlennost'." *Iskusstvo i khudozhestvennaia promyshlennost',* 20 (1900), 439–40.

M.Z.i G.I. [Ministerstvo zemledeliia i gosudarstvennykh imushchestv]. *Obzor deiatel'nosti zemstv po kustarnoi promyshlennosti (1865–92),* 2 vols. St. Petersburg, 1897.

Ocherk deiatel'nosti Ministerstva Gosudarstvennykh imushchest po razvitiiu i ulushcheniiu kustarnoi promyshlennosti v 1880–1890 godakh. St. Petersburg, 1890.

Otchety i issledovaniia po kustarnoi promyshlennosti, 6 vols. St. Petersburg, 1892–1902.

Spravochnyi ukazatel' kustarnykh proizvodstv i kustarnykh masterov, 5 vols. St. Petersburg, 1892–1902.

Tekushchaia statistika kustarnykh promyslov. St. Petersburg, 1897.

Maikov, L. "Materialy dlia izucheniia kustarnoi promyshlennosti i ruchnogo truda v Rossii." *Statisticheskii vestnik russkoi imperii,* vol. 3. St. Petersburg, 1872.

Makarenko, N. *Shkola Obshchestva pooshchreniia khudozhestv 1839–1914.* Petrograd, 1914.

Meshcherskii, A. A., and K. P. Modzalevskii. *Svod materialov po kustarnoi promyshlennosti Rossii,* 16 vols. St. Petersburg, 1874.

Miklashevskii, S. *Itogi poluvekovykh tolkov o pomoshchi kustariu.* St. Petersburg, 1891.

Noakovskii, S. "O protsvetanii khudozhestvennoi promyshlennosti." *Mir iskusstva,* 11–12 (1899), 132–3.

Norden, Julius. "Kunstgewerbe in Russland." *Westermann's Monats-Hefter,* 92 (September 1903), 804–12.

"Novyi zemskii dom v gorode Poltave." *Niva,* 18 (1908), 836–67, 840.

Obzor deiatel'nosti zemstv po kustarnoi promyshlennosti, 3 vols. St. Petersburg, 1913–16.

Ocherk deiatel'nosti Ministerstva gosudarstvennykh imushchestv po razvitiiu i ulushcheniiu kustarnoi promyshlennosti v 1880–1890 godakh. St. Petersburg, 1890.

Plotnikov, M. A. *Kustarnaia promyshlennost' Nizhe-Novgorodskoi gubernii.* Nizhnii Novgorod, 1894.

Pogosskaia, A. L. *Fellowship in Work.* London, n.d.

Idealy truda kak osnova schastlivoi zhizni. Kaluga, 1914.

"Kustari na vserossiiskoi vystavke." *Novoe slovo,* 1 (1896), 1–19; 2 (1896), 1–19.

Pogruzov, A. D. *Kustarnaia promyshlennost' v Rossii. Ee znachenie, nuzhdy i vozmozhnoe budushchee.* St. Petersburg, 1901.

Polovtsova, E. I. "O stepeni vmeshchatel'stva intelligentnykh khudozhnikov v narodnom tvorchestve." *Trudy vserossiiskogo s'ezda khudozhnikov dekiabr 1911–ianvar 1912,* vol. 2, 263–75. Petrograd, 1912.

Ponomarev, N. V. *L'Industrie domestique et rurale en Russie: les koustari.* Paris, 1900.

Kratkii obzor kustarnoi promyshlennosti. Illiustrirovannoe izdanie. St. Petersburg, 1896.

Obzor kustarnykh promyslov Rossii. St. Petersburg, 1902.

Vystavka igrushek v St. Peterburge: Kustarnoe proizvodstvo igrushek v Rossii. St. Petersburg, 1890.

Prilezhaev, A. V. *Chto takoe kustarnoe proizvodstvo?* St. Petersburg, 1882.

Rerikh, N. K. "Iskusstvo i arkheologiia." *Iskusstvo i khudozhestvennaia promyshlennost',* 3 (1899), 185–94; 4 (1899), 251–66.

"Nekhudozhestvennost' nashikh khudozhestvennykh magazinov." *Iskusstvo i khudozhestvennaia promyshlennost',* 11 (1899), 914–17.

Root, N. F. "Khudozhestvenno-promyshlennoe obrazovanie v Rossii i za granitsei." *Zodchii,* 34, no. 1 (3 January 1910), 1–7.

"Vospitanie i razvitie prikladnogo iskusstva i khudozhestvennoi promyshlennosti v Rossii." *Trudy pervogo Vserossiiskogo s'ezda khudozhnikov 1911–12,* vol. 2, 218–25. Petrograd, 1912.

Root, N. F., ed., *Ezhegodnik Ministerstva Torgovli i Promyshlennosti po voprosam khudozhestvenno-promyshlennogo obrazovaniia.* Kiev, 1913.

Russkoe narodnoe iskusstvo na vtoroi vserossiiskoi kustarnoi vystavke v Petrograde v 1913 godu. Petrograd, 1914.

Rybnikov, A. *Kustarnaia promyshlennost' i sbyt kustarnykh izdelii.* Moscow, 1913.

Salmond, Wendy R. "Design Education and the Quest for National Identity in Late Imperial Russia: The Case of the Stroganov School." *Studies in the Decorative Arts,* 1, no. 2 (Spring 1994), 2–24.

Sheremetev, P. S. *O russkikh khudozhestvennykh promyslakh. Doklad chlena kustarnogo soveta Moskovskogo gubernskogo zemstva.* Moscow, 1915.

"Shkola narodnogo iskusstva." *Niva,* 6 (1914), 114–15.

Sinzheimer, G. P. G. "The Russian Kustar Industries, 1830–1928." Ph.D. diss., Seattle University, 1963.

Smirnova, O. "What They Are Capable Of." *Free Russia,* 20, no. 4 (1899), 27–9; no. 5 (1899), 41–3.

Spravochnyi ukazatel' kustarnykh proizvodstv i kustarnykh masterov, 5 vols. St. Petersburg, 1892–1902.

Svod materialov po khudozhestvennoi promyshlennosti v Rossii. St. Petersburg, 1894.

Tourneur, N. "Kustarny." *Art World* (October 1917), 80–2.

Trepke, Basilius. *Kustargewerbe im Gouvernement Poltawa.* Berlin, 1910.

Trudy kommissii po issledovaniiu kustarnoi promyshlennosti v Rossii, 17 vols. St. Petersburg, 1879–90.

Trudy s'ezda deiatelei po kustarnoi promyshlennosti v Sankt Peterburge (1 ianvaria–15 fevral'ia), vol. 1, pts. 1 and 2. St. Petersburg, 1910.

Tugan-Baranovskii, M. "Koe-chto o kustariakh i kustarnykh deiateliakh." *Rech',* 20 March–2 April 1913, 1, 4; 22 March 1913, 1, 4.

V. V. [Vorontsov, V. P.]. *Ocherki kustarnoi promyshlennosti v Rossii.* St. Petersburg, 1886.

Vallance, Aymer. "Russian Peasant Industries." *International Studio,* 28, no. III (May 1906), 241–8.

Voronova, L. N. "Evgeniia Ivanovna Pribyl'skaia. (Put' khudozhnika)." *Khudozhestvennye promysly RSFSR,* 53–7. Moscow, 1950.

"Vserossiiskaia kustarno-promyshlennaia vystavka." *Niva,* 11 (1902), 215, 217–18; 12 (1902), 234–6; 13 (1902), 255–7.

Wright, Richardson. "Mostly Moujik: A Glimpse of the Russian Artel and Kustarnui." *Catholic World,* 102 (November 1915), 216–23.

Zhestianikov, M. "Tvorchestvo derevni." *Solntse Rossii,* 17 (April 1913), 12–14.

Bilibin, I. Ia. "Narodnoe tvorchestvo severa." *Mir iskusstva,* 11 (1904), 303–18.

"Russkoe dereviannoe severnoe zodchestvo." *Trudy vserossiiskogo s'ezda khudozhnikov,* vol. 2, 106–7. St. Petersburg, 1912.

Bobrinskii, A. A. *Narodnye russkie dereviannye izdeliia. Predmety domashnogo khoziaistva i otchasti tserkovnogo obikhoda,* 6 vols. Moscow, 1910–16.

Russian Peasant Art. New York, 1922.

Dalmatov, K. *Russkaia vyshivka.* Moscow, 1889.

Dudoreva, V. *Biser v starinnom rukodelii.* Moscow, 1923.

Evdokimov, Ivan. *Sever v istorii russkogo iskusstva.* Vologda, 1921.

Golyshev, I. A. *Atlas risunkov so starinnykh prianichnykh dosok Viaznikovskogo uezda. Vladimirskoi gubernii.* Mstera, 1874.

Hilton, Alison. "The Peasant House and Its Furnishings: Decorative Principles in Russian Folk Art." *Journal of Decorative and Propaganda Arts,* 11 (Winter 1989), 10–29.

Russian Folk Art. Bloomington, Ind., 1995.

Holme, Charles, ed. *Peasant Art in Russia.* London, 1912.

Il'in, M. "Russkaia natsional'naia arkhitektura i problema russkogo stilia." *Arkhitektura SSSR,* 9 (1945), 32–3.

Kruglova, O. V. *Russkaia narodnaia rez'ba i rospis' po derevu.* Moscow, 1974.

Kulzhinskii, S. K. *Opisanie kollektsii narodnykh pisanok.* Moscow, 1899.

Lossky, B. "The Popular Arts in Russia and Their Revival." *Apollo,* December 1973, 454–9.

Materialy po istorii krest'ianskoi promyshlennosti. Moscow, 1950.

Nekrasov, A. I. *Russkoe narodnoe iskusstvo.* Moscow, 1924.

Nekrasova, M. A., and K. A. Makarova, eds. *Problemy narodnogo iskusstva.* Moscow, 1982.

Orshanskii, L. G. *Igrushki. Stat'i po istorii, etnografii i psikhologii igrushek.* Moscow, 1923.

Ovsiannikova, E. B. "Iz istorii pervykh vystavok lubka." *Sovetskoe iskusstvoznanie '20,* 423–44. Moscow, 1986.

Pomerantseva, E. V., ed. *Russkaia khudozhestvennaia promyshlennost' vtoroi poloviny XIX–XX vekov.* Moscow, 1965.

Popova, O. S. *Russkie narodnye khudozhestvennye promysly.* Moscow, 1984.

Prokopev, D. V. *Khudozhestvennye promysly gorkovskoi oblasti.* Gorky, 1939.

Pronin, A., and B. Pronin. *Russian Folk Arts.* New York, 1975.

Rabotnova, I. P. *Russkoe narodnoe kruzhevo.* Moscow, 1956.

Russkie khudozhestvennye promysli vtoroi poloviny XIX–XX vekov. Moscow, 1965.

Sbornik ornamentnykh ukrashenii, pocherknutykh iz grecheskikh i drevne-russkikh rukopisnykh knig so X po XVI vekov vkliuchitel'no. Material dlia promyshlennogo risoval'shika. Moscow, 1874.

Shakhovskaia, S. N., ed. *Dessins d'anciennes broderies en Russie.* Moscow, 1888.

Simakov, N. *Russkii ornament v starinnykh obraztsakh khudozhestvenno-promyshlennogo proizvodstva.* Moscow, 1882.

Skvortsov, A. I. *Russkaia narodnaia propil'naia rez'ba.* Leningrad, 1984.

Sobolev, N. *Naboika v Rossii.* Moscow, 1912.

Russkaia narodnaia rez'ba po derevu. Moscow-Leningrad, 1934.

Russkii ornament: Kamen', derevo, keramika, zheleza, stenopis, naboika. Moscow, 1948.

Sobolevskii, N. *Khudozhestvennye promysli Podmoskov'a.* Moscow, 1948.

Sobranie russkoi stariny kniaginei V. P. Sidamon-Eristovoi i N. P. Shabel'skoi. Vyshivki i kruzheva, vol. 1. Moscow, 1910.

Sokolov, S., and I. Tomskii. *Narodnoe iskusstvo severa Rossii.* Introduction by Igor Grabar. Moscow, 1924.

Sokolov, V. I. *Narodnaia rez'ba po derevu.* Moscow, 1947.

Stasov, V. V. *Russkii narodnyi ornament,* vol. 1: *Shit'e, tkani, kruzheva.* St. Petersburg, 1872.

Trenev, D. K., and N. P. Syreishchikov. *Ornamenty drevnerusskogo iskusstva,* vol. 1. Moscow, 1904.

Vasilenko, V. M. *Narodnoe iskusstvo. Izbrannye trudy o narodnom tvorchestve X–XX vekov.* Moscow, 1974.

Vishnevskaia, V. M. *Russkaia narodnaia rez'ba i rospis po derevu.* Moscow, 1956.

D. WORKS ON ABRAMTSEVO, SOLOMENKO, SERGIEV POSAD, AND TALASHKINO

i. Abramtsevo

A. S. [A. Sementsev]. *Elena Dmitrievna Polenova. 1850–1898.* Moscow, 1902.

"Abramtsevo S. I. Mamontova." *Iskra,* 1 (1916), 8.

Arenzon, E. P. "Ot Kireeva do Abramtseva." *Panorama iskusstv 6,* 359–82. Moscow, 1983.

Arzumanova, O. I., et al. *Muzei-zapovednik Abramtsevo.* Moscow, 1984.

Beloglazova, N. M. *Abramtsevo. Gosudarstvennyi istoriko-khudozhestvennyi i literaturnyi muzei-zapovednik.* Moscow, 1981.

"Pervaia russkaia khudozhnitsa—skazochnitsa E. D. Polenova." *Detskaia literatura,* 1 (1968), 59–61.

Polenov i E. Polenova v Abramtseve. Leningrad, 1980.

Bocharov, G. N. "Deiatel'nost' khudozhnikov po vozrozhdeniiu narodnogo iskusstva. Abramtsevo i Talashkino." *Russkaia khudozhestvennaia kul'tura 1895–1907,* vol. 1, 341–52. Moscow, 1968.

Borisova, E. A. "Arkhitektura v tvorchestve khudozhnikov Abramtsevskogo kruzhka." *Khudozhestvennye protsessy v russkoi kul'ture vtoroi poloviny XIX veka,* 137–82. Moscow, 1984.

Borok, N. [N. V. Polenova]. "E. D. Polenova." *Mir iskusstva,* 18–19 (1899), 97–120.

Dain, Galina. "Istoriia i problemy promysla: Abramtsevo-Kudrinskaia rez'ba." *Dekorativnoe iskusstvo SSSR,* 6/355 (1987), 40–2.

"E. D. Polenova." *Niva,* 16 (1903), 308–11.

Galueva, Evgeniia. "'Prianichnaia' rez'ba—trudnoe delo." *Dekorativnoe iskusstvo SSSR,* 10/299 (1982), 18–21.

Grover, S. R. *Savva Mamontov and the Mamontov Circle, 1870–1905: Art Patronage and the Rise of Nationalism in Russian Art.* Ann Arbor, Mich., 1971.

Iunge, E. F. "Vospominaniia ob E. D. Polenove." *Russkaia starina,* 150 (June 1912), 536–45.
Jacobs, M. "In the Nest of the Gentry (Abramtsevo. Russia)." *The Good and Simple Life,* 111–18. Oxford, 1985.
Kiseleva, E. G. *Dom na Sadovoi.* Moscow, 1986.
Kopshitser, M. I. *Mamontov.* Moscow, 1972.
Lazarevskii, I. "Iskusstvo i zhizn'. (Posmertnaia vystavka E. D. Polenovoi)." *Arkhitekturnyi muzei,* 1 (1903), 9–13.
Lobanov, V. *Viktor Vasnetov v Abramtseve.* Moscow, 1928.
Makarov, K. "Keramika Vrubelia." *Dekorativnoe iskusstvo SSSR,* 12/289 (1981), 31–4.
Mamontov, S. S. "Mamontovskii kruzhok." *Stolitsa i usad'ba,* 23 (1914) 4–9.
Mamontov, V. S. *Vospominaniia o russkikh khudozhnikakh. (Abramtsevskii khudozhestvennyi kruzhok).* Moscow, 1950.
Manin, V. *Sovetskiie khudozhniki v Abramtseve.* Moscow, 1976.
Masalina, N. V. "Tserkov' v Abramtseve. (K istorii postroiki)." *Iz istorii russkogo iskusstva vtoroi poloviny XIX–nachala XX vekov,* 47–58. Moscow, 1978.
Osugin, G. "Zdes' russkikh dukh . . . K 125-letiiu so dnia rozhdeniia S. I. Mamontova." *Literaturnaia gazeta,* 150 (20 December 1966), 2.
Pakhomov, N. I. *Abramtsevo.* Moscow, 1939.
Dom-muzei Abramtsevo. Moscow, 1978.
Pakhomov, N. I., ed. *M. A. Vrubel 1856–1910. Vystavka maiolikovykh rabot i teatral'nykh eskizov k stoletiiu so dnia rozhdeniia.* Moscow, 1957.
Paston, Eleonora. "100 let nazad v Abramtseve." *Dekorativnoe iskusstvo SSSR,* 1/254 (1979), 34–40.
Peacock, Netta. "Dan' pamiati Eleny Polenovoi." *Iskusstvo i khudozhestvennaia promyshlennost',* 18 (1900), 411–13.
"Das russischer Dorf auf der Pariser Weltausstellung." *Dekorative Kunst,* 6 (September 1900), 480–8.
"The New Movement in Russian Decorative Art." *Studio,* 13 (May 1901), 268–76.
Polenova, N. V. *Abramtsevo. Vospominaniia.* Moscow, 1922.
Sakharova, E. V. *Elena Dmitrievna Polenova. 1850–1952.* Moscow, 1952.
Sakharova, E. V., ed. *V. D. Polenov. E. D. Polenova. Khronika sem'i khudozhnikov.* Moscow, 1964.
Samarin, I. "V. A. Serov v Abramtseve." *Iskusstvo,* 3 (1965), 63–6.
Smirnov, L. P. *Kustarnoe delo v Abramtseve.* Moscow, 1928.
Stasov, V. V. "E. D. Polenova." *Iskusstvo i khudozhestvennaia promyshlennost',* 13 (October 1899), 3–49.
Sternin, G. Iu. "Abramtsevo: Ot yusad'by' k ydache,'" *Russkaia khudozhestvennaia kul'tura vtoroi poloviny XIX–nachala XX vekov,* 184–208. Moscow, 1984.
Suprin, Liudmila. "K 125-letiiu E. D. Polenovoi." *Dekorativnoe iskusstvo SSSR,* 6/223 (1976), 38–40.

ii. Solomenko

[Bartram, N.]. "Pamiati N. Ia. Davydovoi." *Izvestiia,* 20 April 1926.
Mikhailov, V. *Khudozhniki tambovskogo kraia. Istoricheskii ocherk razvitiia izobrazitel'nogo iskusstva na Tambovshchine.* Leningrad, 1976.

"Nataliia Iakovlevna Davydova." *Vestnik promyslovoi kooperatsii,* 5/34 (1926), 75.
Peacock, Netta, "A Log House Dining Room in Russia." *The Artist,* 24 (January–April 1899), 1–7.
Salmond, Wendy. "The Solomenko Embroidery Workshops." *Journal of Decorative and Propaganda Arts,* 5 (Summer 1987), 126–43.

iii. The Moscow Kustar Museum and Sergiev Posad

Bartram, N. D. *Igrushka—radost' detei.* Moscow, 1912.
"O vozmozhnosti vozrozhdeniia v igrushke narodnogo tvorchestva." *Apolon,* 2 (1912), 55–64.
Bartram, N. D., ed. *Igrushka. Ee istoriia i znachenie. Sbornik statei.* Moscow, 1912.
Bartram, N. D., and I. E. Ovchinnikova, eds. *Muzei igrushki.* Leningrad, 1928.
Benua, A. N. "Igrushki." *Apolon,* 2 (1912), 49–54.
"Russkie narodnye igrushki v otkrytkakh." *Otkrytoe pis'mo,* 1 (1904–5), 200–10.
Bill, Valentine T. "The Morozovs." *Russian Review,* 14, no. 2 (April 1955), 109–16.
Bocharov, G. N. "Moskovskii kustarnyi muzei i deiatel'nost' N. D. Bartrama." *Russkaia khudozhestvennaia kul'tura kontsa XIX–nachala XX vekov (1908–1917),* vol. 4, 462–72. Moscow, 1980.
Evdokimov, I. *Russkaia igrushka.* Moscow, 1925.
Istoricheskii ocherk ekonomicheskikh meropriiatii moskovskogo gubernskogo zemstva. Moscow, 1895.
Iunitskii, P. *Moskovskii kustarnyi muzei.* Moscow, 1911.
Ivanova, N. "O sozdanii muzeia narodnogo iskusstva." *Sbornik trudov NIIKhP,* 5 (Moscow, 1972), 7–19.
Kustarnoe delo moskovskogo zemstva. Ukazatel' moskovskogo otdela II-oi Vserossiiskoi kustarnoi vystavki v S.-Peterburge 1913 goda. Moscow, 1913.
"Kustarnoe tsarstvo v Sergievskom posade." *Niva,* 39 (1907), 640–4.
"Kustarnye kooperativy Sergieva posada." *Vestnik kustarnoi promyshlennosti,* 1 (1915), 32–4.
Mozhaeva, E., and A. Kheifii. *Matreshka. Russkii suvenir.* Moscow, 1969.
N. D. Bartram. *Izbrannye stat'i. Vospominaniia o khudozhnike.* Moscow, 1979.
Orlov, A. S. *Kustarnaia promyshlennost' moskovskoi gubernii i sodeistvie kustariam so storony zemstva, raznykh uchrezhdenii i chastnykh lits.* Moscow, 1913.
Ovchinnikova, E. S. "N. D. Bartram." *Sbornik trudov NIIKhP,* 5 (Moscow, 1972), 65–72.
Rozova, L. K. "Rabota muzeia [narodnogo iskusstva] po propagande narodnogo iskusstva." *Sbornik trudov NIIKhP,* 5 (Moscow, 1972), 36–48.
Salmond, Wendy. "Reviving Folk Art in Russia: The Moscow Zemstvo and the Kustar Art Industries." In *Art and the National Dream: The Search for Vernacular Expression in Turn-of-the-Century Design,* edited by Nicola Gordon Bowe, 81–98. Dublin, 1993.
"Studio Talk. Moscow." *Studio,* 45, no. 180 (February 1912), 333–5.
Tseretelli, N. M. *Russkaia krest'ianskaia igrushka.* Leningrad, 1933.
Vvedenskii, D. I. *Gorod Sergiev, Moskovskaia gubernia. Mel'kie promysli: opyt statistiko-ekonomicheskogo issledovaniia.* Moscow, 1927.
U sergievskogo igrushechnika. Moscow, 1926.
Zviagintseva, E. A., et al., eds. *Moskva. Putevoditel'.* Moscow, 1915.

iv. Talashkino

Abramova, A. *Talashkino.* Smolensk, 1950.
Zhizn' khudozhnika, Sergeia Maliutina. Moscow, 1978.
Belogortsev, I. "I. E. Repin v Smolenske." *Rabochii put',* 152 (5 August 1949).
"Talashkino." *Smolenskii almanakh* 4. Smolensk, 1948.
Breshko-Breshkovskii, N. "V skazochnom tsarstve." *Novyi mir,* 23 (1905), 266–8.
Cain, K. R. "Talachino: A Home for Russian Folk Art." *The Craftsman,* 27, no. 1 (October 1914), 92–6.
Danilovich, C. de. "Art décoratif russe ancien. Collection de la Princesse M. Tenichef." *L'Art décoratif,* 17 (April 1907), 136–45.
"Exposition d'art russe moderne à Paris," *L'Art et les artistes,* 34 (January 1908), 480–4.
"Snegourotchka à l'Opéra Comique," *L'Art décoratif,* 18 (June 1908), 226–31.
"Talashkino: Princess Tenisheva's School of Russian Applied Art." *International Studio,* 32, no. 126 (August 1907), 135–9.
Diagilev, S. P. "Neskol'ko slov o S. V. Maliutine." *Mir iskusstva,* 4 (1903), 157–60.
Exposition d'art russe moderne. Paris, 1907.
Frolov, A. "Tvorcheskoe sodruzhestvo N. Rerikha i V. Frolova." *Iskusstvo,* 8 (1970), 62–8.
Golynets, G. *S. V. Maliutin.* Leningrad, 1974.
Jamot, P. "Le Salon d'Automne: L'exposition de l'art russe." *Gazette des beaux arts,* 37 (1 January 1907), 43–53.
Kudriavtseva, T. "Bufet po proektu S. V. Maliutina raboty Talashkinskikh masterskikh." *Soobshcheniia Gosudarstvennogo Ermitazha,* 51 (1986), 25–7.
Makovskii, S. K. *Talashkino. Sbornik statei.* Introduction by N. K. Rerikh. St. Petersburg, 1905. (Published in French translation as S. K. Makovsky, *Talachkino. L'art décoratif des ateliers de la Princesse Tenichef* [St. Petersburg, 1906]).
Mourey, Gabriel. "L'Art populaire russe." *Art et décoration,* 14 (August 1903), 237–46.
Objets d'art russes anciens, faisants partie des collections de la Princesse Marie Tenichev exposés au Musée des Arts Décoratifs du 10 mai au 10 octobre 1907. Paris, 1907.
Osokin, V. N., et al. *Khudozhniki zemli smolenskoi.* Leningrad, 1967.
Pastukhova, Z. I. "Talashkino—pamiatnik narodnym talantam." *Po smolenshchine,* 77–85. Moscow, 1985.
Rerikh, N. K. "Zakliatoe zver'e. (Emali kn. M. K. Tenishevoi)." *Niva,* 18 (1909), 310–11, 338.
Roche, D. *Broderies des paysannes de Smolensk executées sous la direction de la princesse Marie Tenichef.* Paris and Chicago, n.d.
"Une exposition d'art russe à Paris." *Art et décoration,* 22 (1907), 173–90.
"L'Exposition d'art russe ancien au Musée des Arts Décoratifs," *Gazette des beaux arts,* 38 (1 September 1907), 250–64.
"Un milieu d'art russe Talashkino." *Revue de l'art ancien et moderne,* 7 (1907), 223–7.
Rogozina, M. G. "Kollektsiia fotografii I. F. Barshchevskogo." *Pamiatniki otechestva,* 2 (1983), 153–6.
Rybchenkov, B., and A. Chaplin. *Talashkino.* Moscow, 1973.
Sargent, Irene, trans. "The Racial Art of the Russians." *The Craftsman,* 5, no. 1 (October 1903), 43–51.

Sazonova, K. K. "S. V. Maliutin v Talashkine." *Problemy razvitiia russkogo iskusstva,* 3 (1972), 80–9.
Sbornik. Khudozhestvennye promysly RSFSR, 53–57. Moscow, 1950.
Serebrennikov, A. P. *Kratkii istoricheskii ocherk gosudarstvennogo istoriko-etnograficheskogo muzeia v Smolenske, osnovennogo Tenishevoi.* Moscow, 1926.
Suprun, L. Ia. "Grani talanta. O dekorativno-prikladnykh rabotakh S. V. Maliutina." *Iskusstvo,* 4 (1971), 38–43.
"Neopublikovannye dokumenty S. V. Maliutina." *Ocherki po russkomu i sovetskomu iskusstvu,* 344–53. Leningrad, 1974.
"Talashkino." *Studio,* 32 (October 1907), 328–30.
Tarovatyi, N. "Muzei russkoi khudozhestvennoi stariny kn. Tenishevoi v Smolenske." *Iskusstvo,* 5–7 (1905), 156–67.
Tenishev, V. N. *Programma etnograficheskikh svedenii o krest'ianakh tsentral'noi Rossii,* 2d ed. Smolensk, 1898.
Tenisheva, M. K. *Vpechatleniia moei zhizni.* Paris, 1933; reprint, St. Petersburg, 1991.
Zelinskii, A. "Zabytyi pamiatnik russkogo iskusstva." *Iskusstvo,* 3 (1962), 61–2.
Zhuravleva, L. "Dalos' mne eto ne bez raboty." *Prometei,* 1 (1987), 65–79.
"Kak vspominaetsia Talashkino." *Dekorativnoe iskusstvo SSSR,* 3 (1981), 27–31.
"Kollektsiia rabot A. P. Zinov'ieva v sobraniiakh smolenskogo muzeia zapovednika i muzeia narodnogo iskusstva v Moskve." *Pamiatniki kul'tury. Novye otkrytiia 1982,* 469–77. Leningrad, 1984.
"M. A. Vrubel v Talashkine." *Muzei 2,* 57–61. Moscow, 1981.
"Redkie fotografii." *Panorama iskusstv 3,* 367–73. Moscow, 1980.
Russkoe narodnoe reznoe i raspisnoe derevo v sobranii Smolenskogo gosudarstvennogo ob'edinennogo istoricheskogo i arkhitekturno-khudozhestvennogo muzeia-zapovednika. Moscow, 1985.
"K stoletiiu so dnia rozhdeniia M. K. Tenishevoi." *Materialy po izucheniiu smolenskoi oblasti,* 7 (Moscow, 1970), 360–74.
"Talashkinskie balalaiki." *Dekorativnoe iskusstvo SSSR,* 12 (1981), 46–7.
Teremok. Moscow, 1974.

E. KUSTAR INDUSTRIES IN THE SOVIET PERIOD

L'Art décoratif et industriel de L'URSS. Edition du comité de la section de l'URSS à l'Exposition internationale des arts décoratifs. Paris, 1925, 27–47. Moscow, 1925.
Bakushinksii, A. V. *Iskusstvo Palekha.* Moscow-Leningrad, 1934.
Feigin, V. *Kustarnaia promyshlennost' SSSR.* Moscow, 1927.
Guliaev, V. A. *Russkie khudozhestvennye promysly 1920-kh godov.* Moscow, 1983.
Khant, David. "Sovetskii Niurenberg. Ocherk." *Krasnaia Nov',* 7 (1932), 149–56.
Khudozhestvennye promysly RSFSR. Spravochnik. Moscow, 1973.
Lunacharskii, A. V. *Stat'i ob iskusstve.* Moscow-Leningrad, 1941.
Malyshev, S. M., ed. *Spravochnik kustarnoi promyshlennosti goroda Moskvy i moskovskoi gubernii.* Moscow, 1927.
Maslenitsyn, S. I. "Polozhitel'nyi opyt raboty khudozhnika-professionala s masterami narodnykh promyslov." *Problemy narodnogo iskusstva,* 82–93. Moscow, 1982.

Maslennikov, N. N., and A. V. Bakushchinskii. *Russkie khudozhestvennye laki. Khudozhestvennye arteli: Palekh i drugie.* Moscow, 1933.

Narodnoe iskusstvo SSSR v khudozhestvennykh promyslakh, vol. 1. Moscow-Leningrad, 1940.

Nekrasova, M. A., and S. M. Temerin, eds. *Sovetskoe dekorativnoe iskusstvo. Materialy i dokumenty 1917–1932. Narodnye khudozhestvennye promysly,* 2 vols. Moscow, 1986.

Orshanskii, L. G. *Khudozhestvennaia i kustarnaia promyshlennost' SSSR 1917–1927.* Leningrad, 1927.

P. "Krest'ianskoe tvorchestvo Ukrainy." *Iskusstvo,* 2–3 (1923), 115–16.

Pervaia sel'skokhoziaistvennaia i kustarno-promyshlennaia vystavka SSSR, 1923. Sputnik po vystavke. Moscow, 1923.

Pervaia vserossiiskaia konferentsiia po khudozhestvennoi promyshlennosti. Avgust, 1919. Moscow, 1920.

Petrov, G. *Promyslovaia kooperatsiia i kustar',* 2d ed. Moscow, [1920].

Prokopev, D. V. *Khudozhestvennye promysli Gorkovskoi oblasti.* Gorky, 1939.

Rybnikov, A. A. *Mel'kaia promyshlennost' i ee rol' v vosstanovlenii russkogo narodnogo khoziaistva.* Moscow, 1922.

Shatrov, M. N. *Kirovskie kustari.* Kirov, 1938.

Sheremeteva, M. E. "Tarusskaia artel' vyshival'shchits." *Sovetskaia etnografiia,* 2 (1953), 179–88.

Soviet Russian Art and Handicraft Exposition. Exhibition Catalogue. Grand Central Palace, New York, February 1929. New York, 1929.

Vaulin, P. K. "K voprosu ob iskusstve v promyshlennosti." *Iskusstvo kommuny,* 13 (2 March 1919), 2.

Volkova, M., and V. Vishnevskaia. "Perestroika v promyslakh ne sdelala i polshaga." *Dekorativnoe iskusstvo SSSR,* 10/359 (1987), 13–14.

Voronov, V. S. *O krest'ianskom iskusstve.* Moscow, 1972.

Vsekopromsoiuz. *Al'bom pletenykh kruzhev. Raiony: Vologodskii, Eletskii, Riazanskii.* Moscow, 1928.

[Vserossiiskii soiuz promyslovoi kooperatsii]. *Kustarno-promyslovaia kooperatsiia v sisteme narodnogo khoziaistva SSSR. Sbornik materialov k XIV s'ezda RKPSb.* Moscow, 1925.

[Vsesoiuznaia konferentsiia po problemam razvitiia sovremennogo narodnogo iskusstva v svete postanovleniia TsKPSS]. *O narodnykh khudozhestvennykh promyslakh.* Moscow, 1976.

Vyshivka, tkachestvo. Fotoal'bom obraztsov eksportnogo stolovogo bel'ia. Moscow, 1934–5.

Index

Page numbers in italic type indicate illustrations.

Abramtsevo. *See also* Polenova, Elena; Mamontova, Elizaveta Grigorievna; Store of Russian Works
carpentry workshop 23–4, 66, 144–5, 188, 193, 194, 197, 204, 208, 218n. 62, 247n. 4; commercial activities of 35–9, 194, 228n. 22; critical reaction to 44–5, 78, 99–100, 105, 115, 117, 167, 177, 182, 218n. 63; designs for 26–8, 30–5, 39–45, 66, 174, 202, 219n. 80; and exhibitions 43, 95, 171, *175*, 199, 228n. 62, 240n. 1, 247n. 4; goals of 21–2, 43–5, 120, 218n. 55; graduates of 43, 88, 195, 203
estate 21–2, 159, 188, 191, 192; ceramics workshop at 39, 69, 217n. 38; church at 23, *24*, 34, 39, 193; girls workshop at 27; museum of folk art at *25*, 193, 218n. 58; studio at 22, *23*
Abramtsevo Ceramics Factory 66, 170, 195, 200, 216n. 26
Alexandra Fedorovna, tsarina 54, 95, 170, 201
All-Russian Arts and Industries Exhibition: (Moscow, 1882) 52, 82, 113, 194, 220n. 2; (Nizhnii Novgorod, 1896) 43, 57–9, 85, 98, 119, 198, 225n. 64, 229n. 35; (Kiev, 1913) 114, 177, 207
All-Russian Congress of Russian Artists: second (St. Petersburg, 1911–12) 167–8, 206, 243n. 80
All-Russian Congress of Workers in the Kustar Industries: first (1902) 93–4, 146, 200, 230n. 45; second (1910) 146–7, 171, 205, 226n. 79; third (1913) 169
All-Russian Kustar Exhibition: first (St. Petersburg, 1902) 43, 93–6, 154, 175, 200, 224n. 58, 229nn. 36, 44, 230n. 45; second (St. Petersburg, 1913) 12, 169–77, 207, 244nn. 85, 89
All-Russian Society against Luxury and Extravagance 178
All-Russian Union of Zemstva 183, 246n. 129
Andreev's Great-Russian Balalaika Orchestra 117, 234n. 8
Andrianov, S. N. 121
Antokolskii, Mark 60
Apollo (*Apolon,* magazine) 108, 132
Archipenko, A. 181
Art and Industrial Art (*Iskusstvo i khudozhestvennaia promyshlennost',* magazine) 69, 70, 198, 225n. 70
artels 114, 205, 207, 208, 209, 215n. 14, 218n. 63, 227n. 14, 232n. 97
Averintsev, I. 209
Avinova, M. Iu. 203

Bakhrushin Brothers Trade School 204
Baksheev, Vasilii 130
Bakst, Lev 71, 141
Ballets Russes 13, 141, 161, 225n. 71
Barshchevskii, Ivan 50, 133, 134, 138, 140, 198, 234n. 16, 236n. 42, 238nn. 65, 79, 239n. 90
Bartram, Nikolai 87, 104–9, *107,* 113, 146, 155, 169, 188,196, 201, 203, 208, 210, 231nn. 75, 79, 247n. 6
Baumgarten, E. 78, 100
beading 56, 223n. 39
Beketov, V. V. 131, 140, 142, 238n. 79
Bem, Elizaveta 61, 95, 167, 224n. 50, 241n. 21, 243n. 79

Benois, Aleksandr 71, 96, 105, 108, 115, 118, 124, 141, 147, 199, 225n. 70, 228n. 18, 230n. 61, 233n. 5
Bilibin, Ivan 88, 141, 147, 154, 168, 176, 199, 202, 204, 230n. 61, 238n. 82
Bobrinskaia, Countess S. A. 202
Bogorodskoe (Vladimir province) 10, 106–8, 146, 169, 207
Boguslavskaia, Ksenia 179, 180, *181*, 246n. 124
Bol'shoe Krasnoe training workshop (Kostroma province) 101
Bondarenko, Ivan 97
Borovenets naboika workshop (Novgorod province) 72
Borutskii, Vladimir 99, 230n. 58, 232n. 97, 243n. 58
Botkin, M. P. 95
Brailovskii, Lev 99
Briagin, N. I. 150
Brik, Osip 209
Buslaev, Fedor 220n. 9
Butovskii, Viktor 28, *29*, 30, 191, 217n. 41, 220n. 9, 221n. 18
Bykovo-Gora workshops (Tambov province) 47, 165, 171, 194

carpets 151–2, 153–4, 159–60, 173, 174, 206, 245n. 98
Cathedral of St. Vladimir (Kiev) 41, 159, 194, 217n. 37
Chekhonin, Sergei 168, 170, 207, 244n. 92
Chekhov, Anton 87, 224n. 62, 228n. 25
Cherchenko, V. I. 157, *157*
Chief Administration of Agriculture and Land Tenure (G.U.Z.i Z., 1905–17) 112–13, 126, 145, 147, 149, 151, 155, 166, 169, 170, 171, 172, 175, 176, 203, 206, 207
Children's Education (toystore) 37, 86, 106, 194, 228n. 22
Chistiakov, Pavel 24–5, 158
Chokolova, Ekaterina 51, 65–6, 96, 197, 199, 224n. 58
Congress of Trade and Industry (Nizhnii Novgorod, 1896) 58
Constructivism 13, 237n. 57
cooperative movement. *See* artels

Dal', Lev 220n. 9
Dal', Vladimir *30*, 192
Danilovich, C. de 142
Darmstadt Art Colony 97, 230n. 52
Davydova, Natalia Iakovlevna 55–6, *56*, 64–5, *65*, 72, 74, 96, 97, 99, *99*, 117, 121, 145, 155, 174, 188, 189, 199, 204, 208, 210, 219n. 80, 222n. 36, 224n. 62, 244n. 89
Davydova, Natalia Mikhailovna 158, 169, 179, 181, 182, 184, 189, 203, 222n. 36, 242n. 45
Davydova, Sofia 48–9, 193, 194, 196, 220n. 10, 244n. 89
Degtiar School for Weaving Instructors (Poltava province) 151, 157
Diagilev, Iurii 72, 226n. 79
Diagilev, Sergei 70–2, 78, 115, 122–4, 140–2, 161, 198, 203, 219n. 67, 225nn. 66, 71, 75, 233n. 2, 235n. 25, 240n. 97
Ditrikh, Maria 168
D. M. C. Press 183, 246n. 130
Dobuzhinskii, Mstislav 168
Dobychina Gallery (Petrograd) 180
Dolgoruka, Princess S. P. 47, 193, 203
Dolivo-Dobrovol'skaia, A. G. 151–2, 159, 244n. 89
Dolivo-Dobrovol'skaia, V. A. 194
Doré Gallery (London) 103, 202
Durnovo, A. N. 154–5, *155*, 162, 171, 203, 226n. 79
dyes 55, 128–9, 153, 159–60, 236nn. 40, 42, 43

Egorov, Evdokim 26, 217n. 34
Egorova, Maria 197, 217n. 34, 223n. 43
Ekaterinburg School of Industrial Art 101, 243n. 80
Elenovka workshop (Kiev province) 159, 236n. 48
Elizaveta Fedorovna, Grand Duchess: and famine relief 54; fate of 188; kustar outlet of 54, 125, 178, 196, 222n. 28, 235n. 31; as patron 74, 90, 97, 159, 199, 200, 229n. 34, 230n. 52
Emancipation Act (February 1861) 16–17, 173–4, 245n. 103
embroidery 46–8, 51, 54, 128–9, 133, 141, 152, 169, 178, 221n. 19
Empire style 110, 173–4, *175*
Erenburg, Natalia 244n. 84
Exhibition of Architecture and Industrial Art in the New Style (Moscow, 1902–3) 78, 93, 97–100, 201, 247n. 4
Exhibition of Contemporary Decorative Art (Moscow, 1917) 182–3
Exhibition of Russian and Finnish Artists (1898) 198, 235n. 25
Exposition d'Anvers (Antwerp, 1894) 57, 197, 223n. 42

Exposition des Arts Décoratifs (Paris, 1925) 13, 189, 210, 242n. 55, 247n. 8
Exposition d'Art Russe Moderne (Paris, 1907) 140, 141, 204, 239nn. 90, 96
Exposition Universelle: (Paris, 1867) 191; (Vienna, 1873) 192; (Paris, 1889) 37, 74, 195; (Paris, 1900) 70, 74–7, 85, 90–3, 117, 154, 199, 228n. 26, 235n. 30; (Paris, 1904) 102; (Milan, 1906) 102, 154, 203; (Turin, 1911) 206
Exter, Aleksandra 158, 160–1, 168, 179, 180, 184–5, 189

Fabergé firm 178, 237n. 55
factory industry: compared with kustar industry 17–18, 83–4, 96, 104, 143, 214n. 5, 227n. 5; effects on peasant life of 27, 47–8, 51–2, 108, 214n. 5, 215n. 23; growth of 57–8, 81–2; training for 231n. 62
famine (1891–2) 53–4, 90, 195, 196
Filippov, N. A. 244n. 89
Fire Bird motif 59–61, *60*, 68, 75, 122, 126, 225n. 64, 235nn. 23, 34
First All-Russian Conference on Industrial Art (1919) 209
First Exhibition of the Contemporary Decorative Art of Southern Russia (1915–16) 179–80, 208
Flenovo. *See* Talashkino
Fokine, Mikhail 169
folk art: collecting of 29–30, 34–5, 217n. 45, 231n. 75; decline of 51–2; definitions of 173, 174–6, 184–6; exhibitions of 82, 168–9; influence of 243n. 62; as source for kustar art 27, 28–30, 84, 113–14, 122, 126, 133, 166–7, 168–9, 171, 173; study of 48–9, 82, 217n. 40
Fomin, Ivan 97–9, *98*, 230n. 51

GAKhN (State Academy of Arts and Sciences) 187, 209
Gallen-Kallela, Axel 123–4, 235n. 25
Gal'nbek, N. A. 50, 202, 230n. 61
Gartman, Viktor 22, *23*, 29, 192, 221n. 18
Genke, Nina 179
Glavkustprom 187, 188, 210
Globa, Nikolai 131, 209
Gogol School of Industrial Art (Mirgorod) 101, 157, 198, 247n. 3
Golitsyn, Prince F. S. 244n. 89
Golitsyn, Prince S. M. 145, 200
Golitsyna, Princess L. V. 204
Goloushev, Sergei 106
Golovin, Aleksandr 66–9, 72, 74, 78, 90, 97, 117, 141, 198, 199, 224n. 60
Golubev, V. V. 134, 238n. 69
Goncharova, Natalia 168
Gorkii, Maksim 52
Gorozhanka workshop (Voronezh province) 51, 66, 72, 171, 197
Grabar, Igor 55, 69, 77, 79
Gudim-Levkovich, Iu. N. 159
Guro, Elena 77–8
G.U.Z.i Z. *See* Chief Administration of Agriculture and Land Tenure

Iakimchenko, A. G. 110, 225n. 64
Iakunchikova, Maria Fedorovna 46, 54–7, 59, 66, 74, 76, 77, 145, 152, 155, 174, 189, 196, 198, 204, 205, 210, 211, 222nn. 31, 32, 34, 35, 223n. 39, 224n. 62, 225nn. 64, 74, 244n. 89
Iakunchikova-Weber, Maria Vasil'evna 55, 62, 70, 72–3, *73*, 74–6, *75*, 88, 92, 121, 197, 199, 200, 202, 211, 222n. 31, 223n. 43, 224n. 60, 225nn. 74, 75, 228n. 26, 229n. 41, 237n. 50
Iashvil, Princess Natalia 158–9, 169, 242n. 48
icons 106–7, 150–1, 168, 169, 199, 206
Ideal Home Exhibition (London, 1913) 154, 162, 207
Imperial Russian Geographical Society 19, 192, 193
Imperial Women's Patriotic Society 149
Institute of Industrial Art and Kustar Industry 210
Iunge, Ekaterina 63
Ivanov, Sergei 65, 88
IZO Narkompros 187, 188, 208, 209, 247n. 2

Kablukov, Nikolai 161
Kandinskii, Vasilii 88, 129
Karpovich, Vladimir 98
Kashkarova, Zinaida 106
Kaznachaeva, S. P. 47, 193, 220n. 2
Kekushev, Lev 97, 230n. 51
Khanenko, Varvara 158, 159, 169, 236n. 48, 244n. 89
Khanykova, L. P. 196
Khokhloma ware. *See* Semenov ware
Khotkovo woodworking artel (Moscow province) 218n. 63
Kiev Kustar Society 157–61, 179, 204, 205, 207
Kiev Museum of History and Antiquities 159
Kiev School of Art 160
Kliuchevskii, Vasilii 34, 218n. 49
Kliun, Ivan 181

Konenkov, Sergei 106
Korin, Aleksandr 130
Korovin, Konstantin 55, 58, 74, 78, 88, 90, 91–2, 95, 97, 98, 113, 119, 141, 154, 197, 199, 229nn. 34, 35, 230n. 51
Koustari Russes (Paris) 102–3, 203
Krichevskii, V. G. 156, *156*
Kudrino woodworking workshop (Moscow province) 171, 203, 209, 218n. 63
Kulbin, Nikolai 168, 244n. 85
kustar industries: and art education 100–1, 105, 145–50, 171–2, 231n. 62; in Central Asia 172–3, 245n. 98; decline of 16–20; definition of 15–16, 94–5, 213n. 1, 214nn. 1, 2, 4, 230n. 45; exports of 78, 85, 89, 102–4, 161–5, 188, 200, 243nn. 58, 68, 245n. 114; government support of 80–2, 144–5, 169, 172–3, 218n. 55; importance of 19–20, 47–8; income from 111–12, 220n. 2, 236n. 48, 245n. 102; markets for 84, 95, 104–5, 161–2; and the middleman 17, 36, 49, 83, 126; in the Soviet era 187–9, 208–10; statistics on 20, 81, 113, 176, 229n. 33, 230n. 45, 232nn. 88, 91, 236n. 48, 240n. 3, 245n. 109; and World War I 177–9, 183. *See also* factory industry; folk art; women's crafts
Kustar Museum (St. Petersburg) 72, 95, 126, 147, 154, 196, 226n. 79

lace 7, 48, 49–51, *50*, 54, 164, 168, 193, 197, 203, 221nn. 14, 18, 19
Lalique, René 72, 199
Lamanova, Nadezhda 179, 189, 209, 242n. 55
Lange, Boris 106
Larionov, Mikhail 168
Last Futurist Exhibition of Paintings 0.10 (Petrograd, 1915) 180
Lemercier Gallery (Moscow) 179, 208
Lentulov, Georgii 179
Levitan, Isaak 65, 87, 227n. 13, 228n. 25
Ligachev (Moscow province) 231n. 64; drawing school 101, 201; School for Instructors in Art Joinery 2, *4*, 149, 171, 206
Lindeman, Agnessa 106, 147, 231n. 77
lubki 53, 87, 106–7
Lugovskaia-Diagileva, Tatiana 72, 226n. 80
Lukomskii, Georgii 13, 167, 177, 244n. 84
Lunacharskii, Anatolii 209
L'vova, Princess A. V. 47, 194, 223n. 41
L'vova, Princess M. A. 196
Lyskovo workshop (Nizhnii Novgorod province) *155*

Mackintosh, Charles Rennie 3, *5*, 97
Mackintosh, Margaret 230n. 53
Maiakovskii, Vladimir 167, 180, 183
Makovskii, Konstantin 52
Makovskii, Sergei 131–3, 202, 237nn. 58, 59
Maleev, Egor 152–3, *153*
Malevich, Kazimir 159, 168, 179, 180
Maliavin, Filipp 152, 241n. 27
Maliutin, Sergei 86, *86*, 92, 110, *111*, 115, 117, 118–24, *121*, *123*, 130–1, 155, 189, 199, 200, 201, 228n. 22, 231n. 85, 234nn. 12, 13, 16, 17, 235n. 29, 237n. 53
Mamontov, Anatolii Ivanovich 119, 211, 228n. 22, 232n. 92
Mamontov, Savva Ivanovich 21, 26, 57–8, 66, 71–2, 90, 100, 116, 117, 118–19, 192, 197, 198, 199, 208, 211, 217n. 38, 227n. 13, 229n. 35
Mamontova, Elizaveta Grigorievna (née Sapozhnikova) 21–4, 26–7, 35, 36, 37, 38, 41–2, 43, 57, 83, 192, 193, 204, 211, 240n. 1
Mamontova, Maria 37, 86, 119, 194, 211, 228n. 22, 232n. 92
Mamontov circle 76, 193, 194, 195, 216n. 26, 233n. 2
Mamontov Private Opera Company 21, 39, 119, 194, 217n. 38, 218n. 65, 234n. 12, 235n. 20
Mariinskii Lace School (St. Petersburg): creation and curriculum of 49–51, 151, 194, 221n. 17; evacuation of 208; exhibitions involving 171, 195, 204; teachers trained at 55, 66, 72, 153, 174, 196, 197, 245n. 104
Matiushin, Mikhail 181
matreshka doll 10, 85–7, 92, 103, 119, 228n. 21, 232n. 87
Matveev, Boris 91–2
Men'kov, M. I. 181
Menter, Sofi 115
Ministry of Agriculture and State Domains (M.Z.i G.I.) 93, 100, 101, 154, 197, 198, 200, 225n. 77, 227nn. 1, 13
Ministry of Finance 20, 37, 49, 100–1, 192, 200, 201
Ministry of State Domains 49, 80, 195, 197, 220n. 2, 227n. 1
Mishonov, A. P. 235n. 33
Moll', Countess A. A. 164–5, 195

Morozov, Savva Timofeevich 83, 227n. 13
Morozov, Sergei Timofeevich 38, 83–4, 87, 90, 97, 104, 113, 152, 223n. 39, 227nn. 13, 14, 228n. 25, 232n. 98, 243n. 58, 244n. 89
Morris, William 2, 128, 139, 163, 213n. 2, 236n. 43, 239n. 86, 243n. 62
Moscow Art Theater 83, 224n. 62, 227n. 13
Moscow Association of Artists 26, 59, 66, 73, 78, 197, 198, 199, 223n. 48, 224n. 58
Moscow History Museum 34, 88, 105, 118, 135, 200, 243n. 80
Moscow Kustar Museum 114, 142, 194, 195, 205, 227n. 13, 229n. 33, 247n. 4; and Abramtsevo 36, 38, 43, 145, 195; artistic policies of 87, 88, 90, 92–3, 108–9, 111–12, 113, 223n. 39; and Bartram 106–9, 113; building of 112–13, *112*, 201, 227n. 14, 232n. 92; and exhibitions 108, 203, 206, 228n. 22; and sales 102–4, 111–12, 125, 152, 178; in the Soviet era 188, 208, 209, 210
Moscow provincial zemstvo 10–11, 36, 75, 81–4, 102, 117, 157, 193, 194, 195, 201, 202, 203, 205, 223n. 39, 229n. 33
Moscow provincial zemstvo assembly 83, 84, 174, 193, 229n. 33
Moscow School of Painting, Sculpture, and Architecture 26, 55, 66, 105, 118, 130, 167, 208
Mosekust 189
Mosvneshtorg 187
Mukhina, Vera 189
Munch, Edvard 62
Muraevnia workshop (Riazan province) 47, 220n. 3
Museum of Russian Antiquities (Smolensk). *See* Talashkino
M.Z.i G.I. *See* Ministry of Agriculture and State Domains

naboika 27, 56, *56*, 72, 113, 152–3, *153*, 168, 223n. 39
Narbut, Georgii 168
Naryshkina, Princess A. N. 47, 95, 126, 194, 200, 223nn. 41, 42, 244n. 89
neo-Russian style 66, 72–3, 78, 96–7, 114, 124, 152, 229n. 41, 230n. 52. *See also style russe moderne*
neo-Ukrainian style 157
NEP (New Economic Policy) 187, 246n. 1
Neradovskii, P. I. 150
Nesterov, Mikhail 141, 159, 203
New Society of Architects 97, 230n. 51
Nicholas II, tsar 170, 198, 201, 222n. 25, 233n. 6
Nizhnii Novgorod provincial zemstvo 81, 155, 171
Novosiltsova, E. E. 221n. 18
Novyi Torzhok. *See* Torzhok

Obrist, Hermann 2, *3*
Olbrich, Joseph Maria 97, 230n. 51
Old Russian style 6, 27, 28, 29, 102, 110, 111, 113, 127, 147, 166, 216n. 27, 217n. 43
Olga Aleksandrovna, Grand Duchess 170, 206
Omega Workshop 162, 180, 243n. 62
"On Measures for Assisting Kustar Industry" (1919 decree) 208–9, 247n. 2
Orlov, K. V. 99, 188, 208, 209, 247n. 4
ornament 9, 39–41, 57–65, 73–4, 77–8, 97–8, 148, 152, 159, 172, 200, 220n. 10
Orshanskii, Lev 105
Ovchinnikov, Iakov 121
Oveshkov, I. I. 106

Pavilion of the Far North. *See* All-Russian Arts and Industries Exhibition (Nizhnii Novgorod, 1896)
Peacock, Netta 69, 76, 225n. 66
Pestel', Vera 181
Peter, Duke of Oldenburg 169, 170
Podgoretskaia, A. G. 95
Pogosskaia, Anna K. 127–8, 130, 138–9, 236n. 40, 238n. 82
Pogosskaia, Anna N. 127, 138–9, 162, 200, 202, 235n. 34, 236nn. 38, 43, 238n. 81, 239nn. 84, 86, 243n. 58
point de Moscou. *See* lace
Poiret, Paul 162
poker-work (pyrogravure) 10, 87–8, 109, 111, 229n. 26
Pokrovskii, Vladimir 141, 147, 204
Polenov, Vasilii 23, 24, 26, 32, 211, 217nn. 35, 45, 222n. 35, 235n. 20
Polenova, Ekaterina 183
Polenova, Elena 24–6, 193, 197, 198, 199, 211, 219nn. 72, 75, 222n. 32, 223n. 48, 224nn. 50, 60, 237n. 50; and the Abramtsevo carpentry workshop 26–7, 28, 30, *32, 33,* 39–42, *40,* 43, 88, 174, 185, 188, 197, 218n. 55, 61, 219n. 67; critical opinion on 77–8, 100; and Iakunchikova dining room 66–9, *67, 68;* fairy-tale illustrations by 33–4, 59–61, *60,* 195, 203; influence of 55, 64–6, 105–6, 117, 118, 122, 136, 152, 154, 160; and ornament *2,* 61–4, *65,* 77–8; as a painter

Polenova, Elena (*cont.*)
62–3, *63*, 72, 73–4, 194; and Solomenko 59–60, 64–5, 77–8, 225n. 64; and the World of Art group 70–4, 225n. 66
Polenova, Natalia 34–5, 37, 43, 217nn. 35, 45, 222n. 35
Polovtsova, E. N. 168, 197, 203, 205
Poltava provincial zemstvo 156–7, 159, 171, 194, 202, 242n. 53
Popova, Liubov 181
Popova, Vera 179
populism 17, 18, 20, 176, 215n. 15
Porai-Koshits, A. E. 152–3, 244n. 89
Posrednik (publishing company) 221n. 22
Prakhov, Adrian 133, 134, 142, 147–8, 171–2, 234n. 16
Prakhov, Nikolai 158, 242n. 50
Pribyl'skaia, Evgenia 160, 169, 179, 183–5, 189, 210, 242n. 55
Prokhudin-Gorskii, S. M. 244n. 93
Puni, Ivan 179, 180

Raevskaia, M. G. 151
Réau, Louis 186, 235n. 29
Repin, Il'ia 23, 36, 115, 118, 121, 217n. 34, 224n. 56
Riabushkin, Nikolai 65, 88
Roche, Denis 128, 129, 135–6, 238n. 79
Rodnik. *See* Source, The
Roerich, Nikolai 132–6, *137*, 141, 142, 168, 200, 201, 203, 204, 237n. 59, 238n. 79, 241n. 21, 244n. 84
Romanov tercentenary (1913) 12, 169, 244n. 88
Root, Nikolai 146, 174, 207
Ropet, Ivan 22, 29, 192, 193, 195, 224n. 50
Rostov enamels 207, 244n. 92
Rozanova, Ol'ga 179, 181, 182, *182*, 208, 247n. 3
Ruskin, John 2, 8, 10, 148, 163, 186, 213n. 2
Russian Export House 164, 207
Russian Grain Society 168, 206, 244n. 81
Russian Industrial Art Society (St. Petersburg) 168, 230n. 61, 243n. 80

Samokish, N. S. 156, 241n. 40
Samusov, A. P. 235n. 33
School of Folk Art (St. Petersburg) 149–51, *151*, 171, 206, 241n. 19, 244n. 81
School of the Society for the Encouragement of the Arts 25–6, 28, 106, 170, 201, 203, 221n. 13, 231n. 77, 233n. 1
Semechkina, Tatiana 74
Semenovka workshop (Kursk province) 105, 196
Semenov ware (Nizhnii Novgorod province) 154, 208
Semigradova, Anastasia 160
Sergiev Posad (Moscow province) 75, 84, 86–7, 88–90, 142, 152, 171, 196, 202, 227n. 14; artels at 114, 205, 207; art joinery workshop at 109–13, 201; and Exhibition of Architecture and Industrial Art in the New Style 97–9; new designs for 85–91, 93, 106–7, 108–9; and Paris 1900 Exposition Universelle 90–2, 93; in the Soviet era 188, 209, 247n. 5; toy workshop at 112, 114, 169, 207; 228n. 26, 232n. 88, 245n. 114. *See also* matreshka doll; poker-work; toys
Serov, Valentin 38, 55, 90, 121, 197, 222n. 34, 229n. 35, 235n. 20
Shabel'skaia, Natalia 48, 52–3, 74, 76, 90, 105, 196, 203, 222n. 25
Shakhovskaia, Princess M. A. 196
Shakhovskaia, Princess M. N. 47, 193, 196, 223nn. 41, 42
Shaliapin, Fedor 36
Shchukin, Petr 105, 197
Shchuko, V. 147
Shchusev, Aleksei 141, 147, 204
Shekhtel, Fedor 97, 155, 230n. 51, 241n. 37
Sheremetev, Count P. S. 166, 174
Sheremeteva, Countess M. F. 200, 244n. 89
Sheremeteva, Countess M. O. 126
Shevchenko, Aleksandr 168
Shneider, Varvara 74, 95, 150, 152, 241n. 21, 244n. 89
Shtiglits School of Industrial Art 50, 101, 154, 205, 208, 216n. 33, 243n. 80
Sizov, Vladimir 133, 134
Skoptsy workshops (Poltava province) 160, *161*, 171, 179–80, 183, 184, 206, 242n. 55
Sobachko, Ganna 184
Sobko, Nikolai 70, 134, 225n. 70, 241n. 21
Société des Artistes Russes (Paris) 57, 168, 207
Society for the Dissemination of Industrial Art Education 198, 227n. 13
Society for the Encouragement of the Arts 29, 134, 193, 194, 198, 200, 201, 206, 218n. 52, 243n. 80
Society for the Improvement of Popular Labor 59, 164, 206
Society for the Renaissance of Artistic Rus 208, 227n. 13
Society in Aid of Manual Labor 125, 127, 149, 200

Soiuz (Union) Association 43, 199, 219n. 81
Sokolov, Vladimir 87–8
Solomenko embroidery and carpet workshops 46, 54, 64, 66, 69–70, 72, 75, 78, 92, 145, 165, 171, 174, 180, 182, 189, 199, 201, 202, 205, 223n. 43, 245n. 104. *See also* Davydova, Natalia Iakovlevna; Iakunchikova, Maria Fedorovna; Polenova, Elena
Sosnovitsy School for Instructors in Spinning and Weaving (Tver province) 176, 198
Source, The (Rodnik) 125–7, 131, 138, 140, 201, 203, 235nn. 31, 34, 239n. 89
Stasov, Vladimir, 5, 11, 30, 33, 39–40, 42, 48–9, 52–3, 59, 60, 70, 72, 74, 82, 134, 192, 195, 219n. 67, 220n. 9, 225n. 70, 238n. 67
Stasova, Nadezhda 216n. 32
Statute on Industrial Art Training Institutions (1902) 100–1
Stelletskii, Dmitrii 115, 121
Stepanova, Varvara 189
Store of Russian Works (Moscow) 38–9, 57, 178, 195
Stroganov School of Technical Drawing 106, 191, 196, 198, 199; and art education 101, 148, 216n. 33; design competitions of 227n. 13, 230n. 52; exhibitions participated in 202, 205, 207; and neo-Russian style 114, 131, 152, 230n. 52; publications of 28, 29; students of 157, 221n. 18, 231n. 78, 237n. 55, 247n. 3; after 1917 208
style russe moderne 3, 4, 73, 97, 114, 167. *See also* neo-Russian style
Subbotin, P. I. *4*
Sudeikin, Sergei 168
Sudzha district carpet workshops (Kursk province) 153–4, 173, 205, 237n. 48
Sultanov, N. V. 30, 238n. 65
Sunki embroidery workshop (Kiev province) 158–9, *160, 161,* 171, 203. *See also* Iashvil, Princess Natalia
Suprematism 246n. 125
Supremus group 180–3
Supremus (magazine) 181–2
Suslov, V. 59, 198, 200, 234n. 16, 238n. 65
Sviatopolk-Chetvertinskaia, Princess Ekaterina 116, 119, 140, 195, 233n. 5
Syrkin, M. G. 237n. 62

Talashkino: Church of the Holy Spirit at 119–20, 199, 234nn. 14, 16, 238n. 70; dye shop 127 9, 236n. 42; embroideries 117, 128–9, 141, 142; estate 115–17, 122, 127, 142–3, 167, 189, 198, 201, 202, 235n. 23; Museum of Russian Antiquities 133, 135, 140, 206, 209; and the 1905 Revolution 136–8, 139–40; school at 116–17, 137–8, 195, 233n. 6; theater at 120, 200, 235n. 20; workshops 118–21, 142, 155, 163–4, 170, 199, 201, 204, 207, 208, 235n. 33, 238n. 79. *See also* Tenisheva, Princess Maria; Source, The
Tarkhanova, Princess E. P. 178
Tarusa Embroidery Artel 189, 210, 247n. 8
Tchaikovsky, Petr 158, 242n. 46
Teliakovskii, Evgenii 106
Tenishev, Prince Viacheslav 116, 196, 198, 201, 233n. 4
Tenisheva, Princess A. D. 195, 236n. 48
Tenisheva, Princess Maria 196, 202, 207, 226n. 80, 233nn. 1, 2; attitudes toward 115, 143, 186; and balalaika orchestra 234nn. 8, 15; and Church of the Holy Spirit 119–20, 234n. 14; as a collector 133–5, 200, 206, 233n. 5; and Diagilev 140–2; embroideries by *125*, 129, *130;* exhibitions organized by 124–5, 140, 141, 201, 204; and Maliutin 118–20, 121, 124; and the 1905 Revolution 136–7, 139–40; after 1917 188, 210; objectives of 116–18, 120–1, 128, 135, 237n. 58; and the Pogosskaias 127, 130, 138–9, 236n. 38; and Roerich 134–5; and Vrubel 119, 133, 234n. 15; and *World of Art* 71, 117, 126, 198, 233n. 2; and Zinov'ev 130–1
Tian-Shanskaia, Ol'ga 220n. 3, 241n. 21
Tiffany and Co. 72, 199, 200
Tolstoi, Lev 163
Torzhok (Tver province) 95, 152, 197, 244n. 92
Toy Museum (Moscow) 209, 247n. 6
toys 84, 102–3, 105–9, 169, 195, 228nn. 18, 19, 229n. 41, 231n. 78, 245n. 102. *See also* matreshka doll
Traubenburg, Rausch von, Konstantin 141, 204
Tret'iakov, Pavel Mikhailovich 26, 36, 211
Trofimov, V. P. 153, 205
Troitse-Sergieva Lavra 21, 27, 84, 88, 99, 188
Tsionglinskii, Ian 118
Tuchkova, Sofia (née Iakunchikova) 211, 222n. 32
Tugan-Baranovskii, Mikhail 18, 176, 214n. 5
Tugendkhol'd, Iakov 168–9
Turkestan Exhibition (Tashkent, 1890) 172–3
Tver provincial zemstvo 81, 223n. 39

Udal'tsova, Nadezhda 179, 181, 182, *182*
Union des Artistes Russes (Paris) 202

Union of Russian Women 178
Urusova, Princess M. A. 47, 194, 223n. 41

Vallance, Aymer 162
Vasil'eva, Maria 179
Vasilkovskii, S. I. 156, 241n. 40
Vasnetsov, Apollinarii 88, 111, 136, 232n. 86
Vasnetsov, Viktor, 5, 23, *24*, 26, 36, 41, 54, 70, 78, 111, 121, 124, 136, 145, 159, 176, 194, 199, 202, 210, 217n. 37, 230n. 61, 235n. 20
Vatagin, V. A. 106
Vaulin, Petr 66, 247n. 3
Verbovka workshop (Kiev province) 158, 159, *160*, 171, 179–80, 182, 184, 203, 222n. 36, 242n. 46
Viatka provincial zemstvo 155, 196, 197
Viollet-le-Duc, Eugène-Emmanuel 193, 220n. 9
VKhUTEMAS (Higher Art and Technical Workshops) 209
Vologda provincial zemstvo 103, 162, 194, 202
Voloshin, Maksimilian 62–3, 76, 136
Vol'ter, A. A. 209, 210
Vornoskov, Vasilii 88–9, *89*, 184, 188, 203, 229n. 28
Vrangel, Baron, N. N. 244n. 84
Vrubel, Mikhail 38–9, 41, 58, 59, 66, 74, 78, 117, 118, 119, 126, 133, 155–6, 199, 230n. 51, 234nn. 15, 16, 235n. 20
Vyshnii Volochok School for Instructors in Spinning and Weaving (Tver province) 176, 198

Walcot, William 97
Witte, Count Sergei 57
women's crafts (*zhenskie promysly*) 9, 16, 46–7, 48, 53, 145, 149, 157–8, 165–6, 178, 189, 196, 236n. 48. *See also* beading; carpets; embroidery; lace
woodworking 22, 27–8, 154, 216n. 24
World of Art: exhibitions 66, 72, 73, 74, 117, 180, 199, 200, 224n. 58, 225nn. 71, 75, 229n. 36, 235n. 30; group 57, 73, 140, 147, 181–2, 201, 219n. 67, 225n. 75, 230n. 52, 233n. 2
World of Art (*Mir iskusstva,* magazine) 9, 70, 71, 72, *73*, 76, 77, 96, 117, 119, 122, 126, 198, 202, 225nn. 66, 71
World's Columbian Exposition (Chicago, 1893) 43, 197, 220n. 10, 222n. 25, 223n. 41, 228n. 22

Zabelin, Ivan 34, 52, 105, 118, 192, 218n. 48
Zagorsk. *See* Sergiev Posad
Zakharevich, A. F. 121
Zelenkov, Egor 43, 219n. 78
zemstvo system 81, 227n. 3
Zholtovskii, Ivan 97
Zinov'ev, Aleksei, 130, 131–2, *132*, 135, 140, 142, 208, 237nn. 53, 55, 238n. 79
Zozovo workshops (Kiev province) 159, 171